Encyclopedia of Air Pollution

Edited by **Raven Brennan**

R **C**ALLISTO
REFERENCE

New York

Published by Callisto Reference,
106 Park Avenue, Suite 200,
New York, NY 10016, USA
www.callistoreference.com

Encyclopedia of Air Pollution
Edited by Raven Brennan

International Standard Book Number: 978-1-63239-174-2 (Hardback)

Printed in the United States of America.

Contents

Preface

This book has been a concerted effort by a group of academicians, researchers and scientists, who have contributed their research works for the realization of the book. This book has materialized in the wake of emerging advancements and innovations in this field. Therefore, the need of the hour was to compile all the required researches and disseminate the knowledge to a broad spectrum of people comprising of students, researchers and specialists of the field.

The crucial problem of air pollution has always been of great concern. It has been a major environmental problem and an issue of global interest for many years. High concentrations of air pollutants due to several anthropogenic actions influence the quality of the air. This book will help you in filling the gaps existing in the fields of air quality monitoring, modelling, exposure, health and control, and will be of great help to graduates, professionals and researchers. It presents a variety of monitoring techniques of air pollutants, their predictions and control. It also includes case studies explaining the exposure and health implications of air pollutants on human beings in various countries around the world.

At the end of the preface, I would like to thank the authors for their brilliant chapters and the publisher for guiding us all-through the making of the book till its final stage. Also, I would like to thank my family for providing the support and encouragement throughout my academic career and research projects.

Editor

Air Pollution in Mega Cities: A Case Study of Istanbul

Selahattin Incecik and Ulaş Im
[1]Istanbul Technical University, Department of Meteorology, Maslak, Istanbul
[2]University of Crete Department of Chemistry, Environmental Chemical
Processes Laboratory (ECPL) Voutes, Heraklion, Crete
[1]Turkey
[2]Greece

1. Introduction

A megacity is defined by the United Nations as a metropolitan area with a total population of more than 10 million people. This chapter provides a brief introduction to the air pollution in megacities worldwide. This is an extensive topic and brings together recent comprehensive reviews from particular megacities. We have here highlighted the air quality in megacities that are of particular relevance to health effects.

The main objective of this chapter is to enhance our understanding of the polluted atmosphere in megacities, with respect to the emission characteristics, climate, population and specific meteorological conditions that are leading to episodes. Therefore, the chapter will provide state-of-the-art reviews of air pollution sources and air quality in some selected megacities, particularly Beijing, Cairo, Delhi and Istanbul. Furthermore, a detailed analysis of emission sources, air quality, mesoscale atmospheric systems and local meteorology leading to air pollution episodes in Istanbul will be extensively presented.

The world population is expected to rise by 2.3 billion, passing from 6.8 billion to 9.1 billion in between 2009 and 2050 (UN Report, 2010). Additionally, population living in urban areas is projected to gain 2.9 billion from 3.4 to 6.3 billion in this period. However, during the industrial revolution years, only about 10% of the total population lived in the cities. As an example, in 1820, which is the beginning times of the United States (US) transformation from rural to urban, the great majority of the population lived in rural areas of US (about 96%) (Kim, 2007). Today, according to UN Report the world population in urban areas has reached to 50.5%. In other words, half of the world's population are concentrated in the cities. However, distribution of urban population in the world is not evenly. A significant diversity in the urbanization levels can be seen in different regions of the world. About 75% of the inhabitants of the more developed regions lived in urban areas in 2010, whereas this ratio was 45% in the less developed regions. It is expected that urbanization will continue to rise in both more developed and less developed regions by 2050 with about 86% and 69%, respectively. These developments have created new physical, social and economic processes in the cities. For example, uncontrolled urban sprawl has leaded the rising of environmental problems due to high traffic volume, irregular industry, and low quality housing, etc. Massive urbanization in the cities due to the better job opportunities and challenges in the urban areas began first in Europe and then in other regions of the world, particularly in Asia. At this point, urbanization levels have led to a new classification and a concept- megacity- which is usually defined as a metropolitan area with a total population in excess of 10 million inhabitants. Megacities are highly diverse in the world, spanning from Paris (France), Los Angeles, New York City (USA) in developed countries to Delhi (India), Dhaka (Bengladesh) and Lagos (Nigeria) in developing countries. In today's developing countries, megacities exhibit the highest levels of pollution and therefore, in the studies of the anthropogenic impact on atmospheric composition, have become of primary importance, particularly those having high traffic volumes, industrial activities and domestic heating emissions.

The United Nations Environment Programme Urban Environment (UNEP-UE) unit expressed that more than 1 billion people are exposed to outdoor air pollution annually and the urban air pollution is linked to up to 1 million premature deaths and 1 million pre-native deaths each year. Additionally, UNEP presented the cost of urban air pollution with approximately 2% of GDP in developed countries and 5% in developing countries, respectively. In addition, the UNEP/Global Environmental Monitoring System (GEMS) reported that rapid industrialization, burgeoning cities, and greater dependency on fossil fuels have caused increasing production of harmful pollutants, creating significant health problems in most urban cities. The serious air quality problems, specifically inverse health effects, have been experienced in megacities of both developing and developed countries. due to the exposure to high concentrations of particular matter (PM), nitrogen oxides (NO_x), ozone (O_3), carbon monoxide (CO), hydrocarbons (HC) and sulfur dioxide (SO_2) depending on country's technology level. Especially, exposure to eleveted levels of particular matter and surface ozone causes loss of life-expectancy, acute and chronic respiratory and cardiovascular effects. Furthermore, damage to the ecosystem biodiversity by excess nitrogen nutrient is an important consequence of pollution.

In the beginning of the 2011, The European Commusion released a paper about the current policy efforts and the expected results to maintain a hard line against countries that are yet to comply with EU air quality legistlation limiting fine particulate matter ($PM_{2.5}$) concentrations. WHO (2009) concluded that megacities have faced particularly health impact by transportation, governance, water and sanitation, safety, food security, water and sanitation, emergency preparedness, and environmental issues. Furthermore, Baklanov (2011) recently shared results of the EU MEGAPOLI project (Megacities: Emissions, Impact on Air Quality and Climate, and Improved Tools for Mitigation and Assessment), which focuses on the multiple spatial and temporal scales from street to global levels and vice versa. The project addresses megacities with air quality and climate having complex effects on each other. Another EU-funded project CityZen (Megacities: Zoom for the Environment) also focused on impact of megacities on their environment and climate and vice versa from local to global aspects using long-term ground and satellite observations as well as regional and global modeling.

2. Megacities and air quality

As of 2011, there are 26 megacities in the world such as Tokyo, Guangzhou, Seoul, Delhi, Mumbai, Mexico City, New York City, Sao Paulo, Istanbul and other sixteen , eight of which exceeds 20 million. Fig. 1 presents the population in megacities world-wide with their continents. The four of the megacities are located at the South Hemisphere. Fifteen megacities are located at the tropical and humids-subtropical regions. This characteristic is important due to the growing evidence of the climate–health relationships posing increasing health risks under future projections of climate change and that the warming trend over recent decades has already contributed to increased morbidity and mortality in many regions of the world (Patz, et al., 2005). A total of 14 megacities, corresponding to more than of 50% of the total megacities, are located at the Asia continent, particularly in south and east parts. In this very dynamic region of the world, there are significant increases in industrialization and urbanization enhanced the urban population growth and economic development. This also leads to drastic increases in energy consumption and pollutant emissions in these regions. As an example, China is a rapid developing country with an urban population rate that increased from 19.6% to 46% within the last three decades.

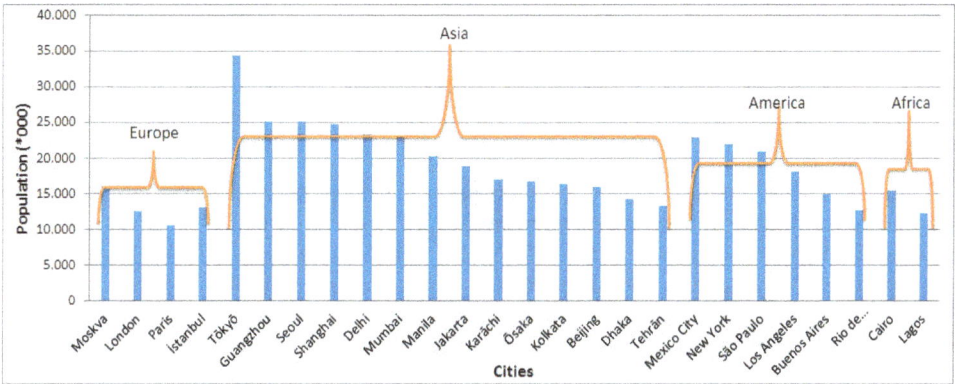

Fig. 1. Populations of the megacities with respect to their continents.

According to "China's blue paper", urban population ratio will reach 65% by 2030 in China. In recent years, a remarkable increase in the number of studies for air quality in China has been conducted (Kai et al., 2007; Chan and Yao, 2008; Wu et al., 2008; Fang et al., 2009; Wang et al, 2010;Zhu et al., 2011; Jahn et al., 2011). As an example, Chan and Yao (2008) extensively discussed the urbanization and air quality characteristics in Beijing, Shanghai and cities in Pearl River Delta (PRD) which is the mainland of China's leading commercial and manufacturing region covering Guangzhou, Shenzhen and Hong Kong. They noticed that in spite of the much attention to reduce emissions through effective control measures, particulate pollution is still severe in megacities of China. Among them, Guangzhou, which is the fourth largest city in China, is the main manufacturing hub of the PRD. In this city, the major industries are located in this industrial zone. In an earlier study by Kai et al. (2007) Air Pollution Index (API) values of Guangzhou were compared with the values of Shanghai and Beijing. The API for Guangzhou is higher than those of Beijing and Shanghai indicating that TSP was the prominent pollutant accounting for 62% of the major share in Guangzhou

(Zhou et al., 2007). In order to improve the air quality in Guangzhou, several new strategic efforts have been planned and established in industry and transportation sectors. Examples of the new control measures in transportation are; the metro line, which was opened in 1997, bus rapid transit system, hybrid buses, and design of low-emission zones in busy traffic areas. In a very recent study, Zhu et al. (2011) investigated the transport pathways and potential sources of PM_{10} in Beijing.

On the contrary to the classical air pollution events in megacities that are above mentioned, Los Angeles, USA (34°03′N; 118°15′W) which is in a large basin surrounded by the Pacific Ocean to the west and several mountain peaks to the east and south, and having a population of over 18 million, remains the most ozone-polluted region in the country. The Los Angeles region, which has a subtropical Mediterranean climate, enjoys plenty of sunshine throughout the year. The frequent sunny days and low rainfall contribute to ozone formation and accumulation as well as high levels of fine particles and dust in Los Angeles. The city area has the highest levels of ozone nationwide, violating federal health standards with an average of 137 days a year. The population growth, dependence on private motor vehicles, and adverse natural meteorological conditions can lead the episodic air quality levels in this area.

2.1 The general characteristics of air pollution and emission sources in megacities

Air pollution in urban areas comes from a wide variety of sources. The sources responsable for high emission loads are grouped into several sectors such as transport, domestic commercial and industrial activities for anthropogenic sources and NMVOCs from biogenic sources. Transport sector includes mainly motor vehicles, trains, aircraft, ship and boats while industry and domestic activities include fuel combustion including wood, coal, and gas for heating and production. Besides, biogenic (natural) emissions include NO_x and VOC emissions from vegetation and soils (Guenther et al., 2006). Today, urban air quality is a major concern throughout the world. Molina (2002) indicated that the quality of the air we breathe is fundamental to the quality of life for the growing millions of people living in the world's burgeoning megacities and deteriorating urban air quality threatens the public health. Furthermore, airborne emissions from major urban and industrial areas influence both air quality and climate change. This challenge is particularly acute in the developing world where the rapid growth of megacities is producing atmospheric pollution of unprecedented severity and extent. Mage et al. (1996) reviewed the difficulties in finding solutions to the air pollution in the megacities. Baldasano et al. (2003) examined the air quality for the principal cities in developed and developing countries. According to the study, the current state of air quality worldwide indicates that SO_2 maintains a downward tendency throughout the world, with the exception of some Central American and Asian cities, whereas NO_2 maintains levels very close to the WHO guideline value in many cities. However, in certain cities such as Kiev, Beijing and Guangzhou, the figures are approximately three times higher than the WHO guideline value. In the Asian databases consulted, only Japan showed really low figures. Surface ozone levels presents average values that exceed the selected guideline values in all of the analysis by regions, income level and number of inhabitants, demonstrating that this is a global problem with consequences for rich and poor countries, large and medium cities and all the regions.

Gurjar et al. (2008) examined the emissions and air quality pertaining to the megacities. He and his colleagues ranked megacities in terms of their trace gas and particle emissions and ambient air quality, based on the newly proposed multi-pollutant index (MPI) which considers the combined level of the three criterion pollutants (TSP, SO_2 and NO_2) in view of the World Health Organization (WHO) guidelines for air quality. Simulations of the export of air pollution from megacities to downwind locations via long-range transport (LRT) have shown different transport patterns depending on the megacity location: in the tropics export is occurring mostly via the free troposphere, whereas at mid and high latitudes it occurs within the lowest troposphere (Lawrence et al., 2007). Butler & Lawrence (2009) simulated small impacts of megacities on the oxidizing capacity of the atmosphere and larger on reactive nitrogen species on global scale. They also pointed out the need of parameterization of the sub-grid effects of megacities. Butler et al. (2008) analyzed different emission inventories and found substantial differences in emission's geographical distribution within countries even if the country total emissions are the same. They also reported large differences in the contribution of various sectors to the total emissions from each city.

Table 1 presents the megacities with their location, climate type, major emissions and critical air quality parameters. As seen in Table 1, there is a significant geographical variation in domestic heating emissions. Specifically, particulate matter is a major problem in almost all of Asian and Latin cities. In all of the megacities, emissions from the motor vehicles are a major contributor of harmful pollutants such as nitrogen oxides and particulate matter. The most important source for the classical pollutants such as particular matter, sulfur dioxide, nitrogen oxides, carbon monoxide, and volatile organic compounds are combustion of fossil fuels.

3. Megacity of Beijing

Beijing (39°54'N; 116°23'E), the capital of China, has completed its third decade of economic development known as the Economic Reform and Open Policy starting in 1978, and is a rapidly developing megacity with a 16 million population (Fang et al., 2009). As it's the capital city, Beijing continues to experience substantial growth in population, economic activity, business, travel and tourism. The city is situated at the northern tip of the roughly triangular North China Plain, which opens to the south and east of the city. Mountains to the north, northwest and west shield the city and northern China's agricultural areas from the desert steppes. Beijing has been experiencing severe anthropogenic air pollution problems since 1980s due to the significant energy consumption depending on developments of the city. Furthermore, natural sources have a significant impact on the city environment such as dust transport from the northern parts of the city. This leads to polluted smog covering the city as a thick blanket under specific meteorological conditions.

3.1 Climate

Beijing is in a warm temperate zone and has a typical monsoon–influenced humid continental climate with four distinct seasons. It is usually characterized by hot and humid summers and dry winters.

Megacity	Lat ; Lon	Popul. (million)	Area (km²)	Climate Type	Major Emission Source(s)	Critical Air Quality Parameters
Beijing (CN)	39°54'N; 116°23'E	16.0	16,800	Monsoon-influenced humid continental	Domestic heating, traffic,industry, dust,biomass burning	PM_{10}, $PM_{2.5}$, NO_2, O_3
Buenos Aires (ARG)	34°36'S; 58°22'W	15.0	4,758	Humid subtropical	Motor vehicles, industry	CO, NO_x
Cairo (EGY)	30°3'N; 31°13'E	17.2	86,370	Hot and dry desert	Industry, motor vehicles,dust transport	PM_{10}, $PM_{2.5}$
Delhi (IND)	28°36'N; 77°13'E	23.0	1,483	Humid subtropical	Motor vehicles, industry	PM_{10}, $PM_{2.5}$, NO_2
Dhaka (BNG)	23°42'N; 90°22'E	14.6	360	Hot, wet and humid tropical	Industry,road dust, open burning	PM_{10}, $PM_{2.5}$, SO_2
Guangzhou (CN)	23°08'N; 113°16'E	12.7	7,434	Humid subtropical	Industry, motor vehicles, power generation	PM_{10}, NO_2
Istanbul (TUR)	41°01'N; 28°58'E	13.2	5343	Mediterranean	Motor vehicles, industry	PM_{10}, CO, NO_x
Jakarta (IN)	6°12'S; 106°48'E	18.0	740	Hot and humid tropical wet and dry	Motor vehicles, Industry	PM_{10}, NO_2, O_3,CO
Karachi (PK)	24°51'N; 67°0'E	17.0	3,527	Arid	Industry, Motor vehicles	PM_{10} (TSP), CO, NO_2
Kolkata (IND)	22°34'N; 88°22'E	15.6	1,480	Tropical wet and dry	Domestic, Motor vehicles, Waste	PM_{10},NO_2
Lagos (NGR)	6º 25'N;3º 23'E	12.0	999	Tropical savanna	Industry, Motor vehicles,	SO_2, PM_{10}, NO_2
London (UK)	51°30'N; 0°7'W	12,6	1572	temperate	Mostly traffic	PM_{10}, NO_2
Los Angeles (USA)	34°3'N;118°15'W	17.7	1302	Subtropical-Mediterranean	Motor vehicles, petroleum rafinery, power generation	O_3, NO_x
Manila (PHP)	14°35'N;120°58'E	20.8	638	Tropical savanna/ tropical monsoon climate	Power generation, industry motor vehicles	PM_{10}, SO_2

Megacity	Lat ; Lon	Popul. (million)	Area (km²)	Climate Type	Major Emission Source(s)	Critical Air Quality Parameters
Mexico City (MEX)	19°26'N; 99°08'W	21,1	1,485	Subtropical high land	Motor vehicles	PM_{10}, O_3, BC (soot), NO_2
Moscow (RUS)	55°45'N 37°37'E	16.0	1,081	Humid continental	Motor vehicles, Industry, Power generation	SO_2, PM_{10}, NO_2, CO
Mumbai (IND)	18°58'N; 72°49'E	23.0	603	Tropical wet and dry	Industry, Motor vehicles, Power plants, Domestic, land fill open burning, road dust	PM_{10}, PAH, hazardeous chemicals
New York City (USA)	40°43'N;74°00'W	19.0	1,214	Humid subtropical	Motor vehicles	$PM_{2.5}$, O_3
Osaka (JPN)	34°41'N; 135°30'E	16.6	222	Humid subtropical	Motor vehicles	PM_{10}, $PM_{2.5}$, NO_2, O_3
Paris (F)	48° 51' N;02° 21' E	11.8	14,518	Western European oceanic climate	Motor vehicles	PM_{10}, $PM_{2.5}$, O_3, NO_2
Rio de Janeiro (BR)	22°54'S; 43°11'W	14.4	4,557	Tropical savanna climate / tropical monsoon	Motor vehicles	PM_{10}, NO2
Sao Paulo (BR)	23°33'S;46°38'W	22.0	7,944	Monsoon-influenced humid subtropical climate	Industry, Motor vehicles	PM_{10}, BC, O_3
Seoul (KR)	37°34'N;126°58'E	25.0	605	Humid subtropical / humid continental climate	Motor vehicles	PM_{10}, NO_2, O_3
Shanghai (CN)	31°12'N; 121°30'E	24.7	6,340	Humid subtropical	Industry, Motor vehicles, Dust transport	PM_{10}, NO_2
Tehran (IRN)	35°41'N ;51°25'E	13.4	1,274	Semiarid continental	Industry, Motor vehicles	CO, NO_2, PM_{10}

Table 1. List of megacities and their characteristics.

3.2 Air pollution sources

The major anthropogenic emission sources in Beijing are domestic heating, traffic and industry. Coal dominated energy structure is also one of the major causes of air pollution in Beijing (Hao, et al., 2007). As an example to the total emissions from energy production sector of 16.0 GWh/y, Hao et al. (2007) calculated a 102,497 t/y of SO_2, 60,567 t/y of NO_x and 11,633 t/y of PM_{10}. Domestic heating in Beijing usually starts in mid-November and ends in the following March and it is the major source for SO_2 in the winter season (Chan and Yao, 2008; Hao et al., 2005). Furthermore, industrial emissions emitted from Shijingshan region, located west of Beijing, are significant sources of particulate matter in Beijing. Beijing experiences a serious urban sprawl which has been claimed to be a major factor leading to the need for long-distance travel, congestion in the city centre and private vehicle usage problem (Zhao, et al., 2010; Deng & Huang, 2004). The number of cars in Beijing has grown rapidly and reached to 4.76 million vehicles in 2011, up from 1.5 million in 2000 and 2.6 million in 2005, according to official statistics provided by the municipal transportation authorities. Particulate matter emitted from motor vehicles and re-suspension of road dust are also likely contributors to PM_{10} pollution in the city (Song et al., 2006). Last but not least, natural dust originating from the erosion of deserts in northern and northwestern China results in seasonal dust storms that plague the city.

3.3 Air quality in Beijing

Beijing is party to the Standard Ambient Air Quality Standards (GB 3095-1996), which sets limits for SO_2, CO, PM_{10} and nitrogen dioxide (NO_2). The Chinese air quality standards set separate limits for different types of areas such as Class I, II and III based on physical characteristics of the region such as natural conservation areas and special industrial areas. Beijing is designated as a Class II area, which applies to residential, mixed commercial/ residential, cultural, industrial, and rural areas.

TSP and SO_2 have been the major pollutants in China for a long time due to the fossil fuel burning from power plants, industry and domestic heating. However, in recent years, the Chinese Government have planned to reveal a major environmental plan to help managing air pollution and is expected to include efforts to reduce pollution through new regulations and strategies including taxes and investments in this field. As an example, energy-related measures include fuel substitution and flue gas desulfurization facilities, which were built at the coal fired power plants, control measures such as energy efficiency and fuel use and dust control improvements (Hao et al. 2007). Initiatives have also been implemented in Beijing. The use of natural gas has been increased four-fold from 2000 as a result of efforts made to replace coal fired boilers and family stoves to use natural gas, and coal heating with electrical heating. As a result of these initiatives, from 1990 to 1999, the annual average TSP concentration in 100 major cities decreased by 30% to 256 μg/m3 and it remained almost constant from 1999 to 2003 (Sinton et al., 2004), despite an overall decrease of 30% in total energy consumption from 1997 to 2002. Fang et al. (2009) examined the air quality management in China and the changing air quality levels with their reasons. The results of the new strategies in Beijing are seen in Fig.2. In 1998, Beijing started its phased intensive control program to fight air pollution. PM_{10} concentrations increased by around 10% from 2003 to 2006 because of the increase in coal-fired boiler emissions, construction activities and dust storms (United Nations Environment Program, 2007). The annual PM_{10} concentrations in Beijing decreased from 162µg/m³ in 2006 to 141µg/m³ in 2007 despite an increase in energy consumption. The recent measurements of PM_{10} indicates lower levels such as 123µg/m³ in 2008; 120µg/m³ in 2009, respectively. One of the reasons of the high level of PM_{10} is residential coal-combustion in Beijing. SO_2 emissions from residential coal-combustion in Beijing were increased from 68,800 tons in 2003 to 85,100 tons in 2005. The expansion of the urban areas and the increase in SO_2 emissions led to increased particulate sulfate concentrations, which resulted in higher PM_{10} levels. However, SO_2 levels in the city are decreasing (Fang et al., 2009). Furthermore, Beijing experiences high PM_{10} pollution during spring dust storms. Zhu et al. (2011) showed that the typical wind speed of such dust storms is approximately 7 m/s or more, and the sand and dust sources are located about 1000-2000 km northwest of Beijing. According to Zhu et al. (2011), dust storms can reach to Beijing within 3 days.

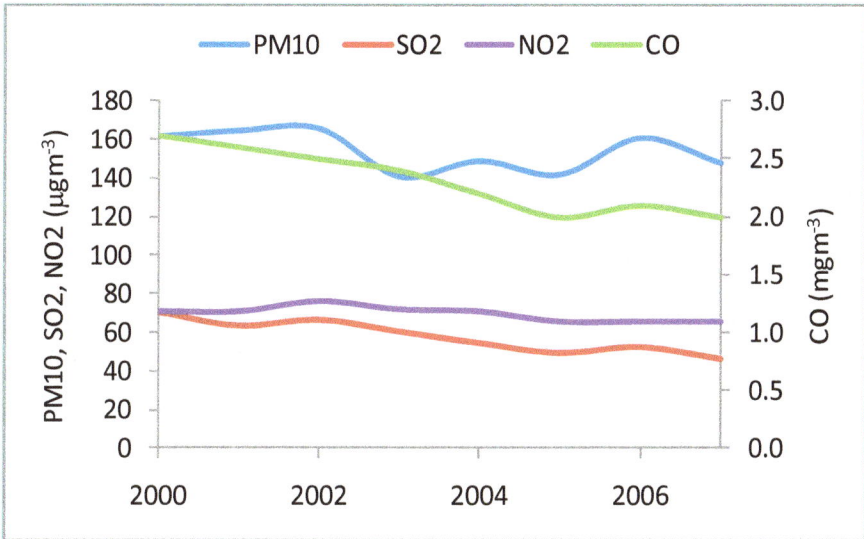

Fig. 2. Variations of annual-mean PM_{10}, SO_2, NO_2 (µgm-3) and CO (mgm-3) levels in Beijing (2000-2008) *Air quality standard for PM_{10} is 100 µg/m3 and WHO guideliness is 20µg/m³.

Recently, heavy industries have been gradually replaced by less polluting industries in Beijing. Besides, several possible activities are planned on major polluting industries such as the closure of cookery units at coke and chemical plants as well as closure of cement, lime and brick plants. The, transport sector is also a major contributor to Beijing's air quality. In this sector, stringent vehicle emission standards have been established. The Beijing municipal government will also implement traffic control measures such as the improvement of fuel quality to meet the new emission standards and to ease the city's traffic congestion.

4. Megacity of Cairo

Greater Cairo (30o3'N; 31o13'E), which is the capital of Egypt, is the largest city in Africa and located in northern Egypt, to the south of the delta in the Nile basin. The Greater Cairo consists of Cario, Giza and Kalubia, and has a population of about 17 million inhabitants. The city which about one-third of Egypt's population and 60% of its industry and is one of the world's most densely populated cities. Gurjar, (2009) and Decker et al., (2000) reported that Cairo's population is confined in 214 km² area making it the most densely populated megacity. The urbanization and industrialzation have increased very rapidly in Greater Cairo, particularly in the second half of the last century.

4.1 Climate

The climate in Cairo and along the Nile River Valley is characterized by a hot, dry desert climate. Wind storms can be frequent, bringing Saharan dust into the city during the spring. Abu-Allaban et al. (2002; 2007 and 2009) reported that wind speeds in wintertime is weaker than during summer, implying a lower ventilation of the area during winter that could favor pollutant accumulation in the vicinity of the sources. Additionally, Safar and Lebib (2010) indicated that the arid climate in the city causes a persistent high background PM level in the Cairo area.

4.2 Air pollutant emissions

The air pollution in Cairo is a matter of serious concern. The major emissions in the city come from industry and motor vehicles which cause high ambient concentrations of PM, SO_2, NO_x and CO. There are over 4.5 million cars on the streets of Cairo according to recent records. The relative contribution to particulate pollution from different economic activities is shown in Table 2. As seen in the table, the major contribution to the particulate load is urban solid waste burning by 30%. Transport and industry follows the solid waste by 26% and 23%, respectively. Kanakidou et al. (2011) summarized a comprehensive overview of the actual knowledge on the atmospheric pollutant sources, and the levels in the Eastern Mediterranean cities including Cairo. The annual sectoral distribution of pollutants are calculated for Cairo based on 2005 year as the reference year. Road transport and residential activities are important sources of PM and responsible for almost 35.9% and 53.4% of the PM_{10} emissions, respectively. Industrial activities have a major part for the SO_2 with 71.5% (Kanakidou et al., 2011). Furthermore, a typical black cloud appears over Cairo and rural regions of the Nile Delta in every fall due to biomass burning. It is found that this event is a major contributor to the local air quality. Molina and Molina (2004) explained the black cloud in Cairo during fall season based on the open burning of agricultural waste (mostly burning of rice harvest by farmers to clear fields for the next harvest in rural areas of Nile Delta). Additionally, traffic, industrial emissions and secondary aerosols was attributed to the black cloud events. Prasat et al. (2010) indicated the long range transport of dust at high altitudes (2.5–6 km) from Western Sahara and its deposition over the Nile Delta region. New evidence of the desert dust transported from Western Sahara to Nile Delta during black cloud season and its significance for regional aerosols have potential impact on the regional climate. Recently, the European Investment Bank approved a 90,000 EURO grant in order to investigate methods to reduce the burning of rice straw, which is thought to be one of the potential sources of these clouds (UN and League of Arap States Report, 2006). In addition to all, Cairo is the only city in Africa having a metro system (about 42 km length and carrying 60,000 passengers per hour in each direction).

Contribution	Solid Waste	Transport (fuel)	Industry (non-fuel)	Industry (fuel)	Agricultural Residues
(%)	30	26	23	9	6

Table 2. Relative contributions to particulate pollution from different economic activities in Cairo (EEPP-Air, 2004).

4.3 Air quality in Cairo

Air quality in Greater Cairo is a major concern to the Government of Egypt, particularly with regard to adverse health impacts. According to Country Cooperation Strategy (CCS) for WHO and the Egypt Report (2010), particulate matter and lead pollution have been recognized as the most significant pollutants threatening health in Cairo causing at least about 6,000 premature deaths annually and about 5,000 excess cancer cases over the lifetimes of current Cairo residents. Under this framework, PM_{10} is the most critical air quality problem in Egypt. Windblown dusts particles were significantly contributed the PM levels in Cairo and surrounding areas (EEPP-Air, 2004). A comprehensive national air quality monitoring system has been recently established in Egypt as part of Environmental Information and Monitoring Program and implemented with support from the Danish Government. The monitoring system has been operational for the measurements of common air pollutants such as SO_2, NO_2, CO, O_3 and PM_{10}. This is carried out by 42 monitoring stations throughout the country with one-third of them located in Cairo.

In recent two decades, several studies are published in the literature about the air pollution in Cairo (Zakey& Omran, 1997; Abu-Allaban et al., 2007; Favez et al., 2008, Zakey et al., 2008; Mahmoud et al., 2008; Safar & Labip, 2010). These studies include both anthropogenic and natural contributions to air quality in Cairo. As an example, Zakey & Omran (1997) showed that the dust and sand storms frequently occur specifically in spring and autumn and hot desert cyclones known as the "Khamasin" depressions pass over the desert during spring months. Hot and dry winds can often carry the dust and sand particulates to the city and they increase PM levels in the Cairo atmosphere. A source attribution study was performed by Abu-Allaban et al. (2007) where they used the chemical mass balance receptor modeling order to examine the sources of PM_{10} and $PM_{2.5}$ in Cairo's ambient atmosphere. They found that major contributors to PM_{10} included geological material, mobile source emissions, and open burning. $PM_{2.5}$ tended to be dominated by mobile source emissions, open burning, and secondary species. Favez et al. (2008) examined the seasonality of major aerosol species and their transformations in Cairo. Mahmoud et al. (2008) investigated the origins of black carbon concentration peaks in Cairo atmosphere. Seasonal and spatial variation of particulate matters was examined in Cairo by Zakey et al. (2008). They indicated that the highest recorded PM_{10} values were found in industrial and heavy traffic locations. The annual mean $PM_{2.5}$ and PM_{10} are observed to be 85 and 175 µg/m³, respectively) due to the traffic emissions and burning of waste materials. Recently, Safar&Labip (2010) investigated health risk assessments of PM and lead in Cairo. They showed that due to the arid climate, there is a persistent high background PM level in this area. This is one of the reasons of the high daily PM_{10} levels that is above the air quality limits in the country. From the meteorological conditions view, Cairo has very poor dispersion characteristics. Irregular settlements, layout of tall buildings and narrow streets create a bowl effect in the city environment. The high levels of lead were recorded in the major Egyptian cities. Safar&Labip (2010) explained that the lead levels in Cairo are among the highest in the world. The maximum annual average concentration of lead is found in the Shoubra Kheima industrialized region due to the lead smelters in this area during the period of 1998 through 2007. The highest annual average Pb levels recorded were 26.2 and 25.4 $\mu g/m^3$ at the Shoubra Kheima and El Sahel monitoring stations, respectively, during the baseline year (October 98 to September 99). The annual average Pb levels have been gradually decreased when the lead smelters in the area were closed and moved to the industrial area of Abou Zaabal. CCS (2010) report indicated that lead was completely phased out from petrol distributed in Cairo, Alexandria and most of the cities of Lower Egypt in late 1997, and consequently, lead concentration in the atmosphere of Cairo city centre and residential areas gradually decreased. Surface ozone levels were also examined in Cairo. Gusten et al. (1994) studied the ozone formation in the Greater Cairo in 1990. The peak values of 120 ppb and daily mean value of 50 ppb throughout the year indicate a substantial contribution of photochemistry to the ozone content of the atmosphere. It is estimated that the ozone is produced predominantly over the industrial area in the north and in the centre of Cairo and transported southward by the prevailing northerly winds. Contrary to many urban areas in Europe and in North America, fairly high average ozone levels of 40 ppb are observed during the night throughout the spring and the summer. This may imply that health hazards and crop damage may be higher in the greater Cairo area than in Central Europe. Recently, Mi (2009) examined the diurnal, seasonal and weekdays-weekends variations of ground level ozone concentrations in an urban area in Greater Cairo (Haram, Giza). He found that the daytime (8-h) mean values of wintertime and summertime O_3 were 44 ppb and 91 ppb, respectively. Besides, he reported that the highest levels of NO_x were found in winter. The concentrations of O_3 precursors (NO and NO_2) in weekends were lower than those found in weekdays, whereas the O_3 levels during the weekends were high compared with weekdays.

In order to enhance the air quality in Cairo atmosphere, there are several efforts under the new strategy by the Government such as switching to natural gas in industrial, residential and transport sectors. Policies to remove old fleet of vehicles from the streets and to promote public transport especially through the expansion of underground metro, enhancement of solid waste management, banning of the open air burning of solid waste are also among the major strategies.

5. Megacity of Delhi

Delhi (28°36'N; 77°13'E) is the second largest metropolitan in India. The name Delhi is often used to refer both to the urban areas near the National Capital Territory of Delhi and New Delhi, the capital of India, which is surrounded by other major urban agglomerations of adjoining states such as Haryana and Uttar Pradesh (National Summary Report, 2010). The National Capital Territory of Delhi is spread over an area of 1484 km^2. There are three local bodies in Delhi. Municipal Corporation of Delhi which is the major one has an area of 1397 km^2 and the two small areas are New Delhi Municipal Committee and Delhi Cantonment Board. Its population has increased from 9.4 million in 1991 to 18.9 million in 2010. Presently, about 30% of the population lives in squatter settlements. The number of industrial units in Delhi in 1951 was approximately 8,000. By 1991 this number had increased to more than 125,000. The vehicular population has increased phenomenally, from 235,000 in 1975 to 2,629,000 in 1996, and closed to touch 6 million in 2011. In 1975 the vehicular population in Delhi and Mumbai was about the same; today Delhi has three times more vehicles than Mumbai.

5.1 Climate

The climate of Delhi is a monsoon-influenced humid subtropical climate with an extremely hot summer, and cold winters. Delhi has relatively dry winters and has a prolonged spell of very hot weather. Delhi usually experiences surface inversions and heavy fog events during the winter season. This leads to restriction of dilution of the emissions from specifically motor vehicles and episodic events in Delhi. In December, reduced visibility leads to disruption of road, air and rail traffic. Molina & Molina (2004) explained that during summer, large amounts of wind-blown dust carried by strong westerly winds from the Thar Desert result in elevated PM levels. These dust storm periods are followed by the monsoon season (July to mid-September), which is the least polluted season due to the heavy monsoon rains that wash out the pollutants.

5.2 Air pollution sources

The major air pollution sources in Delhi are motor vehicles and industry. The number of industrial units in Delhi increased from 8,000 in 1951 to 125,000 in 1991 while automobile vehicles increased from 235,000 in 1975 to 4,5 million in 2004 (Government of India, 2006). The vehicular pollution contributes 67% of the total air pollution load (approx.3 Mt per day) in Delhi and its sharing is rapidly growing (Narain & Krupnick, 2007). The 25% of air pollution is generated by industry and coal based thermal power plants. The three power plants in Delhi generate approximately 6,000 Mt of fly ash per day. Industrial effluent load is about 320 Mt per day. Municipal solid waste generation is also estimated to be 5,000 Mt per day.

5.3 Air quality in Delhi

Delhi is the fourth most polluted megacity in the world. Air quality in Delhi is poor and airborne concentrations of major air pollutants frequently exceed National Ambient Air Quality Standards (NAAQS) set by India's Central Pollution Control Board (CPCB) (The Ministry of Environment and Forests, National Ambient Air Quality Standards 2009) A number of studies have analyzed air pollution data for Delhi. Delhi's annual mean PM_{10} concentration is highest among major Asian cities, and was between three and four times the Indian standard during 2001–2004 (HEI, 2004). A summary of PM_{10}, $PM_{2.5}$, NO_2, and SO_2 levels for different stations (background, residential, industry and kerbside) and different seasons (winter, post monsoon, and summer) in Delhi is given in Table 3. As seen in Table 3, almost at all locations and in all seasons, standards of PM_{10} and $PM_{2.5}$ have been exceeded (except for the industrial area). Even the background locations are highly polluted because these locations also fall within the city area and are impacted from the city emissions. PM_{10} contribution is mainly originated from heavy duty diesel vehicles.

Pollutant /Season	Background			Residential			Industry			kerbside		
	Wint	Post-Mon	Sum	Wint	Post-Mon	Sum	Wint	Post-Mon	Sum	Wint	Post-Mon	Sum
PM_{10}	355	300	232	505	671	81	546	781	229	451	941	337
$PM_{2.5}$	-	-	131	301	-	30	197	314	52	306	361	107
SO_2	8	15	8	14	18	78	85	77	11	20	20	12
NO_2	31	33	25	73	88	29	159	142	60	109	121	47

Table 3. Average air quality levels ($\mu g/m3$) in Delhi at background, residential, industry and kerbside areas for winter, post monsoon and summer periods of 2007 (Nat.Rep., 2010).

It can also be seen that in terms of PM levels, Delhi shows highest air pollution levels during post-monsoon. Observations at Delhi show much higher variability according to the characteristics of the monitoring station area. NO_2 levels are exceeded at the residential area sites in Delhi (35%), NO_2 levels generally exceed the ambient air quality standards at kerbside locations, particularly during winter and post monsoon seasons at Delhi by 85 – 95%.This analysis shows that PM problem is severe and NO_2 is the emerging pollutant that requires immediate planning to control its emissions. NO_2 is mainly contributed by man-made sources such as vehicles, industry and other fuel combustion activities. Delhi exhibits high percentage of NO_2 from energy production owing to presence of power plants. Fig. 3 presents the trend of annual variation of Respirable Suspended Particulate Matter (RSPM or PM_{10}), SO_2 and NO_2 in Delhi in 2001-2008. The annual average RSPM and NO_2 concentrations are increasing in Delhi while SO_2 values are declining due to the low sulfur fuel use in power plants. Other fuels consumed in domestic use and, in some cities, for vehicles, are LPG and CNG respectively.

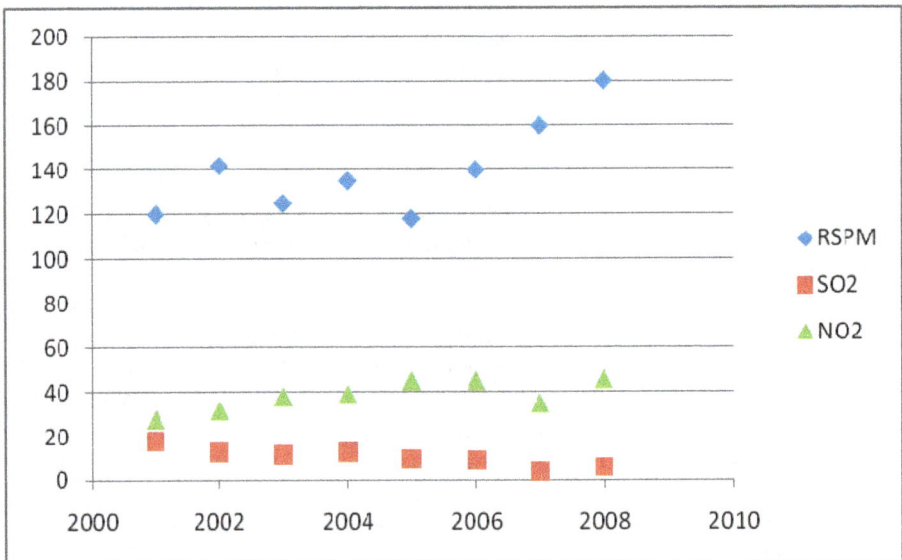

Fig. 3. Trends in annual average concentrations of RSPM, SO_2, and NO_2 in residential areas of Delhi (National air quality standard is 60 for RSPM ,NO_2 and SO_2).

6. Megacity of Istanbul

6.1 Introduction

Air pollution problem in Istanbul has received wide public attention since the 1980s and has remained the focal point among Turkey's environmental problems. In the late 1980s and beginning of 1990s, Istanbul has experienced significant particulate matter and sulfur dioxide episodes due to the fossil fuel burning for domestic heating and industry. Following the fuel switching policy, particulate matter concentrations and sulfur dioxide levels were gradually decreased in the city. However, today the city is facing specifically secondary particulate matter and NO_x problems depending on the emission sources. In this part of this chapter, firstly, a general description of the topography and meteorology of Istanbul, emission sources, and the spatial and temporal variations of the pollutants, particularly PM_{10} and SO_2, in the city will be provided that will be followed by a description of the recent emission inventory preparations. Surface ozone and its precursors in the city are also discussed. Additionally, meteorological characteristics leading to air pollution episodes will be extensively presented.

Istanbul is one of the significant and historically ancient megacities in the world with 13.2 million inhabitants and has an area of 5343 km2 covering 39 districts. The city is the center of industry, economics, finance and culture in Turkey. Istanbul generates about 55% of Turkey's trade and produces 27.5% of Turkey's national product. This lovely city has been associated with major political, religious and artistic events for more than 2,000 years. Recently, historical areas of the city have been added to the list of UNESCO World Heritage.

6.2 Topography and climate

Istanbul is located at 41oN and 29oE and is in the NW of Turkey's from the Black Sea to Marmara. The Bosphorus channel separates the city to Asian and European parts on the direction of NNE/SSW (Fig.4). The Bosphorus connects the Sea of Marmara to the Black Sea. The city also encompasses a natural harbor known as the Golden Horn in the northwest. The historic peninsula in the European part of the city is built on seven hills and surrounded by historical city walls. There are two significant hills in the city. The highest points in Istanbul are Yakacık (420m) and Aydos (537m) hills in Kartal province and Camlıca hill (288m) nearby the Bosphorus, on the Asian sides.

Istanbul has Mediterranean climate in temperate zone with four distinct seasons. Summer months (June-July-August) are relatively dry and warm while winter months (December, January, February) are mild and rainy. The lowest monthly average temperature is 6.5oC in January and the highest monthly average temperature is 22.7oC in July. Domestic heating in Istanbul usually starts in early-November and ends in the late March and early-April. There are also around 124 rainy days with a total precipitation of 843 mm. Most of the rain comes in winter season. Istanbul is humid and monthly average RH is 75%. RH exceeds 80% in most winter months. Besides, irradiation is strong with average daily values in summer by approximately 21 MJ/m2, and sunshine duration is 2460 hours annually. The prevailing wind directions are north-easterly and south-westerly in winter, and northerly in summer, especially when the Etesian system controls the weather in the region (Unal et al., 2000). However, the prevailing wind direction varies from north northeast to northeast and south southwest in the winter and varies from north northeast, northeast and east-northeast in the summer with moderate wind speeds. The urban area in Istanbul has continuously expanded to the suburbs.

Fig. 4. A map of Istanbul with air quality measurement stations.

6.3 Emission sources

Emission sources in Istanbul have shown significant changes over time. While today, traffic, industrial processes, domestic heating, road dust and biogenic emissions are the most significant emission sources, domestic heating and industry was the major emission sources before two decades ago in the city.

Istanbul has experienced a complicated period from an ancient metropolis to a sprawling megacity, growing from just over 2 million inhabitants in 1970 to 13.2 million today. However, due to the migration from the other cities its population has increased more than six fold between 1970 and 2010 (Tayanc et al., 2009). Rapid urbanization and development of society and economy, with the increased migration from the less developed regions of the country at the end of the 1980s caused a significant increase in the population and an expansion of the built-up areas in the city. Not only the rapid increase of the urban population due to influx of the people from other cities, but also the establishment of many small and medium sized industries in and around Istanbul, caused many environmental problems. Atimtay and Incecik (2004) extensively discussed the period of 1970-1980s. The industrial as well as the domestic heating mostly burned fuel oil to produce energy before 1970's. Following the energy crises of 1970's, due to the tremendous increase in the oil prices, the preferred fuel in the city was coal. The coal used was mostly local Turkish lignite with high in sulfur and ash content, but low in calorific value. In those days there was not any regulation on air pollution control. The First Environmental Law was accepted in 1983 and the first Air Pollution Control Regulation was entered in force in 1986. The very active period involving the banning of lignite, natural gas agreement, establishing of gas distribution company (IGDAS) and starting the operations for infrastructure followed the regulation. IGDAŞ started to distribute natural gas on January 1992, first on the Asian side of the city and then expanded to the metropolitan area (Atimtay&Incecik, 2004). Today 95% of the total gas is used for domestic heating and industry purposes.

As one of the major emission sources, traffic was the secondary source in 1980s. Following 1980s the number of motor vehicles on Istanbul's streets has increased even faster than its population growing since 1980. It was only about 0.3 million cars in 1980, but now it is closing three million (about 2.72 million vehicles are registered based on the 2010 figures). Every day more than 700 new cars enter to the Istanbul streets. About 60% of these vehicles are operated by gasoline and 40% by diesel. PM_{10} emissions on traffic originate from diesel vehicles rather than the gasoline powered ones. Besides, most of the heavy vehicles, such as buses and commercial vehicles in transport and construction sectors, are powered by the diesel system. Furthermore, following the 1999, liquefied petroleum gas has been widely used in traffic. Istanbul Chamber of Industry reported and Im et al., 2006; Kanakidou et al. (2011) used that the low quality solid and liquid fuels with high sulfur content, natural gas and LPG are the most commonly used fuel types in the industrial activities including textile, metal, chemical, food and other industries. Istanbul is on the two transit motorways passing over the city connections between Europe and Asia on the east-west direction by the Bosphorus and Fatih Sultan Mehmet Bridges. The total length of roads and highways has significantly increased from 1980s within three decades while the number of vehicles has increased. The city traffic much more depends on the two bridges in the city. The number of vehicles crossing the two bridges has increased approximately 25% from 2001 (120,000 vehicles) to 2010 (150,000 vehicles) (KGM, 2011). Hence, traffic emissions are elevated significantly within the past decade. Recently, Turkey adopted the Euro 4 standards to reduce vehicular emissions of air pollutants in the beginning of 2009. Istanbul has a metro with 16 km long in the European side and it is still expanding. On the Asian side, a new metro construction continues and is scheduled to be opened in 2013. Furthermore, a Bus Rapid Transit (BRT) system established in 2009 connecting the both side of the city from far west (Avcılar) to Sogutlucesme (Kadikoy) with about 30 km. Istanbul will be connected at soon with an underwater tunnel. The construction is currently in progress and underwater rail tunnel (Marmaray) will be open in 2013.

There are three industrial zones in the city for small and medium-scale industrial operations. They are located at both sides of the city. There are two busy international airports. One is located at the European part (Atatürk Airport) and the other is Sabiha Gokcen Airport at the Asian part (Fig. 4). Atatürk International Airport is ranked 34th in the busiest airports of the world with a passenger traffic of 32.1 million. Sabiha Gokcen Airport serves many domestic and some low-fare international flights, with about 104,000 total aircraft movements annually. Istanbul experiences frequently dust transport coming from the Sahara. Karaca et al. (2009), Celebi et al. (2010) and Kocak et al. (2011) showed that air parcels arriving in Istanbul in the spring months are mainly from the Sahara. Nevertheless, even limited amounts coal is consumed for domestic heating in winter months; it causes significant particular matter pollution problems at some residential areas in the city where mostly illegal squatter settlements are located.

6.4 Emission inventory preparations

Emission inventory has always been a significant problem in Istanbul. In recent decade, there were several efforts on the emission inventory preparations for Istanbul

Metropolitan area. In 2005, EMBARQ and Istanbul's Directorate General of Environmental Protection launched an ambitious initiative to reduce air pollution in the city considering reducing transport emissions. In addition, Istanbul Metropolitan Municipality has developed an "Istanbul Air Quality Strategy Report" emissions inventory of air pollutants (IBB, 2009). The local emissions inventory was based on a European Union project in the scope of the EU Life Program has been implemented in a partnership of Istanbul Metropolitan Municipality and Dokuz Eylül University (Elbir, 2010). In this project, local emission inventory was prepared with 1-hour temporal and 1-km spatial resolution within an area of 170 km by 85 km centered at the metropolitan area of Istanbul. In a systematic way, the emission sources are broadly categorized as point, line and area sources, covering industrial, vehicular and domestic sources respectively. Five major pollutants consisting of particulate matter (PM_{10}), sulfur dioxide (SO_2), carbon monoxide (CO), non-methane volatile organic compounds (NMVOCs) and nitrogen oxides (NO_x) emitted through these sources were identified (IBB, 2009). As seen in Table 4, industry is the most polluting sector for SO_2 contributing to about 83% of total emissions while domestic heating is the most polluting sector for PM_{10} contributing to 51% of total emissions. Traffic is also the most polluting sector for NO_x, and CO emissions with the contributions of 89% and 68%, respectively.

	Emissions (tons/year)				
	PM_{10}	SO_2	NO_x	NMVOC	CO
Industry	7630	58458	9394	117	1714
Domestic Heating	13631	10983	7014	18451	123510
Traffic	5200	1016	158000	38500	270000
Total	26461	70467	154408	56968	270000

Table 4. Sectoral emissions in Istanbul (IBB, 2009).

Furthermore, in recent years, there are important efforts on higher resolution (in high spatial (2×2 km^2) and temporal resolutions) detailed emission inventories of anthropogenic sources studies. Markakis et al. (2012) developed high resolution emission inventory for 2007 reference year as annual sectoral distribution of pollutants; and cited by Im et al. (2010 and 2011) and Kanakidou et al, (2011). The sectoral distributions of these emissions is presented in Table 5. Im et al. (2010) showed that wintertime PM_{10} episodes can be explained almost entirely by the local anthropogenic sources. Kocak et al. (2011) confirmed this finding by the PMF analysis conducted on the chemical composition data provided by Theodosi et al. (2010). The emissions are spatially allocated on the cells using a grid spacing of 2 km over the Istanbul. The pollutants considered are NO_x, CO, SO_x, NH3, NMVOCs, PM_{10} and $PM_{2.5}$. The NMVOCs emissions are chemically speciated in 23 species based on Olivier et al. (2002) and Visschedijk et al. (2005) source sectoral profiles (CARB, 2007). PM_{10} emissions were chemically speciated in organic and elemental carbon, nitrates, sulfates, ammonium and other particles. According to Markakais et al. (2012) and Kanakidou et al. (2011), CO and NO_x are the major emissions for road transport by 83.1% and 79.4%, respectively. Industry has also major contributer for PM_{10} by 64.9% and SO_2 by 23.2%, respectively. However, shipping emissions in Istanbul is gradually increasing. Because,

Bosphorus Istanbul has a busy ship traffic. Recently, there are about 60,000 ships passing yearly from the Bosphorus channel (Kesgin and Vardar, 2001; Deniz and Durmusoglu, 2008; Incecik et al., 2010). Kanakidou et al. (2011) presents significant contribution of shipping to SO_2 emissions by 17.6% and NO_x by 9.5 and PM_{10} by 3.1%.

	CO	NO_x	SO_2	NMVOC	PM_{10}
Combustion Residential	10.8	2.1	14.7	2.6	7.1
Industry	3.7	2.4	23.2	0.5	64.9
Fuel Extraction/Distribution			2.3		0.1
Solvents Use				29.8	
Road Transport	83.1	79.4	2.3	44.8	17.4
Off-road machinery		2.8	4.1	0.4	3.9
Maritime	0.3	9.5	17.6	0.6	3.1
Waste	0.7			20.4	1.7
Energy	0.7	3.2	35.6	0.2	1.8
ALL (ktons)	437	305	91	77	61

Table 5. Sectoral distribution (%) of annual anthropogenic emissions in Istanbul (Kanakidou et al., 2011).

6.5 Air quality in Istanbul

6.5.1 Monitoring network

The first air pollution monitoring network was designed to measure SO_2 and suspended particulate matter (TSP) concentrations in the air in 1985 with 7 stations in the city by the Ministry of Health (Hifzisihha Institute). The network provided daily SO_2 and TSP measurements. In this network, daily values of TSP concentrations were measured by the reflectivity method and SO_2 concentrations were measured with the West-Geake method. In mid-1989, 10 more stations were established in the city. In 1998, the Greater Istanbul Municipality established a separate air pollution monitoring network in the city. The new network system consists of 10 measuring sites located in residential, commercial and industrial parts of the city. The new network measures air pollution parameters on hourly basis. Conventional parameters of PM_{10}, which is replaced with TSP and SO_2, are measured in all stations. CO, NO_x are only measured in four stations. Surface ozone was measured at two stations in 1999-2005 and then continued with one station. Now, O_3 is being measured at two stations in the city. However, understanding of air quality in megacities requires high spatial resolution of the monitoring network. For this purpose, in frame of a new European Project (SIPA), about 36 stations including urban, semi urban and rural are being launched in Istanbul and Marmara areas.

6.6 Temporal and spatial variation of air quality

The first measurements showed that Istanbul has a serious air pollution problem due to domestic heating, industry and traffic (Incecik, 1996; Tayanc, 2000; Topcu et al.,2001). The pollution was more severe especially in the regions of densely populated settlements. Uncontrolled expansion of the city in 1980s also caused severe air pollution problems in certain areas. City regulations banning usage of less efficient, poor-quality lignite for heating purposes was inadequate. Fig.5 presents the first results of the measurements as annual average SO_2 and TSP concentrations in the city. As can be seen in the figure, both SO_2 and TSP levels were much higher than the air quality standards and WHO guidelines.

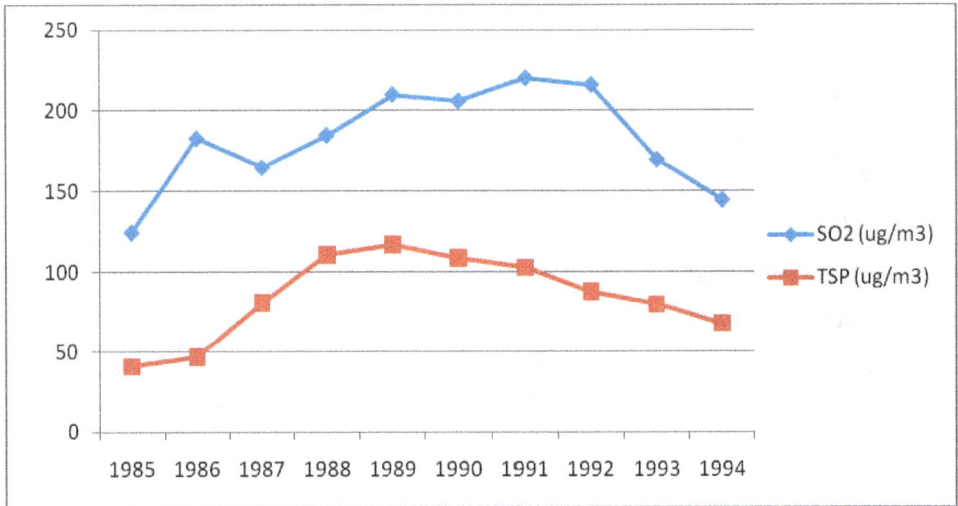

Fig. 5. Annual mean values of daily SO_2 and TSP concentrations in Istanbul for the polluted period 1985-1994.

Incecik (1996) extensively examined the air quality levels and atmospheric conditions leading to air pollution episodes in winter months of 1985-1991 periods. Incecik (1996) showed that anticyclonic pressure patterns and lower surface wind speeds lead to unfavorable conditions for air pollution potential over the city. Poor quality fuel used in the city was the major reason of the dramatic levels of SO_2 and TSP in the winter months. Furthermore, he examined the ratios of January/July SO_2 and TSP concentrations during the 1985-1994 periods. The ratios for SO_2 an TSP presented and increase from 1.4 to 13.6 and 2.3 to 8.8, respectively in the study period. Besides, it is shown that the higher pollutant concentrations in cool seasons might also be related to the generally stable atmosphere, which limits the volume of air into which emissions from the local emissions are dispersed.

The relationship between occurrence of the intense episode days and other meteorological parameters such as inversions and light winds are seen in many cities. In Istanbul, inversions form by radiative cooling at night or as a result of subsidence in anticyclones. Table 6 gives the characteristics of inversions recorded from radio sounding station at Goztepe-Istanbul in the period of November 1989 – February 1990. There are almost 16 episodes occurring in this period and the days with surface or elevated inversions almost coincide with the episode days.

	Time (LST)	Surface inversion (day)	Elevated inversion (day)	Mean thickness of surface inversion (m)
November	0200	10	12	273
	1400	2	13	408
December	0200	13	13	242
	1400	2	23	145
January	0200	9	2	280
	1400	0	9	-
Feb	0200	9	13	126
	1400	3	16	171

Table 6. Frequency and depth of inversions in Istanbul during November 1989-February 1990 (Incecik, 1996).

According to Table 3, the heights of the night time surface inversions in 48% of the cases are below 250m. Furthermore, it is found that during the episodic period, calms are about 20% or more of the total episodic periods which are probably associated with weak synoptic scale pressure gradient at the surface atmosphere in Istanbul.

In Istanbul, natural gas usage has just started in January 1992 in the Asian part of the city. Hence, the city of Istanbul, particularly in European side, has faced many significant episodic air quality problems due to the poor quality lignite usage and weak dispersion conditions in winter periods up to mid 1990s. As an example, one of the dramatic air pollution episodes had been experienced in 1993 winter. The catastrophic values of the SO_2 and TSP were measured to be 4070 and 2662 $\mu g/m3$, respectively, on 18th of January 1993 in the European part of the city (Batuk et al., 1997). The stagnant anticyclonic pressure conditions continued for two days in the city and Governor of Istanbul announced the emergency situation in the city particularly for sensible people such as older people and babies. The schools were closed for two days following the decision in the city. Fig. 6 gives the surface synoptic map for 18 January at 00Z. As it can be seen from this figure, a very strong anticyclonic pressure system was established over the Central Mediterranean and South Europe and consequently the Balkan Peninsula up to the west part of the Asia Minor and Black Sea Balkan Peninsula. Such synoptic conditions favor the formation of an anticyclonic subsidence inversion, which results in residential and industrial stack plumes capping specifically in the European side of the city. Fig.7 indicates the 850 hPa map indicating geopotential heights and temperatures. This map indicates very strong temperature advection over Istanbul.

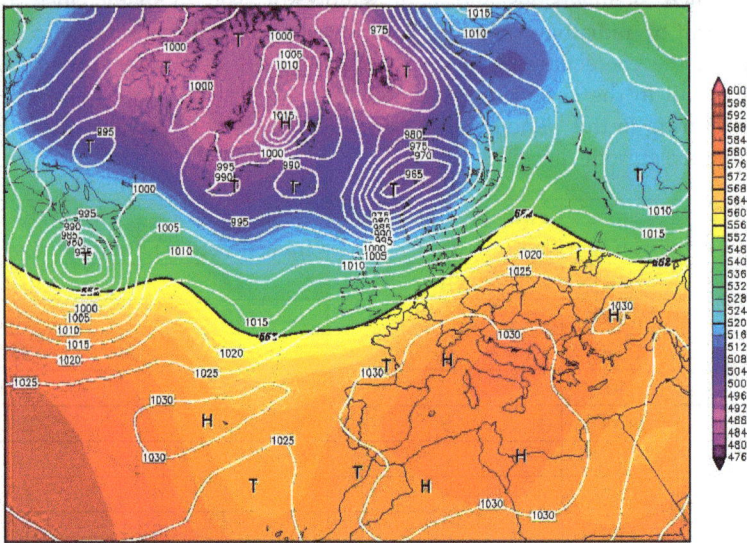

Fig. 6. 500 hPa and surface map on 18th Jan. 1993 00Z.

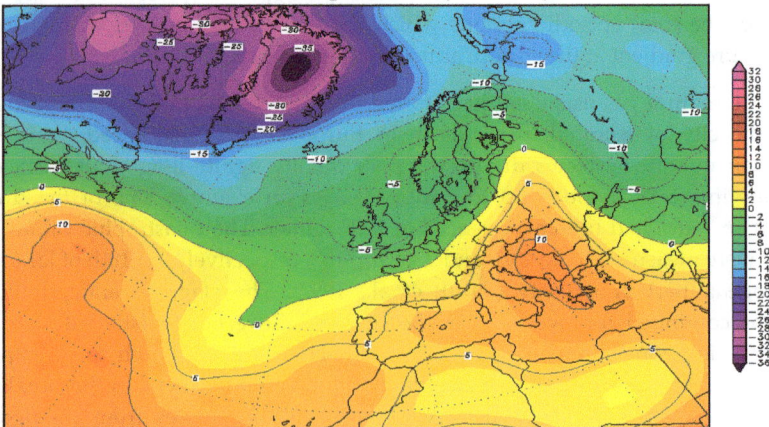

Fig. 7. 850 hPa Geopotential and temperature map indicating the major episode in Istanbul on 18th January 1993, 00Z.

A ridge covers this area and in some cases a significant warm advection are observed on 17th and 18th of January. As seen in Figures 6 and 8, high pressures were observed at surface as well as aloft, and the lower troposphere becomes strong stable. During the night of January 17, a surface temperature inversion was formed. On January 17, the first day of the episode, the surface pressure was rather high over the entire Balkans and Eastern Mediterranean. The light surface winds blew from southern directions, weak at noon and from SW at night (Fig.9). The thermodynamic structure of the lower atmosphere during the episode shows that during the night of 17 January, a strong surface inversion was formed with a depth of 230 m more than 5° C in strength as result of warm advection. On that night surface atmospheric pressure was 1028 hPa and surface winds measured at the Goztepe meteorological stations from 17th January 00Z to 19 January 12 Z were light (~1 m/s) throughout the night and daytime period. around 1 m/s were at the south and southwest

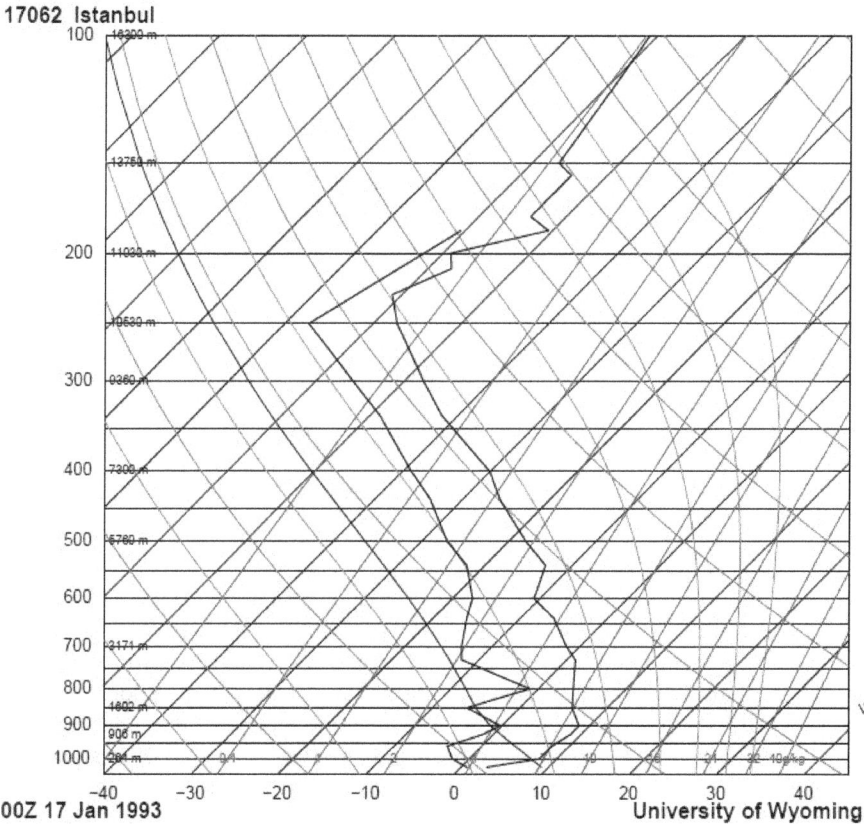

Fig. 8. Skew T log P diagram at 00Z on 17th Jan 1993.

directions. The pollutants (SO$_2$ and TSP) were accumulated during the following day due to the stagnant weather conditions. Very dramatic concentrations of SO$_2$ and TSP (4070 and 2662 µg/m3) on 18th January 1993 were measured to be associated with very strong surface inversion and stagnant conditions during the poor quality coal usage in the European and west parts of Istanbul. The episode ends on January 20, when a strong low pressure system moves over the Black Sea.

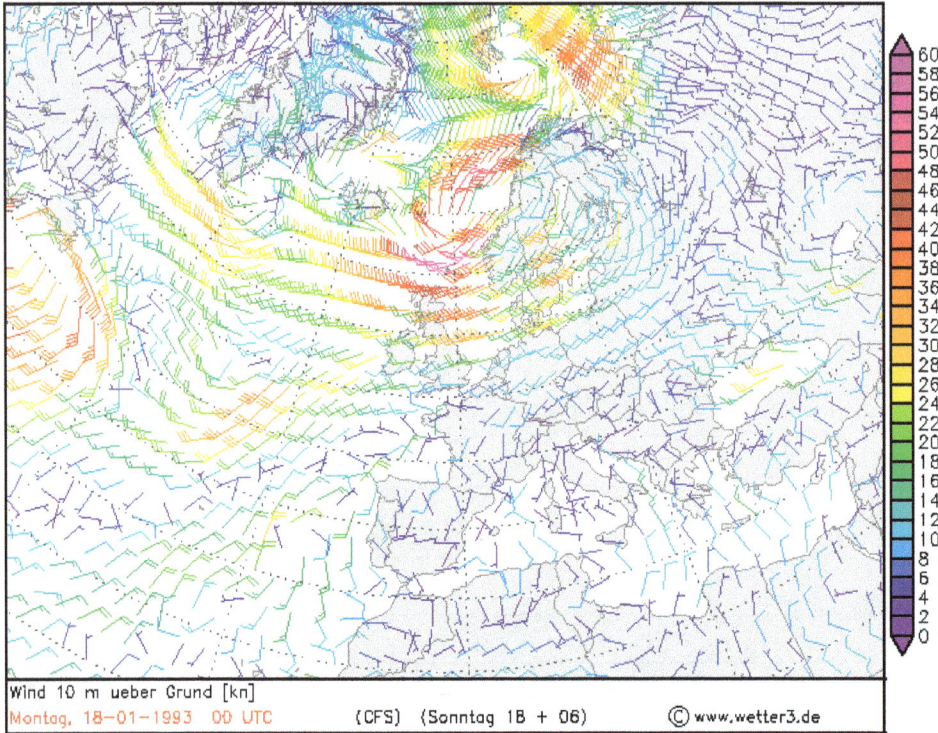

Fig. 9. Surface winds at 10m on 18 January 1993 at 00Z.

The hourly concentrations of PM_{10} were examined by several researchers. As an example, Karaca et al. (2005) interpreted the analyses for monthly average variations of PM_{10} concentrations in Istanbul. The numerical results of the study indicate that cyclonic behavior of the time series of PM_{10} concentrations occurs in winter and summer times with the effect of prevailing meteorological conditions. According to Kindap et al. (2006) Istanbul had hourly PM_{10} levels observed from monitoring sites that were in excess of $300\mu g/m^3$ at several locations in the beginning of 2000s. Attributing to predominantly westerly winds at this period it is investigated that long-range transport is effective on elevating PM levels in Istanbul. Following the 2000s, air quality levels for Istanbul present different characteristics based on the changing of emission sources in the city. Fig. 10 presents the temporal variation of the annual average concentrations of some critical pollutants following fuel switching period in the city. In this figure, the hourly PM_{10}, SO_2 and NO_x, concentrations reported over the five year period (1 January 2005 through 31 December 2010). As seen in the figure sulfur dioxide concentrations in the city remained below the air quality standards whereas PM_{10} and NO_2 concentrations exceeded the air quality standards.

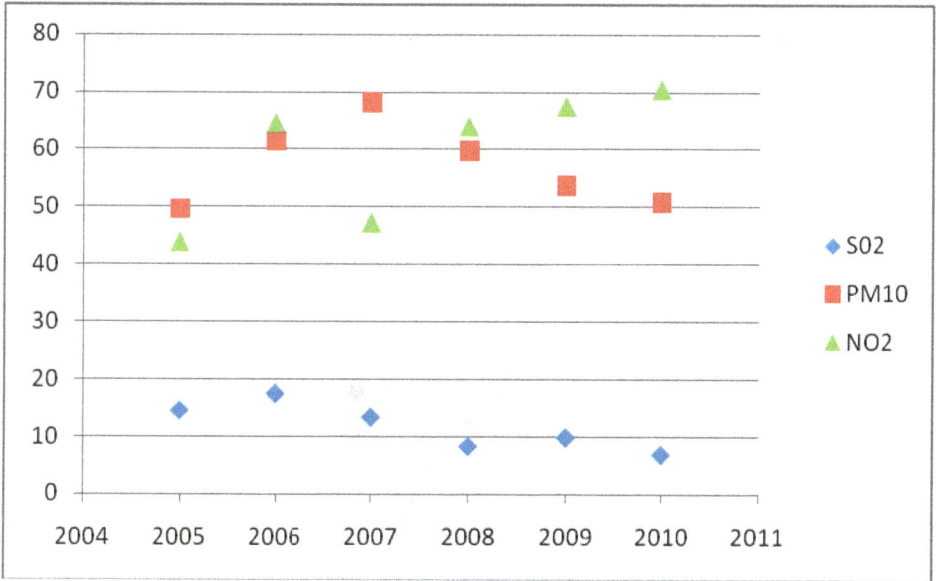

Fig. 10. Temporal variation of annual average PM_{10}, SO_2 and NO_2 concentrations ($\mu g/m3$) in Istanbul.

In a recent study, Unal et al., (2011) assessed the PM_{10} data in ten monitoring stations considering the EU PM_{10} standards, and to identify air-monitoring sites with similar pollutant behavior characteristics by means of hierarchical cluster analysis and to explore possible pollutant sources in such clusters. PM_{10} concentration averages over the monitoring sites under the influence of urban traffic and residential heating greatly exceeded the EU's daily ambient air quality standard. As an example, PM_{10} data for the monitoring stations of the metropolitan air quality monitoring network show that about 32% of the total days in 6 years (2005-2010) daily PM_{10} concentrations exceed the limit value of 50 $\mu g/m3$ within the city. Istanbul yields an urban average PM_{10} of 58 $\mu g/m3$. The monthly means for all stations vary in between 32-68 $\mu g/m3$ in summer and 43-87 $\mu g/m3$ in winter. Unal et al., (2011) indicated that Kartal (Asian site) Esenler and Alibeykoy (European sites) (Fig.4), where urban areas with high traffic and industrial activity, were seriously polluted by PM_{10}. Many people who live in the Kartal region might be exposed to a much higher PM_{10} level than the rest of the population. Topography or the barriers in the city environment of the Kartal might support the accumulation of the concentrations. These results indicate that higher the PM_{10} concentration, the greater the risk of premature mortality from heart and lung disease to the residents living in these locations. Celebi et al. (2010) found similar PM_{10} results for these locations (Fig.11).

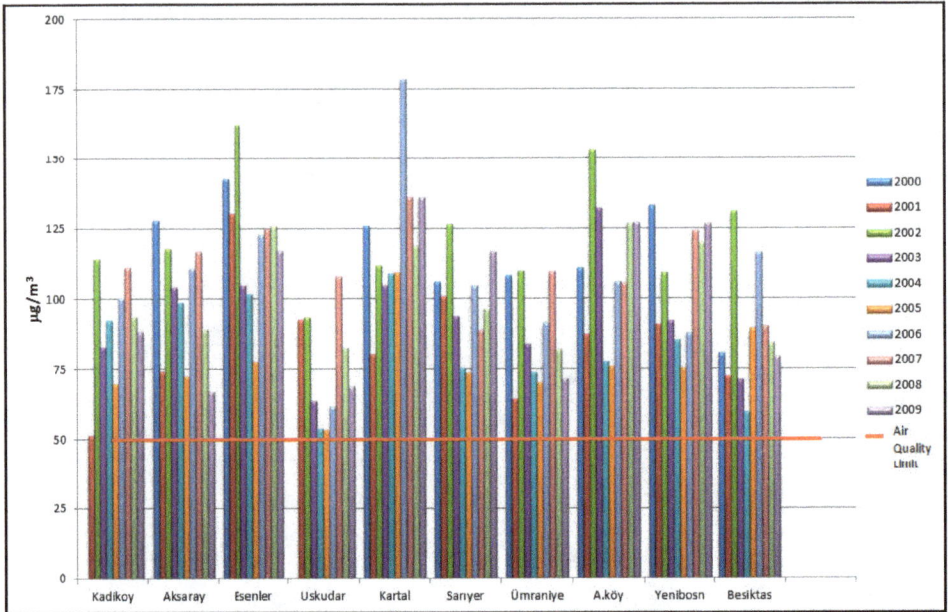

Fig. 11. Annual variation of PM_{10} hourly concentrations in Istanbul citywide (Celebi et al., 2010).

The first complete chemical composition data for PM_{10} levels in Istanbul has been provided by Theodosi et al. (2010). Daily PM_{10} samples were collected at the Bogazici University Campus and major ionic species as well as metals and organic and elemental carbon were measured between November 2007 and June 2009. Fig. 12 shows the temporal variation of major chemical species in Istanbul during the sampling period and Fig.13 shows the contribution of aerosol species to PM_{10} concentrations. The seasonal variations of metallic elements revealed that elements of mainly natural origin peak during spring, associated with natural processes such as wind flow (Sahara dust transport), whereas elements associated with human activities peak during winter, due to domestic heating, traffic-related and industrial emissions. The organic to elemental carbon ratio indicates that the organic carbon is mostly primary and that the elemental part is strongly linked to traffic. During winter additional sources like household heating contribute to the total carbon loadings. The water-soluble organic to organic carbon ratio is characteristic for an urban area, demonstrating a higher ratio in the summertime, mostly due to the large fraction of secondary (oxidized and more soluble) organic species.

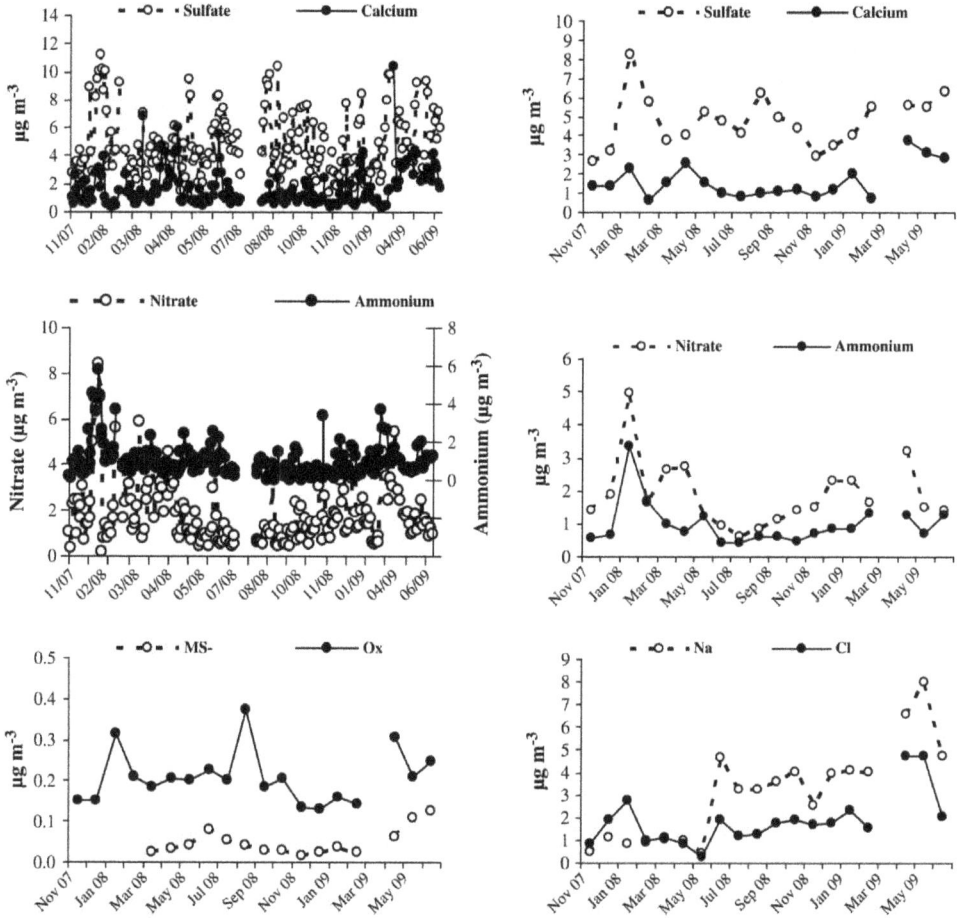

Fig. 12. Temporal distribution of major aerosol components in Istanbul (Theodosi et al., 2010).

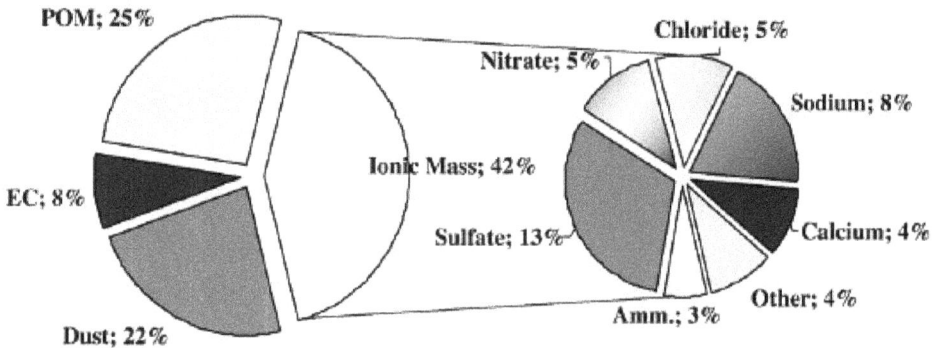

Fig. 13. Annual relative contribution of aerosol species to PM$_{10}$ mass (Theodosi et al., 2010).

Im et al. (2010) simulated the high PM_{10} levels observed during 13 to 17th of January, 2008 and showed that high resolution modeling with updated anthropogenic emissions can successfully reproduce the wintertime episodes associated with local anthropogenic emissions. Figure 14 shows the OC/PM_{10} ratios simulated by Im et al. (2010) which clearly indicate the nature of PM levels in the urban parts of the city. Around the emissions hot spots, which are located along the two sides of the Bosphorus, the ratios are calculated to be highest (~0.40). Figure 15 shows the origin of crustal materials from Sahara based on Kocak et al. (2011). As seen in the figure, Algerian, Libyan and Tunisian deserts are the important sources of natural dust affecting Istanbul, as well the north-eastern parts of Black sea region. The analysis has been conducted based on the extensive PM_{10} chemical composition data from November 2007 to June 2009 in Istanbul (Theodosi et al., 2010). Further analysis conducted by Kocak et al. (2011) clearly showed the potential impacts of Istanbul on its surroundings (Fig.16). The results show that Istanbul is under influence of several sources including Balkans and Eastern European countries. On the other hand, Istanbul pollution influences western Black Sea, Balkan counties, Levantine Basin and north-eastern Africa countries.

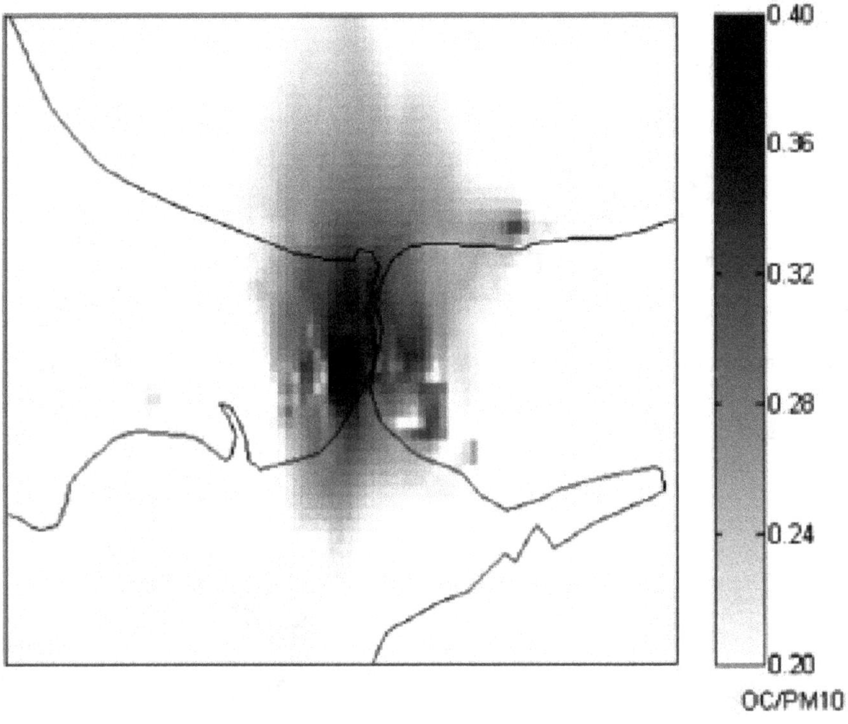

Fig. 14. Spatial distributions OC/PM_{10} mean ratio averaged over the 5-day period between 13th to 17th of January, 2008 (Im et al., 2010).

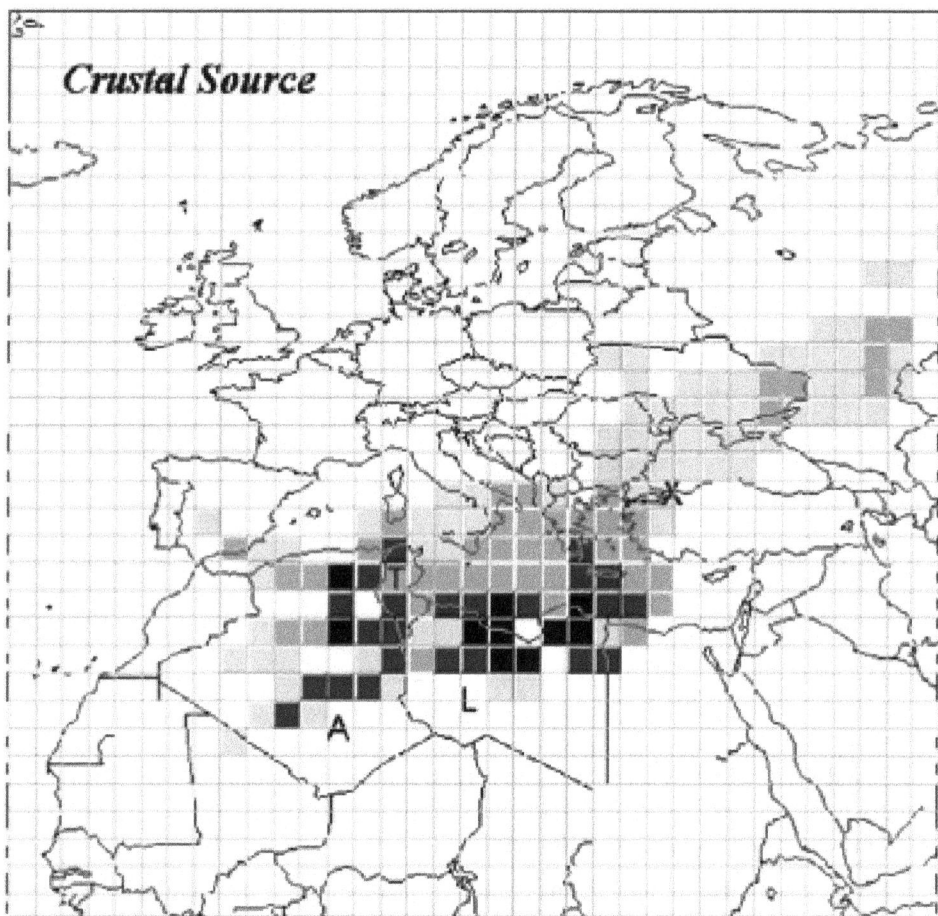

Fig. 15. Distribution of crustal source. X: Sampling site (Istanbul), A: Algeria, L: Libya and T: Tunisia Kocak et al. (2011).

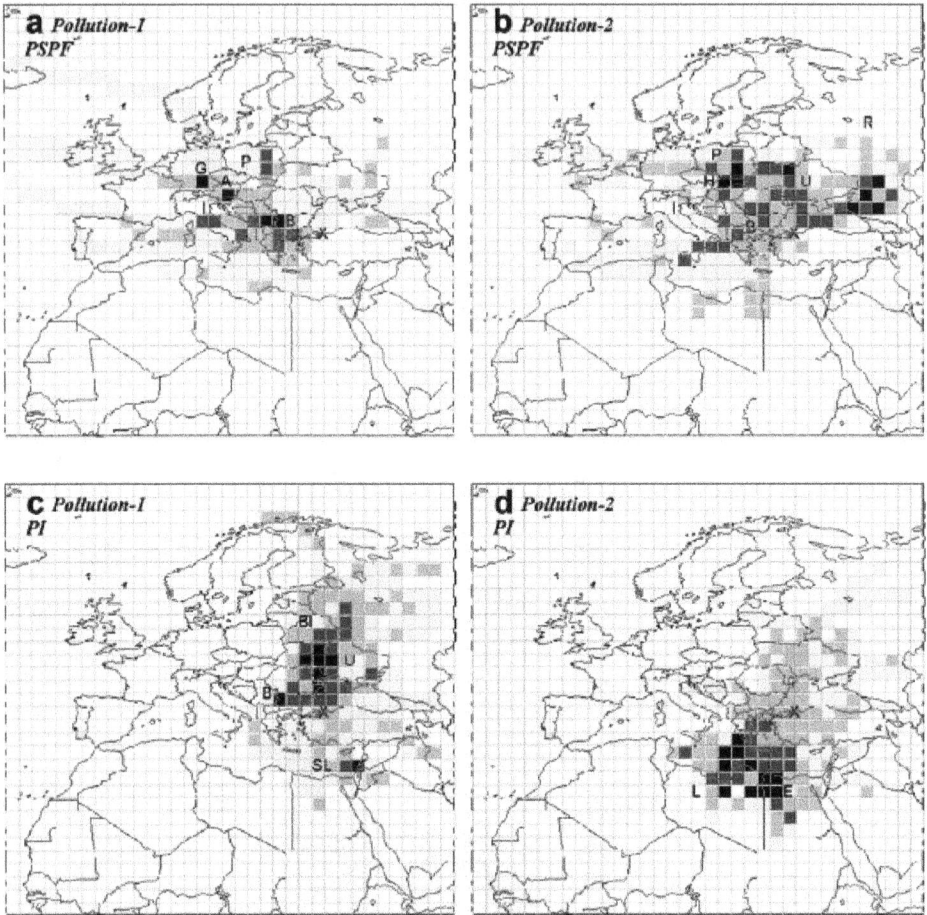

Fig. 16. Sources influencing Istanbul for a) wintertime primary sources, b) summer secondary and natural sources, and potential impacts of Istanbul on c) wintertime primary sources and d) summertime secondary and natural sources Kocak et al. (2011).

6.6.1 Surface ozone variations in Istanbul

Surface ozone is a secondary pollutant produced by a series of complicated photochemical reactions involving NO_x and HCs in the present of intense solar radiation. The first measurements of surface ozone at two sites in Istanbul (Kadikoy and Aksaray as seen in Fig.1)) were studied by Topcu & Incecik (2002 and 2003). They showed that ozone levels do not show yet episodic values in the city. But, when meteorological conditions are favorable such as when Istanbul and its surrounding region were dominated by an anticyclonic pressure system, ozone levels becomes high. During conducive ozone days, southerly and south-westerly winds with low speeds (<1m/s) influence Istanbul. Fig 17 presents the time serious of the daily peak ozone concentrations in between 2001-2006. Im et al., (2008) evaluated the highest ozone concentrations in Istanbul. They observed in summer periods having sunny days and maximum temperatures above 25 °C, and the episodes were mainly characterized by south-westerly surface winds during the day and north-easterly surface winds during the night. A modeling study conducted by Im et al. (2011a) showed that a NO_x-sensitive ozone chemistry is more pronounced in the northern parts of the city that are characterized by forested areas with high biogenic VOC emissions. Furthermore, the high anthropogenic NO_x emissions in the urban parts leads to response of ozone more to changes in NO_x, agreeing with the findings of Im et al. (2008).

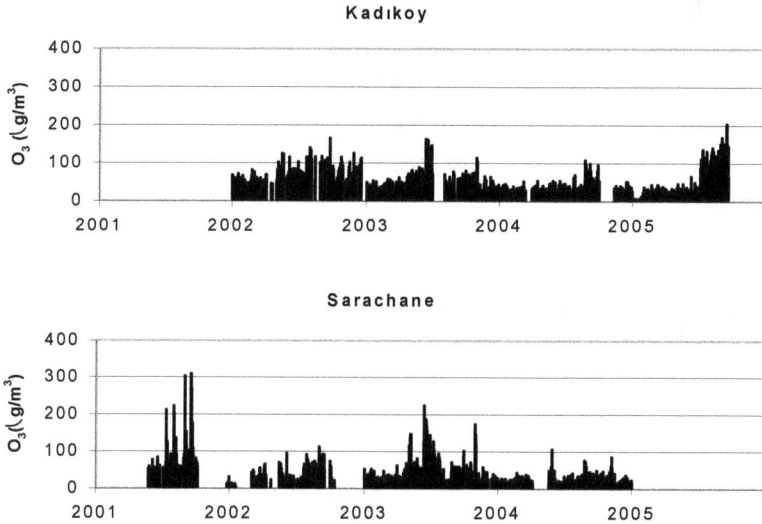

Fig. 17. Daily peak ozone concentrations measured in Aksaray (Sarachane) and Kadikoy.

Furthermore, an assessment of the wind field simulations for a case study in explaining the ozone formation mechanism over Istanbul is performed by Anteplioglu et al., 2003. In this study, meteorological conditions favorable for high ozone concentrations appear when Istanbul and the surrounding region are dominated by an anticyclonic pressure system. During the ozone favorable days, south and south-westerly winds with low wind speed influence Istanbul. In addition to the examining of the available monitoring air quality data in the city, surface ozone concentrations and NO, NO_2 were measured first time at three different sites one is on the island (Prince's Island or Büyükada) as background site; semi urban site, Kandilli just over the Bosphorus and as a nearby major high way site (Goztepe DMO). Figures 18a,b and 19a,b present the time serious of the data for the major high way and background stations (Incecik et al., 2010). The assessment of meteorological variables has shown that the production and destruction of the surface ozone was highly related to temperature and wind speeds. In urban and rural areas emissions of NO decrease ozone concentrations in the absence of solar radiation due to the reaction O_3, NO, NO_2 and NO_x. Conversely, O_3 concentrations show comparatively less diurnal variability in rural areas due to the absence of high NO_x emission sources.

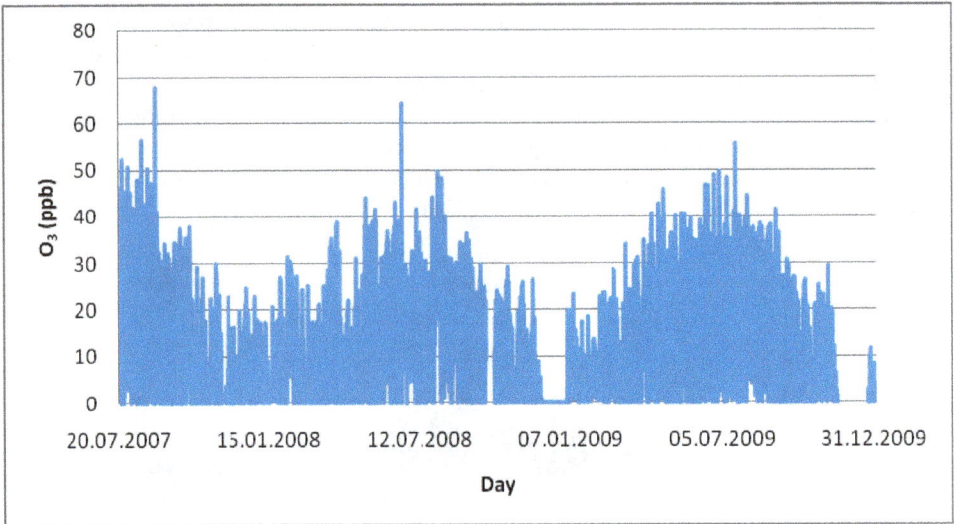

Fig. 18a. Hourly ozone concentrations for Goztepe (on major highway) air quality station, from July 20, 2007 to December 31, 2009.

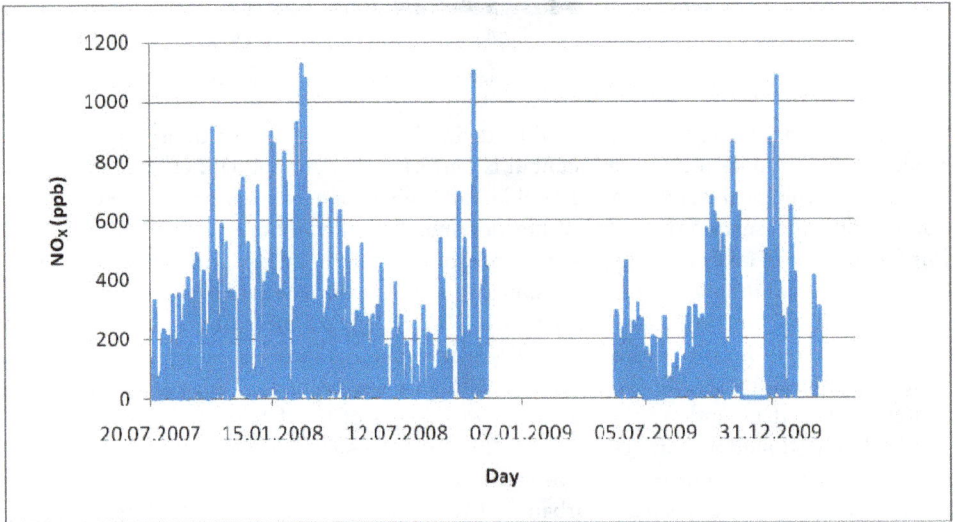

Fig. 18b. Hourly NO_x concentrations for Goztepe (on major highway) air quality station, from July 20, 2007 to December 31, 2009.

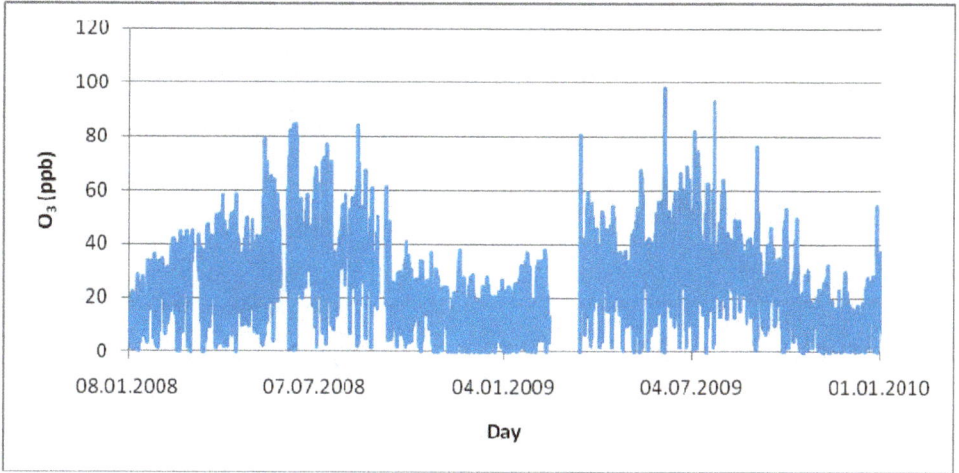

Fig. 19a. Hourly ozone concentrations for Princes's Island Air Quality Station, from January 9, 2008 to 31 December, 2009.

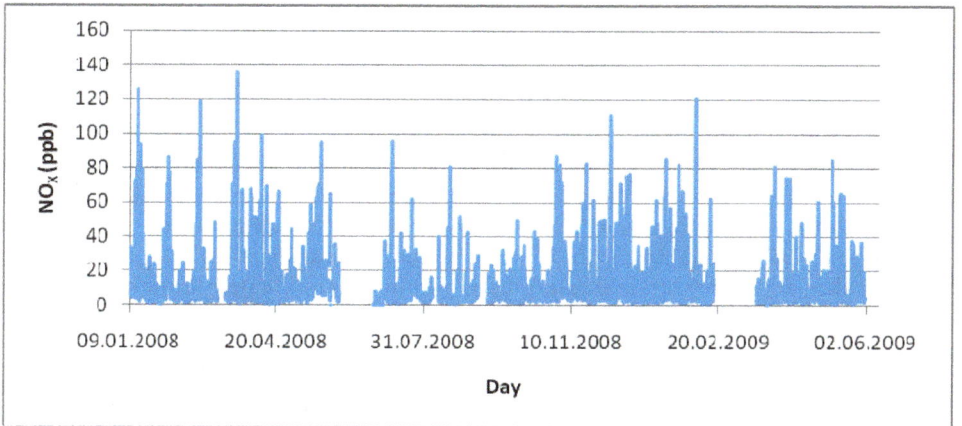

Fig. 19b. Hourly NO_x concentrations for Princes's Island Air Quality Station, from January 9, 2008 to 31 December, 2009.

Im et al. (2011b) simulated the summertime ozone concentrations in the Eastern Mediterranean, evaluating the contribution of the physical and chemical process to the simulated ozone levels. The results showed that for Istanbul, due to the high NO_x emissions, chemistry is a sink by 34% in the surface layer and 45% in the whole PBL. By horizontal (28%) and vertical transport (22%), Istanbul receives ozone. For the precursors of ozone (NO_x and VOCs), Istanbul is a sink in terms of transport, leading to transport of these species to downwind locations to produce ozone. Fig.20 shows the simulated surface ozone and NO_x concentrations in the area. As seen in the Fig. 20a, Istanbul is characterized by low ozone concentrations (~19 ppb) due to the high NO_x (Fig.20b)).

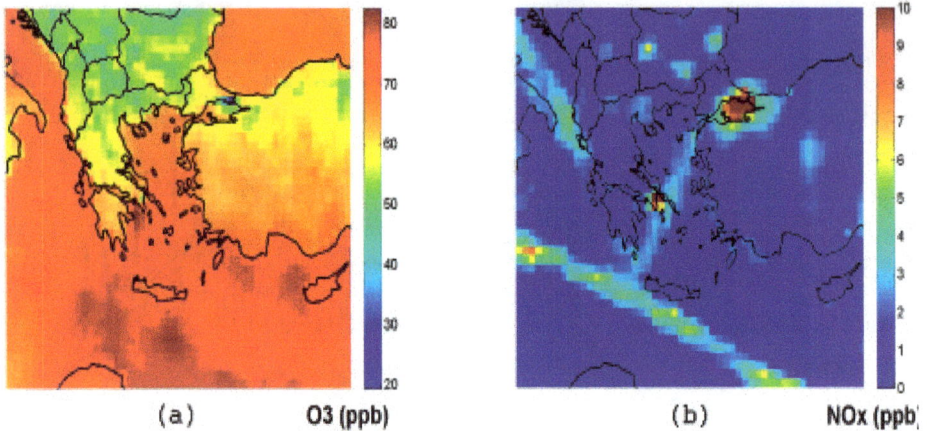

(a)   O3 (ppb)   (b)   NOx (ppb)

Fig. 20. Simulated levels of surface a) ozone and b) NO_x mixing ratios in Istanbul averaged over 1-15 June, 2004 (Im et al., 2011b).

6.7 Concluded remarks for Istanbul

As a summary, Istanbul had experienced many episodic air pollution events. One is very dramatic happened in January 1993. It is similar to a small scale London smog event. Following these events, Istanbul Greater Municipality established several strategically programs in Istanbul for air quality management in the beginning of 1990s. The first action was about the strict control program for the use of poor quality lignite in domestic heating and industry. Then fuel switching program was established in the city in 1992. It was first started in Asian parts of the city and then gradually expanded to the citywide. Now, about 95% of the city has natural gas. On the other hand, coal combustion is still a leading factor of the particulate emissions at residential heating even if it has a limited consumption in the city. Due to the high natural gas prices in the country, in recent years, some parts of the urban areas use coal instead of natural gas even if they are equipped with natural gas system in their houses. Furthermore, periodical motor vehicle inspection stations established at international standards in late 2008. However, traffic is still the most polluting sector for NO_x, NMVOC and CO emissions. Surface ozone levels are not elevated but depending on the increasing number of the motor vehicles, ozone potential is developing in the city. Industry is the most polluting sector for both sides of the city. Finally, shipping emissions over the Bosphorus in Istanbul is becoming serious problem. The emissions should be strictly controlled.

7. Conclusions

Megacities in general experience elevated levels of primary pollutants such as particular matter, carbon monoxide and sulfur dioxide whereas they influence the secondary pollutants like ozone in their surrounding regions. Air pollution in megacities has been influenced by many factors such as topography, meteorology, emissions from domestic heating, industry and traffic. The level of air pollution in megacities depends on the country's technology and pollution control capability by air quality improving plans, using cleaner fuels, renewable.

Megacities in developed and developing countries have different emission sources and air quality problems. Most of the megacities in developed countries (Los Angeles, New York, USA; Osaka, Tokyo, Japan and Paris, France) motor vehicles have been the major emission source and PM_{10}, $PM_{2.5}$, and NO_2 were the critical air quality parameters whereas, industry, motor vehicles and residential heating were major emissions in developing countries. In addition to the emission sources, the critical air quality parameters in megacities of developing countries are PM_{10}, $PM_{2.5}$, NO_2, SO_2 and CO. Dust transport leads to serious air quality and visibility problem in some megacities in developing countries such as Beijing and Cairo.

Rapid urbanization has resulted in increasing urban air pollution in major cities, especially in the developing countries. Over 90% of air pollution in cities of these countries is attributed to vehicle emissions brought about by high number of older vehicles coupled with low fuel quality. Increased population density causes new risks for the residents through deteriorated environment and social problems due to the intense and complicated interactions between economic, demographic, social political and ecological processes in megacities.

Finally, in megacities of the developing world, in order to enhance quality of life, city planning needs to adopt new visions and innovating management tools. One of the key public concerns in megacities is transportation. High population density and high motor vehicle rates need to be improved. Transport policies should consist of multiple strategies particularly in developing countries.

8. Acknowledgements

The authors acknowledge Özkan Çapraz, Dr Huseyin Toros and Melike Celebi for their help with graphics of this chapter.

9. References

Anteplioğlu, U.; Topçu, S. & İncecik, S. (2003). An Application of a Photochemical Model for Urban Air Shed in Istanbul, *Journal of Water, Air & Soil Pollution: Focus*, Vol.3(5-6), pp. 55-66.
Abu-Allaban, M.; Gertler, A.W. & Lowenthal, D.H. (2002). A Preliminary Apportionment of the Sources of Ambient PM_{10}, $PM_{2.5}$ and VOCs in Cairo. *Atmospheric Environment*, Vol.36, pp. 5549- 5557.
Abu-Allaban, M.; Lowenthal, D.H.; Gertler, A.W. & Labib, M. (2007). Sources of PM_{10} and $PM_{2.5}$ in Cairo's Ambient Air. *Environmental Monitoring and Assessment*, Vol.133, pp. 417-425.
Abu-Allaban, M.; Lowenthal, D.H.; Gertler, A.W. & Labib, M. (2009) . Sources of Volatile Organic Compounds in
Cairo's Ambient Air. *Environmental Monitoring and Assessment*, Vol.157, pp. 179–189.

Atimtay, A. & İncecik, S. (2004). Air Pollution Problem in Istanbul and Strategically Efforts in Air Quality Management, *13th World Clean Air Congress*, 22-27 Aug.2004, London.

APCR, Air Pollution Country Report, Rep. of Turkey, Min. of Health, Refik Saydam Center for Hygiene (1998).

Baklanov, A. (2011). Megapoli Project Megacities: Emissions, Urban, Regional and Global Atmospheric Pollution and Climate Effects, and Integrated Tools for Assessment and Mitigation THEME FP7-ENV-2007.1.1.2.1: Megacities and Regional Hot-spots Air Quality and Climate Collaborative Project (Medium-Scale Focused Research Project) Grant Agreement no.: 212520.

Baldasano, J.M.; Valera, E. & Jimenez, P. (2003). Air Quality Data from Large Cities, *Science of the Total Environment*, Vol.307, pp. 141 – 165.

Batuk, D.N.; Gürsoy, E.; Ertut, H.; Erdun, H. & Incecik, S. (1997). Analysis of SO_2 and TSP Under Mesoscale Weather Conditions in Istanbul, Turkey, *Environ. Res. Forum*, Vol 7-8,pp 73-83.

Butler, T.M.; Lawrence, M.G.; Gurjar, B.R.; van Aardenne, J.; Schultz, M. & Lelieveld, J. (2008). The Representation of Emissions from Megacities in Global Emission Inventories, *Atmos. Environ.*, Vol.42, pp. 703-719.

Butler, T.M. & Lawrence, M.G. (2009). The Influence of Megacities on Global Atmospheric Chemistry: a Modeling Study, *Environ. Chem.*, Vol.6, pp. 219-225.

Cairo Air Improvement Project, (2004). USAID/Egypt, Office of Environment Contract 263-C-00-97-00090-00

Celebi, M.; İncecik, S. & Deniz, A. (2010). Investigation of PM_{10} Levels in Urban Atmosphere of a Megacity: Istanbul, *Better Air Quality Conference*, BAQ2010, 9-11 Nov. 2010. Singapore.

CCS Report , (2010). Country Cooperation Strategy for WHO and Egypt 2005-2009, WHO Regional Office for the Eastern Mediterranean.

Chan, C.K. & Yao, X. (2008). Review: Air Pollution in Megacities in China, *Atmospheric Environment*, Vol.42, pp. 1-42.

Decker, E.H.; Elliot, S.; Smith, F.A.; Blake, D.R. & Rowland, F.S. (2000). Energy and Material Flow Through the Urban Ecosystem. *Annual Review of Environment and Resources*, Vol.25, pp. 685 – 740.

Deng, F.F. & Huang, Y. (2004). Uneven Land Reform and Urban Sprawl in China: the Case of Beijing. *Progress in Planning*, Vol.61, pp. 211–236.

Deniz, C. & Durmusoglu, Y. (2008). Estimating Shipping Emissions in the Region of the Sea of Marmara, Turkey. *Science of the Total Environment*, Vol.390, pp. 255-261.

EEPP-Air, (2004). Egyptian Environmental Policy Program, Planning for Integrated Air Quality Management. Final Report | June, Chemonics International Inc. 1133 20t h Street, NW Washington, DC 20036 / USA. USAID/Egypt, Office of Environment Contract 263-C-00-97-00090-00

Elbir, T. (2008). A GIS Based Decision Support System for Urban Air Quality Management in the City of İstanbul (LIFE06-TCY/TR/000283)

Elbir, T.; Mangir, N.; Kara, M. & Ozdemir, S. (2010). Development of a GIS-based Decision Support System for

Urban Air Quality Management in the City of Istanbul, *Atmospheric Environment*, Vol.44, pp. 441-454.

Fang, M.; Chan, C.K. & Yao, X. (2009). Managing Air Quality in a Rapidly Developing Nation, China, *Atmospheric Environment*, Vol.43, pp. 79-86.

Farber, R.J.; Welsing, P.R. & Rozzi, C. (1994). PM_{10} and Ozone Control Strategy to Improve Visibility in the Los Angeles Basin, *Atmospheric Environment*, Vol.28, pp. 3277-3283.

Favez, O.; Cachier, H.; Sciare, J.; Alfaro, S. C.; El-Araby, T. M.; Harhash, M. A. & Abdelwahab , M. M. (2008). Seasonality of Major Aerosol Species and Their Transformation in Cairo, *Atmospheric Environment*, Vol.42, pp. 1503-1516.

Government of India (2006). Government Report.

Guenther, A.; Karl, T.; Harley, P.; Wiedinmyer, C.; Palmer, P.I. & Geron, C. (2006). Estimates of Global Terrestrial Isoprene Emissions Using MEGAN (Model of Emissions of Gases and Aerosols from Nature, *Atmospheric Chemistry and Physics Discussions*, Vol.6, pp. 107 – 173.

Gurjar, B. R.; van Aardenne, J. A.; Lelieveld, J. & Mohan, M. (2004). Emission Estimates and Trends (1990-2000) for Megacity Delhi and Implications. *Atmospheric Environment*, 38 (33), 5663-5681. doi: 10.1016/j.atmosenv.2004.05.057

Gurjar, B. R. & Lelieveld, J. (2005). New Directions: Megacities and Global Change. *Atmospheric Environment*, 39, 391-393. doi:10.1016/j.atmosenv.2004.11.002

Gurjar, B. R; Butler, T. M.; Lawrence, M. G. & Lelieveld, J. (2008). *Evaluation of Emissions and Air Quality in Megacities, Atmospheric Environment* , Vol.42, pp. 1593-1606.

Gusten, H.; Heinrich, G.; Wappner, J.; Abdel-Aal, M.M.; Abdel-Hay, F.A.; Ramadan, A.B.; Tawfik, F.S.; Ahmed, D.M.; Hassan, G.K.Y.; Cuitas, T.; Jeftic, J. & Klasinc, L. (1994). Ozone Formation in The Greater Cairo Area, *Science of The Total Environment*, Vol.155, pp. 285 – 295.

Hao, J.; Wang, L.; Shen, M.; Li, L. & Hu, J. (2007). Air Quality Impacts of Power Plant Emissions in Beijing, *Environmental Pollution*, Vol.147, pp. 401 – 408.

HEI (2004). International Scientific Oversight Committee, 2004.

HEI, (2010). Health Effects Institute Special Report 18, 2010.

IBB, (2009). Istanbul Municipality Report.

Im, U.; Tayanc, M. & Yenigün, O. (2008). Interaction Patterns of Major Photochemical Pollutants in Istanbul, Turkey, *Atmospheric Research*, Vol.89, pp. 382-390.

Im, U.; Markakis, K.; Unal, A.; Kindap, T.; Poupkou, A.; Incecik, S.; Yenigun, O.; Melas, D.; Theodosi, C. & Mihalopoulos, N. (2010). Study of a Winter PM Episode in Istanbul Using High Resolution WRF/CMAQ Modeling System. *Atmospheric Environment*, Vol.44, pp. 3085-3094.

Im, U.; Poupkou, A.; Incecik, S.; Markakis, K.; Kindap, T.; Unal, A.; Melas, D.; Yenigun, O.; Topcu, S.; Odman, M.T.; Tayanc, M. & Guler, M. (2011a). The Impact of Anthropogenic and Biogenic Emissions on Surface Ozone Concentrations in Istanbul. *Science of the Total Environment*, Vol.409, pp. 1255-1265.

Im, U.; Markakis, K.; Poupkou, A.; Melas, D.; Unal, A.; Gerasopoulos, E.; Daskalakis, N. & Kanakidou, M. (2011b). The Impact of Temperature Changes on Summer Time Ozone and Its Precursors in the Eastern Mediterranean. *Atmospheric Chemistry and Physics*, Vol.11, pp. 3847-3864.

Incecik, S. (1996). Investigation of Atmospheric Conditions in Istanbul Leading to Air Pollution Episodes, *Atmospheric Environment*. Vol.30, pp. 2739-2749.

Incecik, S.; Im, U.; Yenigün, O.; Odman, M.T.; Kindap, T.; Topcu, S.; Tek, A. & Tayanç, M. (2010). Simulations of Meteorological Fields and Investigation of Air Pollution in

Istanbul and Kocaeli Regions. Turkish Scientific and Technical Research Council Project No: 105Y005 (in Turkish)

Jahn, H.J.; Schnider, A.; Breitner, S.; Eibner, R. & Wendisch, M. (2011). Particulate Matter Pollution in the Megacities of the Pearl River Delta, China-A Systematic Literature Review and Health Risk Assessment, *Int.J. Hygiene and Environmental Health*, Vol.214, pp. 281-295.

Karaca, F.; Anil, I. & Alagha O. (2009). Long-range Potential Source Contributions of Eisodic Aerosol Events to PM_{10} Profile of a Megacity, *Atmospheric Environment*, Vol.43, pp. 5713 – 5722.

Kai, Z.; You-Hua, Y.; Qiang, L.; Ai-Jun, L. & Shao-Lin, P. (2007). Evaluation of Ambient Air Quality in Guangzhou, China, *Journal of Environmental Sciences*, Vol.19, pp. 432–437.

Kanakidou, M.; Mihalopoulos, N.; Kindap, T.; Im, U.; Vrekoussis, M.; Gerasopoulos, E.; Dermitzaki, E.; Unal, A.; Koçak, M.; Markakis, K.; Melas, D.; Kouvarakis, G.; Youssef, A.F.; Richter, A.; Hatzianastassiou, N.; Ebojie, A. F.; Wittrock, F.; von Savigny, C.; Burrows, J. P.; Weissenmayer, A. L. & Moubasher, H. (2011). Review: Megacities as Hot Spots of Air Pollution in the East Mediterranean, *Atmospheric Environment*, Vol.45, pp. 1223-1235.

Kesgin, U. & Vardar, N. (2001). A Study on Exhaust Gas Emissions from Ships in Turkish Streets. *Atmospheric Environment*, Vol.35, pp. 1863-1870.

Khoder, M. (2009). Diurnal Seasonal and Weekdays-Weekends Variation of Ground Level Ozone Concentrations in an Urban Area in Grater Cairo, *Environmental Monitoring and Assessment*, Vol.149, pp. 349 – 362.

Kim, S. (2007). Immigration, Industrial Revolution and Urbanization in the United States, 1820-1920: Factor Endowments, *Technology and Geography*, NBER Working Paper No. 12900.

Kindap, T.; Unal, A.; Chen, S.H.; Odman, M.T. & Karaca, M.(2006). Long-range Aerosol Transport from Europe to Istanbul, Turkey, *Atmopheric Environment*, Vol.40, pp. 3536 – 3547.

Kocak, M.; Theodosi, C.; Zarmpas, P.; Im, U.; Bougiatioti, A.; Yenigun, O. & Mihalopoulos, N. (2011). Particulate matter (PM_{10}) in Istanbul: Origin, source areas and potential impact on surrounding regions. *Atmospheric Environment*, 45, 6891-6900.

Lawrence, M.G.; Butler, T.M.; Steinkamp, J.; Gurjar, B.R. & Lelieveld, J. (2007). Regional Pollution Potentials of Megacities and Other Major Population Centers, *Atmos. Chem. Phys.*, Vol.7, pp. 3969-3987.

Mage, D.; Ozolins, G.; Peterson, P.; Webster, A.; Orthofer, R.; Vandeweerd, V. & Gwynne, M. (1996). Urban Air Pollution in Megacities of the World. *Atmospheric Environment*, Vol.30, pp. 681-686.

Mahmoud, K.F.; Alfaro, S.C.; Favez, O.; Abdel Wahab, M.M. & Sciare, J. (2008). Origin of Black Carbon Concentration Peaks in Cairo, *Atmospheric Research*, Vol.89, pp. 161-169.

Markakis, K., Im, U., Unal, A., Melas, D., Yenigun, O. & Incecik S. (2012). Compilation of a GIS based high spatially and temporally resolved emission inventory for the Greater Istanbul Area. *Atmospheric Pollution Research*, Vol. 3, pp. 112-125.

Markakis, K.; Im, U.; Unal, A.; Melas, D.; Yenigun, O. & Incecik, S., (2009). A Computational Approach for the Compilation of a High Spatially and Temporally Resolved

Emission Inventory for the Istanbul Greater Area. In: *7th International Conference on Air Quality Science and Application*, 24–27 March 2009, Istanbul.

Markakis, K.; Poupkou, A.; Melas, D.; Tzoumaka, P. & Petrakakis, M. (2010). A Computational Approach Based on GIS Technology for the Development of an Anthropogenic Emission Inventory for Air Quality Applications in Greece, *Water, Air & Soil Pollution*, Vol.207, pp. 157–180.

Molina, M.J. & Molina, L.T. (2002). Air Quality in the Mexico Megacity: An Integrated Assessment, Springer, New York.

Molina M.J, Molina L.T, (2004). Megacities and Atmospheric Pollution, *Journal of Air &Waste Management Assoc.* Vol.54, pp. 644-680.

National Summary Report, (2010). Air Quality Monitoring, Emission Inventory and Source Apportionment Study for Indian Cities, Central Pollution Control Board.

Narain, U. & Krupnick, A. (2007). The Impact of Delphi's CNG Program on Air Quality, RFF Discussion Paper No 07 – 06.

Patz, J.; Campbell-Lendrum, D.; Holloway, T. & Foley, J. A. (2005). Impact of Regional Climate Change on Human Health, *Nature*, Vol.438, pp. 310-317.

Prasad, A. K.; El-Askary, H. & Kafatos, M. (2010). Implications of High Altitude Desert Dust Transport from Western Sahara to Nile Delta During Biomass Burning Season , *Environmental Pollution*, Vol.158, pp. 3385-3391.

Safar, Z. S. & Labib, M. W. (2010). Assessment of Particulate Matter and Lead Levels in the Greater Cairo Area for the Period 1998–2007, *Journal of Advanced Research*, Vol.1, pp. 53-63.

Song, Y.; Zhang, M. & Cai, X. (2006). PM_{10} Modeling of Beijing in the Winter, *Atmospheric Environment*, Vol.40, pp. 4126-4136.

Tayanc, M. (2000). An Assessment of Spatial and Temporal Variation of Sulfur Dioxide Levels over Istanbul, *Env.Poll.*, Vol. 107, pp. 61-69.

Theodosi, C.; Im, U.; Bougiatuoti, A.; Zarmpas, P.; Yenigun, O. & Mihalopoulos N. (2010). Aerosol Chemical Composition Over Istanbul. *Science of the Total Environment*, Vol.408, pp. 2482-2491.

Topcu, S.; Incecik, S. & Unal, Y.S. (2001). The Influence of Meteorological Conditions and Stringent Emission Control on High TSP Episodes in Istanbul , *Environ. Sci. and Pollut. Res.*, Vol.10, pp. 24-32.

Topcu, S. & İncecik, S. (2002). Surface Ozone Measurements and Meteorological Influences in the Urban Atmosphere of Istanbul, *Int.Journal of Environment and Pollution*, Vol.17, pp. 390-404.

Topcu, S. & İncecik, S. (2003). Characteristics of Surface Ozone Concentrations in Urban Atmosphere of Istanbul: A

Case Study, *Fresenius Environmental Bulletin*, Vol.12, pp.413-417.

UN and League of Arap States Report, (2006). Air Quality and Atmospheric Pollution in the Arab Region, UN Economic and Social Commission for Western Asia & League of Arab States.

UN (2009). United Nations Department of Economic and Social ffairs/Population Division. World Urbanization Prospects.

Unal, Y.S.; Incecik, S.; Borhan, Y. & Mentes, S. (2000). Factors Influencing the Variability of SO_2 Concentrations in Istanbul, *Journal of the Air & Waste Management Association*, Vol.50, pp. 75-84.

Unal, Y.S.; Deniz, A.; Toros, H. & Incecik, S. (2011). Influence of Meteorological Factors and Emission Sources on Spatial and Temporal Variations of PM_{10} Concentrations in Istanbul Metropolitan Area, *Atmospheric Environment*, Vol.45, pp. 5504-5513.

UNEP United Nations Environment Program, Urban Environment Unit.

United Nations Environment Programme/Global Environmental Monitoring System. Urban Air Pollution; Environmental Library No.4, Oxford, England, 1991.

Visschedijk, M.; Hendriks, R. & Nuyts, K. (2005). How to Set up and Manage Quality Control and Quality Assurance, *The Quality Assurance Journal*, Vol.9, pp. 95 – 107.

Wang, H.; Fu, L.; Zhou, Y.; Du, X. & Ge, W. (2010). Trends in Vehicular Emissions in China's Mega Cities from 1995 to 2005, *Environmental Pollution*, Vol.158, pp. 394–400.

WHO, (2008). Air Quality and Health, Fact Sheet N°313

WHO, (2009). Center for Health Development, World Health Organization Centre for Health Development (WHO Kobe Centre – WKC)

Wu, Z.; Hu, M.; Lin, P.; Liu, S.; Wehner, B. & Wiedensohler, A. (2008). Particle Number Size Distribution in the Urban Atmosphere of Beijing, China, *Atmospheric Environment*, Vol.42, pp. 7967-7980.

Zakey, A.S.; Abdelwahab, M.M.; Pattersson, J.J.C.; Gatari, M.J. & Hallquist, M. (2008). Seasonal and Spatial Variation of Atmospheric Particular Matter in a Developing Megacity, the Greater Cairo, *Egypt Atmosfera*, Vol.21, pp. 171-189.

Zhao,P (2010). Sustainable Urban Expansion and Transportation in a Growing Megacity: Consequences of Urban Sprawl for Mobility on the Urban Fringe of Beijing, *Habitat International*, Vol.34, pp. 236–243.

Zhou, K.; Ye, Y-H.; Liu, Q.; Liu, A-J. &Peng, S-L. (2007). Evaluation of Ambient Air Quality in Guangzhou, China, *Journal of Environmental Sciences*, Vol.19, pp. 432-437.

Zhu, L.; Huang, X.; Shi, H.; Cai, X. & Song, Y. (2011). Transport Pathways and Potential Sources of PM_{10} in Beijing, *Atmospheric Environment*, Vol.45, pp. 594-604.

2

Air Pollution Monitoring Using Earth Observation & GIS

Diofantos G. Hadjimitsis, Kyriacos Themistocleous and Argyro Nisantzi
Cyprus University of Technology
Cyprus

1. Introduction

Air pollution has received considerable attention by local and global communities (Wald et al, 1999). Many major cities have established air quality monitoring stations but these stations tend to be scarcely distributed and do not provide sufficient tools for mapping atmospheric pollution since air quality is highly variable (Wald et al, 1999). Satellite remote sensing is a valuable tool for assessing and mapping air pollution as satellite images are able to provide synoptic views of large areas in one image on a systematic basis due to the temporal resolution of the satellite sensors. Blending together earth observation with ground supporting campaigns enables the users or governmental authorities to validate air pollution measurements from space. New state-of-the-art ground systems are used to undertake field measurements such as Lidar system, automatic and hand held sun-photometers etc. The rise of GIS technology and its use in a wide range of disciplines enables air quality modellers with a powerful tool for developing new analysis capability. Indeed, thematic air pollution maps can be developed. Moreover, the organization of data by location allows data from a variety of sources to be easily combined in a uniform framework. GIS provides the opportunity to fill the technical gap between the need of analysts and decision-makers for easy understanding of the information.

This Chapter presents an overview of how earth observation basically is used to monitor air pollution through on overview of the existing ground and space systems as well through the presentation of several case studies.

2. Relevant literature review

The use of earth observation to monitor air pollution in different geographical areas and especially in cities has received considerable attention (Kaufman and Fraser, 1983; Kaufman et al., 1990; Sifakis and Deschamps, 1992; Retalis, 1998; Retalis et al., 1998; Sifakis et al., 1998; Retalis et al., 1999; Wald and Balleynaud, 1999; Wald et al., 1999; Hadjimitsis et al., 2002; Themistocleous et al., 2010; Nisantzi et al. 2011; Hadjimitsis et al., 2011). All the studies have involved the determination of aerosol optical thickness (AOT) either using indirect methods using Landsat TM/ETM+, ASTER, SPOT, ALOS, IRS etc. images or MODIS images in which AOT is given directly. The superior spatial resolution of Landsat and SPOT enable several researchers to develop a variety of methods based on the radiative transfer equation, atmospheric models and image-based methods. Indeed, Kaufman et al. (1990) developed an algorithm for determining the aerosol optical thickness (using land and water dark targets) from the difference in the upward radiance recorded by the satellite between a clear and a hazy day. This method assumes that the surface reflectance between the clear day and the hazy day images does not change. Sifakis and Deschamps (1992) used SPOT images to estimate the distribution of air pollution in the city of Toulouse in France. Sifakis and Deschamps (1992) developed an equation to calculate the aerosol optical depth difference between one reference image (acquired under clear atmospheric conditions) and a polluted image. Their method was based on the fact that, after correction of solar and observation angle variations, the remaining deviation of apparent radiances is due to pollutants. Retalis (1998) and Retalis et al. (1999) showed that an assessment of the air pollution in Athens could be achieved using the Landsat TM band 1 by correlating the aerosol optical thickness with the acquired air-pollutants. Moreover, Hadjimitsis and Clayton (2009) developed a method that combines the Darkest Object Subtraction (DOS) principle and the radiative transfer equations for finding the AOT value for Landsat TM bands 1 and 2. Hadjimitsis (2009a) used a new method for determining AOT through the use of the darkest pixel atmospheric correction over the London Heathrow Airport area in the UK and the Pafos Airport area in Cyprus. Hadjimitsis (2009b) developed a method to determine the aerosol optical thickness through the application of the contrast tool (maximum contrast value), the radiative transfer calculations and the 'tracking' of the suitable darkest pixel in the scene for Landsat, SPOT and high resolution imagery such as IKONOS and Quickbird.

Satellite remote sensing is certainly a valuable tool for assessing and mapping air pollution due to their major benefit of providing complete and synoptic views of large areas in one image on a systematic basis due to the good temporal resolution of various satellite sensors (Hadjimitsis et al., 2002). Chu et al. (2002) and Tang et al. (2004) reported that the accuracy of satellite-derived AOT is frequently assessed by comparing satellite based AOT with AERONET (AErosol RObotic NETwork – a network of ground-based sun-photometers) or field based sun-photometer. The determination of the AOT from the Landsat TM/ETM+ imagery is based on the use of radiative transfer equation, the principle of atmospheric correction such as darkest pixel method (Hadjimitsis et al., 2002). Since MODIS has a sensor with the ability to measure the total solar radiance scattered by the atmosphere as well as the sunlight reflected by the Earth's surface and attenuated by atmospheric transmission, the AOT results found by MODIS were compared with the sun-photometer's AOT results for systematic validation (Tang et al., 2004). For the MODIS images, despite its low pixel resolution, aerosol optical thickness values were extracted for the 550 nm band. MODIS data have been extended correlated with PM10 over several study areas (Hadjimitsis et al., 2010; Nisanzti et al., 2011) with improved correlation coefficients but with several limitations. PM2.5 are also found to be correlated against MODIS data using new novel improved algorithms (e.g Lee et al., 2011) that considers the low spatial resolution of MODIS with future positive premises. Themistocleous et al. (2010) used Landsat TM/ETM+ and MODIS AOT data to investigate the air pollution near to cultural heritage sites in the centre of Limassol area in Cyprus. Nisantzi et al. (2011) used both particulate matter (PM10) device, backscattering lidar system, AERONET and hand-held sun-photometers for developing PM10 Vs MODIS AOT regression model over the urban area of Limassol in Cyprus.

It is expected that future satellite technologies will provide data with finer spatial and temporal resolutions and more accurate data retrievals (Lee et al., 2011). Moreover, the advanced capability of discriminating by aerosol species in satellite technologies will further contribute to health effect studies investigating species-specific health implications (Lee et al., 2011).

3. Aerosol Optical Thickness – Aerosol Optical Depth (AOD)

The key parameter for assessing atmospheric pollution in air pollution studies is the aerosol optical thickness, which is also the most important unknown element of every atmospheric correction algorithm for solving the radiative transfer equation and removing atmospheric effects from satellite remotely sensed images (Hadjimitsis et al, 2004). Aerosol optical thickness (AOT) is a measure of aerosol loading in the atmosphere (Retails et al, 2010). High AOT values suggest high concentration of aerosols, and therefore air pollution (Retalis et al, 2010). The use of earth observation is based on the monitoring and determination of AOT either direct or indirect as tool for assessing and measure air pollution. Measurements on PM10 and PM2.5 are found to be related with the AOT values as shown by Lee et al. (2011), Hadjimitsis et al. (2010), Nisantzi et al. (2011).

"Aerosol Optical Thickness" is defined as the degree to which aerosols prevent the transmission of light. The aerosol optical depth or optical thickness (τ) is defined as 'the integrated extinction coefficient over a vertical column of unit cross section. Extinction coefficient is the fractional depletion of radiance per unit path length (also called attenuation especially in reference to radar frequencies)'. (http://daac.gsfc.nasa.gov/data-holdings/PIP/aerosol_optical_thickness_or_depth.shtml).

4. PM10

Particulate matter (PM10) pollution consists of very small liquid and solid particles floating in the air. PM10 particles are less than 10 microns in diameter. This includes fine particulate matter known as PM2.5. PM10 is a major component of air pollution that threatens both our health and our environment.

5. Satellite imagery

Several remotely sensed data are used to provide direct or indirect air or atmospheric pollution measurements. MODIS satellite imagery provides direct AOT measurements; however, Landsat TM/ETM+, ASTER, SPOT or other satellites provide indirect measurements of AOT through the application of specific techniques.

5.1 MODIS

The Moderate Resolution Imaging Spectroradiometer (MODIS) onboard NASA's Terra spacecraft performs near-global daily observations of atmospheric aerosols. The MODIS instrument measures upwelling radiance in 36 bands for wavelengths ranging from 0.4 to 14.385 μm and has a spatial resolution of 250m (2 channels), 500m (5 channels), and 1 km (29 channels) at nadir. (Kaufman et al, 1997; Tanré et al, 1997, 1999; Remer et al, 2005 ; Vermote and Saleous, 2000). The aerosol retrieval makes use of 7 of these channels (0.47 – 2.12 μm) to retrieve aerosol boundaries and characteristics. MODIS has one camera and measures radiances in 36 spectra bands, from 0.4 µm to 14.5 µm with spatial resolutions of 250m (bands 1-2), 500 m (bands 3-7) and 1000 m (bands 8-36). Daily level 2 (MOD04) aerosol optical thickness data (550 m) are produced at the spatial resolution of 10x10 km over land, using the 1 km x 1 km cloud-free pixel size. MODIS aerosol products are provided over land and water surfaces. Typical MODIS satellite image with raw AOT values over Cyprus is shown in Figure 1.

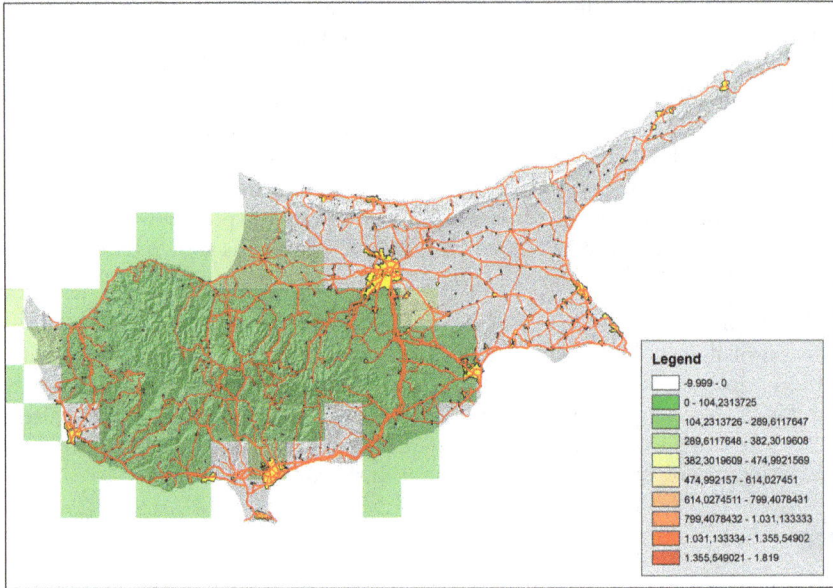

Legend

	-9.999 - 0
	0 - 104,2313725
	104,2313726 - 289,6117647
	289,6117648 - 382,3019608
	382,3019609 - 474,9921569
	474,992157 - 614,027451
	614,0274511 - 799,4078431
	799,4078432 - 1.031,133333
	1.031,133334 - 1.355,54902
	1.355,549021 - 1.819

Fig. 1. MODIS AOT satellite image (direct measurements of AOT).

5.2 ASTER

ASTER is an advanced multi-spectral imager that was launched onboard NASA's Terra spacecraft in December 1999. ASTER covers a wide spectral region with 14 bands from visible to thermal infrared with high spatial, spectral and radiometric resolutions. The spatial resolution varies with wavelength: 15m in the visible and near-infrared (VNIR), 30m in the short wave infrared (SWIR), and 90m in the thermal infrared (TIR).

5.3 LANDSAT TM/ETM+

The Landsat Thematic Mapper (TM) is an advanced, multi-spectral scanning, Earth resources sensor designed to achieve higher spatial resolution, sharper spectral separation, improved geometric fidelity, and greater radiometric accuracy and resolution (Guanter, 2006). TM data are scanned simultaneously in 7 spectral bands. Band 6 scans thermal (heat) infrared radiation, the other ones scan in the visible and infrared. The Enhanced Thematic Mapper-Plus (ETM+) on Landsat-7 that was launched on April 15, 1999 is providing observations at a higher spatial resolution and with greater measurement precision than the previous TM

6. Ground measurements for supporting remote sensing measurements

As both direct or indirect retrieval of the aerosol optical thickness from satellites require ground validation, several resources (equipments, systems) are available to support such outcomes. The authors provide an overview of the existing systems which are used to support the air pollution measurements from space. Such systems are available at the Cyprus University of Technology-The Remote Sensing Laboratory.

6.1 Automatic sun-photometer (AERONET)

The CIMEL sun-tracking photometer (Fig.2) measures sun and sky luminance in visible and near-infrared. Specifically, it measures the direct solar radiance at 8 wavelengths (340nm, 380nm, 440nm, 500nm, 675nm, 870nm, 1020nm, 1640nm) and sky radiance at four of these wavelengths (440nm, 670nm, 865nm, 1020nm) providing information for determination of aerosol optical properties such as size distribution, angstrom exponent, refractive index, phase function and AOT. The CIMEL sun-photometer is calibrated according to program of AERONET specifications. An example of how measurements of AERONET aerosol optical depth (AOD) retrievals compared with the AOD derived from satellite MODIS data are correlated are shown in Figure 3. Such regression models are essential in order to test and validate the AOD retrieved from satellites with ground measurements.

Fig. 2. CIMEL sun-photometer (Cyprus University of Technology Premises: Remote Sensing Lab).

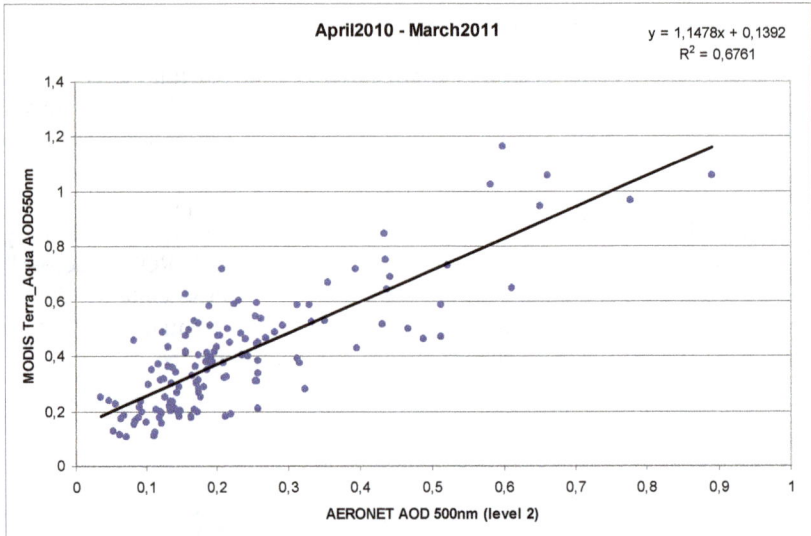

Fig. 3. Correlation between MODIS AOD and AERONET AOD (level 2) for Limassol, April 2010 - March 2011.

6.2 Hand-held sun-photometer

The MICROTOPS II is a hand-held multi-band sun photometer (fig. 4). The sun-photometer measures the aerosol optical thickness (440 nm, 675 nm, 870 nm, 936 nm and 1020 nm bands) and precipitable water (936nm and 1020nm bands) through the intensity of direct sunlight. The sun-photometer consists of five optical collimators (field of view 2.5 degrees) while the internal baffles eliminate the reflections occurred within the instrument. In order to extract the aerosol optical thickness the Langley method was used.

Such hand-held sun-photometers are useful to assess the effectiveness of the AOT measurements derived from satellites. For example, Figure 5 shows the results obtained from applying a linear regression model concerning the satellite AOT retrievals and ground-based Microtops derived AOT for more than two years measurements in the area of Limassol centre area in Cyprus.

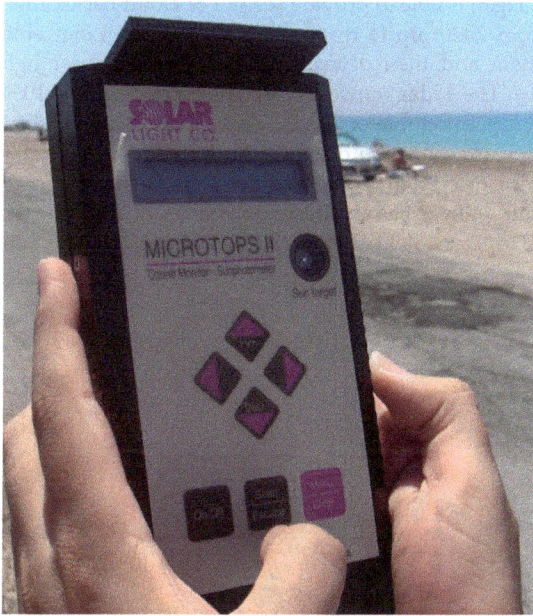

Fig. 4. View of hand-held sun photometer (MICROTOPS II sun-photometer).

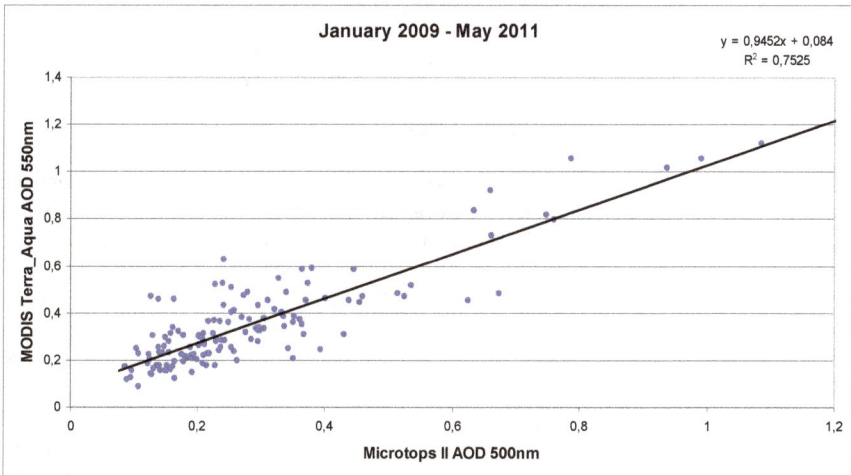

Fig. 5. Correlation between MODIS and Microtops AOD for Limassol, January 2009 - May 2011.

6.3 LIDAR

The Lidar system (see Fig.6) is able to provide aerosol or cloud backscatter measurements from a height beginning from 200m up to tropopause height. The Lidar emits a collimated laser beam in the atmosphere and then detects the backscattered laser light from atmospheric aerosols and molecules. The Lidar transmits laser pulses at 532 and 1064 nm simultaneously and co-linearly with a repetition rate of 20 Hz. Three channels are detected, with one channel for the wavelength 1064 nm and two channels for 532 nm. One small, rugged, flash lamp-pumped Nd-YAG laser is used with pulse energies around 25 and 56 mJ at 1064 and 532 nm, respectively. The primary mirror has an effective diameter of 200 mm. The overall focal length of the telescope is 1000 mm. The field of view (FOV) of the telescope is 2 mrad.

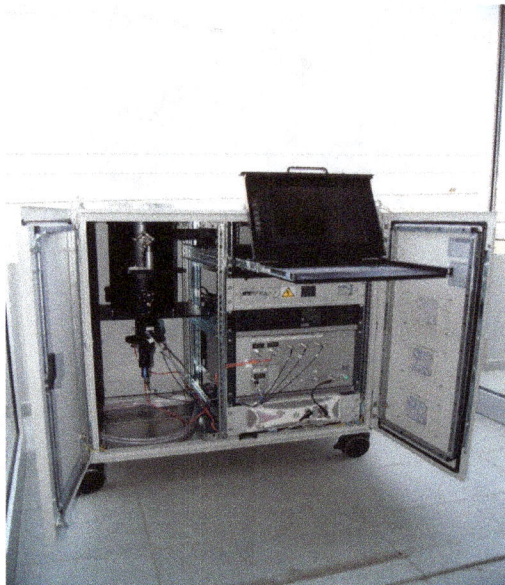

Fig. 6. Light Detection and Ranging, Lidar System.

Using the two different wavelength of lidar we can extract aerosol optical properties such as color index (CI), aerosol backscatter and extinction coefficient, Angstrom exponent, depolarization ratio and AOT, which are all useful parameters for characterizing the type and the shape of particles within the atmosphere. Moreover the lidar system can be used for the detection of clouds and the provision of the Planetary Boundary Layer (PBL) height in near-real time. The data that was extracted after the processing of lidar signals is shown in Fig. 7 and 8. Such data are important for supporting the findings from satellite imagery and through systematic measurements a local atmospheric model can be used for future applications. Moreover, air pollution episodes can be fully supported and explained. According to Fig. 7 and 7, the PBL extends from the surface up to 1600m and there are also two layers above the PBL, the first are centred around 3300m and the second at 3800m. Additionally there are multiple layers within the PBL and it can be distinguished not only by the different colours of temporal evolution of atmosphere (fig.7) but from the peaks presents in the signal of backscatter coefficient (fig.8).

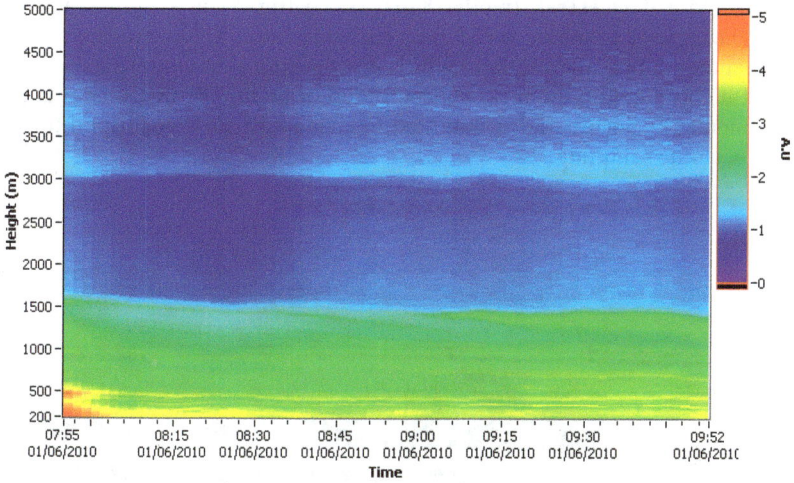

Fig. 7. Temporal evolution of the dust layers over Limasol on 1st of June 2010: Profile from the Lidar system.

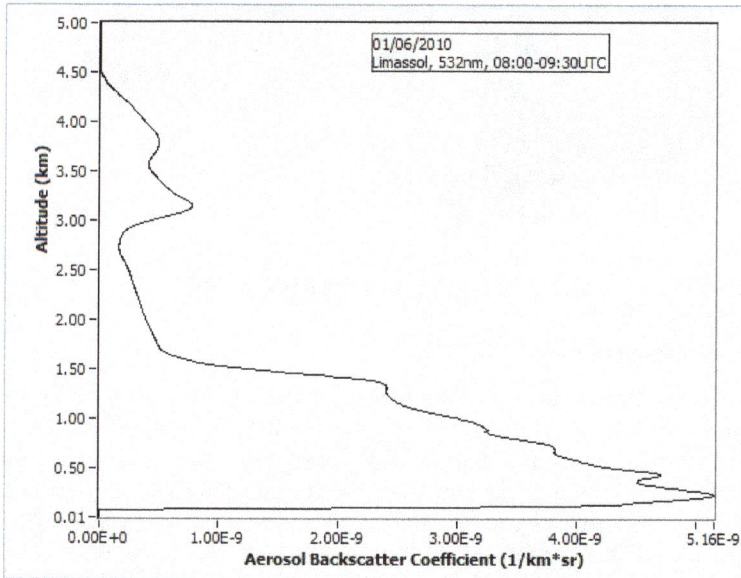

Fig. 8. The vertical distribution of the backscatter coefficient on 1 June 2010, Limassol.

6.4 PM10 device

Several PM10 devices are available to measure particulate matter. Such devices can be used to develop regression models between PM10 measurements with AOT values obtained from satellites,which assists in the systematic monitoring of PM10 from space.

For example, the TSI DustTrak (model 8520/ 8533) (fig. 9) is a light scattering laser photometer that is used to measure PM mass concentrations. The instrucment specifically measures the amount of scattering light which is proportional to volume concentrations of aerosols and it could obtain the mass concentration of them. The instrument is placed in the Environmental Enclosure and is mounted to a standard surveyor tripod for allowing reliable and accurate sampling.

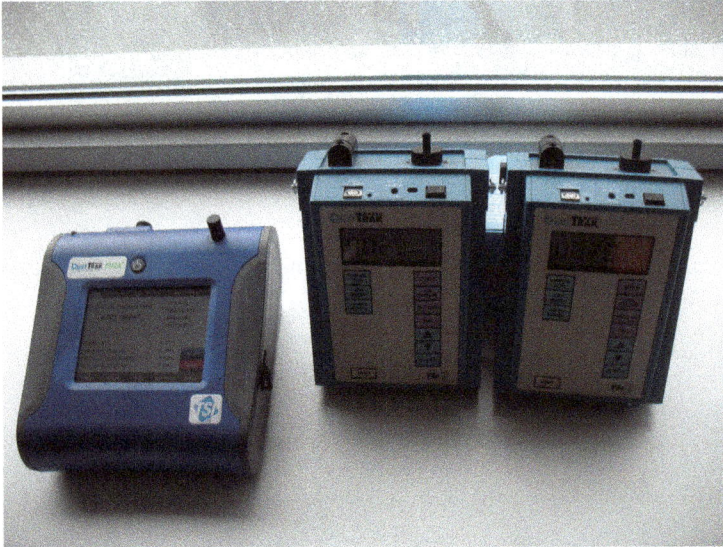

Fig. 9. PM10 devices (for example, Dust Trak model 8533/ 8520).

6.5 Field spectroradiometers

Several spectroradiometers are available to support the effective removal of atmospheric effects from satellite imagery such as the GER-1500 and SVC HR-1024 field spectroradiometers (see Fig.10). Spectroradiometers are used to retrieve the ground reflectance of various standard calibration targets. The GER1500 field spectroradiometer are light-weight, high performance covering the visible and near-infrared wavelengths from 350 nm to 1050 nm while the SVC HR-1024 provides high resolution field measurements between 350 nm and 2500 nm. The application of an effective atmospheric correction algorithm by using the retrieved ground reflectance values in conjunction with the use of the radiative transfer equation provides an indirect method for determining the aerosol optical thickness. Field spectroradiometric measurements are acquired over dark targets such as asphalt areas (see Fig.10) so as to retrieve the AOT values through the application of atmospheric correction algorithms over satellite imagery such as Landsat TM/ETM+, ASTER, SPOT, ALOS, IRS etc.

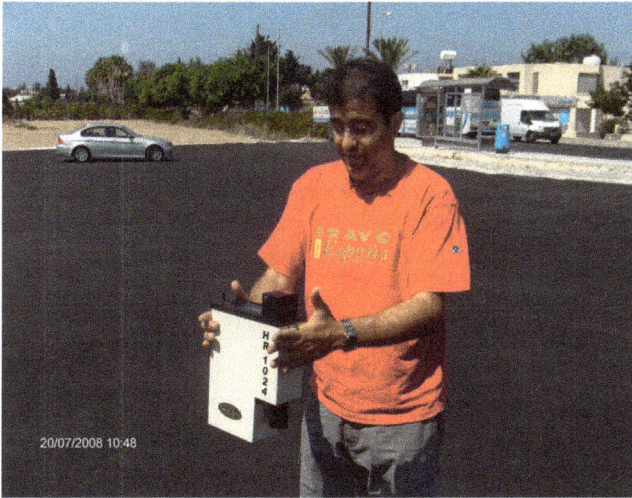

Fig. 10. Field spectroradiometric measurements over dark asphalt target.

7. Case studies

7.1 Case study 1: Retrieval of aot through the application of rt equation and atmospheric correction algorithm

Hadjimitsis et al. (2010) provides several examples of how Lidar and sun-photometers measurements can support the AOT values found from the Landsat TM/ETM+ satellite images. AOT values are derived from Landsat TM/ETM + images using the darkest pixel atmospheric correction method and radiative transfer equation.

The basic equations used to retrieve AOT values is described by Hadjimitsis and Clayton (2009) has been fully applied. The revised method presented by Hadjimitsis et al. (2010) differs from the traditional DOS (Darkest Object Subtraction) method in the following ways: The method incorporates the true reflectance value which is acquired from in-situ spectro-radiometric measurements of selected pseudo-invariant dark-targets such as dark water bodies or asphalt black surfaces (fig.9); the method combines both the basic principles of the darkest object subtraction and radiative transfer equations by incorporating in the calculations the aerosol single scattering phase function, single scattering albedo and water vapour absorption (i.e. Relative Humidity) (values as acquired from several in-situ measurements).

The retrieved target reflectance is given by equation 1:

$$\rho_{tg} = \frac{\pi . \left(L_{ts} - L_P \right)}{t(\mu) \uparrow . E_G} \tag{1}$$

where

ρ_{tg} is the target reflectance at ground level

L_{ts} is the at-satellite radiance (integrated band radiance measured in $W/m^2/sr$)

L_p is the atmospheric path radiance (integrated band radiance measured in $W/m^2/sr$)

E_G is the global irradiance reaching the ground

$t(\mu)\uparrow$ is the direct (upward) target-sensor atmospheric transmittance

For a dark object such as a large water reservoir the target reflectance (at ground level, $\rho_{tg} = \rho_{dg}$ is very low but is not zero. From this large reservoir the darkest pixel will be seen at-satellite to have radiance $L_{ts} = L_{ds}$. Therefore equation 1 can be re-written as:

$$\rho_{dg} = \frac{\pi.(L_{ds} - L_P)}{t(\mu)\uparrow.E_G}$$
(2)

where

ρ_{dg} is the dark target reflectance at ground level

L_{ds} is the dark target radiance at the sensor in $W/m^2/sr$

The aerosol optical thickness was calculated using equation 3 as given by Hadjimitsis and Clayton (2009) [3]:

$$L_{pa} = \omega_a \left\{ \frac{(E_O.\cos(\theta_0).P_a)}{4\pi(\cos(\theta_0) + \cos(\theta_v))} \right\}.$$
$$\left\{ 1 - \exp\left[-\tau_a \left(1/\cos\theta_0 + 1/\cos\theta_v \right) \right] \right\}.t_{H_2O}\uparrow.t_{O_3}\uparrow.$$
$$\exp\left[-\tau_r \left(1/\cos\theta_0 + 1/\cos\theta_v \right) \right]$$
(3)

The algorithm as presented by Hadjimitsis and Clayton (2009) required the following parameters:

- ground reflectance values for the selected ground dark target either zero or any value up to 5 % until the AOT value gets maximum (from spectro-radiometric measurements, suitable water target, for example for water dams values of 0 to 5 % for TM band 1)
- the at-satellite reflectance value of the darkest pixel or target (from the image)
- and the aerosol characteristics (e.g. aerosol scattering albedo) unless you will make assumptions based on previous studies .eg for urban areas aerosol scattering albedo=0.78

By running equation (3) and (2), AOT value was found for Landsat TM band 1.

The first example indicates the AOT values as measured by CIMEL sun-photometer for the whole day, 20/4/2010 as well for the whole month (see Figures 10 and 11). For the Landsat ETM+ image acquired on 20/4/2010, the AOT was found to be 1.4 after the application of the darkest pixel atmospheric correction i.e indirect method of retrieving AOT values. Such value complies with the AOT value found during the satellite overpass for that day as shown from Figure 11. Figure 12 shows that the daily AOT values for April 2010 from AERONET, the 20th of April 2010 has the highest value.

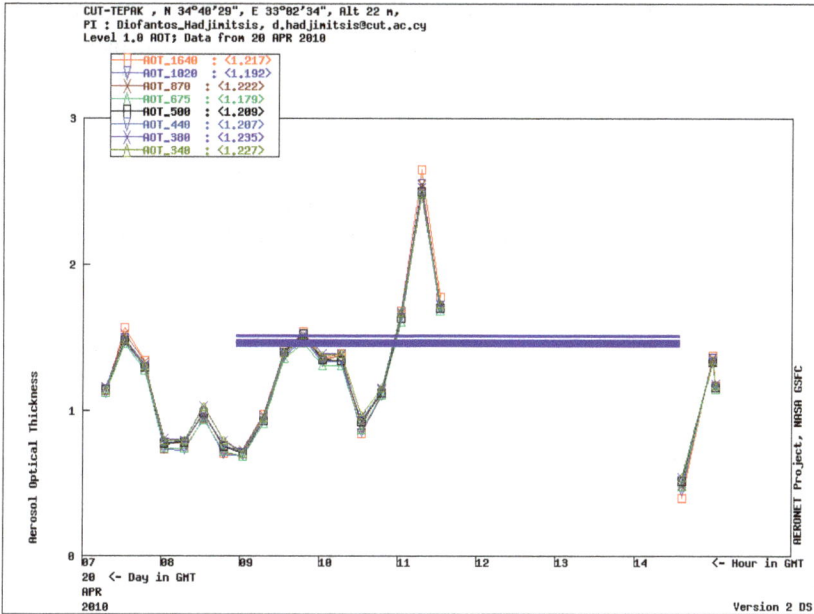

Fig. 11. Daily distribution of AOT by AERONET for the 20th of April 2010. (Hadjimitsis et al., 2010).

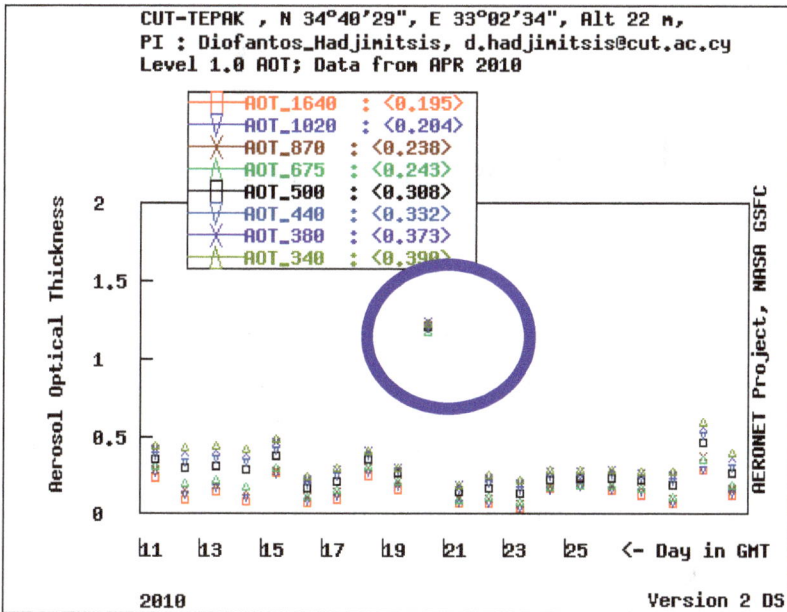

Fig. 12. Monthly distribution of AOT (11-30/April/2010) by AERONET (Hadjimitsis et al., 2010).

Hadjimitsis et al. (2010) provided another example for the May 2010 (27th and 28th of May 2010) measurements both from LIDAR and AERONET. An assessment of the existing atmospheric conditions was done using Lidar system and CIMEL sun-photometers for the 27th and 28th of May 2010 in which dust scenarios were occurred. By considering the values given by the CIMEL sun-photometer as well the support given by the Lidar regarding the dust event, the user assessed air pollution through the application of the atmospheric correction algorithm. By using three months of measurements with both CIMEL sun-photometer and LIDAR, it was found that the mean daily value of Aerosol Optical Thickness exceeded 0.7, while the highest value of AOT during the day was 1.539 (Figure 13).

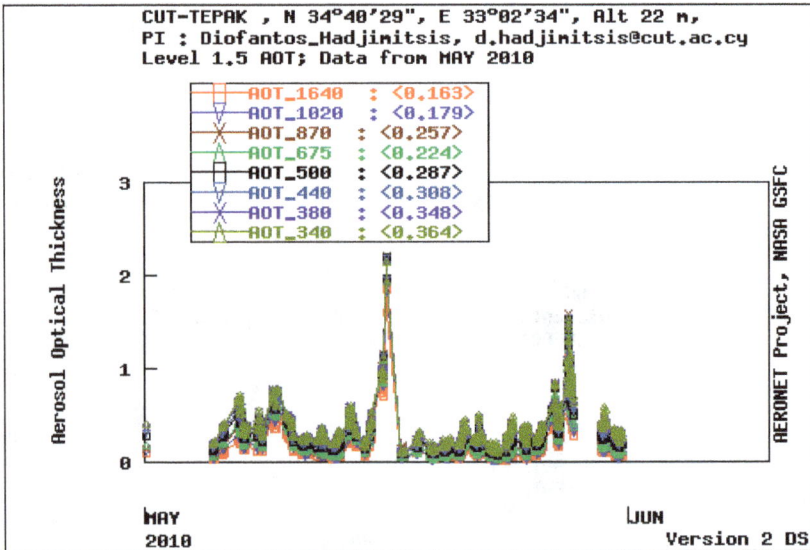

Fig. 13. Monthly distribution of AOT (May) by AERONET (Hadjimitsis et al., 2010).

The measurements from LIDAR showed the previous day a distinct layer in a height between 2 and 3 km (Fig.14). Thus the dust layer in a specific altitude above the Planetary Boundary Layer (PBL) followed the high AOT values a day after as prescribed the CIMEL sun-photometer from AERONET. Morevoer, according to Hadjimitsis et al. (2010), in order to verify dust transport to Cyprus, the NOAA Hysplit Model was used, which provided three-days air-mass back-trajectories in three different levels within the atmosphere. The relevant 72-h back-trajectory analysis for air-masses ending over Limassol, Cyprus at 00:00 UT on 28 May is presented in Figure 15. It is remarkable that for 28 May all air-mass trajectories ending over Limassol at levels between 1km and 4km originated from Saharan at levels 0.5-4 km.

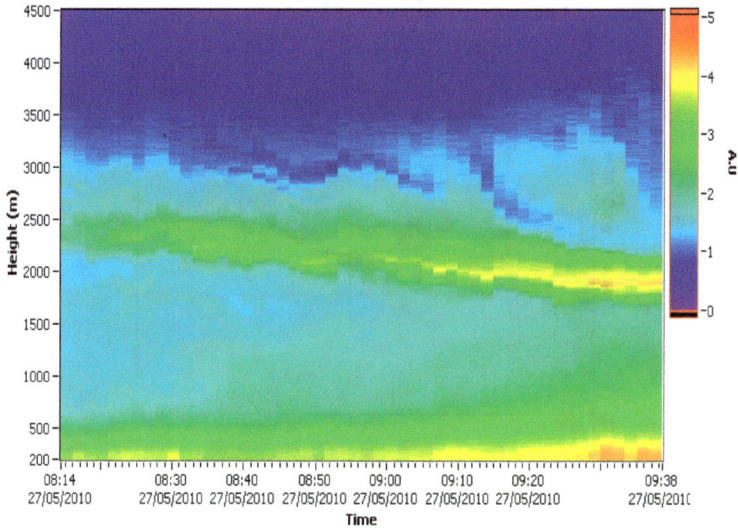

(a)

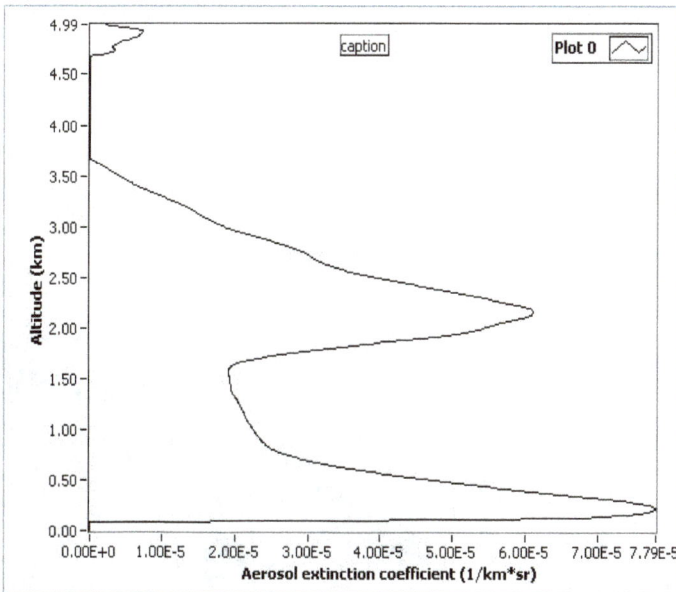

(b)

Fig. 14. (a)Temporal evolution of a dust layer at an altitude of 1.9-2.6 km over the area of Limassol (b) vertical distribution of the aerosol extinction coefficient in a case of high aerosol loading (27/05/2010) (Hadjimitsis et al., 2010).

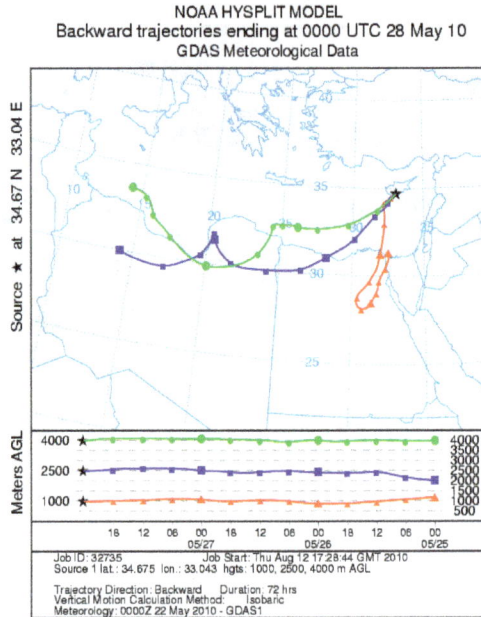

Fig. 15. 72-h air-mass back-trajectories ending over Limassol, Cyprus at 00:0 UT on 28 May 2010 (Hadjimitsis et al., 2010).

By extending the same procedure over a large area of interest, AOT maps can be produced using Landsat TM.ETM+ imagery as shown in Fig. 16

Fig. 16. AOT map from Landsat TM images using indirect method such as the application of atmospheric correction (Themistocleous et al., 2010).

7.2 Case study 2: Retrieval of aot from modis and development of regression model between aot vs pm10

Nisantzi et al. (2011) show how MODIS AOT data can be correlated with PM10 over an urban area in Limassol, Cyprus. Based on the developed regression models, such model can be used for systematic monitoring of air pollution such as PM10 over the urban area of Limassol. Nisantzi et al. (2011) made a comparison between MODIS AOD at 550nm and AERONET AOD (automatic-sun photometer as shown in Fig.2). The number of measurements are 136 and the correlation coefficient according to coefficient of determination ($R^2 = 0.67$) is R=0.822 (fig.16). These results indicate that MODIS-derived AOD values can be used in case of no AERONET AOD will be available and vice versa. Similarly MODIS AOD compared with Microtops II (See Fig.3) AOD at 500nm and the retrievals for 141 data shows a better correlation R=0.867 (R^2=0.7525) for more than two years measurements (fig.18).

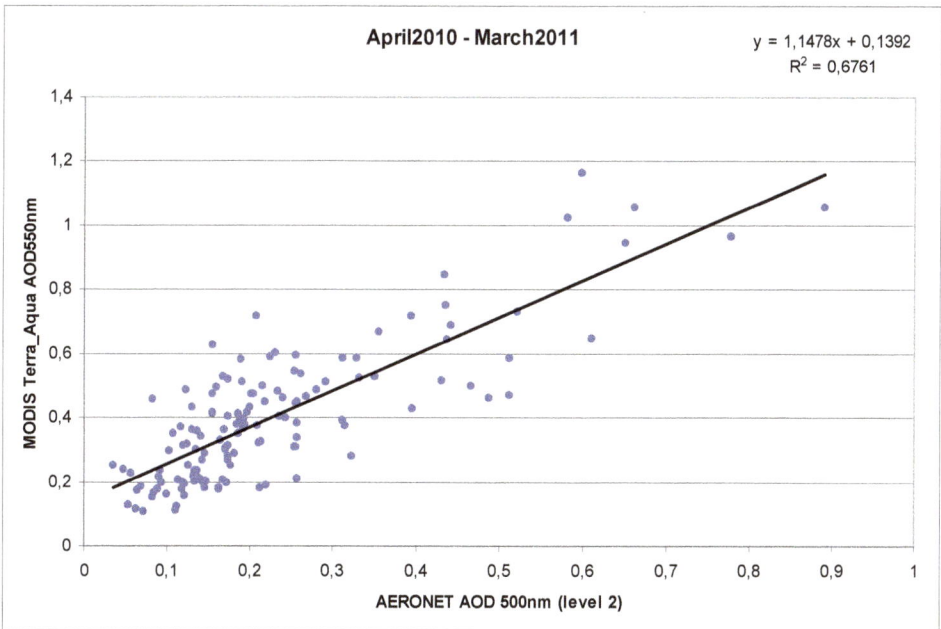

Fig. 17. Correlation between MODIS AOD and AERONET AOD (level 2) for Limassol, April 2010 - March 2011. (Nisantzi et al., 2011).

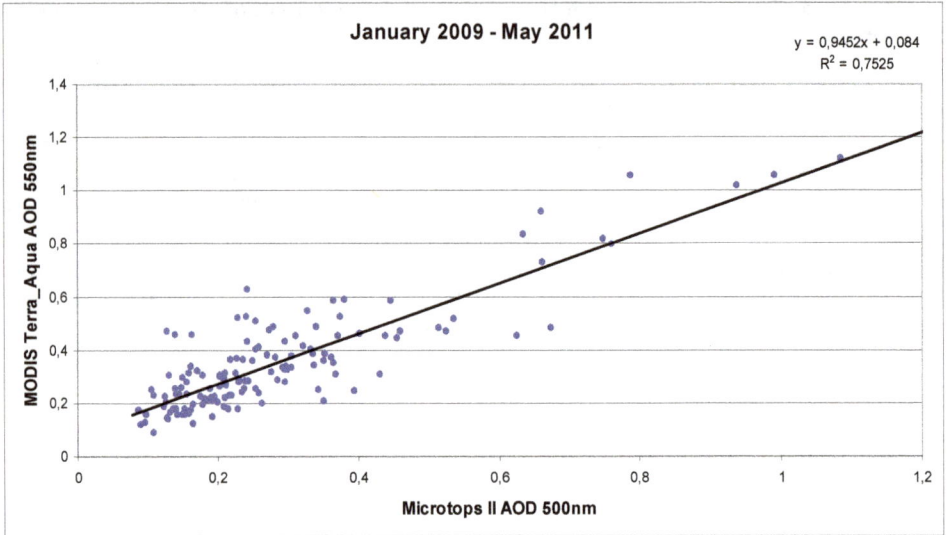

Fig. 18. Correlation between MODIS AOD and Microtops AOD for Limassol, January 2009 - May 2011 (Nisantzi et al., 2011).

In order to develop a regression model between MODIS data and PM10, Nisantzi et al. (2011) attempted several correlations between PM10 and AOT sun-photometer prior to the MODIS correlation in order to be more accurate. Nisantzi et al. (2011) acquired PM_{10} data (using the DUST TRAK -Figure 9) every 10 minutes and AOD measurements for the CIMEL sun photometer at the same time, with a possible deviation 5 minutes. The correlation is shown in fig.19.

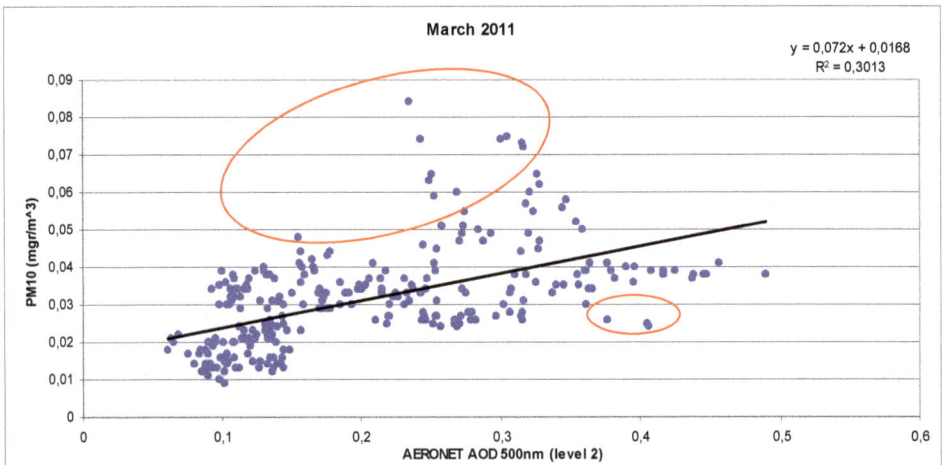

Fig. 19. Correlation between PM_{10} and AERONET AOD for Limassol area, March 2011 (Nisantzi et al., 2011).

Nisantzi et al. (2011) considered the effect of dust in the field campaign measurements. By taking into consideration the measurements within the red cycles it was found that these measurements corresponded to days with dust layer above the ABL. Also examined the lidar measurements (see Figure 6) and by using backward trajectories for the three dust events (3/3/2011-5/3/2011, 16/3/2011-18/3/2011 and 28/3/2011-31/3/2011) occurred within this period. Nisantzi et al. (2010) excluded the data for these days and the correlation increased as shown in Fig. 20 (n=231, R=0.68). By excluding the measurements with aerosol layer above the ABL from both sun photometer and MODIS data, the correlation between the two variables PM0 Vs. AOD (MODIS & Sun photometer) increases and the plots are shown in figures 21, 22.

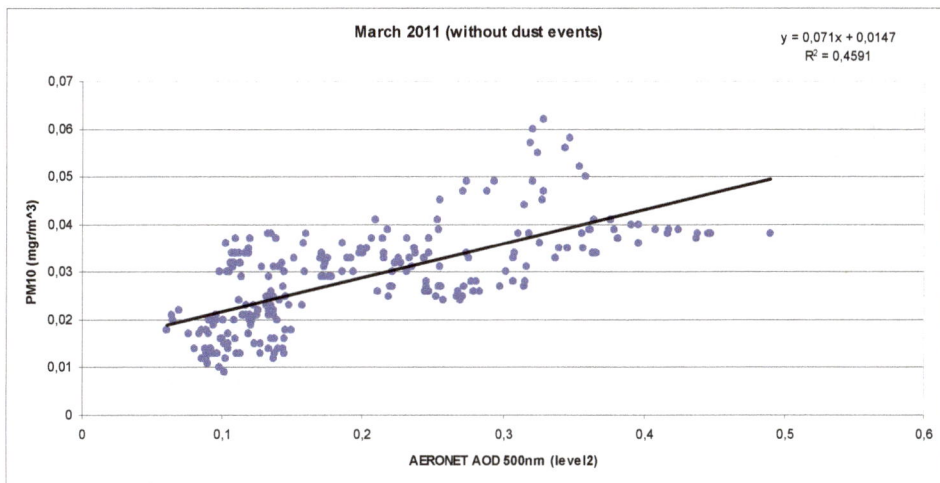

Fig. 20. Correlation between PM_{10} and AERONET AOD (level 2) for Limassol area for March 2011 after excluding days with dust layers (Nisantzi et al., 2011).

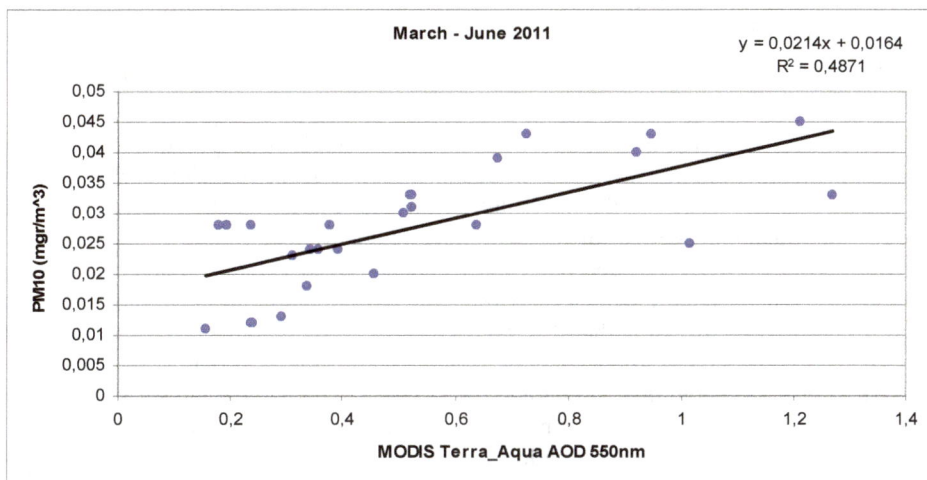

Fig. 21. Correlation between PM_{10} and MODIS-derived AOD (Nisantzi et al., 2011).

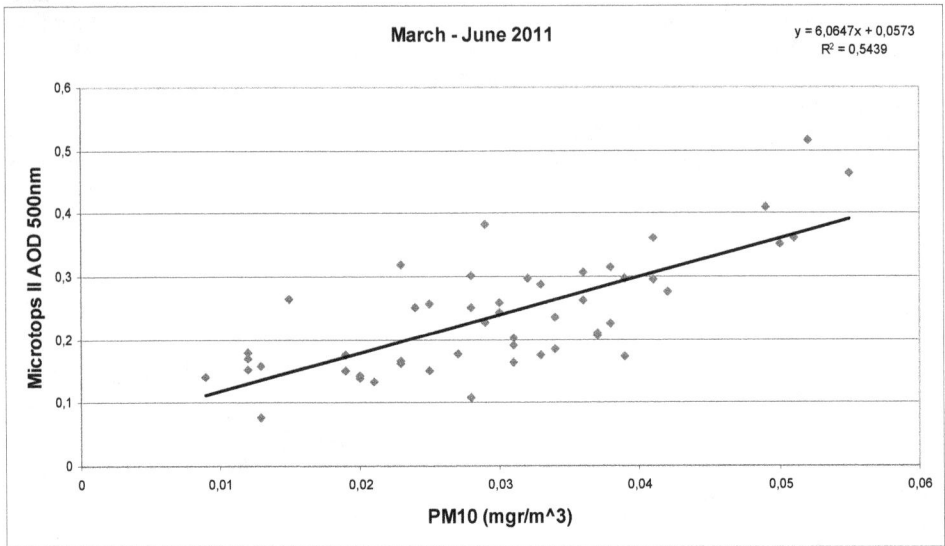

Fig. 22. Correlation between PM_{10} and Microtops II-derived AOD (Nisantzi et al., 2011).

8. GIS

GIS has the advantage of the high power of analysing of spatial data and handling large spatial databases. Indeed, in air pollution there are a large amount of data that GIS can be used for their handling. Data that is used for air pollution studies is air pollutants, wind direction, wind speed, traffic flow, solar radiation , air temperature, mixing height etc. The integration of both GIS and remote sensing provides an deal efficient tool for the air pollution monitoring authorities (Zhou, 1995)

With GIS, remote sensing data can be integrated with other types of digital data, such as air pollution measures. Merging satellite remote sensing and GIS tools provides a quick and cost effective way to provide an improved qualitative assessment of pollution. GIS is a tool that can be used for assessing the air pollution through the use of AOT values retrieved directly from satellite imagery or in-situ sun-photometers, from air pollution measurements including CO_2, CO, SO_2, PM10 and other environmental data. The integration of the above information can be inserted into a GIS software which can monitor and map high risk areas resulting air pollution. By using the AOT values retrieved from satellite, a thematic map can be developed through the application of Kriging algorithm in order to indicate high-polluted areas. Based on the determined AOT over an area of interest, and through interpolation, thematic maps can be generated using colour themes showing the levels of the AOT and/or pollution. Such information can be used as a tool for decision-makers to address air quality and environmental issues more effectively. Figure 23 shows a typical example of air pollutants over Cyprus using a GIS.

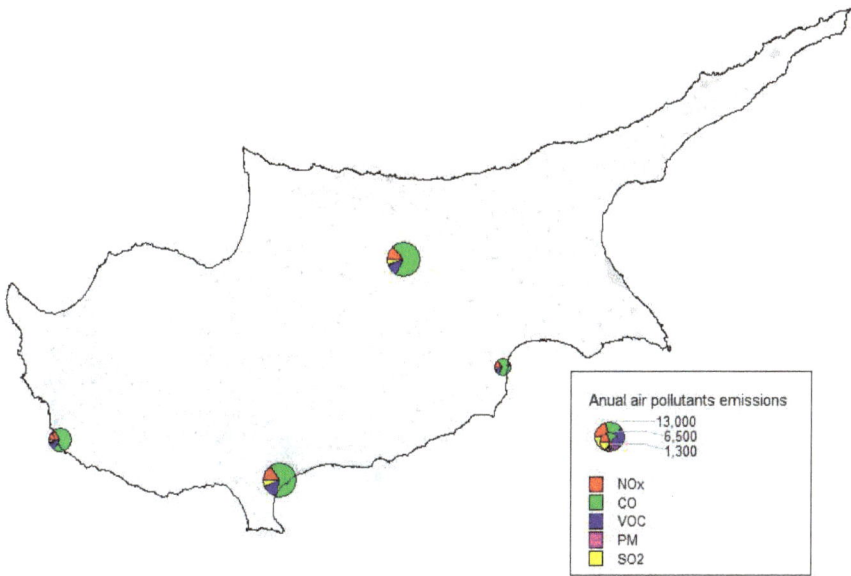

Fig. 23. GIS indicating air pollutant emissions in Cyprus.

Several studies showed the importance of using both GIS tools as well as satellite remotely sensed imagery to view and analyze the concentration of air pollutants and linkages with land cover and land use (Hashim and Sultan, 2010; Weng et al; 2004; 2006).

8.1 Case study

After determining the AOT using both MODIS (direct method) and Landsat TM/ETM imagery (indirect method), AOT thresholds were established based on PM10 measurements acquired over studies shown in section 7.1 and 7.2. Using the developed regression models shown in study area 7.2, thresholds of AOT values were obtained using the relevant limit, which is $50\mu g/m^3$ as prescribed by the European Union. Then , a GIS map was developed using the Kriging algorithm to identify areas with PM10 measurements above the acceptable threshold in order to monitor and map high risk areas due to air pollution (figure 24) by blending together satellite imagery and GIS. The GIS was conducted by dividing the area of interest using grid cells. The study found that by using the AOT values from MODIS or Landsat, a GIS map can be produced which can show high-polluted areas as shown in Figure 24. Themistocleous (2010) developed the fast atmospheric correction algorithm and the simplified image based AOT retrieval based on RT equation for GIS modeling which was used to create a thematic map of AOT values over Limassol.

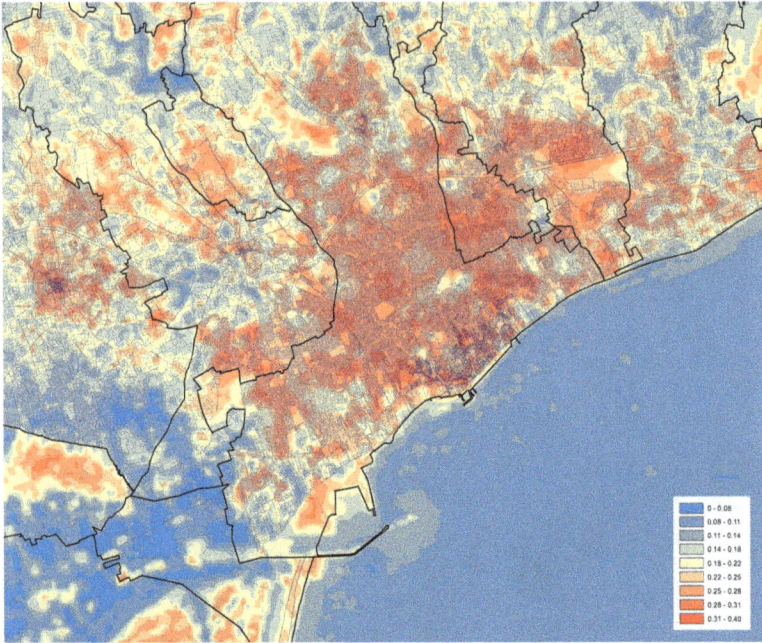

Fig. 24. GIS Thematic map using Krigging algorithm to display AOT levels in the Limassol area.

9. Conclusion

Earth observations made by satellite sensors are likely to be a valuable tool for monitoring and mapping air pollution due to their major benefit of providing complete and synoptic views of large areas in one snap-shot. Blending together GIS and remote sensing, AOT values can derived over a large area of interest systematically. GIS is used also to map air pollutants based on ground and satellite-derived AOT values.

10. Acknowledgment

The authors acknowledge the Cyprus University of Technology (Research Committee) for their funding ('MONITORING' internal funded project) & research activity funds and the Cyprus Research Promotion Foundation for their funding ('AIRSPACE' research project). Special thanks are given to the Remote Sensing Lab of the Cyprus University of Technology.

11. References

Chu, D. A., Kaufman, Y. J., Ichoku, C., Remer, L. A., Tanré, D., Holben, B. N. (2002). *Validation of MODIS aerosol optical depth retrieval over land*, Geophys. Res. Lett., 29(12), pp. 8007.

Hashim B. and Sultan A. (2010) Using remote sensing data and GIS to evaluate air pollution and their relationship with land cover and land use in Baghdad City, *Earth Sciences 2 (2010) // 120-124*

Hadjimitsis, D.G. (2009a). *Aerosol Optical Thickness (AOT) retrieval over land using satellite image-based algorithm*, Air Quality, Atmosphere & Health- An International Journal, 2 (2), pp. 89-97 DOI 10.1007/s11869-009-0036-0

Hadjimitsis, D.G. (2009b) *Description of a new method for retrieving the aerosol optical thickness from satellite remotely sensed imagery using the maximum contrast value principle and the darkest pixel approach*, Transactions in GIS Journal 12 (5), 633−644.

Hadjimitsis, D.G., Clayton, C.R.I (2009). *Determination of aerosol optical thickness through the derivation of an atmospheric correction for short-wavelength Landsat TM and ASTER image data: an application to areas located in the vicinity of airports at UK and Cyprus.* Applied Geomatics Journal. 1, pp. 31--40

Hadjimitsis, D.G., Retalis A., Clayton, C.R.I. (2002). *The assessment of atmospheric pollution using satellite remote sensing technology in large cities in the vicinity of airports.* Water, Air & Soil Pollution: Focus, An International Journal of Environmental Pollution 2, pp. 631−640.

Hadjimitsis, D.G., Clayton C.R.I., & Hope V.S. (2004). *An assessment of the effectiveness of atmospheric correction algorithms through the remote sensing of some reservoirs*, International Journal of Remote Sensing, Volume 25, 18, 3651-3674. DOI: 10.1080/01431160310001647993

Hadjimitsis, D. G., Nisantzi, A., Themistocleous, K., Matsas, A., Trigkas, V. P. (2010). *Satellite remote sensing, GIS and sun-photometers for monitoring PM10 in Cyprus: issues on public health,* Proc. SPIE 7826, 78262C (2010); doi:10.1117/12.865120

Holben, B.N., Eck T.F., Slutsker I., Tame D., Buis J.P., Setzer A., Vermote, E., Reagan, J.A., Kaufman, Y., Nakajima, T., Lavenu F., Jankowiak, I., Smimov, A. (1998). *AERONET – A federated instrument network and data archive for aerosol characterization*, Rem. Sens. Environ. 66, pp. 1−16.

Kaufman, Y.J., Fraser, R.S., Ferrare, R.A. (1990) *Satellite measurements of large-scale air pollution: methods*, Journal of Geophysics Research 95, pp. 9895−9909.

King, M.D., Kaufman, Y.J., Tanré, D., Nakajima, T. (1999). *Remote Sensing of Tropospheric Aerosols from Space: Past, Present, and Future*, Bulletin of the American Meteorological Society 80, pp. 2229--2259

Knobelspiessea, K. D., Pietrasa, C., Fargiona, S. G., Wanga, M., Frouine, R., Millerf, M. A., Subramaniamg, A., Balchh, W. M. (20040. *Maritime aerosol optical thickness measured by handheld sun photometers*, Remote Sensing of Environment 93, pp.87−106.

Lee, H. J., Liu, Y., Coull, B. A., Schwartz, J., and Koutrakis, P. (2011). *A novel calibration approach of MODIS AOD data to predict PM2.5 concentrations*, Atmos. Chem. Phys., 11, 7991-8002, doi:10.5194/acp-11-7991-2011, 2011.

Nisantzi A., Hadjimitsis D.G., Aexakis D. (2011) Estimating the relationship between aerosol optical thickness and PM10 using lidar and meteorological data in Limassol, Cyprus, SPIE Remote Sensing 2011, Prague Sept.2011 (in press)

Raju, P.L.N. (2003). Fundamentals of Geographical Informatin Systems. *Proceedings of Satellite Remote Sensing and GIS Applications in Agricultural Meteorology* 7-11 July 2003, Dehra Dun, India, p.p. 103-120.

Retalis, A., Cartalis, C., Athanasiou, E. (1999) *Assessment of the distribution of aerosols in the area of Athens with the use of Landsat TM.* International Journal of remote Sensing 20, pp. 939−945.

Retalis, A.: Study of atmospheric pollution in Large Cities with the use of satellite observations: development of an Atmospheric correction Algorithm Applied to

Polluted Urban Areas, Phd Thesis, Department of Applied Physics, University of Athens (1998).

Retalis, A.; Hadjimitsis, D.G. Chrysoulakis, N.; Michaelides, S.; Clayton, C.R.I. (2010). *Comparison between visibility measurements obtained from satellites and ground*, Natural Hazards and Earth System Sciences Journal, 10 (3) p.p. 421-428 (2010) doi:10.5194/nhess-10-421-2010

Sifakis, N., Deschamps, P.Y. (1992) *Mapping of air pollution using SPOT satellite data.* Photogrammetric Engineering and Remote Sensing, 58, 1433 – 1437.

Sivakumar, V., Tesfaye, M., Alemu, W., Moema, D., Sharma, A., Bollig, C., Mengistu, G. (2006) *CSIR South Africa mobile LIDAR- First scientific results: comparison with satellite, sunphotometer and model simulations*, South Africa Journal of Science 105, pp. 449 – 455.

Schott, J.R. (2007). *Remote Sensing: the image chain approach*. Oxford, UK. Oxford University Press.

Tang, J., Xue, Y., Yu, T., Guan, Y. (2004) *Aerosol optical thickness determination by exploiting the synergy of Terra and Aqua MODIS*. Remote Sensing of Environment 94, pp. 327 – 334.

Themistocleous K., Nisantzi A., Hadjimitsis D., Retalis A., Paronis D., Michaelides S., Chrysoulakis N., Agapiou A., Giorgousis G. and Perdikou S. (2010) *Monitoring Air Pollution in the Vicinity of Cultural Heritage Sites in Cyprus Using Remote Sensing Techniques*, Digital Heritage, Lecture Notes in Computer Science, 2010, Vol. 6436/2010, 536-547, DOI: 10.1007/978-3-642-16873-4_44. M. Ioannides (Ed.): EuroMed 2010, LNCS 6436, pp. 536–547, 2010, © Springer-Verlag Berlin Heidelberg 2010.

Tsanev, V. I., Mather, T. A. (2008) *Microtops Inverse Software package for retrieving aerosol columnar size diustributions using Microtops II data*, Users Manual.

Tulloch, M., Li, J. (2004) *Applications of Satellite Remote Sensing to Urban Air-Quality Monitoring: Status and Potential Solutions to Canada*, Environmental Informatics Archives, 2, pp. 846 – 854.

Wald, L., Basly, L., Balleynaud, J.M. (1999) *Satellite data for the air pollution mapping. In: 18th EARseL symposium on operational sensing for sustainable development*, pp.133--139 Enschede, Netherlands.

Wang, J., Christopher, S. A. (2003) *Intercomparison between satellite derived aerosol optical thickness and PM2.5 mass: Implications for air quality studies*, Geophys. Res. Lett., 30 (21), pp. 2095 (2003). http://daac.gsfc.nasa.gov/data-holdings/PIP/aerosol_optical_thickness_or_depth.shtml

Zhou, Q. (1995). The integration of GIS and remote sensing for land resource and environment management. *Proceedings of United Nation ESCAP Workshop on Remote Sensing and GIS for Land and Marine Resources and Environmental Management*, 13-17 February, 1995. Suva, Fiji, p.p. 43-55.

Weng, Q., Lub, D., Schubring, J. (2004) Estimation of land surface temperature – vegetation abundance relationship for urban heat island studies, Remote Sensing of Environment, v.89 (4), p. 467-483,

Weng, Q., Yang, S (2006) Urban air pollution patterns, land use and thermal landscape: an examination of the linkage using GIS Environmental Monitoring and Assessment, v.117 (4), p.463-489.

3

Urban Air Pollution

Bang Quoc Ho
Institute of Environment and Resources,
Vietnam National University in HCM City
Vietnam

1. Introduction

Over the last 60 years, the urban population has increased at an incredible pace. According to the statistical documents of the United Nations Environment Program (UNEP, 2006), in 1900 the world only had 15 cities having a population of 1 million, whereas in 1950 the world had 83 cities having a population with more than 1 million and today there are more than 350 cities having a population more than 1 million. The population living in urban areas is about 50% of the world's population and the population living in urban areas will continue to increase rapidly in the future. In 2005, Asia had 50% of the most populous cities in the world. The urbanization process is a consequence of the explosion of the industrialization and automation process world-wide. People are attracted by high rates of economic growth in urban areas because there is more employment, educational opportunities and a better quality of life.

However, the urbanization process creates high density of street network, building, population and other activities (industry, etc.). These activities are relation with the high consumption of fossil fuel, such as people in urban areas uses more energy for cooking, air-conditioning, transportation, etc., and industry uses energy for production (Zarate, 2007). Consequently, these activities of high energy consumption emit a large amount of pollutants into the atmosphere which brings many environmental problems, for example, air, water and noise pollution as well as waste management. Among them, air pollution is one of the most serious environmental problems in urban areas. The World Health Organization (WHO) (WHO, 2005) has estimated that urban air pollution causes the death of more than 2 million people per year in developing countries, and millions of people are found to be suffering from various respiratory illnesses related to air pollution in large cities.

Therefore, the urban air quality management should be urgently considered in order to protect human health. Up to now, developed countries have made extensive efforts to improve the air quality through reducing emissions, such as: using cleaner energy, applying new air quality regulations, moving the industrial activities to the developing countries, etc. These efficient strategies at global scale are to move to developing countries. Air quality in developing countries has deteriorated considerably, thus exposing millions of people to harmful concentrations of pollutants because in developing countries the urban air quality management has not been adopted for a variety of difficulties.

2. Status of the air quality in large cities

A vast majority of the urban and suburban areas in the world is exposed to conditions which exceed air quality standards set by WHO. Especially, the large cities in developing countries have the highest air pollution levels. Some researches show that the emissions from Asian cities will rise and this will continue to have an impact on hemispheric background ozone level as well as global climate (Gurjar et al., 2005). In general the cities in developed countries have the concentrations of air pollutants lower than the cities in developing countries. A lot of measurements of air pollutants (such as O_3, SO_2, NO_2, TSP and PM10) in the world have been done; some of them are shown in the table 1.

Data in Table 1 has been extracted for the period 1999 – 2000 (only the data of Bogotá and HCMC were for the year 2005). Data were not complete for all cities because often measurements are not available for the same year.

Data has been used from urban stations to represent the overall pollution levels of the city. In the case of NO_2 concentrations (standard limit =40 μ g.m-3, WHO), the high concentrations for both the developed and developing cities occurred where the main source of emission is road traffic. However the road traffic emissions in developed countries are less dangerous than developing countries because they use modern vehicular combustion and emission control technology, clean fuel and more public transport. While in developing countries the out-dated vehicle technologies are used and few public transports are used. As results most cities in developing countries such as: Shanghai, Delhi, Jakarta, Beijing, HCMC, etc contain a relatively high level of air pollution. Especially, the maximum NO_2 concentrations in several cities of China are approximately three times higher than the WHO guideline value.

In the case of total suspended particles concentrations (standard limit =90 μ g.m-3, WHO), particle are normally related to SO_2 concentrations because TSP and SO_2 are emitted from burning coal for industrial activities. In general, the highest concentrations of TSP and SO_2 are found in developing cities, exceeding 300 μ g.m-3 where industrialization rate is rapid.

Particle matters (PM10) with diameters less than 10 μ m are very dangerous for respiratory human system. PM10 is originated from industrial and traffic sources, the values of PM10 are normally highest in developing cities where they use much old vehicle and diesel fuel.

The guideline value for the average annual value of O_3 does not exist. In some countries they apply the guideline for the maximum daily hourly O_3 at ground-level (standard limit = 180 μ g.m-3, TCVN) to manage air quality. So, Table 1 shows the maximum daily hourly O_3 concentrations. The highest O_3 concentrations are found in Mexico City (546 μ g.m-3), followed by Sao Paulo (403 - 546 μ g.m-3). The lowest maximum hourly concentrations of O_3 are found for the European cities where the lowest photochemical reaction occurs.

The tendency concentration of pollutants in the worldwide is to reduce because the local government and international organizations impose strong restriction laws and have the air quality management program. However, in developing countries the concentration of pollutants is tendency to increase because they develop too fast and release more pollutants and because they lack air quality management tools.

City	Population[a]	O_3	TSP[b]	PM10	SO_2	NO_2
Tokyo, JP	33.4		49		18	68
Seoul, KR	23.1		84		44	60
Mexico, MX	22.0	546	201	52	46	55
New York, US	21.8	272		24	26	70
Bombay, IN	21.1		240		33	39
Delhi, IN	20.8		415		24	41
Sao Paulo, BR	20.3	403	53		18	47
Shanghai, CN	18.6		246		53	73
Los Angeles, US	17.9	225		39	9	66
Jakarta, ID	16.9		271			
Osaka, JP	16.6		43		19	63
Cairo, EG	15.8				69	
Calcutta, IN	15.4		375		49	34
Manila, PH	15.2				32	
Karachi, PK	14.6					
Dacca, BD	13.6					
Buenos Aires, AR	13.5		185			20
Moscow, RU	13.4		100			80
Beijing, CN	12.4		377		90	122
Rio de Janeiro, BR	12.2		60		50	40
Bogota, CO	8.4	348		58	40	39
HCMC, VN	6.3	247	260	79.6	44	34
WHO standard[c]		160[d]	90	20	50	40

Source: Baldasano et al. (2003); Erika (2007); ADB (2006) and HEPA (2006).
[a] Population expressed in millions, 2005.
[b] TSP = Total suspended particles.
[c] WHO standard for PM10 was manly issued in 2005, the rest in 2000. Source: WHO (2000) and WHO (2005).
[d] WHO standard for O_3 (maximum 1-h concentration). Source: Molina and Molina, 2001

Table 1. Air quality in large cities of the world. Almost data reported correspond to the mean annual concentration in μ g.m-3 (only O_3 is reported in maximum 1-h concentration).

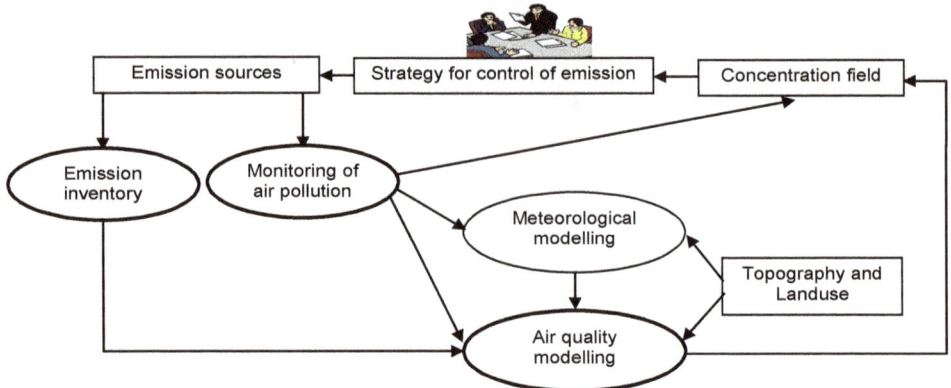

Fig. 1. Air quality management system for urban areas.

3. Air quality management

Air pollutions in cities are very complex because of several factors contributing to deterioration of the air quality in cities. These factors include: (1) a large amount of emission sources (traffic, industrial, residential, natural, etc.); (2) meteorological processes (wind components, temperature, moisture content, solar radiation, etc.); (3) chemical transformations (chemical reactions, dry depositions, etc.).

The design of effective abatement strategies for reduction emission becomes very difficult if we take into account the socio-economical problems. The population growth leads to increase economical as well as industrial activities which can not inhibited due to the development needs. These difficulties could be solved using an improved technology to reduce the pollution. However, the implementation of an improved technology is very costly. As an example it is quoted that it costs 7 billions US$ to install the catalytic converters in all new vehicles which are bought each year in United States (US) (Clappier., 2001).

The design of effective abatement strategies for reducing emission requires a good understanding of all factors related to air pollutions. Over past 20 years, a multitude of different urban AQM systems have been developed by scientists and environmental authorities (Moussiopoulos, 2004). These tools have been used very well to study air pollution and to propose efficient abatement strategies.

Many tools are integrated together in AQM system. These main tools which are presented in Figure 1 include: monitoring of air pollution, emission inventory (EI), air quality modelling and meteorological modelling. The current state of the art of these tools is presented in the next sections.

3.1 Monitoring of air pollution

Monitoring of air pollution is one of the most important tools in AQM. It helps us to understand the status of air pollution levels and to understand the evolution of air pollution. Monitoring gives us the information on emissions sources, because monitoring could be done on roadside for evaluating the traffic emissions and in industrial park for evaluating the industrial emission sources. Nowadays, almost all large cities in the world have installed the monitoring system in order to inform the level of pollution exposure to the population. Another important role of monitoring of air pollution is to validate the model which is used to simulate air quality.

There are different methods in use to monitor the air pollution including: automatic, semi-automatic and manual methods. Automatic methods use the equipments which can measure directly the pollution and can be moved anywhere for monitoring of air pollution. Example, the equipments manufactured by Environmental S.A, France, just to mention some, monitors automatically CO, NO_2, NO, O_3, PM10, etc. On the other hand, semi-automatic methods involve collecting air quality samples at the selected sites by placing the equipments there, samples are collected and then these samples are transported to laboratory for analysis. Such methods are used, for example, for collecting BTEX (Benzene, Toluene and Xylene) sample by collecting samples for a continuous duration of 6h and then transporting the samples to laboratory for analysis by Gas Chromatography (GC). The third category of methods called manual methods involves collecting the samples manually as is done for the case of CO monitoring.

Among these, automatic method is the best one, because this method allows monitoring real time air pollution. A lot of measurements are made which can be later used these data to study the evolution of pollution in different periods and throughout the year. However, this method is very costly because the equipments are expensive, additionally, there is a need to maintain it regularly and we have trained technicians to operate the devices (Molina and Molina, 2004).

Nowadays, automatic and manual air pollution measurement networks have been installed in the world. In Europe some 1450 measurement stations covering 350 cities all over Europe have been installed, in US over 1000 stations are operated throughout the country by the US Environmental Protection Agency (US-EPA) (Baldasano et al., 2003), in Latin America there are more than 4000 urban monitoring stations (Belalcazar, 2009), in Asia countries with installed networks include China, Japan, Korea, India, Indonesia, Thailand, Vietnam, etc. In addition to these, WHO installed the monitoring networks in over 100 cities around the world since 1990s.

In Vietnam, the information from air pollution measurement networks is used to improve air quality. The air the cities of Vietnam before 1995 was polluted by lead due to traffic activities (several times higher than Vietnam air quality standard). The import of leaded gasoline was banned by Vietnam government, thereafter; the concentrations of lead in atmosphere are reduced and lower than the local air quality standard.

3.2 Emission inventory

Development of EI database is very important to describe the emissions and to manage air quality (Moussiopoulos, 2004; Ranjeet et al., 2008). The information from EIs helps us to understand the emission sources and also the emission fluxes in the study domain. The atmospheric pollutants needed for assessment and management of air quality are SO_2, NOx, CO, VOCs and particle maters (PM). The emission sources are grouped in different categories: mobile source (such as road traffic), area sources (such as agriculture, natural) and point sources (such as industry). Resolution in space of EI depends on the scale of study domain and the availability of information for generating emission inventories (EIs). In general the smaller the region of interest, the finer is the required spatial resolution of the EIs. Resolution in time of EIs must follow the activity pattern of emission sources.

The principal equation for calculation of emissions is combined by two variables:

$$E = e \times A \tag{1}$$

Where, E is the total emission

e is the emission factor (EF)

A is the activity data of emitters

There are two main methods currently used for generating emission inventories: top-down and bottom-up

Top-down approach

A top-down approach starts to estimate the total emissions and uses several assumptions to distribute these emissions in space and time (Friedrich and Reis, 2004).

Strengths: this method is easy to apply because it needs less input information and the time to generate EIs is rapid. This method is particularly appropriate to estimate the inventories at large scale.

Weaknesses: one of the main limitations of this method is that the results of EIs are normally highly uncertain.

Examples of using top-down approach for generating EIs from the literature are:

-   Bond and Streets calculated emissions of Carbonaceous aerosols, black carbon and organic carbon for the entire world by using total fuel consumption of the year and emission factors (g/kg of fuel burned) (Streets et al., 2004).
-   Streets calculated emission of gaseous and primary aerosol in Asia by using parameters such as: energy use, human activities and biomass burning (Streets et al., 2003).
-   Generoso combined the results of a global chemistry and transport model and satellite data to evaluate the emission produced by the Russian fires in 2003. In this research, they used top-down approach to estimate the emissions (Generoso and Bey, 2007).

Bottom-up approach

The bottom-up methodology is based on a source oriented inquiry of all activity and emission data needed for describing the behavior of a single source. It starts to evaluate the spatial and temporal repartition of the parameters used to calculate the emissions (Friedrich and Reis, 2004).

Strengths: this approach is more accurate than the top-down approach to evaluate the spatial and temporal repartition and is more appropriate for the small scale (city scale and lower).

Weaknesses: The main limitation of this method is that it needs a large amount of input data for generating EIs. Sometime, the information can even not be collected in the cities of developing countries. The time of generating EIs of this approach is longer than top-down approach

Examples of using bottom - up approach for generating EIs from the literature are:

- Erika combined the traffic fluxes obtained from a traffic model and emission factors per vehicles to calculate the traffic emissions in Bogota, Colombia (Erika, 2007).
- Molina measured emissions of industry and used GPS to locate the position of factory on the map to calculate the industrial emissions for Mexico City (Molina et al.2001).
- Brulfert calculated the EIs for Maurienne and Chamonix valley, France (G.Brulfert et al. 2004).
- Mattai calculated the EIs for London, England (J. Mattai et al. 2002)

The bottom-up approach is in general impossible due to lack of available information, while top-down approach might lead to poor accuracy level for urban EIs. So several researchers used top-down approach to generate EIs for point sources and area sources because data for these sources are not easy to access, while they used bottom-up approach to generate EIs for road traffic sources. An example from the literature is:

Sturnm generated EIs for urbanized triangle Antwerp-Brussels-Ghent which is located in Northwest Europe. He used the top-down approach for generating EIs for point sources (based on statistic data and emission factors from literatures) and area sources (based on collective data per km2). However for road traffic emissions he used bottom-up approach using road traffic emission measurements and an urban traffic flow model (Moussiopoulos et al., 2004).

Combination of top-down and bottom-up approaches

In the cities of developing countries, the data for generating EIs for traffic sources are generally not available, so it is difficult or even impossible to use bottom-up approach for generating EIs. In this chapter, we present a new road traffic emission model which combines top-down and bottom-up approaches (Bang et al., 2011).

The strengths and weaknesses of the combination of the two approaches are as follows:

Strengths: By using this method (1) the computation time is rapid; (2) less input data is needed to generate EIs; (3) we have control on the uncertainty in EIs due to input data; (4) we can generate EIs both for developed cities and developing cities and (5) the accuracy in the results of EIs is improved.

Weaknesses: the limitation of this method is that (1) we have to follow many steps to generate EIs and (2) we have to study the input parameters which generate most uncertainty in EIs.

3.3 Modeling of air pollution

The European Directive on evaluation ambient air quality (EU-N°96/62/CE 1996) requires the use of modeling tools to define high pollutant concentration zones. Modeling tools use mathematical formula to simulate all the atmospheric processes over various time and space scales. These processes are complex and nonlinear so that modeling is the only tool which takes into account all the processes.

Concerning the space scales of modeling, there are many scales such as global scale, regional or continental scale, mesoscale (city or country) and microscale (street canyons). There are many mesoscale models that are used to simulate urban air quality, such as METPOMOD, TVM-Chem, CIT, CHIMERE, CMAQ, TAPOM, etc. Input parameters of these air quality models are meteorological conditions, EIs, landuse, topography, boundary and initial conditions.

In recent decades, computer technology has rapidly developed; so many new mesoscale models are developed. With the increased power of computer technology, the time scale of modeling more refined. The new models can now simulate air quality for long periods and on temporal resolution from few hours to few months.

One of the most important functions of air quality modelling is to evaluate the effective of abatement strategies to reduce air pollution in cities. Modeling tools are also used to study the impact of different activities on urban air quality and to evaluate the methodology for generating EIs, such as Erika used TAPOM model to evaluate the accuracy of different methodology for generating EIs for Bogota city, Colombia (Erika et al., 2007).

3.4 Analysis of uncertainties

Uncertainty in emission inventory

One of the most important strengths of the emission inventory is generate the uncertainties of EIs due to the input parameters. There are many methods used to calculate uncertainty. One of these methods called analytical method is as follows:

Emissions are calculated as the combination of different parameters:

$$E = H_1 \times H_2$$

Where, E is the emission

H1 is the parameter to quantify the activities

H2 is an emission factor per unit of activities

H1 and H2 can depend on many other factors like the number of vehicles, road parameters, etc.

When H1 and H2 are simple enough it is possible to compute directly the uncertainties. For example when H1 and H2 are constants:

$$E = (\overline{H}_1 + \Delta H_1) \times (\overline{H}_2 + \Delta H_2)$$

Where, $\overline{H}_1$, $\overline{H}_2$ is the average of H1, H2 respectively

ΔH_1 , ΔH_2 is the uncertainty of H1, H2 respectively.

$$E = (\overline{H}_1 \times \overline{H}_2) + (\overline{H}_1 \times \Delta H_2) + (\overline{H}_2 \times \Delta H_1) + (\Delta H_1 \times \Delta H_2) \text{ and also: } E = \overline{E} + \Delta E$$

In this example the uncertainty on E is equal to:

$$\Delta E = (\overline{H}_1 \times \Delta H_2) + (\overline{H}_2 \times \Delta H_1) + (\Delta H_1 \times \Delta H_2)$$

As we can see in this example the calculation of the emissions is non-linear and generates several uncertainty terms. All these terms rapidly become impossible to calculate analytically when the parameters used in the emission calculation are more complex. In such situation, the most oftenly used approach is the Monte-Carlo method (Hanna et al. 1998). The detailed of Monte-Carlo method is explained as follows:

The Monte Carlo methodology has been used to evaluate the uncertainties in previous air quality studies (Hanna et al., 1998; Sathya., 2000; Hanna et al., 2001; Abdel-Aziz and Christopher Frey., 2004). In the emission inventory model, the EIs are generated as the combination of different input parameters:

$$E = f(H_1, ... H_n) \tag{2}$$

where H_i are the input parameters (i=1, n). H_i can change due to the uncertainties. Each parameter H_i is distributed around an average $\overline{H}_i$ and a standard deviation σ_i.

The Monte Carlo method (Ermakov, 1977) generates for each input parameters a pseudorandom normally distributed numbers η^H ($\sigma = 1$ and mean= 0) which can be used to compute several values of H_i:

$$H_i^k = \overline{H}_i + \eta^H \sigma_i \tag{3}$$

The percentage of standard deviation σ_i could be calculated as: $\tilde{\sigma}_{Hi} = \dfrac{\sigma_i \times 100}{\overline{H}_i}$

The equation (3) becomes: $H_i^k = \overline{H}_i \left(1 + \dfrac{\tilde{\sigma}_{Hi} \eta^H}{100} \right)$

These parameters H_i^k are used to calculate several values of E^k: $E^k = f(H_1^k, ... H_n^k)$. The average value $\overline{E}$ and the standard deviation σ_E are deduced from the distribution of E^k.

The percentage of standard deviation σ_E could be calculated as: $\tilde{\sigma}_E = \dfrac{\sigma_E \times 100}{\overline{E}}$

The classification of standard deviation $\tilde{\sigma}_E$ is based on the standard deviation of the input parameters $\tilde{\sigma}_{Hi}$:

If $\tilde{\sigma}_{Hi}$ is the standard deviation of input parameters due to spatial and temporal repartition, the values $\tilde{\sigma}_E(x, y, t)$ are the standard deviation of emission for all input parameters but in space and time.

If $\tilde{\sigma}_{Hi}$ (the standard deviation of each input parameters) is constant in the entire domain, then the values of $\tilde{\sigma}_{E_{Hi}}$ are the standard deviations of emission for all the domain but for each input parameter.

Examples from the literature for uncertainties in emission inventory and Monte-Carlo application to air quality model are:

- Kuhlwein calculated uncertainties of input parameters in modeling emissions from road transport. The emission inventory was calculated by the working group "urban emission inventories" of GEMEMIS and the working group "Val.3 of Saturn" (Kuhlwein J et al. 2000)
- Hanna evaluated the effect of uncertainties in UAM-V input parameters (emissions, initial and boundary conditions, meteorological variables and chemical reactions) on the uncertainties in UAM-V ozone predictions by using Monte-Carlo uncertainty method in framework of research: "Uncertainties in predicted ozone concentrations due to input uncertainties for UAM-V photochemical grid model applied to July 1995 OTAG domain" (Hanna et al, 2001).

Uncertainty in air quality simulation

Results of numerical simulations are more reliable if the estimation of uncertainties in model prediction is generated. The uncertainties of the air quality model due to input parameters could be generated by using the standard deviation (square root of variance) around the mean of the modeled outputs (Hwang et al., 1998). Up to now, the Monte-Carlo (MC) technique is a brute-force method for uncertainty analysis (Hanna et al., 2000). A simple program was developed (named EMIGEN) by our research team using the MC technique for generating different emission files. The MC technique used in EMIGEN includes different steps: (1) we generate a series of random numbers which follow a normal distribution; (2) the EI results of emission inventory model are used as the input parameters for EMIGEN. The uncertainties of the input parameters for other sources (industrial, residential and biogenic sources), can be calculated by using the available data which is estimated the emissions in developing cities; (3) Running EMIGEN to get one hundred EI files. These hundred EI files are used as input for the air quality model. The uncertainty and the median of pollutants are calculated from 100 MC air quality simulations. The results of pollutants and their uncertainties from the output of the air quality modeling are divided into two pollutant types, primary and secondary pollutants. Their results from 100 MC in using air quality model are used to calculate the uncertainties in process to simulate air pollution.

4. Air quality study over developing countries: A case study of HCMC

4.1 Air quality in HCMC, Vietnam

HCMC is the largest city in Vietnam and is the most important economic center in Vietnam. HCMC became one of 100 rapid economic growth cities in the world (Gale, 2007). The population of city is 6.105 million (8% population of Vietnam), however the city accounts for 20.2 % GDP, 28 % industrial output of Vietnam. HCMC had 28500 factories and 2,895,381 motorcycles. They are the most important sources which contribute to air pollution in HCMC.

The HCMC's government has been set-up a number of air quality monitoring stations around the city for monitoring air pollution due to road traffic and industrial activities. The results are shown in Table 2.

The measurements results show that air quality in HCMC are polluted by TSP, PM and especially O_3. The highest TSP concentrations are found in 2001 about 900 μ g.m-3 (10 times higher than WHO standard). TSP concentrations reduce to the value of 260 μ g.m^{-3} in 2005 (Table 2). The high TSP concentrations in HCMC are related to the construction and traffic activities (Belalcazar., 2009). The mean annual concentration of NO_2 and SO_2 are lower than WHO standard, but their average daily/hourly values regularly exceed the WHO standards. PM10 and O_3 are the most critical pollutants in HCMC, because their concentrations are very high and almost exceeded the WHO standard during the period 2001- 2006. They have high toxic health effects. The measurements show that O_3 concentrations are highest during the midday where the highest photochemical processes are happened. The high O_3 concentrations are related to high VOCs concentrations in HCMC. However, there is very few information on VOCs concentrations which are monitored in the Asian countries due to their complex in measurement. Up to 2007, the first long-term study on the VOCs levels in HCMC was carried out by Belalcazar (Belalcazar., 2009). He measured continuously roadside levels of 17 VOCs species in range C2 - C6 during the dry season (from January to March) of 2007 in HCMC. Table 3 firstly shows VOCs concentrations in HCMC, then compare with available roadside VOCs concentrations of other Asian cities.

Pollutants	2001	2002	2003	2004	2005	2006	WHO standard
TSP	900.0	475.0	486.7	496.7	260.0	-	90
PM10	-	122.5	83.5	61.5	79.6	77.4	20
NO_2	-	-	-	-	34.0	32.1	40
SO_2	-	-	-	-	44.2	28.5	50
O_3	-	-	238	266	247	-	160[a]

Source: HEPA (2004); HEPA (2005) and HEPA (2006).
[a] WHO standard for O_3 (maximum 1-h concentration). Source: Molina and Molina (2001).

Table 2. Air quality in HCMC from 2001 to 2006. Almost data reported correspond to the mean annual concentration in μ g.m-3 (only O_3 is reported in maximum 1-h concentration).

Unfortunately, the air quality standard for all VOCs compounds does not exist. Only some individual VOCs standard limits are defined by WHO based on health effects. Among those, benzene is one of the very toxic VOCs. The measurements of benzene in HCMC show a very high concentration of 14.2 ppbv (annual mean WHO standard - 6ppbv). The VOCs concentrations in HCMC are generally higher than other Asian cities. The benzene concentrations in Hanoi are higher than in HCMC, because benzene in Hanoi is measured at peak hours (short time) when the highest benzene concentration happed during the year.

Up to now, very few researches attempts have been made for improving air quality in developing cities, especially in HCMC. These researches attempts are on air pollution status and effective abatement strategies. The researches on status of air pollution analyzed the data from the monitoring networks to evaluate air pollution status of the city. These results help the government in proposing solutions to reduce air pollution level today and in the future (Hang et al., 1996; DOSTE., 2000).

VOCs	HCMC (Mean)	Changchun (Mean)	Karachi (Mean)	HongKong (Mean)	Hanoi (Mean)
n-Propane	3.7				
Propene	19.5				
n-Butane	22.6				
Trans-2-Butene	5.2				
1-Butene	4				
Cis-2-butene	5.2				
i-Pentane	80.3	14.7	74		
n-Pentane	21.8				
Trans-2-Pentene	16.4				
1-Pentene	4.6				
2-methyl-2-butene	3.8				
Cis-2-Pentene	4.2				
2,3-Dimethylbutane	8.6				
2-Methylpentane	7.7	6.1	39		
3-Methylpentane	43.5				
n-Hexane	91	1.7	71	4.4	
Benzene	14.2	11.9	19.7	8.2	40
total	356.3				

Sources: Belalcazar (2009).

Table 3. VOCs levels in BTH street (HCMC) and comparison with the VOCs levels in other Asian cities (concentrations are in ppbv).

The researches on effective abatement strategies propose solutions to reduce air pollution. It consists mainly of master thesis which is carried out at IER (Institute for Environment and Resources in HCMC). These researches are focused only on some typical aspects of abatement strategies at small scale, due to the lack of the information for research, methodology, etc. Nguyen (2000) proposed some methods to reduce air pollution by traffic in HCMC as well as for developing public transportation system and using unleaded gasoline. Another study proposed technical solutions for reducing air pollution level by traffic activities (Duong, 2004).

In conclusion, both the developed and developing cities have air pollution problem, especially in the developing countries where it is related to high level of air pollution. It's urgent to adapt an air quality management (AQM) system for improving the air quality in these cities. In the next section, the research for improving air quality in HCMC is carried out by using numerical tools.

4.2 Emission inventory of air pollution over HCMC

4.2.1 Traffic source

Methodology

The following Fig. 2 shows an outline of the process of generating an EI for HCMC where the input data is limited and using the EMISENS model (Bang, 2011).

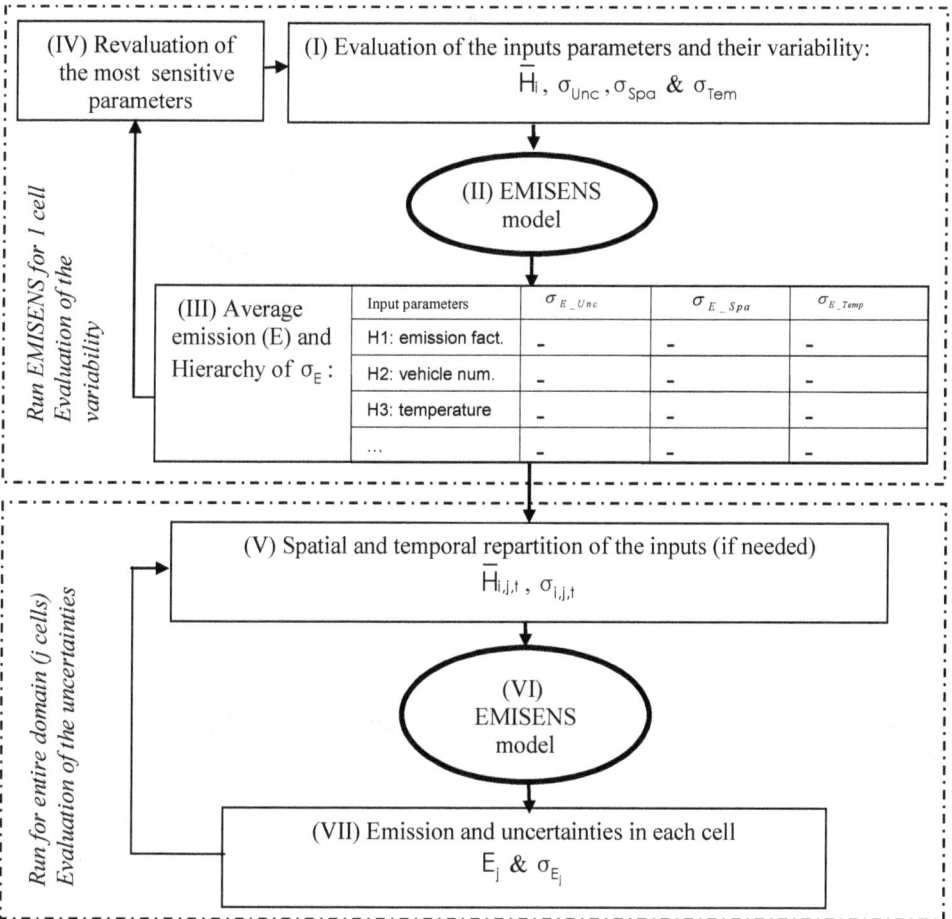

Fig. 2. Research process outline ($\overline{H}_i$ is the average value of input parameters. σ_{Unc}, σ_{Spa} and σ_{Tem} are the standard deviation of the input parameters due to the uncertainties, the spatial and temporal variability, respectively. The σ_{E_Unc}, σ_{E_Spa} and σ_{E_Tem} are the standard deviation of the emissions due to variability of the input parameters. The E_j & σ_{E_j} are the emission and their standard deviation for each cell.

This process can also be applied for other cities than HCMC.

i.   We collected the necessary data: minimum information to compute emission, estimation of the variability of the parameters (use the literature, other EI, existing data from HCMC, etc.).

ii.   We run the EMISENS model with variable parameters in one cell.

iii. Results of model are a list hierarchy of standard deviation (σ_{E_j}) due to input uncertainties.Then, we analyze the results to identify the parameters which generate the maximum of variability in the EI.

iv.   After the determination of the most sensitive parameters, we try to find more information of these parameters and will conduct additional campaigns to reduce the level of uncertainties of these parameters. We rerun the EMISENS model to revaluate the standard deviation of the emissions.

v.   Identification of the parameters that generate significant standard deviation in space and in time and generate a spatial and temporal distribution of these parameters.

vi.   We run the EMISENS model for each cell (run the model with parameters variable in all cells of the grid).

vii. Results of model are the emissions of each pollutant for each cell including its uncertainties (E_j and σ_{E_j}).

Input data

The air quality of HCMC is controlled by HoChiMinh environmental protection agency (HEPA). However, the research on air quality for HCMC is mainly studied by Institute of Environment and Resources (IER), Vietnam national university in HCMC (VNU-HCM). The database for air quality studies in HCMC is still limited because there have not been a lot of research conducted for this city. The only available source of data or other information about this city are the reports on air quality from monitoring activities and some research conducted by master student of VNU-HCM.

Within the framework of this research, several campaigns were organized which focused mainly on the on-road traffic activities (vehicle fleet, traffic flow, etc.) using different methods.

Fig.3 shows the daily average variation of the fleet composition (in %) at 3 Thang 2 street (urban street category).

The fleet is almost dominated by light gasoline vehicles (motorcycles: 92%, cars: 3.46%, light trucks: 2.8% and buses 0.1%). Only 1.1% of the fleet is made of heavy truck diesel vehicles. The fleet distribution can be assumed the same for all streets in HCMC in the urban category. The fleet composition we have determined is also similar to the fleet of HOUTRANS project which was counted for the whole city of HCMC (HOUTRANS, 2004). We organized survey for the characterization of the vehicles (on road). The campaign was organized during the year of 2007 and 2008.

Some results of the campaign are shown in the Fig.4. This figure describes the distribution motorcycle age. We can conclude that motorcycles currently used in HCMC have been produced in recent years (more than 32% of motorcycles are produced less than 1 year and about 94% of motorcycle are produced less than 10 years) and more than 98% are the 4 strokes motorcycles. The main fuel used by motorcycles is gasoline. In spite of the fact, that most of motorcycles used in HCMC are not very old, they are a major producer of pollution. The reason is that they are manufactured by low standard technology (older than EURO II standard). The largest fraction of the motorcycles are imported from China and Taiwan which are very cheap in price and easily meet the requirement of the local people according to their purchasing power.

The survey further revealed that there are more than 25% cars which are less than 1 year old and more than 70% cars which are less than 4 years old. More than 98% cars use the gasoline for fuel and the rest of cars use the diesel oil.

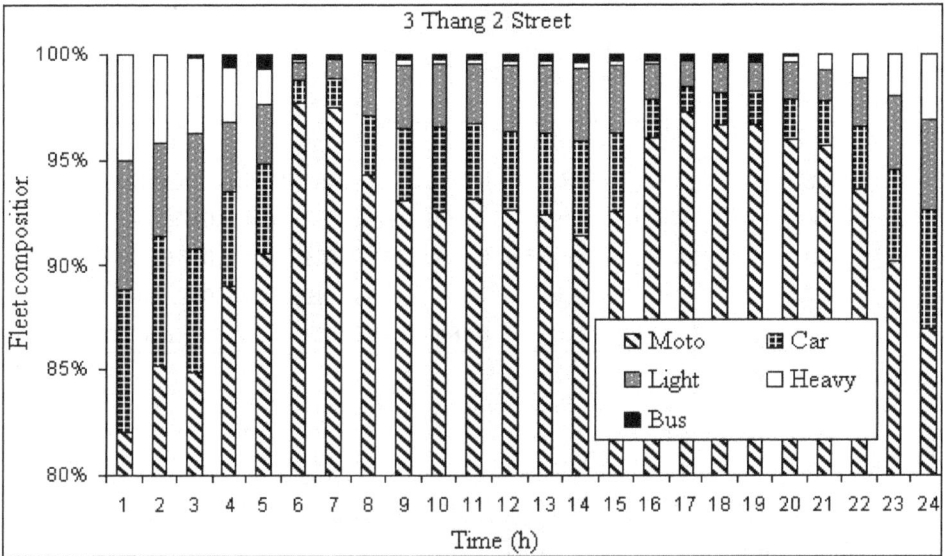

Fig. 3. Daily variations of the fleet composition at 3 Thang 2 street (urban street category).

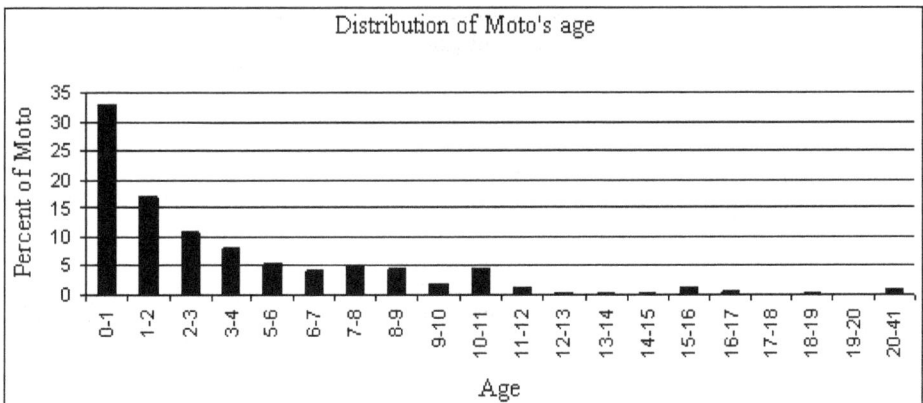

Fig. 4. Distribution of Moto's age in HCMC.

Emission factors (EFs)

The EFs from the three different sources are shown in Table 4. The differences of emissions factors from three sources can partly be explained by the different quality of fuel and the maintenance of vehicles. The motorcycle's engines are not regularly maintained in Vietnam and more than 60% of the motorcycles in Hanoi and in HCMC exceeded the emissions standard (Trinh, 2007).

We evaluated the EFs from the 3 different sources. The main EFs which were used in this study are the EFs from HCMC (NOx, NMVOC and CO). However, there is no data available for the EFs of SO_2 and CH_4 for HCMC. So we used the emission factor of SO_2 from China because the characterizations of China's vehicles are similar to HCMC. For the emission factor of CH_4, we unfortunately had to use the emission factor of CH_4 from Copert IV. The EFs and their uncertainties which were used in this study are shown in Table 4.

Vehicle	NOx[a]	CO[a]	SO_2[b]	NMVOC[a]	CH_4[c]
Motorcycle	0.05±0.02	21.8±8.67	0.03±0.015	2.34±1.17	0.115±0.121
Bus	19.7±5.2	11.1±5.3	1.86±1.08	89.9±33.01	0.077±0.051
Light	1.9±0.9	34.8±15.5	0.05±0.029	15.02±7.36	0.017±0.018
Heavy	19.7±5.2	11.1±5.3	1.86±1.08	89.9±33.01	0.062±0.041
Car	1.9±0.9	34.8±15.5	0.18±0.105	15.02±7.36	0.0031±0.0032

Source:
[a]: the EFs were calculated for HCMC (Belalcazar et al., 2009; Ho et al., 2008)
[b]: the EFs of China (DOSTE, 2001)
[c]: the EFs were calculated from Copert IV.

Table 4. EFs (in g.km-1.vehicle-1) and their uncertainty.

- Result of total emissions:

The total result of traffic emission is presented in Table 5 as follows:

Pollutants	Total emissions [ton h-1]	Total uncertainties (%)
NOx	3.44	19.56
CO	331.4	33.77
SO_2	0.733	27.09
NMVOC	46.24	27.55
CH4	2.04	49.66

Table 5. The average of emission and their total uncertainties of all parameters over HCMC.

Temporal and spatial distribution

Temporal distribution: the vehicle flows from existing traffic counts were analyzed to derive the temporal distribution (Fig.5).

Spatial distribution: the spatial distributions are based on the length of each street category (L_{Is}) in each cell of study domain. The MapInfo software is used to distribute the parameters spatially. As expected, Fig. 6 shows that most of the urban streets are concentrated in the districts of central HCMC.

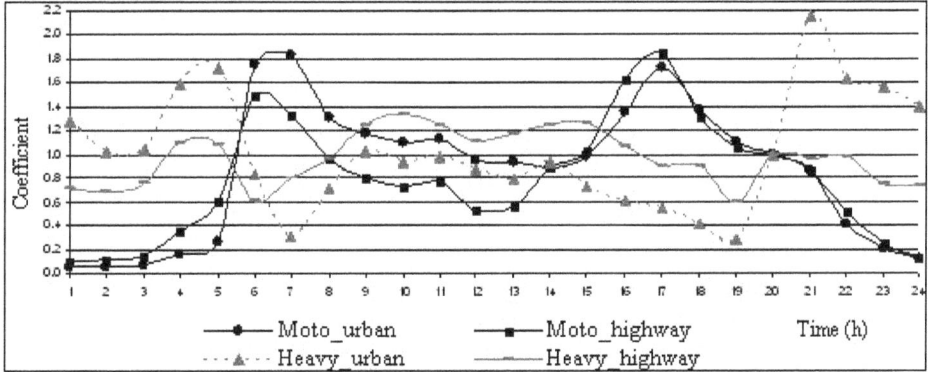

Fig. 5. Normalized hourly distribution of Motorcycles (Moto) and Heavy trucks, per urban and highway street category. A factor of 1.0 is attributed to the hour between 1900LT and 2000LT to both categories of streets, for a working day.

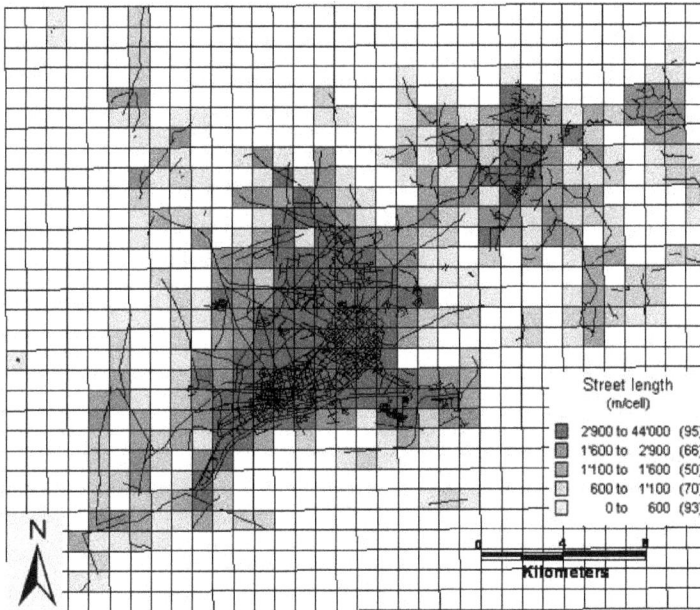

Fig. 6. The network of urban streets category in the study domain. The different colour is the number of kilometres of street length per cell. The numbers in parentheses is the number of cells.

Result of emissions in spatial

Finally, we run the EMISENS model for all cells of study domain. The outputs of the model are the emissions and their uncertainties in each cell (Fig.7 and Fig.8). For each pollutant, the emissions are calculated for each cell per hour. The results of CO in Fig.7 show that most of the emissions occur in the centre of the city where we can see the highest of street density of urban streets. In contrast, in the uncertainty map (Fig.8) the lowest uncertainties occur in urban streets in the city centre. Because we sub-divided the urban streets into three urban street categories, the uncertainty for urban streets is relative small.

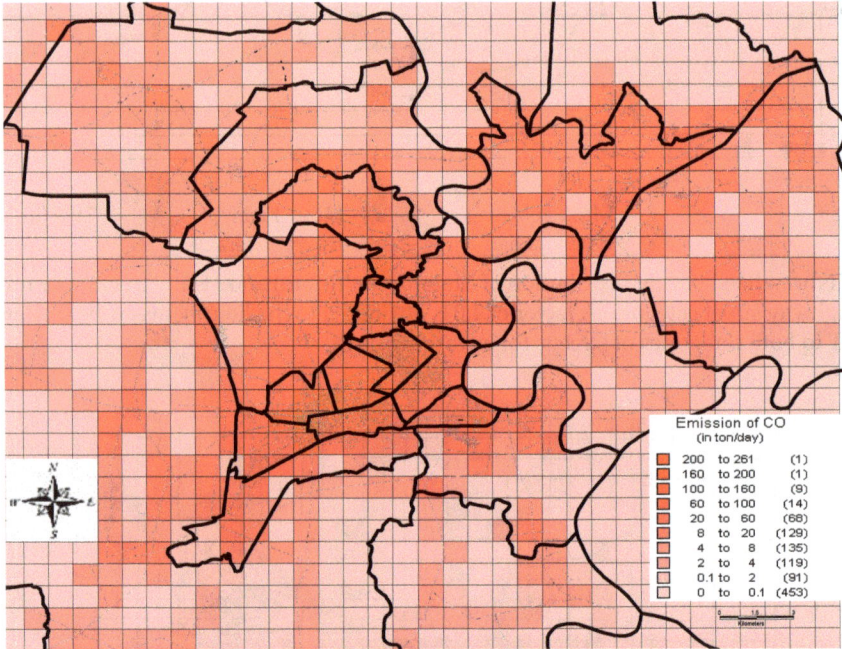

Emission of CO
(in ton/day)

200 to 261	(1)
160 to 200	(1)
100 to 160	(9)
60 to 100	(14)
20 to 60	(68)
8 to 20	(129)
4 to 8	(135)
2 to 4	(119)
0.1 to 2	(91)
0 to 0.1	(453)

Fig. 7. The traffic emission map of CO. The contour of the districts (black colour) and the street network (grey colour). The numbers in parentheses are the number of cells.

4.2.2 Industriel sources

The industry is second most important source of atmospheric pollution in HCMC (Bang et al., 2006; Nguyen et al., 2002). The city spans only 0.6% of the whole country's area but more than 20% industrial producing capacity and 40% industrial output of whole country (Nguyen, 2002) is located in HCMC. The main industries in HCMC are thermo-electricity, cement production, steel lamination and refinement, weaving and dying, food processing and chemistry. The existing information for industries is very poor and an official detailed database does not exist. Most of the industries in HCMC have the following characteristics:

- The factories are old and their operation times are over 20 years.
- They have old technology which is imported from Soviet Union.
- The engines consume much energy and fuel due to old and poor quality engines. The fuel used is of a low quality.

The above list mentions the major characteristics of a majority of the industries in HCMC. In this study, the emissions from industrial sources are calculated by using a top-down approach. Starting from the total emissions in Vietnam (Table 6) and the percentage of distribution of pollutants in different sectors in Vietnam (Fig.9), we estimate the yearly industrial emissions for Vietnam. The industry of HCMC accounts for 20.2% of the total industrial emissions of Vietnam (HCMC statistics, 2003).

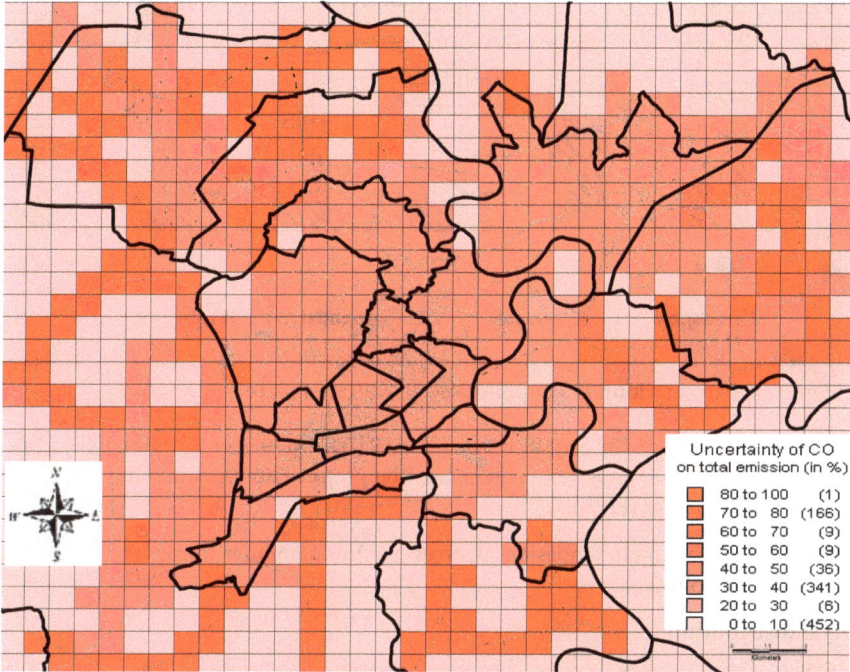

Fig. 8. The uncertainty map of CO. The contour of the districts (black colour) and the street network (grey colour). The numbers in parentheses are the number of cells.

Pollutants	Total Emission (Tg/year)
SO_2	0.193
NOx	0.283
CO_2	169.200
CO	9.248
CH_4	2.907
BIO NMVOC	1.037
ANT NMVOC	1.168
COVNM total	2.205

Sources: Mics-Asia, 2000 (1 Tg = 106 ton).
Note: BIO NMVOC: NMVOC Biogenic; ANT NMVOC: NMVOC Anthropogenic

Table 6. Total emission in Vietnam.

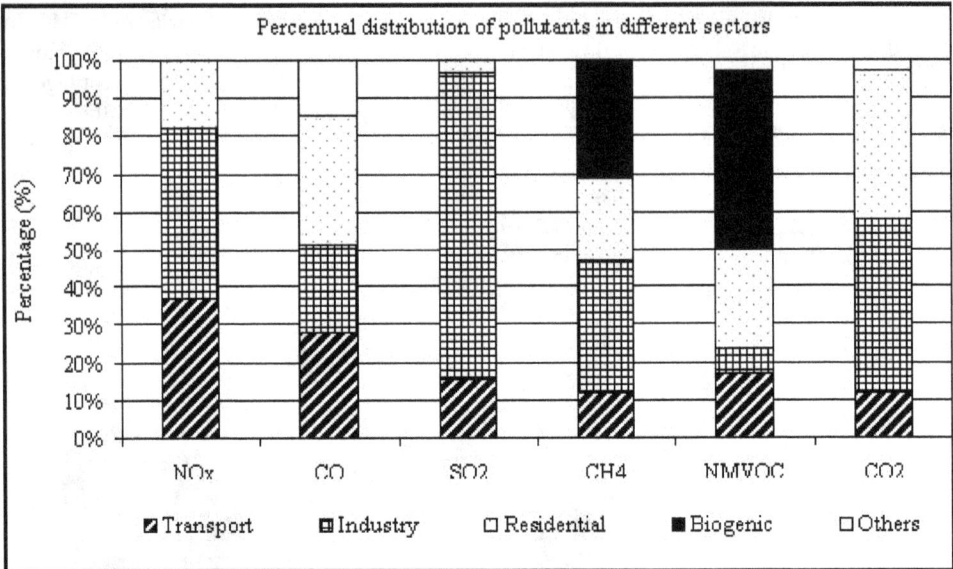

Fig. 9. Percentage of distributions of pollutants in different sectors in Vietnam.

The total emissions of Vietnam are shown in Table 6. These values were estimated by Mic-Asia project from all sources of emission in Vietnam. The project also estimated the contribution of different emission sources (Fig.9)

Fig.9 shows that the main emission sources in Vietnam are transport and industries depending on the pollutant type.

Temporal distribution

The results of the air quality monitoring program in south of Vietnam conducted by the Institute of Environment and Resources (IER, 2006) were used to estimate the monthly, daily and hourly coefficient distribution of industrial emissions in HCMC. The results show that November is the most polluted month because it is the post-rainy season in HCMC. During this time period, all the companies want to complete their already planned targets of products for the specific year. So the companies utilize the maximum of the available resources and run their industries at full capacity to meet their targets. On a weekly basis, Friday is the day of the week with largest emissions as it suffers maxima of pollution at 0900LT in morning and 1400LT in afternoon.

Spatial distribution

The spatial distribution of industrial emission sources is estimated by using the population density in each cell. We also used the GIS software for distributing the emissions spatially.

4.2.3 Residential sources

The main emission sources of residential areas are anthropogenic, which are mainly caused from the gastronomic activity at the residences and restaurants in HCMC. Natural gas is the major source of domestic fuel and is mainly used for cooking purposes. However, there are many small restaurants in HCMC which still use the fossil coal. The pollutants such as SO_2 and CO are mainly produced by the burning of this fossil fuel (Dinh, 2003).

Temporal distribution

We estimated the monthly, daily and hourly coefficients for the temporal distribution of residential emissions in HCMC by using the data of the air quality monitoring program in south of Vietnam conducted by the IER (IER, 2006). The monitored results show that December is the most polluted month of the year. This is because December is the start of hot season in HCMC. Saturday is the most polluted day of the week because of the weekend. Normally people cook more dishes on weekend than normal days of the week. The maximum pollution is measured at 1100LT on Saturday which is corresponds to lunch time in HCMC.

Spatial distribution

The spatial distribution of residential emission sources on each cell of study domain is also estimated using the population density in each cell.

4.2.4 Biogenic sources

Only volatile organic compounds (VOCs) are calculated for biogenic emission sources. The biogenic EI are very important for air quality modelling, because biogenic VOCs contribute significantly to the formation of ozone (Varinou et al., 1999; Rappenglück et al., 2000, Moussiopoulos, 2003). The largest contribution through biogenic emissions is caused by trees, which emit 10 times more biogenic VOCs as compared to smaller plants. The biogenic VOCs are mainly divided in 3 major pollutants: Isoprenoids, Terpenes and other VOCs (EEA, 1999).

Temporal distribution

The emissions of biogenic VOCs depend on the air temperature and the intensity of solar radiation (EEA, 1999; Rappenglück et al., 2000; Moussiopoulos, 2003). So, the estimation of monthly, daily and hourly coefficient for biogenic source in HCMC is based on the results of solar-radiation measurements (from the urban air quality monitoring of HEPA (HEPA, 2006)). The monitored results show that the highest intensity of solar radiation is in the month of April, because the April is in the middle of the hot season in HCMC. Midday is the time when the maximum of radiation intensity is observed.

Spatial distribution

In this research, we estimate the emission for the districts in centre of HCMC. We assume that the percentage of green space is similar every where in the domain. The spatial distribution of biogenic emission sources to each cell of study domain is estimated by using the area of each cell.

4.2.5 Results

Table 7 presents the results of total EI for all emissions sources in HCMC and the percentage of contribution of each emission source towards the total emissions in HCMC. The column "total emissions" represents the total emissions (in ton per day). The last four columns correspond to the percentage of contribution of each emission source on total EI.

Pollutant	Total emissions	Industry (%)	Residential + other (%)	Biogenic (%)	Traffic (%)
NOx	106.27	15.79	6.46	-	77.75
CO	8860.25	5.45	4.78	-	89.77
SO$_2$	1603.37	80.42	18.49	-	1.10
NMVOC	1241.45	1.90	8.33	0.38	89.39
CH$_4$	214.81	27.02	33.19	16.98	22.80

Table 7. Total emissions in HCMC [ton day-1] and the contribution of each emission source on the total emissions in HCMC (in %).

Table 7 shows that the emissions of SO$_2$ (80.42%) from industrial sources are very important. Because the industry in HCMC uses a lot of diesel oil, mazut oil and fossil coal which contain high percentage of sulphur as fuel (Dinh, 2003).

The traffic sources contribute with a high percentage of total emissions (NOx 77.75%; CO 89.77%; and NMVOC 89.39%).

4.3 Modeling of air pollution over HCMC

4.3.1 Models description and set-up

Meteorological

The Finite Volume Model (FVM) model used in this research is a three dimensional Eulerian meteorological model for simulating the meteorology. The model uses a terrain following grid with finite volume discretization (Clappier et al. 1996). This mesoscale model is non-hydrostatic and anelastic. It solves the momentum equation for the wind component, the energy equation for the potential temperature, the air humidity equation for mean absolute humidity and the Poisson equation for the pressure. The turbulence is parameterized using turbulent coefficients. In the transition layer these coefficients are derived from turbulent kinetic energy (TKE, computed prognostically), and a length scale, following the formulation of Bougeault and Lacarrere (Bougeault et al. 1989). In the surface layer (corresponding to the lowest numerical level), in rural areas, the formulation of Louis (Louis et al. 1979) is used. The ground temperature and moisture, in rural areas, are estimated with the soil module of Tremback and Kessler (Tremback et al. 1985). An urban turbulence module in the model simulates the effect of urban areas on the meteorology (Matilli et al. 2002 b). A second module, the Building Energy Model (BEM, Krpo 2009), takes into account the diffusion of heat through walls, roofs, and floors, the natural ventilation, the generation of heat from occupants and equipments, and the consumption of energy through air conditioning systems. The FVM model was developed at the Air and Soil Pollution Laboratory (LPAS) of the Swiss Federal Institute of Technology in Lausanne (EPFL).

For choosing the domain used in the meteorological simulations, we have to take account of their size and especial resolution due to the capacity of computer. The four selected domains (Fig. 10) are domain 1 (resolution of 75 km x 75 km cells 16 x 16 grid points), domain 2 (resolution of 16 km x 16 km cells 33 x 33 grid points), domain 3 (resolution of 5 km x 5 km cells 40 x 40 grid points) and domain 4 (resolution of 1 km x 1km cells 34 x 30 grid points). The domain 4 includes main part of HCMC. The results from nesting procedure for initial and boundary conditions are used over this domain. For the largest domain (domain 1), the initial and the boundary conditions are interpolated from 6-hourly data from National Centers for Environmental Prediction (NCEP). In vertical position the grids extend up to 10,000 m, with 30 levels. The vertical resolution is 10 m for the first level, and then it is stretched up to the top of the domain at 1000 m [(grid stretching ratio equal to 1.2) (Martilli et al. 2002)]. Land use data obtained from the U.S. Geological Survey (USGS) is used as input for the simulations. For obtaining more realistic initial conditions, a pre-run of one day is computed for the meteorological simulations.

Air quality model

The air quality model used for this research is the Transport and Photochemistry Mesoscale Model (TAPOM) (Martilli et al . 2003; Junier et al . 2004) implemented at EPFL and at the Joint Research Centre of Istituto Superiore per la Protezione e la Ricerca Ambientale (ISPRA). It is a transport and photochemistry three-dimensional Eulerian model. It is based on the resolution of the mass balance equation for the atmospheric substances. This equation includes the advection by the mean wind, the vertical diffusion by the turbulence, the chemical transformation by reactions, the dry deposition and the emissions. The chemical transformations are simulated by using the RACM (Stockwell et al. 1997), the Gong and Cho (Gong et al. 1993) chemical solver for the gaseous phase and the ISORROPIA module for inorganic aerosols (Nenes et al. 1998). The transport is solved using the algorithms developed by Collella and Woodward (Collella et al. 1984). Then this algorithm has recently developed by Clappier (Clappier et al. 1998). Now there are a lot of atmospheric models uses this algorithm. The photolysis rate constants used for chemical reactions are calculated using the radiation module TUV which is developed by Madronich (Madronich et al. 1998). In vertical position, the grids extend up to 7300 m with 12 levels. The vertical resolution is 15 m for the first level, and then it is stretched up to the top of the domain at 2000 m (grid stretching factor of 1.2 for lower and 1.6 for upper layers of the grid).

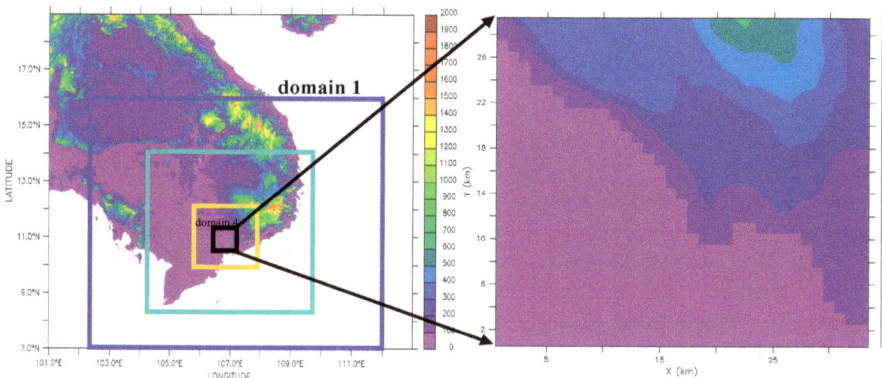

Fig. 10. Topography of South of Vietnam and description of simulated domains (left panel). The central black square (shown in left panel) used for the air quality simulation domain (right panel). (Source: http://edcdaac.usgs.gov/gtopo30/gtopo30.html (online & free downloading)).

Initial and boundary conditions for the photochemical simulations are based on measurements obtained by the Institute of Environment and Resources –VNU (IER-VNU) and the HEPA. Measurements taken from stations located in the surrounding of HCMC. They show 30 ppb of O_3 and very low values of NO and NO_2 (0.19 ppb).

A pre-run of one day with the same emissions (in emission inventory section) and wind fields (in meteorological modeling section) is performed. This calculation provides more realistic initial conditions for the air quality simulations. The air quality simulations are run for the episode of 3 day 6th - 8th February 2006.

4.3.2 Results

Results of meteorological simulations over HCMC

In the morning, the wind direction in HCMC is towards the north-west. At 0600LT (Fig. 11(a)), the wind is influenced by the Trade Winds. At this time, we do not observe the sea breeze phenomenon because it is too weak and the Trade Winds dominate the wind direction in the grid at this time. By 0900LT the wind is stronger and we observe the development of some small converge zones, produced due to the slope winds phenomenon developed in the city. Until 1300LT as shown in Fig. 11(b), the sun light has warmed up the ground rapidly. The slope winds are stronger at that time and air masses come up from the south plateau toward the highland area in the north. Some other air masses come from the east. Three main converge fronts can be perceived in the grid. The wind speed increases strongly and reaches its maximum at 1700LT as shown in Fig .11(c). At this time, the warming of the ground reaches its maximum and the sea breeze phenomenon develops strongly. From 2200LT until the next morning, wind fields are similar to 0600LT as it shown in Fig .11(d).

The measurements taken during the episode are used to validate these wind fields. The results of TSN station (Fig .13) show daily and nightly temperature values (Fig .12(a)). In general, FVM reproduces correctly the variation of the temperature. The results show that during all the day, measured and modeled temperatures are very similar. The model predicts well the time of the day when the sun rises (0700LT) and temperatures start increasing.

The maximum value of temperatures (between 1200LT and 1500LT) is 35.19°C. However it underestimates nightly temperatures, this can be explained by the NCEP nightly temperatures are also underestimated at ground level. These boundary conditions contribute to cool down the borders of the grid, and then the simulations are underestimated. The measurements of TSN station (Fig .12(b)) show very clearly the daily and nightly maximum and minimum wind speed values. Unfortunately, there are very few measurements for meteorology over HCMC area. The TSN station is located in west part of domain. We observed that minimum wind speed values are between 0500LT and 0700LT, when land and sea are coolest. The minimum wind speed was observed at the same time together with a change in the wind direction (Fig .12(c)). The maximum wind speed values observed during the day occur at the same time with the maximum of development of local phenomena. The change in wind direction due to the slope winds cannot be seen clearly, because TSN station is situated towards the west of city where the local phenomena are less notorious. The wind and temperature of simulation in vertical are agreement with the observations.

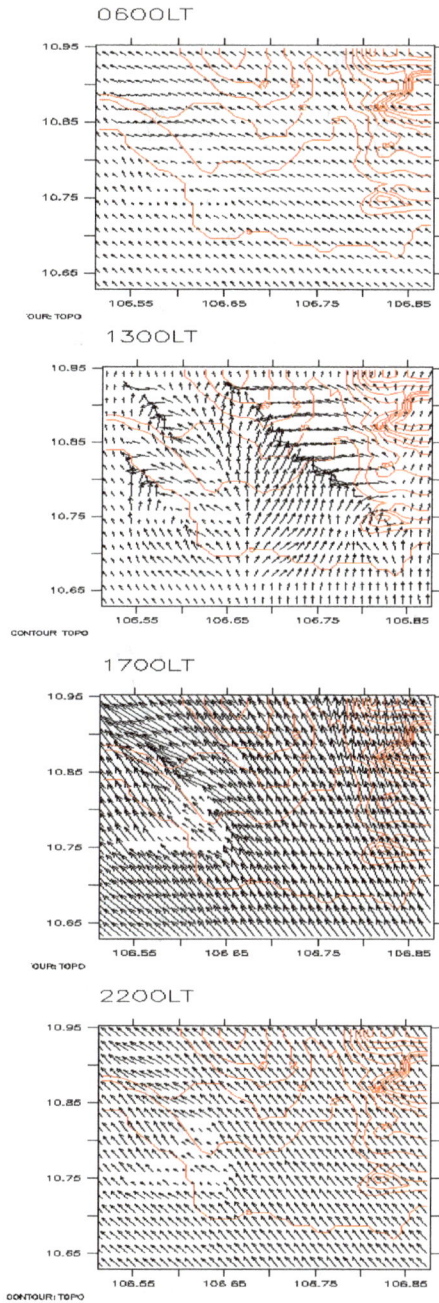

Fig. 11. Wind field results from simulations at ground level for the domain 34 km x 30 km, 7th February 2006. Geographical coordinates of the lower left corner: 10.64°N and 106.52°E. Maximum wind speed is 5.5 m s-1.

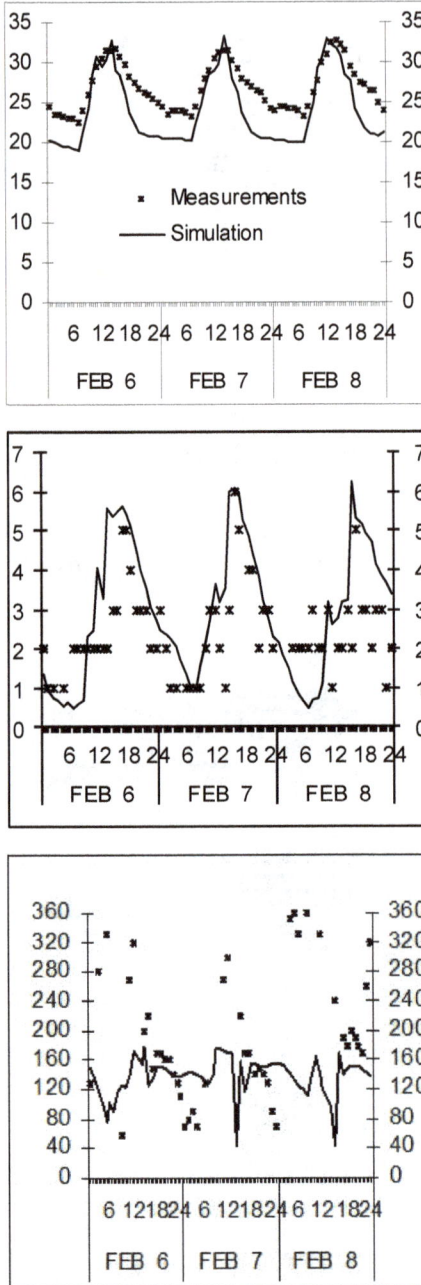

Fig. 12. Comparison between the results of simulated (solid line) and measured (starts). The temperature in °C (a), wind speed in m.s-1 (b), wind direction in degree (c) at ground level in TSN station, 6th – 8th February 2006.

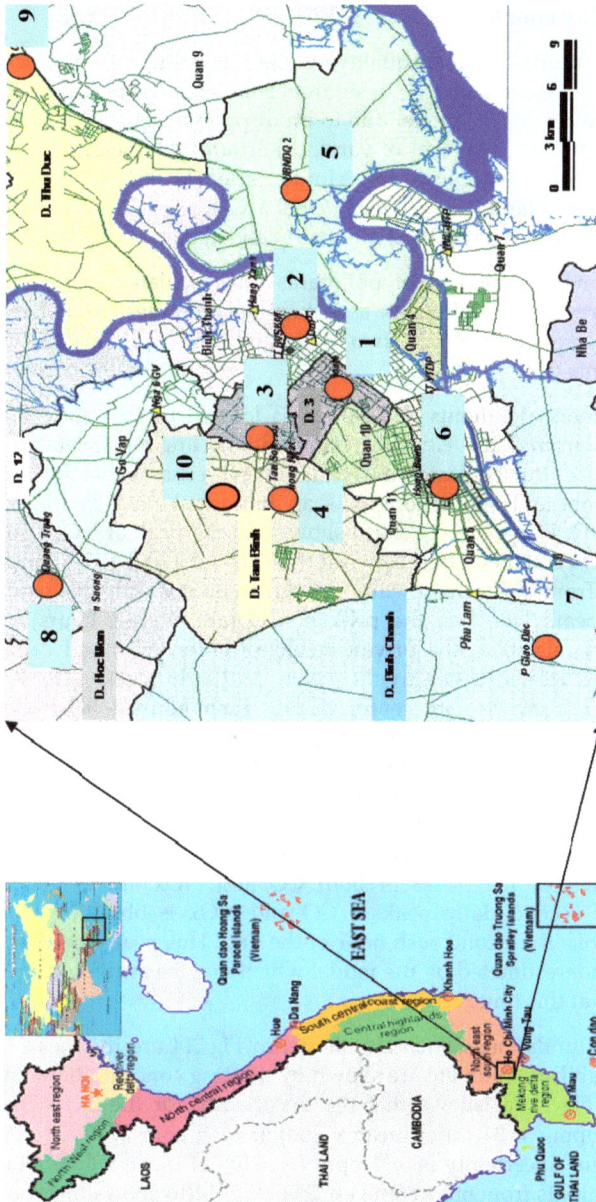

Fig. 13. The map on the left panel is the location of Vietnam. The map on right panel is presenting the location of monitoring stations in domain simulated. Five road-side stations (1. DO; 4. TN; 6. HB; 7. BC; 9. TD) and four urban background stations (2. ZO; 3. TS; 5. D2; 8. QT) for air quality monitoring are located on the map. The meteorological stations are located in TSN airport (station number 10) and Tan Son Hoa site (station number 3). (Source: Library of Institute of Environment and Resources (IER)-Vietnam).

Results of air quality simulations over HCMC

Evaluation of the uncertainty in air quality simulation: results of numerical simulations are more reliable if the estimation of uncertainties in model prediction is generated. The uncertainties of the air quality model due to input parameters could be generated by using the standard deviation (square root of variance) around the mean of the modeled outputs (Hwang et al. 1998). Up to now, the Monte-Carlo (MC) is a brute-force method for uncertainty analysis (Hanna et al. 2000). 100 MC air quality simulations are run for estimating uncertainty in the results of air quality and abatement strategies.

The uncertainty and the median of pollutants are calculated from 100 MC air quality simulations. The results of pollutants and their uncertainties from the output of the air quality modeling are divided into two pollutant types, primary (CO, NOx, etc.) and secondary pollutants (O_3).

Simulation of primary pollutants: TN station is located in the centre of HCMC, and D2 station is located around the city, but both stations are representative for ambient air quality. They are not situated beside the road. Fig .14(a) shows that the concentration of CO (from measurements and simulations) has an important peak in the morning (between 0700LT and 1000LT). However, Fig .14(b) shows that the peak of NOx (from measurements and simulations) is presented later, between 0900LT and 1200LT. The peak is related to the highest emissions from traffic mainly due to the rush hour during this period in HCMC. The peak of NOx appears late than the peak of CO (around 2-3 hours), because the high emission of CO is related to the private vehicle (motorcycles and cars), while the high emission of NOx is related to the trucks (heavy and light trucks). In HCMC, trucks have limited circulation from the city centre during rush hours (600LT-830LT and 1600LT-2000LT). The peak is amplified to a very high concentration due to a low mixing height in the early morning. At this time the temperature of the air masses are still cold, the vertical diffusion of pollutants is very weakly so the pollutants are stored at ground level. The air-monitoring network in HCMC includes nine stations but due to the lack of calibration and maintenance of the equipments, measurements are available only for some day at some stations. The values of the peaks of both CO and NOx are in good agreement with observations. The second daily peak of CO and NOx is observed around 1700LT and 1800LT, because this is a second rush hour of the day. This peak is related to the traffic and it is sometimes underestimated by the model which may be attributed to an overestimation of the wind speed at this time.

The results from simulations which are shown in Fig .14 are the mean values of 100 MC simulations. Probabilistic estimate are shown by plotting concentration enveloped (mean $\pm$ 1 σ) with time. Fig .14(a) shows that the uncertainty for the CO simulated differ by a maximum of 1.8 ppm ($\approx$ 34.4% of mean value) at rush hour 0700 LT – 0800LT on 6th Feb 2006. The minimum uncertainty is 0.01 ppm ($\approx$ 0.5% of mean value) which is observed in the middle of all nights from 6th to 8th Feb 2006. Fig. 14(b) shows that the NOx uncertainty differs by a maximum of 11.28 ppb ($\approx$ 13% of mean value) at the same time of appearance NOx peak. The minimum uncertainty is 0.47 ppb ($\approx$ 5.9% of mean value) which is observed during the middle of night.

Simulation of secondary pollutants: HB station is located in the centre of HCMC (Fig .13). In general the concentration of O_3 in D2 station (Fig .14(c)) is higher than in HB station (Fig .14(d)), because D2 station is situated closer to the O_3 plumes than HB station.

The simulation shows high O_3 levels at the same stations as the measurements do on 7th Feb as shown on Fig .14(c), indicating a good reproduction of the plume position. We can see that the HB station is located closer to the south of the city than the D2 station (Fig .13). This confirms the fact that pollutants are being transported in the northern and north-western direction at 1300LT on 7th Feb. The uncertainty analysis for the air quality model is also studied by running 100 MC simulations. Fig .14(c) and Fig .14(d) show that the uncertainties of O_3 differ by a maximum of 5ppb ($\approx 8.6\%$ of mean value) at 1100LT-1300LT on 6th - 8th Feb 2006 at both stations. The minimum uncertainty of O_3 is 1ppb ($\approx 15\%$ of mean value) at 0700LT – 0900LT on 6th – 8th Feb 2006 at both stations.

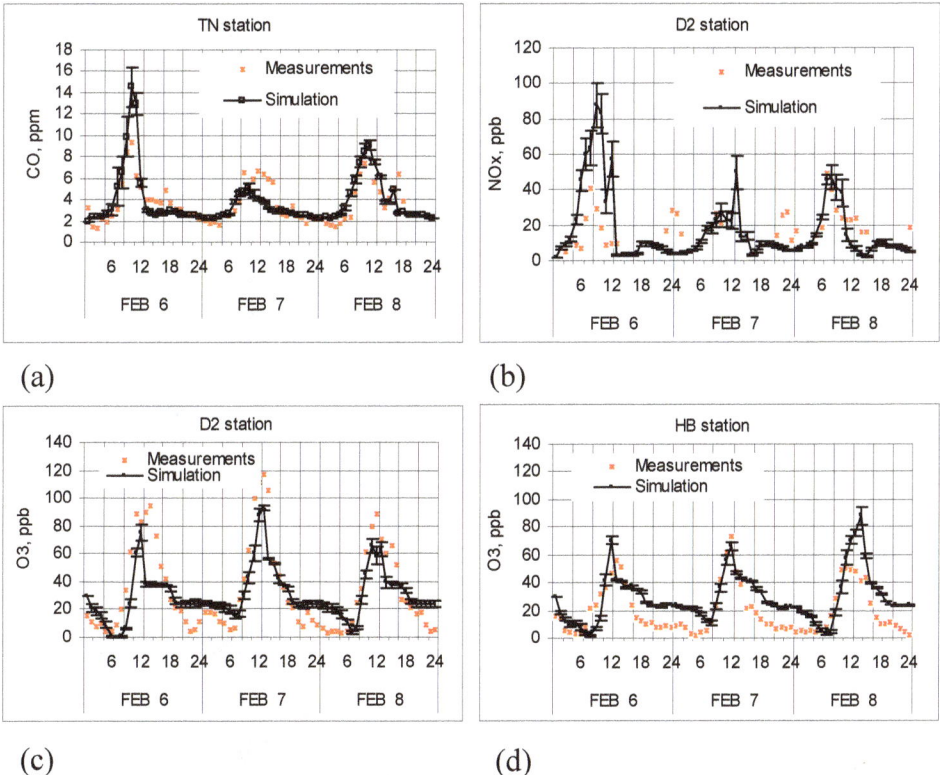

(a)   (b)

(c)   (d)

Fig. 14. Comparison between the results of measurements (stars) and simulation (solid line) during the selected episode (on 6th - 8th Feb 2006) for CO (ppm) at TN station, NOx (ppb) at D2 station and O_3 (ppb) at D2 and HB stations. The uncertainties of CO, NOx and O_3, from 100 MC simulations are presented by 1σ (standard deviation). NOx refers to NO + NO_2.

Spatial distribution of O_3

Figure 15 shows the plume of O_3 developed during the 7th Feb 2006. In the early morning, there are very high concentrations of NOx stored in the centre of the city, which generates O_3 destruction at this location, while Fig .15(a) shows that at 1000LT O_3 is being formed in the neighboring city. At this time, pollutants are pushed to the north-west of city by the wind coming from the south-east. Figure 15(b) shows that until 1300LT, the time with the highest solar radiation, the maximum quantity of O_3 is formed. At this time, wind is divided in three main convergences which divide the plume of O_3 into two different small plumes. Two O_3 maxima are formed at this time on 7th Feb, 140 ppb and 150 ppb, for the northern and north-western parts of the city, respectively. Then, until 1400LT the wind direction is the same wind direction than at 1300LT. However the wind speed is very strong (two times stronger than at 1300LT), which pushes rapidly the O_3 plume to go up to the north and north-west of the city. Fig .15(c) shows that at 1500LT the plumes leave the basin of HCMC through the north and north-west. The O_3 concentrations remain low in the city. Then, at 2000LT the wind direction is the same wind direction than at 1000LT. At 2000LT, there is not solar radiation coming to the earth which prevents O_3 production and promotes the destruction of O_3, especially in the north-western part of the city.

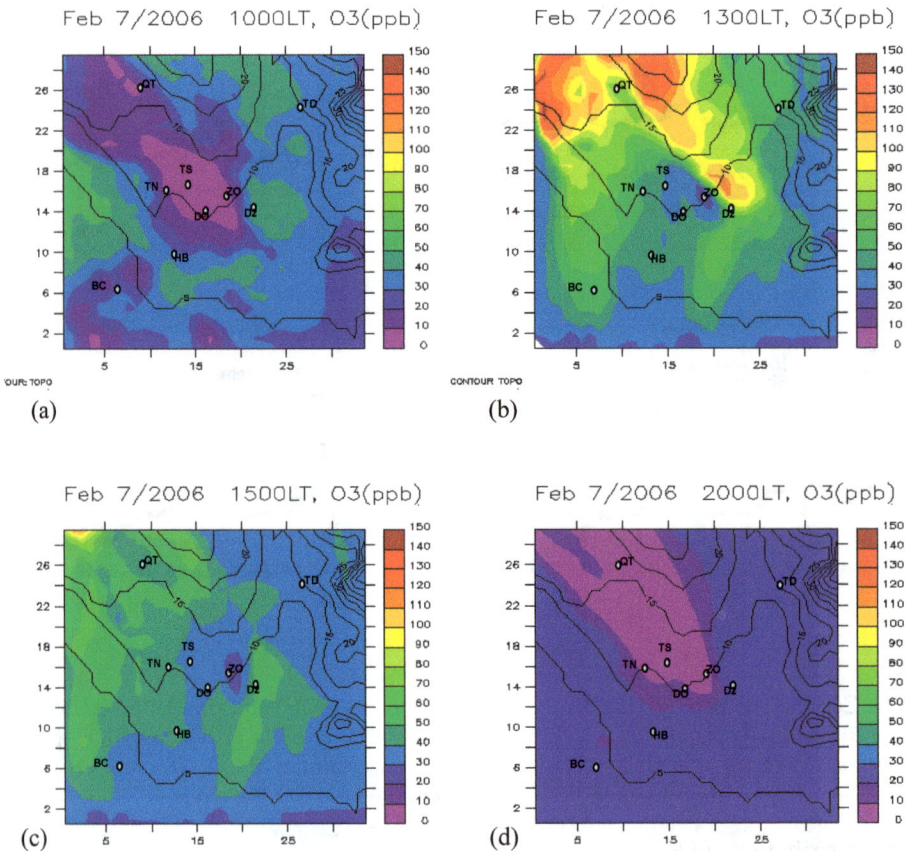

Fig. 15. Map of O_3 concentrations (ppb) at ground level in the domain of 34 km x 30 km, at 10LT00 (upper left panel), 1300LT (upper right panel), 1500LT (lower left panel) and 2000LT (lower right panel), 7th Feb 2006 and measurement stations. The different colours are the O_3 levels.

In conclusion, 6th – 8th February 2006 is a period which is representative for one of the highest O_3 episodes during the dry season of the year in HCMC. The primary pollutants show highest values in centre of city where the highest density of traffic is found. Therefore, the huge part of population in the centre of HCMC is living with unfavorable conditions due to high concentration of primary pollutants. However, in the case of secondary pollutant we can see that it has most favorable conditions for the population living in the centre of HCMC and unfavorable for the population living in the north and north-west of HCMC. Once the models were shown able to reproduce and understand the principal characteristics of pollution in HCMC, it is very useful to study different strategies to reduce pollution for the city in the future.

4.4 Air pollution abatement strategy

For over 15 years, a lot of studies to evaluate air pollution abatement strategies have been carried out by using air quality models (Metcalfe et al. 2002; Palacios et al. 2002; Zarate et al. 2007 and to cite a few). The previous section shows that it is urgent to establish emission control scenarios for HCMC. Over some recently years, the results of air quality monitoring have shown that the pollutants exceeded regularly the standard limits in HCMC due to the emissions from the traffic source (HEPA, 2005; HEPA, 2006). The local government has started to design some emission control plans for traffic in HCMC. The plans are designed for the year of 2015 and 2020. The two emission control plans are named: (1) Emission reduction scenario for 2015 and (2) Emission reduction scenario for 2020.

The main ideas are that in 2015 the HCMC government will perform many activities to control air pollution concerning the road traffic source (Trinh, 2007): (1) controlling the emission of all vehicles (Thang, 2004), (2) the first metro line will be finished at the end of 2014 (Bao du lich, 2008) and (3) HCMC government will add 3000 new buses during 2006-2015 (Tuong, 2005).

For the year of 2020, four metro lines will be constructed (metro system will replace 50% of total motorcycle) (Bao du lich, 2008) and the number of buses will be increased to 4500 during 2006-2020 (Bang et al., 2010).

Spatial distribution of O_3

The impacts of two strategies on the levels of troposphere O_3 in HCMC are shown in Fig .16(a) and Fig .16(b). If HCMC follows the reduction plan: (i) For the year of 2015, the reductions for the HCMC grid area as a whole are 3.3% for CO, SO_2, and CH_4. There is an increase in NOx emissions of 8%. Mean values of O_3 are reduced from 28.5 ppb to 28.0 ppb and the maximum from 150 ppb to 136 ppb on 7th Feb. (ii) For the year of 2020, the reductions for HCMC grid area as a whole are 8.6% for CO and 12.5 % for CH_4, there are increases in NOx, SO_2 and NMVOC emissions of 20.1%, 7.6% and 6.2%, respectively. Mean values of O_3 are reduced from 28.5 ppb to 27.6 ppb and the maximum from 150 ppb to 120 ppb for 7th Feb.

The highest reduction of O_3 concentration is found at the same place of the principal O_3 plume for both abatement strategies. A deeper analysis of this reduction will be discussed by plotting the O_3 concentration variable with respect to time and its uncertainties. In this research, we select some stations where we find the maximum of O_3 reduction, the medium of O_3 reduction and the minimum of O_3 reduction. The MA, D2 and HB stations are chosen for the maximum, medium and minimum reduction zones, respectively (Fig .16 (a)). Their results are shown in the following section.

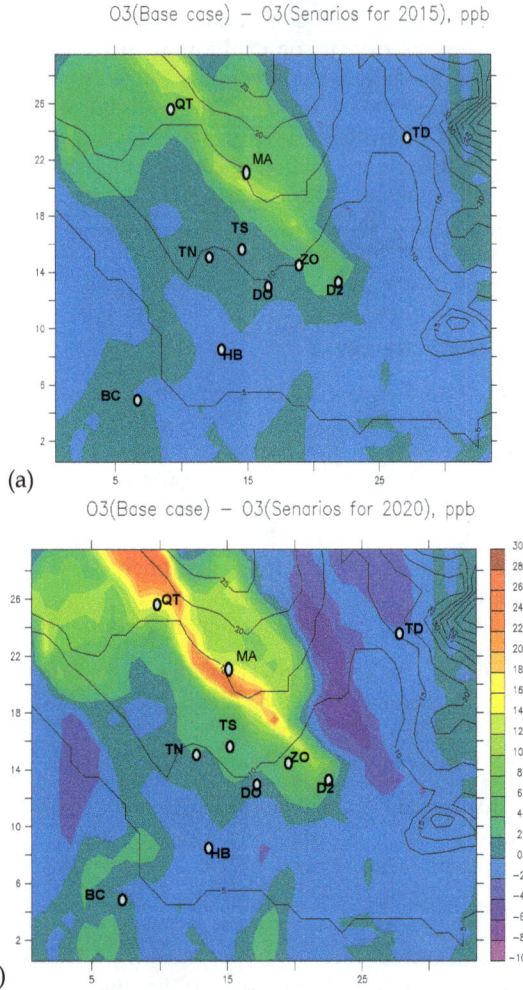

Fig. 16. Effect of two strategies on O_3 concentration (ppb) fields for the 7th Feb 2006 at 1300LT ground level. Figure 16(a) represents the reduction of O_3 concentration in 2015 from the ozone concentration in 2006. Figure 16(b) represents the reduction of O_3 concentration in 2020 from the ozone concentration in 2006. The measurement stations are shown in Fig. 13.

Analysis of O_3 at different measuring stations

Strategy in 2015: Figure 17(a, b, c)-upper panels shows the reduction of O_3 concentrations ($\overline{\Delta O_3}$ in equation 4) in 2015 from the O_3 concentration in 2006 and its uncertainties ($\sigma_{\overline{\Delta O_3}}$ in equation 5) in different stations (we consider the standard deviation as the uncertainty).

The Delta O_3 ($\overline{\Delta O_3}$) and uncertainties of O_3 ($\sigma_{\overline{\Delta O_3}}$) presented on Fig .17 are calculated using equations (4) and (5). These values are calculated from 100 MC simulations for the base case and 100 MC simulations for each strategy.

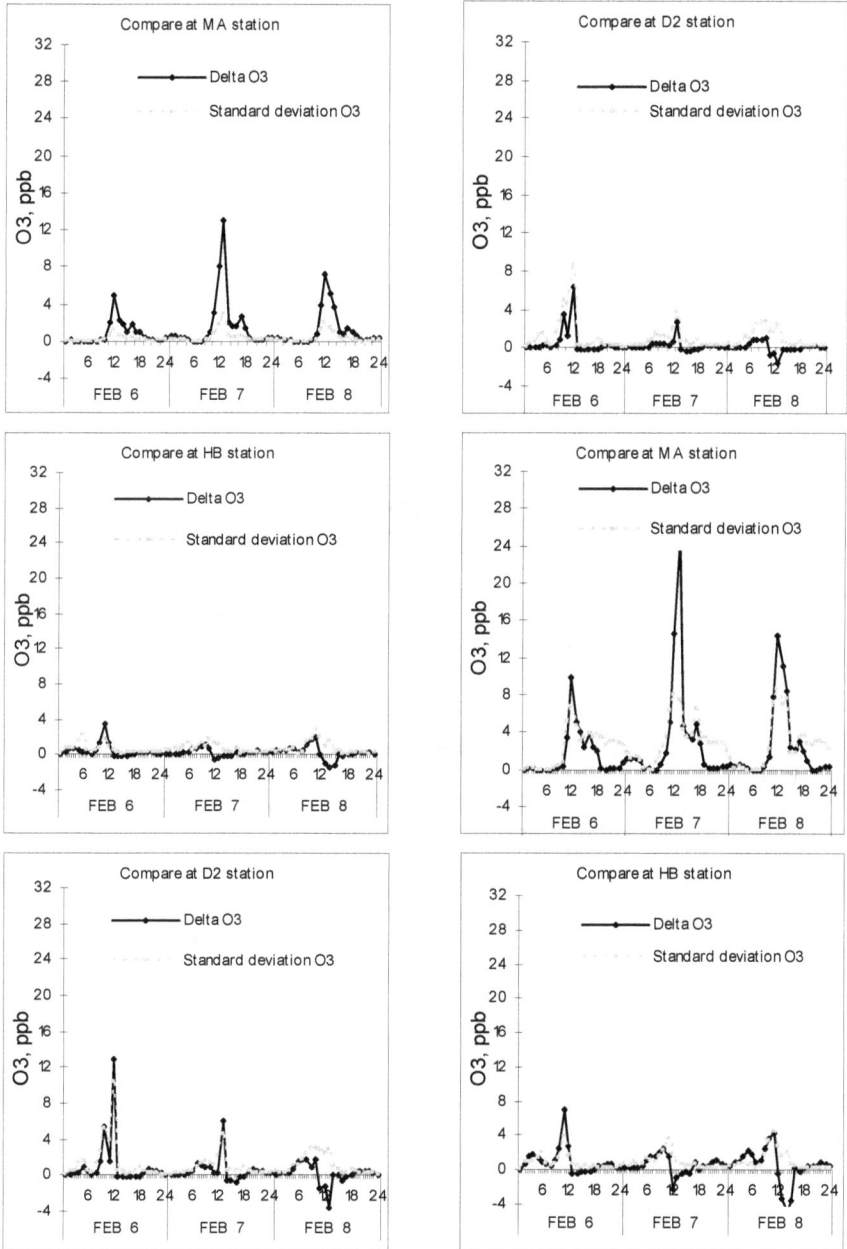

Fig. 17. Reduction of O_3 (in ppb) in 2015 (strategy in 2015; Fig .17 (a, b, c)_upper panels) and in 2020 (strategy in 2020; Fig .17 (d, e, f)_lower panels) from the O_3 concentration in 2006 and its uncertainties. The results are plotted for the first layer near the ground during the selected episode (on 6th - 8th Feb) at MA, D2 and HB stations.

$$\text{Delta } O_3 = \overline{\Delta O_3} = \frac{\sum\limits_{i=1}^{100}(O^i_{3\_Base\ case} - O^i_{3\_2015})}{100} \qquad (4)$$

where: i is the number of simulation

$O^i_{3\_Base\ case}$ is the O_3 concentration of base case (2006) for the ith simulation

$O^i_{3\_2015}$ is the O_3 concentration of strategy in 2015 for the ith simulation,

$$\text{Standard deviation of } O_3 = \sigma_{\overline{\Delta O_3}} = \sqrt{\frac{\sum\limits_{i=1}^{100}(\Delta O^i_3 - \overline{\Delta O_3})^2}{99}} \qquad (5)$$

Where: ΔO^i_3 is the difference in O_3 concentration between the base case and the strategy in 2015 for the ith simulation.

The highest uncertainties of O_3 reduction appear at the same time of the highest O_3 reduction at 1200LT - 1400LT of each day (Fig .17). The highest reduction of O_3 concentration at MA, D2 and HB stations during 6th - 8th Feb are 14ppb, 6.7ppb and 3.9ppb, respectively. However, the highest uncertainties of O_3 reduction at MA, D2 and HB stations are 3ppb, 9ppb and 3.5ppb respectively. Therefore, the uncertainties of O_3 reduction are in general similar to the O_3 reduction. We can not conclude that the change in O_3 concentration is due to the impact of the emission control plan, because the change can probably be due to the impact of uncertainties of input parameters.

For the evolution of primary pollutants of the strategy in 2015, the concentrations of NOx in simulations will increase 7% than those were in 2006. However, the concentrations of CO and CH_4 decrease around 10% and 8%, respectively than those were in 2006. In conclusion, there is very little impact of the emission control plan in 2015.

Strategy in 2020: Figure 17 (d, e, f)-lower panels shows that the highest uncertainties of O_3 reduction ($\sigma_{\overline{\Delta O_3}}$) also appear at the same time of the highest O_3 reduction ($\overline{\Delta O_3}$) at 1200LT - 1400LT of each day. In general, the O_3 concentration in 2020 is lower than the O_3 concentration in 2006. The highest reduction of O_3 concentration at MA, D2 and HB stations during 6th – 8th Feb are 23.5ppb, 13.4ppb and 7.8ppb, respectively. While, the highest uncertainties of O_3 reduction at MA, D2 and HB stations are 8.2ppb, 11ppb and 5.1ppb, respectively. The O_3 reduction is higher than the uncertainty of O_3 reduction in all stations. It means that the change in O_3 concentration is due to the change in emissions from the emission control plan.

For the evolution of primary pollutants of the strategy in 2020, the concentrations of NOx in simulations will increase 16% than those were in 2006. However, the concentrations of CO decrease around 7% than those were in 2006. In conclusion, this emission control plan in 2020 has strong impact on the primary and secondary pollutants in HCMC.

5. Summary and key recommendations

In this chapter, the state of the art for studying urban air pollution was reviewed and discussed. Air in the large cities is polluted, especially in developing cities. Road traffic is the main source of pollution in cities. The environmental managers strongly need for air quality management systems to help them control air quality in the cities. The available air quality management system is discussed. Emission inventory is the most important part in air quality management system as it is the mostly used tool to identify the pollution source. However there are a lot of difficulties to be faced when we generate EIs. In this chapter, we also present an example of generation an EI for developing cities where there are not enough available data. The results of EI are used as input parameters for air quality modeling. Then we present many Air pollution abatement strategies over HCMC as a case study of developing cities.

6. References

- Print Books

Dinh, X.T., 2003. Air pollution, VNU-HCM edition. 399p.

Ermakov, X.M., 1997. Monte Carlo methodology and its relations. Translated by Pham,T.N. Sciences and Technology publishers. 271p.

Molina, L., Molina, M., 2002. Air Quality in the Mexico Megacity. An integrated assessment. Kluwer Academic Publishers, Dordrecht, The Netherlands. ISBN 1-4020-045204.

Rainer, Friedrich., Stefan, Reis., 2004. Emissions of Air Pollutants, Springer. 333p

Moussiopoulos, Nicolas., 2003. Air Quality in Cities (book). Springer, Heidelberg, Germany. ISBN 3-540-00842-x. 298 p.

- Papers in Journals

Bang Q. HO, Clappier, A., 2011. Road traffic emission inventory for air quality modelling and to evaluate the abatement strategies: a case of Ho Chi Minh City, Vietnam. Atmospheric Environment Vol 45, Issue 21 (2011) pp. 3584-3593. ISSN: 1352-2310.

Bang Q. HO, Clappier, A., Golay F., 2011. Air pollution forecast for Ho Chi Minh City, Vietnam in 2015 and 2020. Air Quality, Atmosphere & Health, Volume 4, Number 2, p.145-158. ISSN: 1873-9318.

Bang Q, HO, Clappier, A., Zarate, E., Hubert, V.D.B., Fuhrer, O., 2006. Air quality meso-scale modeling in Ho Chi Minh City: evaluation of some strategies' efficiency to reduce pollution. Journal Science and Technology Development. Vol 9, N° 5, 2006.

Belalcazar, L., Fuhrer, O., Ho. D., Zarate, E., Clappier, A., 2009. Estimation of road traffic emission factors from a long term tracer study in Ho Chi Minh City (Vietnam), atmospheric environment, vol. 43 (2009) 5830–5837.

Bougeault, P., Lacarrere, P., 1989. Parameterization of orography-induced turbulence in a mesobeta-scale model. Monthly Weather Review 117, 1872–1890.

Clappier, A., Perrochet, P., Martilli, A., Muller, F., Krueger, B.C., 1996. A new non-hydrostatic mesoscale model using a control volume finite element (CVFE) discretisation technique. In: Borrell, P.M., et al. (Ed.), Proceedings of the EUROTRAC Symposium '96. Computational Mechanics Publications, Southampton, pp. 527–553.

Clappier, A., 1998. A correction method for use in multidimensional time-splitting advection algorithms: application to two- and three-dimensional transport. Monthly Weather Review 126, 232–242.

Collella, P., Woodward, P., 1984. The piecewise parabolic method (PPM) for gas dynamical simulations. Journal of Computational Physics 54, 174–201.

Gurjar, B.R., J. Lelieveld., 2005. New Directions: Megacities and global change. Atmospheric Environment 39 (2), 391-393.

Generoso, S and Bey, I., 2007: A satellite and model-based assessment of the 2003 Russian fires: Impact on the Arctic region. J. Geophys.Res.,vol 112, D15302

Gong, W., Cho, H., 1993. A numerical scheme for the integration of the gas phase chemical rate equations in three-dimensional atmospheric models. Atmospheric Environment. 27A, 2147–2160.

Hanna, S.R., Chang, J.C., Fernau, M.E., 1998. Monte Carlo estimates of uncertainties in predictions by a photochemical grid model (UAM-IV) due to uncertainties in input variables. Atmos Environ 1998; 32:3619-3628.

Hanna, S.R., Lu, Z., Frey, C.H., Wheeler, N., Vukovich, J., Arunachalam, S., Fernau, M.,Hansen, A., 2000. Uncertainties in predicted ozone concentration due to input uncertainties for the UAM-V photochemical grid model applied to the July 1995 OTAG domain. Atmospheric Environment 35 (5), 891–903.

Hwang, D., Karami, H.A., Byun, D.W., 1998. Uncertainty analysis of environmental models within GIS environments. Computers & Geosciences 24 (2), 119–130.

Ho M.D., Dinh X.T. Estimation of emission factors of air pollutants from the road traffic in HCMC. VNU Journal of Science, Earth Sciences 24 (2008) 184-192

HOUTRANS, 2004. The Study on Urban Transport Master Plan and Feasibility Study in Ho Chi Minh Metropolitan Area. In: Vol. 6, No. 1 Traffic and Transport Surveys. ALMEC Corporation

Junier, M., Kirchner, F., Clappier A. and Hubert, V.D.B., 2004. The chemical mechanism generation program CHEMATA, part II: Comparison of four chemical mechanisms in a three-dimensional mesoscale simulation, Atmos. Environ. 39, 1161-1171

Kuhlwein, J and Friedrich, R., 2000. Uncertainties of modeling emissions from road transport, Atmos. Environ., (2000) 34, 4603-4610.

Louis, J.F., 1979. A parametric model of vertical eddies fluxes in the atmosphere. Boundary-Layer Meteorology 17, 187–202

Martilli, A., Clappier, A., Rotach, M.W., 2002. An urban surface exchange parameterization for mesoscale models. Boundary-Layer Meteorology 104, 261–304.

Martilli, A., Roulet, Y.-A., Junier, M., Kirchner, F., Rotach, M.W and Clappier, A., 2003. On the impact of urban exchange parameterization on air quality simulations: the Athens case, Atmos.Environ.37, 4217-4231.

Metcalfe, S.E., Whyatt, J.D., Derwent, R.G., O'Donoghue, M., 2002. The regionaldistribution of ozone across the British Isles and its response to control strategies. Atmospheric Environment 36, 4045–4055.

Nenes, A., Pandis, S., Pilinis, C., 1998. ISORROPIA: A new thermodynamic equilibrium model for multiphase multi-component inorganic aerosols. Aquatic Geochemistry 4, 123-152.

Palacios, M., Kirchner, F., Martilli, A., Clappier, A., Martin, F., Rodriguez, M.E., 2002. Summer ozone episodes in the Greater Madrid area. Analyzing the Ozone Response to Abatement Strategies by Modeling. Atmospheric Environment 36, 5323–5333.

Rappenglück, B., Oyola, P., Olaeta, I., Fabian, P., 2000. The evolution of Pho-tochemical Smog in the Metropolitan Area of Santiago de Chile. Journal of Applied Meteorology 39, 275-290.

Ranjeet S. Sokhi and Nutthida Kitwiroon, 2008. World Atlas of atmospheric pollution. Anthem Press, first edition. 144p.

Streets.D.G, Bond.T.C, Carmichael. G.R, Fernandes.S.D, Fu.Q, He.D, Klimont.D, Nelson.S.M, Tsai.N.Y, Wang.M.Q, Woo.J.H and Yarber.K.F., 2003: An inventory of gaseous and primary aerosol emissions in Asia in the year 2000. J. Geophys. Res. 108 (D21) (2003), p. 8809 doi:10.1029/2002JD003093.

Streets.D.G, Bond.T.C, Lee.T and Jang.C., 2004: On the future of carbonaceous aerosol emission. J. Geophys.Res.,vol 109,D24212

Varinou, M., Kallos. G, Tsiligiridis, G and Sistla., 1999. The role of anthropogenic and biogenic emissions on tropospheric ozone formation over Greece. Phys. Chem. Earth (C), Vol. 24, No. 5, pp. 507-513, (1999).

Stockwell, W.R., Kirchner, F., Kuhn, M., Seefeld, S., 1997. A new mechanism for regional atmosphere rich chemistry modeling. Journal of Geophysical Research 102, 25847–25879.

Zarate, E., Belalcazar, L.C., Clappier, A., Manzi, V and Hubert V, D, B. 2007. Air quality modeling over Bogota, Colombia: Combined techniques to estimate and evaluate emission inventories. . Atmospheric Environment, 41, 6302–6318.

-   Technical reports

ADB (Asian Development Bank) and Clean Air Initiative for Asian Cities Center. Country Synthesis Report on Urban Air quality Management: Vietnam. Dec., 2006.

Belalcazar, Luis.C. Alternative Techniques to Assess Road Traffic Emissions. Ph.D Thesis of EPFL. N° 4504 (2009).

Clappier, Alain. Modélisation numérique des polluants atmosphériques. 2001. 98p. Cours de troisième année EPF Lausanne.

DOSTE (Department of Science, Technology and Environment of Ho Chi Minh city). Urban transport energy demand and emission analysis – Case study of HCM city. N° 1 (phase II). 2001.

DOSTE. Project on Energy – air pollution in HCMC. 2000. DOSTE - HCMC and ADEME.

Duong. T,M,H. 2004. Proposed technical solutions for diminishing air pollution level by traffic and for analyzing the solution to overcome barriers in application. MSc/IER

European Environment Agency (EEA). 1999. EMEP/CORINAIR. Emission inventory guidebook.

HEPA (Ho Chi Minh environmental protection agency). Report 2006 on air quality in Ho Chi Minh City. December 2006.

HEPA (Ho Chi Minh environmental protection agency)., 2005. Last report of 2005 on inventory of emissions sources for HCMC. December 2005.

IER., 2006. Report annual. Environmental monitoring in South of Vietnam, Zone III. Air quality monitoring program in south of Vietnam, Institute of Environment and Resources (IER), (2006).

Junier, Martin., 2004. Gas phase chemistry mechanisms for air quality modeling: generation and application to case studies. (2004). 112 p. Thèse Doctorat EPF Lausanne, no 2936

Krpo, A., 2009. Development and Application of a Numerical Simulation System to Evaluate the Impact of Anthropogenic Heat Fluxes on Urban Boundary Layer Climate. Ph.D thesis. EPFL.

Kuentz Burchi, C., 1996. Les polluants atmosphériques. Approche toxicologique de l'évaluation des risques. Thèse de Doctorat. Université Louis Pasteur de Strasbourg

Kreinovich, J., Beck, C., Ferregut, A., Sanchez, G. R., Keller, M., Averill., Starks, S.A. Monte-Carlo-Type Techniques for Processing Interval Uncertainty, and Their Potential Engineering Applications College of Engineering and NASA Pan-American Center for Earth and Environmental Studies (PACES), University of Texas, El Paso, TX 79968, USA .

Mics-Asia project. International Institute for Applied Systems Analysis, Laxenburg, Austria. 2000.

Nguyen, T.T.M., 2000. Proposed methods to reduce air pollution by traffic in HCMC: developing public transportation and using lead free gasoline. (2000), MSc/IER-VNU.

Nguyen, T, 2002: Asian Regional Research Program in Energy, Environment and Climate (ARRPEEC). Project phase III, supported by SIDA organization, from 2002 to 2004.

Sathya, V., 2003 Uncertainty analysis in air quality modeling - the impact of meteorological input uncertainties. Thesis N°2318 (2003). EPFL

Statistical yearbook Ho Chi Minh City., 2003. Ho Chi Minh Statistical Office [réf. du Juin 2004].

Statistical yearbook Ho Chi Minh City., 2006. Ho Chi Minh Statistical Office.

Thang, Q. D., 2004. A Vision for Cleaner Emissions from Motorcycles in Viet Nam. Paper presented at the Cleaner Vehicles and Fuels in Viet Nam Workshop, 13-14 May 2004, Hanoi, Viet Nam. Vietnamese Ministry of Transportation and US-EPA.

Trinh, N.G., 2007. Motorcycles do not meet emissions standards should be upgraded or replaced. Conference in: "Control emission from motorcycles in major cities of Vietnam", HCMC. August 2007.

Tremback, C.J., Kessler, R., 1985. A surface temperature and moisture parameterization for use in mesoscale numerical models, Proceedings of Seventh conference on Numerical Weather Prediction, Montreal, Quebec, Canada, June 17-20.

World Health Organization (WHO), 2000. Guidelines for Air Quality. World Health Organization, Geneva.

World Health Organization (WHO), 2005. WHO Air Quality guidelines for particulate matter, ozone, nitrogen dioxide and sulphur dioxide. Global update 2005. World Health Organization, Geneva.

Zarate, E., 2007. Understanding the Origins and Fate of Air Pollution in Bogotá, Colombia. Doctoral thesis N° 3768, EPFL.

-      Papers in Conference Proceedings

Le, T.G.; Dan, G and Nao, I., 2008. Clean Air Initiative, "Air Pollution Blamed as Study Finds Respiratory Illness Hitting HCMC's Children," March 26, 2008.

Nguyen, D.T., Pham, T.T. Air pollution in HoChiMinh City, Vietnam. Conference on: "Better Air quality in Asian and Pacific Rim Cities (BAQ 2002), Dec.2002, Hong Kong.

-      World Wide Web Sites and Other Electronic Sources

EU directive N° 96/62/CE of 27/09/96 concernant l'évaluation et la gestion de la qualité de l'air ambiant.
        http://www.ineris.fr/aida/?q=consult_doc/consultation/2.250.190.28.8.4339

Gale group., 2007. Ho Chi Minh City becomes one of 100 rapid economic growth cities. March, 2007. Ipr strategic business information database - articles.
        http://www.encyclopedia.com/doc/1G1-160479731.html

Madronich, S., 1998. TUV troposphere ultraviolet and visible radiation model, from the Website: http://acd.ucar.edu/models/open/tuv/tuv.html/.

Tuong, L., 2005. Go together by bus. Conference in: "The solutions for reducing the transport congestion for HCMC City in 2020", HCMC. 2005. Available at
        http://vietbao.vn/Phong-su/Di-xe-buyt-thoi-xang-tang-gia/30066850/263/.

4

Monitoring Studies of Urban Air Quality in Central-Southern Spain Using Different Techniques

Florentina Villanueva, José Albaladejo, Beatriz Cabañas,
Pilar Martín and Alberto Notario
Castilla La Mancha University
Spain

1. Introduction

In urban areas, emissions of air pollutants by anthropogenic processes such as traffic, industry, power plants and domestic heating systems are the main sources of pollution (Fenger, 1999). The massive growth in road traffic and in the use of fossil fuels during the last decades has changed the composition of urban air, increasing the frequency of pollution episodes and the number of cities experiencing them. The main pollutants monitored in the atmosphere in these areas are ozone (O_3), nitrogen oxides (NO_x), sulphur dioxide (SO_2), carbon monoxide (CO), aromatic compounds and particulate matter. While CO, NO and aromatic compounds are mainly emitted by traffic, O_3 and NO_2 are originated by photochemical reactions.

The high levels of solar irradiation observed in the Mediterranean countries favour, in general, the enhanced photochemical production of secondary oxidising pollutants, including O_3, nitrogen dioxide (NO_2), and peroxyacetylnitrate (PAN). Amongst these, the O_3 and the nitrogen dioxide (NO_2) are capable of causing adverse impacts on human health and the environment (Lee et al., 1996; WHO, 2000a; Mazzeo and Venegas, 2002, 2004).

Nitrogen dioxide is considered to be an important atmospheric trace gas pollutant not only because of its effects on health but also because (a) it absorbs visible solar radiation and contributes to impaired atmospheric visibility, (b) as an absorber of visible radiation, it could play a potentially direct role in the change in the global climate if its concentrations were to become high enough (WHO, 2000a), (c) it is one of the major sources of acid rain (Tang and Lau, 1999), (d) it is, along with nitric oxide (NO), a chief regulator of the oxidising capacity of the free troposphere by controlling the build-up and fate of radical species, including hydroxyl radicals, and (e) it plays a critical role in determining concentrations of O_3, nitric acid (HNO_3), nitrous acid (HNO_2), organic nitrates such as PAN ($CH_3C(O)O_2NO_2$), nitrate aerosols and other species in the troposphere. In fact, the photolysis of nitrogen dioxide in the presence of volatile organic compounds is the only key initiator of the photochemical formation of ozone and photochemical smog, whether in polluted or unpolluted atmospheres (WHO, 2000a; Varshney and Singh, 2003). Therefore, ozone mainly generated in the photochemical reaction mentioned above, is a secondary pollutant. Ozone is also produced by natural sources (trees and thunderstorms for example).

It has been found out that the photochemical ozone production in urban areas rises with the NOx emissions and is less sensitive to the VOC emissions (Sillman and Samson, 1995). In areas moderately contaminated, ozone sensitivity to emission of nitrogen oxides depends on the season and on the emission rates (Kleinman, 1991).

Hence, monitoring of air pollution is necessary, not only to comply with the environmental directives, but also to provide public information and to measure the effectiveness of emission control polices. It is also important to have a number of measurements from different cities to help understand the role of the different physicochemical processes in the troposphere. In addition to meteorological and irradiance conditions, traffic patterns, volatile organic compounds (VOC)/NOx emissions and industrial activities vary significantly between different locations.

Monitoring of urban air pollutants has been the subject of study in the city of Ciudad Real (in central-southern Spain) using passive and continuous techniques in order to identify, measure and study the daily, weekly and seasonal variation of the concentrations of important tropospheric constituents (Saiz-Lopez et al., 2006, Martín et al., 2010 and Villanueva et al., 2010). Therefore, a review of these studies has been presented in this chapter, where the evolution of air quality during three different periods (July 2000-March 2001; January-December 2007; March-December 2008) is shown and data of O_3, NO_2 and SO_2 are compared.

2. Material and methods: Experimental section

2.1 Area of study

The study was carried out in the urban area of Ciudad Real (Spain). The city has around 65,000 inhabitants and is located in the heart of La Mancha region in central-southern Spain (38.59 N, 3.55 W, at approximately 628 m above sea level) in a fairly flat area, 200 km south of Madrid. With a low presence of industry, traffic is likely to be the most important source of air pollution in this city. Meteorologically, the zone is characterised by very hot and dry summer period with high insulation, variable direction winds and dry and cold winters, conditions that could play an important role in the evolution of the polluting agents.

A DOAS system, passive samplers and a mobile air pollution control station were used to sample, measure and identify the different pollutants. A map of the city is shown in Figure 1, where the situation of the all measurement points can be seen. The passive samplers were placed in town centre (total sampling area of 7.5 km² approximately) with the exception of points 6, 7 and 10 that correspond to rural zones, but in the case of point 10 in the influence zone of a small airport under construction. Industry next to the city is especially scarce, and the zone is mainly surrounded by rural areas. There is a petrochemical complex located in Puertollano at 30 km from southwest of Ciudad Real. This complex would affect to increase the NO_2 and O_3 concentrations in the Ciudad Real atmosphere by transporting processes.

The DOAS instrument was placed at a fixed location inside the campus of the University of Castilla-La Mancha, situated in the Eastern part of the city and downwind of the main traffic of the city (also point 1). The mobile station (point MS) was also placed, at a fixed location located between the town centre and the ring road. This site represents well-mixed urban air conditions near major anthropogenic sources of primary pollutants such as traffic.

The measurements were carried out in different periods of time, the pollutants were monitored by means of DOAS from 21st July 2000 to 23rd March 2001, the passive sampling were carried out from January to December 2007 and the mobile station monitored the pollutants from March to December 2008.

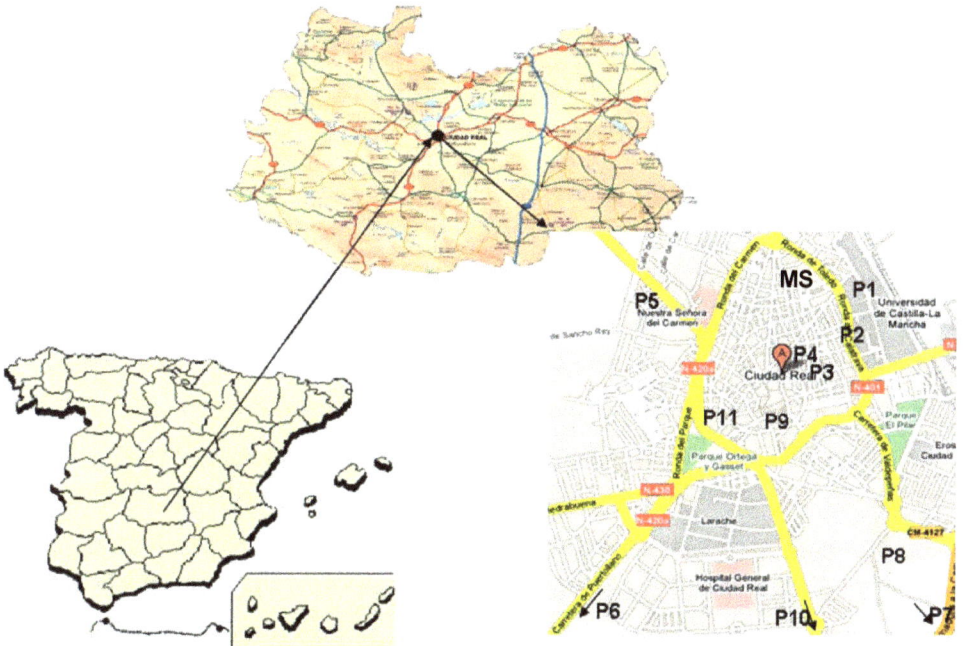

Fig. 1. Situation map of Ciudad Real (Spain) and the sampling points of passive samplers. P1 (DOAS and one passive sampler), MS (Mobile station).

2.2 Description of experimental techniques

2.2.1 DOAS technique

The DOAS principle and instrumentation have proven to be a powerful tool for simultaneously monitoring relevant atmospheric trace gases (Platt, 1994; Plane and Saiz-Lopez, 2005). The use of light paths which range from hundred of meters to several kilometres can avoid the problems of local influences and surface effects, as the measured concentrations are averaged over the light path and are barely influenced by small-scale perturbations. The technique is based on the fact that all trace gases absorb electromagnetic radiation in some part of the spectrum. If the radiation of the appropriate frequency is transmitted through the atmosphere, the absorption features of each molecule in that spectral region allow the identification and quantification of the gas concentration. A commercial DOAS system (OPSIS, model AR 500) was employed to simultaneously monitor the gas concentration of O_3, NO_2, NO, and SO_2 integrated along a light path of 391 m. More experimental details are described in Saiz-Lopez et al., 2006 and only a brief description is made. The system used in this study consists of an emitter (EM 150) and receiver (RE 150) in combination with the AR 500 analyzer. The emitter was located on the roof of the School of Agricultural Studies about 16m above the ground, and the receiver system was installed 12m above street level, on a terrace on the School of Computing Science.

The average altitude of the optical path was 14m above ground level. The 391m light beam crosses over a small parking area and is nearby two interior city roads containing the heaviest traffic in the city. Also, a monitoring station of the National Meteorological Institute was located in the middle of the light path and provided surface meteorological data for the interpretation of the measurements in this study. In the evaluation procedure (Platt, 1994; Plane and Saiz-Lopez, 2005), the wavelength region of the strongest absorption was used for each gas: 200–250 nm (NO), 250–300 nm (O_3), 360–440 nm (NO_2) and 260–340 nm (SO_2). The calibration of the DOAS system was made using the cell reference system provided by the manufacturer (an OPSISCB100 calibration bench with cells of various lengths, a calibration lamp CA 150 and an OPSIS OC 500 ozone calibrator unit). For the pollutants monitored, the manufacturer specifies a detection limit of the order of 1 μg m^{-3} and an accuracy of 2% for NO_2, SO_2 and O_3 and 3%–15% for NO.

2.2.2 Passive sampling technique

Passive samplers are based on free flow (according to the Fick's first law of diffusion) of pollutant molecules from the sampled medium to a collecting medium. These devices can be deployed virtually anywhere, being them useful for mean concentrations determinations. Radiello® samplers (Fondazione Salvatore Maugeri, Padova, Italy) have been employed to measure ambient NO_2 and ozone concentrations. Radiello passive samplers are different from the axial ones because the diffusion process is radial through a microporous cylinder in which a cartridge with adsorbent material is positioned. Consequently, there is a greater diffusion area, which facilitates the reaction between the gas and the collection cartridge (Cocheo et al., 1999), giving an uptake rate of at least two times higher.

Radiello passive samplers were used to describe the evolution of NO_2 and O_3 levels in the urban air of Ciudad Real throughout the four seasonal periods during 2007. The Radiello sampler for measuring nitrogen dioxide and O_3 is based on the principle that NO_2 and O_3 diffuse across the diffusive body towards the absorbing material on the inner cartridge. NO_2 and O_3 in the atmosphere are captured in the sampler as nitrite (NO_2^-) and 4-pyridylaldehyde, respectively. According to the Fick's first law, the quantity of NO_2^- and 4-pyridylaldehyde in the sampler is proportional to the concentration outside the sampler, the diffusion coefficient, the dimensions of the sampler and the sampling time. Concentrations are calculated from the quantity of nitrite or 4-pyridylaldehyde captured in the sampler by means of Eq. 1:

$$C(\mu g m^{-3}) = \frac{m \times z}{A \times t \times D} 10^6 = \frac{m}{Q_k t} 10^6 \tag{1}$$

Where the sampling rate Q_k is constant:

$$Q_k(mL\,min^{-1}) = 60xDx\frac{A}{z} \tag{2}$$

C represents the ambient air NO_2 or O_3 concentration measured by the passive sampler (micrograms per cubic metre), m is the quantity of NO_2 or O_3 captured in the sampler (micrograms) and z, A, t and D denote the diffusion length (centimetres), cross-sectional area (square centimetres), sampling time (minutes) and diffusion coefficient (square centimetres per second), respectively. Exposure time to sample NO_2 and O_3 was in all points 7 days. Details of the extraction and analysis procedure of the samples are given in Martín et al., 2010 according with R&P-Co 2001. In all cases, three unused cartridges belonging to the same lot were treated in the same manner as the samples. Then, it was subtracted by the average blank value from the absorbance values of the samples.

2.2.3 Mobile air pollution control station

A Mercedes Sprinter van was selected as rolling platform for the measuring equipment. The mobile air pollution control station is equipped with a vertical manifold that distributes the sample to all the analyzers, a multipoint calibrator (API, model 700) with an internal ozone calibrator and photometer, a zero air generator (API, model 701) and a meteorological station. Table 1 presents a list of the characteristics of the instruments and measured parameters including meteorological parameters. The methods used to analyze the different air pollutants are those used by the Air-Quality Networks and therefore, defined by the European Directive (2008/50/CE), as reference methods.

The particulate matter analyzer is based on beta radiation absorption technique and has been designated by the American Environmental Protection Agency (EPA) as equivalent method for measuring the PM_{10} fraction of suspended matter (EQPM-0798-122). The calibration of the photometer and flows of the calibrator and analyzers was made at the beginning of the measurement period by the accredited calibration laboratory of the manufacturer. More experimental details are described in Villanueva et al., 2010.

Parameter	Instrument	Time resolution	Detection limit
O_3	API 400A	20 s	0.6 ppb
NOx	API 200A	20 s	0.4 ppb
SO_2, H_2S	API 100A	20 s	0.4 ppb
CO	API 300A	20 s	0.04 ppm
BTX	Syntech spectras GC 955	15 min	0.05 ppb for benzene
PM_{10}	Met One BAM 1020	1 h	1 $\mu g/m^3$
Temperature and humidity	Met One HP043	10 s	-40°C / 0%
Pressure	Met One 091	10 s	0 mbar
Wind direction and speed	Met One 50.5	10 s	0.5 m/s
Solar radiation	Met One 595	10 μs	0 W/m²
Rain	Met One RC15	--	--

Table 1. Monitoring instruments employed in the mobile monitoring platform.

3. Results and discussion

3.1 Measurements of O_3, NO and NO_2

The evolution of NO, NO_2 and O_3 concentration measured using the continuous techniques DOAS (Saiz-Lopez et al., 2006) and the mobile station (Villanueva et al., 2010) for a typical summer day are shown in Figure 2a and 2b, respectively.

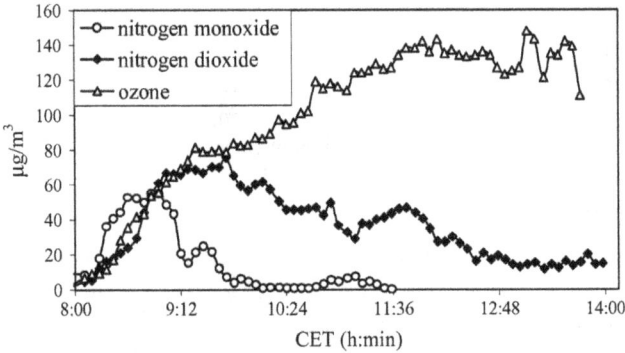

a)

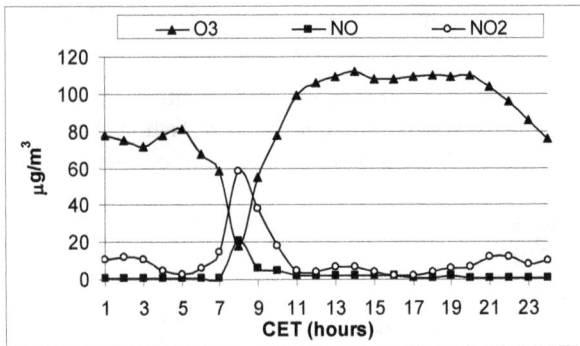

b)

Fig. 2. Evolution of O_3, NO and NO_2, concentrations for a typical summer day in Ciudad Real measured (a) with DOAS (12th August 2000) and (b) with the mobile station (30th July 2008). Time scale is given in CET (Central European Time). CET=UTC+2 in summer; UTC (Universal time coordinated).

It can be seen a typical profile for the three pollutants that is similar in both cases during the time period from 10:00 to 14:00. The maximum of NO concentrations in the early morning (see figure 2b) corresponds with the minimum of O_3. This morning O_3 loss has been previously studied by Zhang et al., 2004. Their results suggested that O_3 concentrations are depressed by the reaction of NO + O_3 → NO_2. During daytime, NO_2, can be quickly photo-disassociated (in a few minutes) and yield O_3. Therefore, as the day progresses, solar radiation levels favour the increase in the O_3 concentrations, having as consequence a decrease in NO and NO_2. After sunset, the cease in solar activity decreases the O_3 concentrations. The diurnal cycles registered in both studies (Saiz-Lopez et al., 2006 and Villanueva et al., 2010) show that the urban atmosphere of Ciudad Real is mainly influenced by road traffic and photochemistry. This behavior was observed in other urban areas of Spain and Europe (Palacios et al., 2002 and Kourtidis et al., 2002).

Figure 3 shows the evolution of NO, NO_2 and O_3 concentrations observed for an autumn day with the analyzers of the mobile station. It can be seen a negative correlation between O_3 and NO_2 as it occurred in summer-time and also an additional peak for NO_2 and NO after sunset which is related to a second rush hour in the traffic circulation. In this case the drop in O_3 concentration must be mainly due to the reaction with NO because of the absence of solar radiation. The same behavior was observed in 2001 with the DOAS system (Saiz-Lopez et al., 2006).

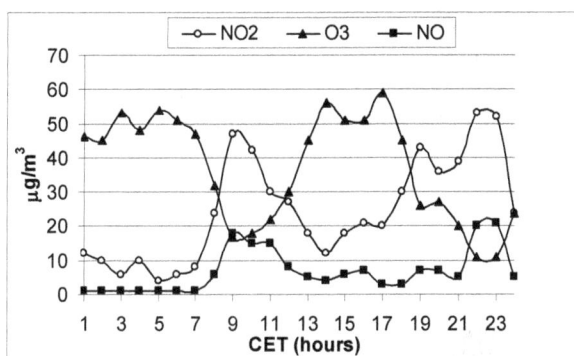

Fig. 3. Example of the evolution of O_3, NO and NO_2 concentrations for an autumn day (12nd December 2008) in Ciudad Real. Time scale is given in CET (Central European Time), being UTC+1 in autumn.

Figure 4 shows the daily averaged mixing ratio of NO_2 and O_3 measured by the DOAS technique (Saiz-Lopez et al., 2006) and figure 5 the weekly variation of O_3 and NO_2 using passive sampling (Martin et al., 2010). Finally, monthly averaged concentrations for NO, NO_2 and O_3 obtained from mobile station (Villanueva et al., 2010) are represented in figure 6 (in all cases during the whole measurements period). As it can be seen, the annual cycle for these trace gases show the largest values of O_3 during summer while NO_2 exhibit maximum value in winter or autumn months. On average, the production of O_3 is reduced in autumn-winter by up to 60% compared to the summer months. Maximum monthly averages of O_3, (in summer) and NO_2 (in autumn-winter) were: with DOAS, 65 and 35 μg m^{-3}, with passive samplers 48 and 24 μg m^{-3} and 82 and 18,6 μg m^{-3} with the analyzers of the mobile station.

The increase of ozone concentration in spring observed in the case of passive samplers, where several sampler points were located outside of downtown, could be due to emitted volatile organic compounds of biogenic origin (BVOCs), which reach the maximum levels in the spring period as a result of the intense activity of plant species and the photochemical associated. In the presence of nitrogen oxides (NOx) and under particular weather conditions (supporting photochemistry), BVOC contribute to the formation in the troposphere of secondary pollutants such as ozone (Atkinson, 2000 and Fuentes et al., 2000). Nevertheless, in rural areas with high levels of BVOCs the formation of ozone may be NOx-limited. In this sense, meteorological conditions and the dominating trajectories of air masses can constitute a source of NOx. Thus, there are several meteorological conditions that may enhance the rise on surface ozone during the spring season. In these months the heating of the ground favors the formation of higher mixing layers, mostly in continental areas such as the studied region, and therefore the photochemical activity inside them is elevated. This increase in O_3 concentration in spring is also observed for the points in which DOAS and the mobile station are located in the urban area.

Regarding NOx, the highest values found in winter and autumn months must be related with the lesser solar radiation and lower temperature which reduce the photochemical activity. Therefore, the oxidation of NO from motor traffic and central heating systems will increase the concentration of NO_2 and it will tend to accumulate due to the less photochemical activity in this period. Also, the reduction of the photochemical production of O_3 will favour the removal of this molecule by reaction with NO to form NO_2 and O_2, which in turn will contribute of the enhancement of the NO_2 values.

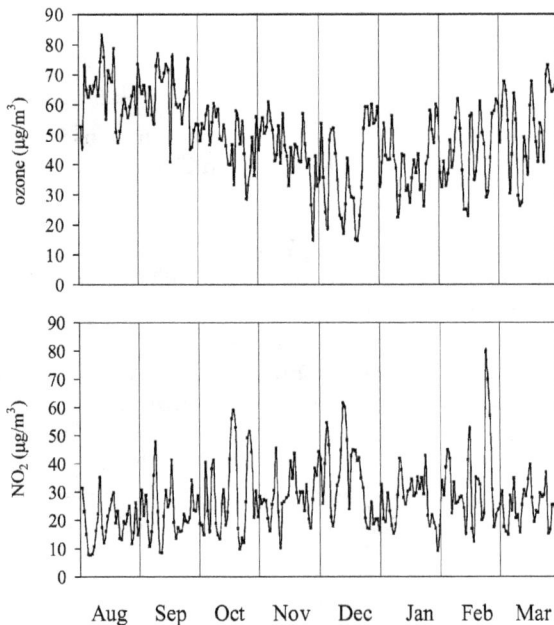

Fig. 4. Daily mean concentration data for O_3 and NO_2 obtained from the DOAS dataset for the whole period of measurements (August 2000–March 2001).

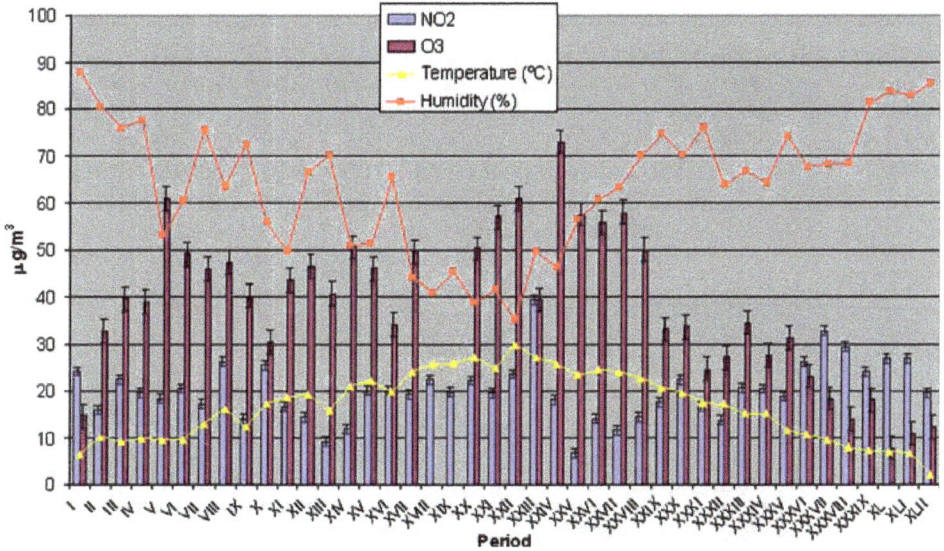

Fig. 5. Evolution of temperature, humidity, NO_2 and O_3 average concentrations of the 11 measured points during the full period (January-December 2007) using passive samplers. Winter (I-V), Spring (VI-XVI), Summer (XVII-XXX) and Autumn (XXXI-XLII).

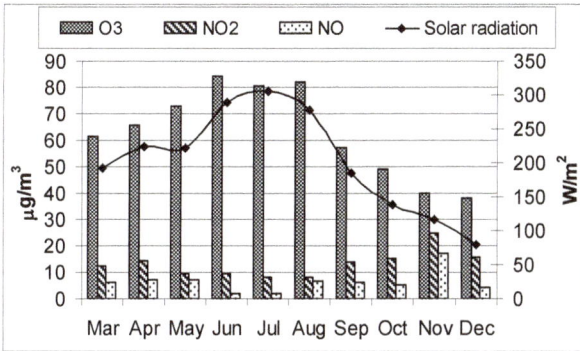

Fig. 6. Monthly means of O_3, NO_2 and NO pollutants measured from daily data (March-December 2008) with the analyzers of the mobile station. Solar radiation has also been plotted.

Mean annual levels for O_3 and NO_2 were 50 and 27 μg m^{-3} with DOAS, 38 and 20.8 μg m^{-3} with passive samplers and 63 and 13 μg m^{-3} for the mobile station. Mean annual levels for O_3 and NO_2 at P1, the point nearest to DOAS and mobile station was 42.5 and 16 μg m^{-3}. The values reported here are generally lower than those observed in previous studies in large urban areas in Spain, such as Madrid (Pujadas et al., 2000; Palacios et al., 2002) and Barcelona (Toll and Baldasano, 2000), where traffic emissions are considerably higher.

One advantage of using passive samplers is the ability to develop a spatial distribution map. In this sense, the O_3 and NO_2 average values obtained in the points P1, P2, P3, P4, P5, P6, P8, P9 and P11 have been used to create the spatial distribution map of these pollutants (Fig. 7) using SurGe Project Manager ver 1.4. As it can be seen, higher values of O_3 correspond to lower values of NO_2 and vice versa. This behaviour is expected considering the inter-conversion of both species according to photochemical cycle of tropospheric O_3 production (Finlayson-Pitts and Pitts, 2000). This same behaviour was also observed in the case of analysis by DOAS and conventional analyzers.

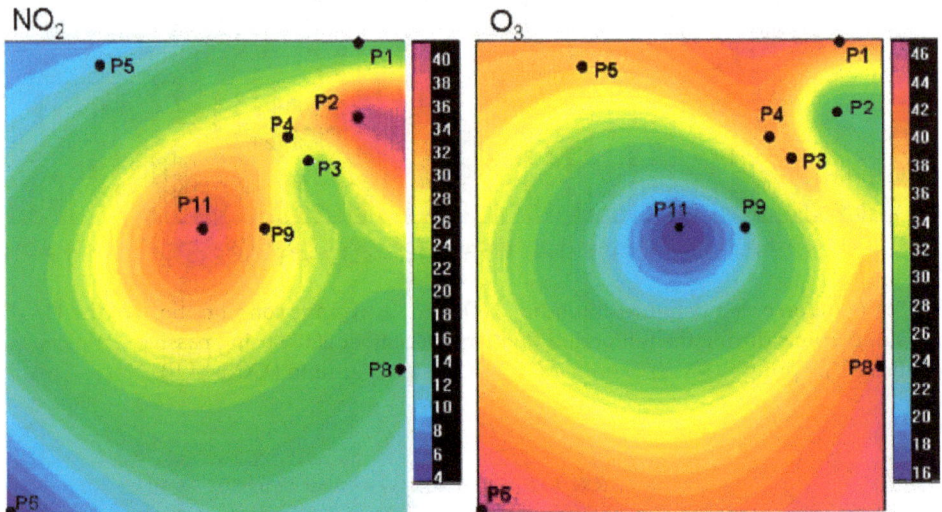

Fig. 7. Maps of O_3 and NO_2 spatial distribution. Units are in $\mu g\ m^{-3}$.

As commented above, meteorological conditions can affect the pollutant transport and distribution. Ozone concentration (in $\mu g\ m^{-3}$) from data obtained by passive samplers, showed a slight negative correlation with relative humidity (RH in %) ($[O_3]=-0.62RH + 88.66$) and a positive correlation with temperature (T in degree Celsius; $[O_3]=0.37T + 2.52$). Periods with higher temperatures and lower humidity usually correspond to higher values of ozone (Martin et al., 2010). In the case of DOAS analysis, the wind speed does not significantly influence the NO_2 observations in winter, by contrast, in summer, higher influence of the wind speed in the NO_2 concentrations is observed with a decreasing trend on NO_2 as the wind speed increases. The higher thermal convection (associated with larger eddy diffusion coefficients) and the faster winds in summer will contribute to a more efficient mixing leading to a higher dependence of the NO_2 values with wind speed as observed by the DOAS (Saiz-Lopez et al., 2006).

Regarding limit and threshold environmental levels values, the European Directive 2008/50/EC for O_3 establishes a population information threshold defined as an hourly average of 180 $\mu g\ m^{-3}$. Results obtained from DOAS in summer of 2000 and from mobile station in summer of 2008, showed that this value was exceeded only one time during summer season in 2008, with a concentration of 191 $\mu g\ m^{-3}$. In addition, the annual mean values of 50, 47 (point P6) and 63 $\mu gm-3$ measured in 2000, 2007 and 2008 exceeded the threshold limit value (40 μg m^{-3} calendar year) which implies damage to materials. The values obtained for NO$_2$ in points P2 and P11 slightly exceeded the annual limit value for the protection of vegetation (30 μg m^{-3}) and human health (40 μg m^{-3}).

3.2 Measurements of SO$_2$

The levels of SO$_2$ in the urban atmosphere of Ciudad Real during August 2000 to March 2001 and March to December 2008 were monitored using the continuous remote-sensing technique DOAS and the SO$_2$ analyzer placed at mobile station. The data obtained are shown in Figure 8 (a) and 8 (b) respectively.

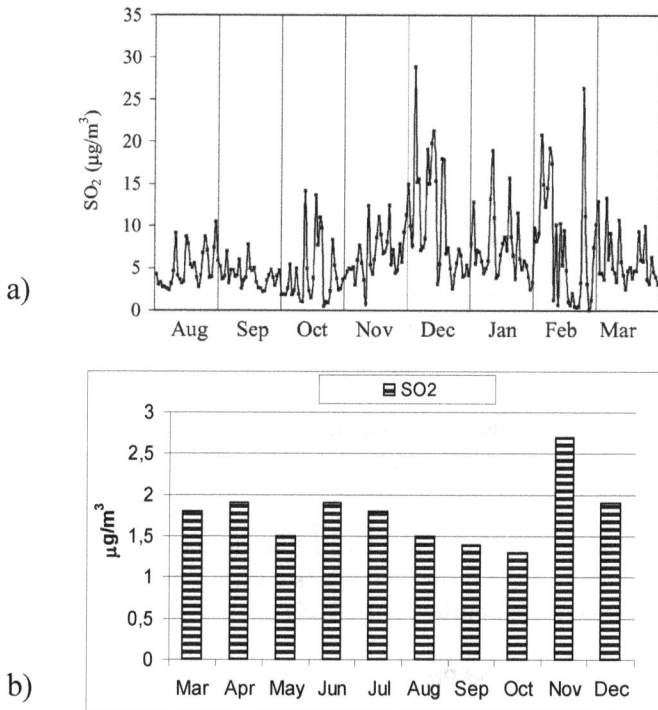

Fig. 8. (a) Daily mean concentration data for SO$_2$ obtained from the DOAS dataset for the entire period of measurements (August 2000–March 2001) and **(b)** monthly mean concentration data obtained from mobile station (March –December 2008).

The daily mean concentration profile obtained with DOAS system shows clearly an increase in the concentration of SO$_2$ during winter being the maximum monthly average of 10 μg m^{-3} in December 2000. SO$_2$ mean values were two times higher in winter than in summer. In the case of mobile station there is not a clear difference throughout the months, the maximum monthly average was 2.7 μg m^{-3} in November 2008 what means a 73% less than the maximum value found in 2000. Mean annual levels were 7 μg m^{-3} measured with DOAS system and 1.8 μg m^{-3} measured with the mobile station.

A diurnal profile of SO_2 for summer and winter time, hourly averaged, is illustrated in Figure 9. The higher SO_2 concentrations in winter were mainly related to house heating and only at a small extent with road traffic. The main difference between them arises in the later part of the day (i.e. 16 h to 22 h). In the case of the January data, the concentration of SO_2 increases up to three times in the early evening hours, when central heating systems (widely employed in this city during the cold winter months) are in greater use. The diurnal variation of SO_2 is also characterised by a secondary small peak observed in the morning, at the rush hour of traffic. In the warm season, only this small peak at the rush hour is observed. This small peak is broader in August indicating a change in the traffic pattern in this vacational month and/or another source of SO_2.

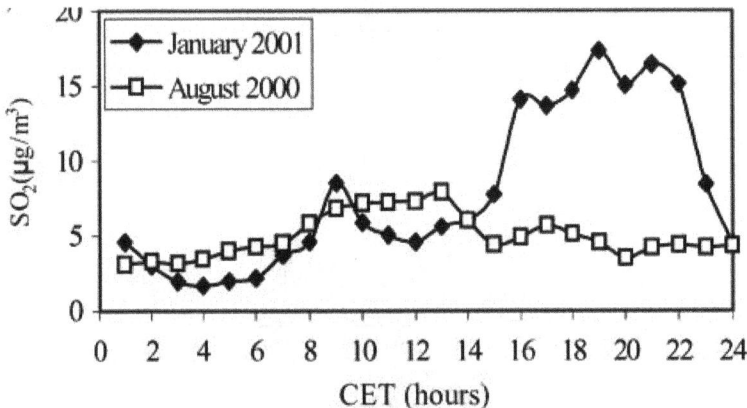

Fig. 9. Mean diurnal variation of the hourly averaged concentration during August 2000 and January 2001. Time scale is given in CET (Central European Time), being UTC+1 in winter and UTC + 2 in summer.

Generally, the levels of SO_2 registered in Ciudad Real during 2000/2001 and 2008 were found to be lower than those encountered in similar studies on other European cities (Gobiet et al., 2000; Vandaele et al., 2002; Kourtidis et al., 2002).

3.3 Measurements of CO, BTX and PM$_{10}$

The concentrations of CO, PM_{10} and benzene, toluene and m,p-xylene (BTX) were reported for the first time in Ciudad Real (Villanueva et al., 2010) using the mobile station.

Figure 10 shows the evolution of BTX, CO and PM_{10} concentrations observed for an autumn day. As above mentioned, the daily evolution of air pollutants reveals a marked daily cycle as in the cases of O_3 and NO_2. The daily evolution of O_3, and NO and their relation with BTX is due to common NO and BTX sources and the photochemical mechanism involving these species (NO, NO_2, BTX, among others) that leads to O_3 formation (Finalyson-Pitts and Pitts, 2000). Two peaks are observed during the day, one in the early morning hours when the emissions from road traffic increase and other after sunset which is related to a second rush hour in the traffic circulation. During summer, only the first peak corresponding to the early morning hours appears as in the cases of NO and NO_2.

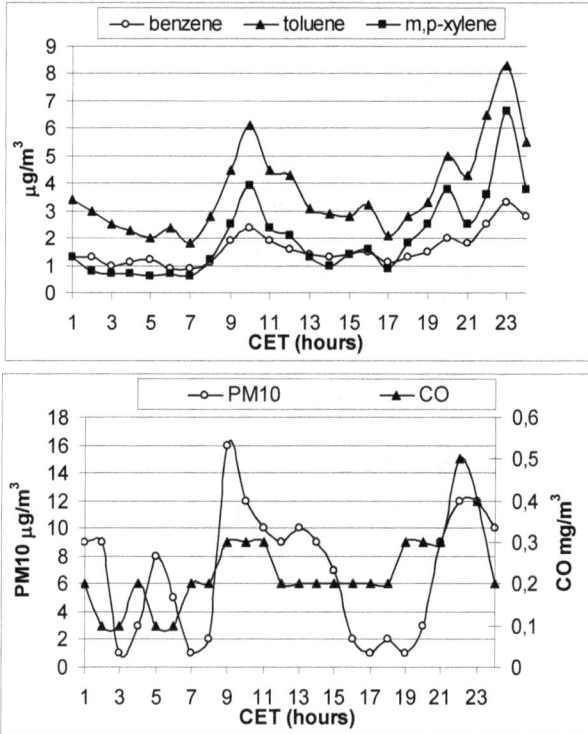

Fig. 10. Example of the evolution of BTX, CO and PM_{10} concentrations for an autumn day (12nd December 2008) in Ciudad Real. Time scale is given in CET (Central European Time), being UTC+1 in autumn.

Figure 11 shows the monthly averaged concentrations for CO, PM10 and BTX during the whole period of measurement. For PM10, there is not a clear trend throughout the months, while for CO the highest values are registered in spring, 0.19 mg m^{-3} and autumn, 0.21 mg m^{-3} while the lowest values are measured during summer, 0.12 mg m^{-3}. For BTX, the highest values are registered in spring (benzene 1.5, toluene 10.8, and m,p-xylene 3.6 μg m^{-3}) and the levels decrease up to reach the minimum values in August (benzene 0.4, toluene 2.6, and m,p-xylene 1 μg m^{-3}). Maximum monthly averages were in μg m^{-3}: 31 (PM10) in August, 2 (benzene) in November, 12 (toluene) in March, 4 (m,p-xylene) in May and 0.3 mg m^{-3} (CO) in November. The mean values of CO, PM10, benzene, toluene and m,p-xylene during the whole measurements period were: 0.2 mg m^{-3}, 20.7 μg m^{-3}, 1.2 μg m^{-3}, 6.4 μg m^{-3}and 2.4 μg m^{-3} respectively.

Regarding BTX, the average level of benzene found in this study during the whole period was 1.2 μg m^{-3} as mentioned above, with the highest concentration reaching 2 μg m^{-3}. This is comparable with the results published by Fernandez-Villarrenaga et al., 2004 concerning benzene levels in Coruña (Spain); the annual mean was 3.4 μg m^{-3} while Cocheo et al., 2000 reported urban concentrations in a few European cities as a mean annual average of 4.4 μg m^{-3} for Antwerp, 3.1 μg m^{-3} for Copenhagen and much higher for the South European cities.

In the present study, the obtained mean concentration was lower, due to the fact that Ciudad Real is not a big city. Toluene and m,p-xylene had concentrations either in the same range or lower than the ones reported for other urban areas, especially North European cities (Derwent et al., 2000).

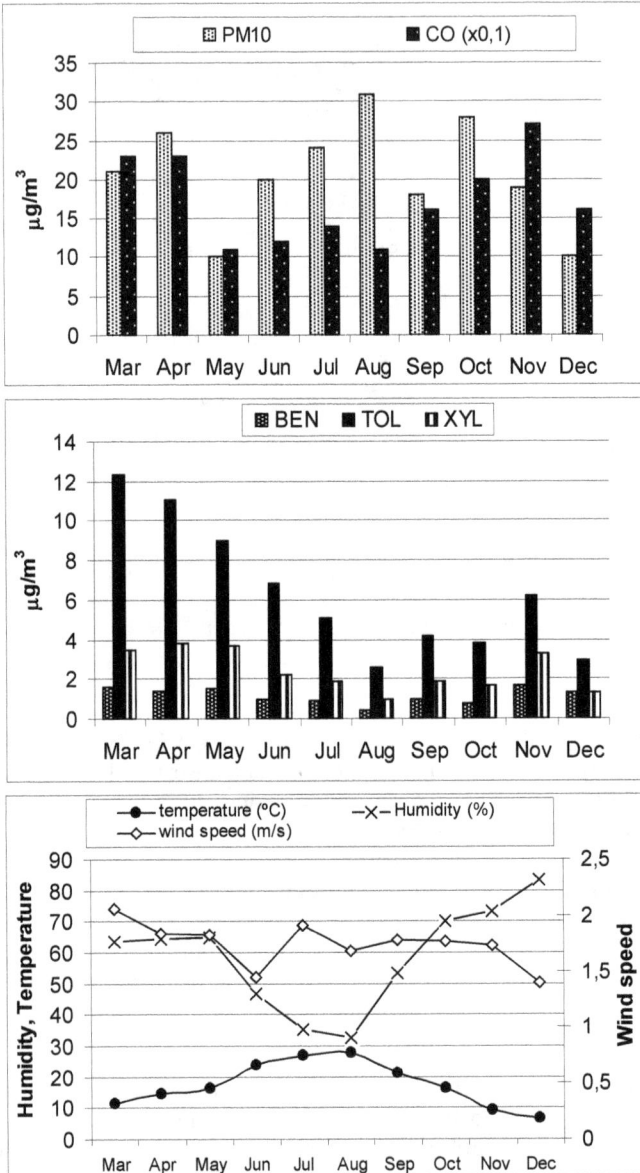

Fig. 11. Monthly means of CO, PM10 and BTX measured from daily data. The meteorological parameters are also presented.

The variation in the particulate matter concentration throughout the study, as mentioned above, has not a clear trend and there is not a clear relation to meteorological parameters such as wind speed that is below 2 m s^{-1} (in the whole period), temperature or humidity (Figure 11). This correlation was observed in the study carried out by Kulshrestha et al., 2009. PM10 concentrations in Ciudad Real must be conditioned for natural processes of the pollutants such as dust intrusion from Sahara desert, fires, etc., together with road activities and re-suspension of road dust. The mean value obtained in this study was 20.7 μg m^{-3}. This value is lower than those reported in other urban areas in Spain (Querol et al., 2008).

Regarding limit and threshold environmental levels values, the European Directive 2008/50/EC defines the threshold to protect the human health as a daily value of 50 μg m^{-3}. This value was exceeded 4 times during the period of study, in all cases during the warm season. The highest daily concentration was 56 μg m^{-3}. On the other hand, since EU has not guide value for toluene and xylenes, we compare our measurements with WHO, 2000b, recommended air guideline for toluene (9-10 ppb for 24 h). The values measured in Ciudad Real are below these limits.

4. Conclusion

We analyse in this study, the air quality of an urban atmosphere, Ciudad Real (in central-southern Spain) through different monitoring techniques. We perform a comparison study on four trace gases (O_3, NO, NO_2 and SO_2) using three different measurement techniques: a commercial DOAS system (Opsis, Sweden), conventional analyzers (API-Teledyne, USA) placed at a mobile station and passive samplers (Radiello, Italy). The results showed a fairly good agreement between the monitoring methods considering the differences in the techniques, the different time period and meteorological situations found during the different years and also the altitude and location of the instruments. In this study, measurement of BTX, CO and PM10 carried out with the mobile station for first time in Ciudad Real have also been included.

Based on the results of the studies made during the three campaigns (July 2000-March 2001, January-December 2007 and March-December 2008) using different techniques it can be concluded that Ciudad Real has high air quality with no significant changes during the three years monitored and where the pollution mainly comes from road traffic and it is influenced by photochemistry. There are not signs that indicate that the pollution from the petrochemical complex located 30 km southwest can significantly affect the air quality of Ciudad Real.

5. References

Atkinson, R. (2000). *Atmospheric chemistry of VOCs and NOx*. Atmospheric Environment 34, (12–14), 2063–2101.

Cocheo, V.; Boaretto, C.; Cocheo, L. & Sacco, P.; (1999). *Radial path in diffusion: The idea to improve the passive sampler performances*. In V. Cocheo, E. De Saeger, D. Kotzias

(Eds.), International Conference Air Quality in Europe. Challenges for the 2000's. Venice, Italy.

Cocheo, V.; Sacco, P.; Boaretto, C.; De Saeger, E.; Perez Ballesta, P.; Skov, H.; Goelen, E.; Gonzales, N. and Baeza Caracena, A. (2000). *Urban benzene and population exposure.* Nature 404, 141–142.

Derwent, R.G.; Davies, T.J.; Delaney, M.; Dollard, G.J.; Field, R.A.; Dumitrean, P.; Nason, P.D.; Jones, B.M.R. & Pepler, S.A. (2000). *Analysis and interpretation of the continuous hourly monitoring data for 26 C2–C8 hydrocarbons at 12 United Kingdom sites during 1996.* Atmospheric Environment 34, 297–312.

Fenger, J. (1999). *Urban air quality.* Atmospheric Environment. 33, 4877–4900.

Fernandez-Villarrenaga, V.; Lopez-Mahia, P.; Muniategui-Lorenzo, S.; Prada- Rodriguez, D.; Fernandez-Fernandez, E. & Tomas, X. (2004). *C1 to C9 volatile organic compound measurements in urban air.* Science of the Total Environment 334-335, 167–176.

Finlayson-Pitts, B. J. & Pitts, Jr., J. N. (2000) Chemistry of the Upper and Lower Atmosphere, Academic Press, New York.

Fuentes, J.D.; Lerdau, M.; Atkinson, R.; Baldocchi, D.; Bottenheim, J.W.; Ciccioli, P.; Lamb, B.; Geron, C.; Gu, L.; Guenther, A.; Sharkey T.D. & Stockwell, W. (2000). *Biogenic hydrocarbons in the atmospheric boundary layer: a review.* Bulletin of the American Meteorological Society , 81, 1537–1575.

Gobiet, A.; Baumgartner, D.; Krobath, T.; Maderbacher, R. & Putz, E. (2000). *Urban air pollution monitoring with DOAS considering the local meteorological situation,* Environmental Monitoring Assessment. 65, 119–127.

Kleinman, L. I. (1991). *Seasonal dependence of boundary layer peroxide concentration: the low and high NOx regimes.* Journal of Geophysical Research, 96, 20721–20733.

Kourtidis, K. A.; Ziomas, I.; Zerefos, C.; Kosmidis, E.; Symeonidis, P.; Christophilopoulos, E.; Karathanassis, S. & Mploutsos, A. (2002). *Benzene, toluene, ozone, NO2 and SO2 measurements in an urban street canyon in Thessaloniki, Greece,* Atmospheric Environment. 36, 5355–5364.

Kulshrestha, A.; Gursumeeran Satsangy, P.; Masih, J. & Taneja, A. (2009). *Metal concentration of PM2.5 and PM10 particles and seasonal variations in urban and rural environment of Agra, India.* Science of the Total Environment 407, 6196-6204.

Lee, D. S., Holland, M. K. & Falla, N. (1996). The potential impact of ozone on materials in the UK. Atmospheric Environment, 30, 1053–1065.

Martin, P.; Cabañas, B.; Villanueva, F.; Gallego, M.P.; Colmenar, I. & Salgado, S. (2010). *Ozone and Nitrogen Dioxide Levels Monitored in an Urban Area (Ciudad Real) in central-southern Spain.* Water, Air & Soil Pollution 208, 305-316.

Mazzeo, N. A. & Venegas, L. E. (2002). *Estimation of cumulative frequency distribution for carbon monoxide concentration from wind-speed data in Buenos Aires (Argentina).* Water, Air and Soil Pollution: Focus, 2, 419–432.

Mazzeo, N. A. & Venegas, L. E. (2004). *Some aspects of air pollution in Buenos Aires City.* International Journal of Environmental Pollution, 22 (4), 365-379.

Palacios, M.; Kirchner, F.; Martilli, A.; Clappier, A.; Martin, F. & Rodrıguez, M. E. (2002). *Summer ozone episodes in the Greater Madrid area. Analyzing the ozone response to abatement strategies by modelling.* Atmospheric Environment. 36, 5323–5333.

Plane J. M. C. & Saiz-Lopez, A. (2005). *UV-visible Differential Optical Absorption Spectroscopy (DOAS), in D.E.* Heard (ed.), Analytical Techniques for Atmospheric Measurement. Blackwell Publishing, Oxford.

Platt, U. (1994). *Differential optical absorption spectroscopy (DOAS), in M.W.* Sigrist (ed.), Air Monitoring by Spectroscopy Techniques. John Wiley, London.

Pujadas, M.; Plaza, J.; Teres, J.; Artinano, B. & Millan, M. (2000). *Passive remote sensing of nitrogen dioxide as a tool for tracking air pollution in urban areas: The Madrid urban pluma, a case of study,* Atmospheric Environment. 34, 3041–3056.

Querol, X.; Alastuey, A.; Moreno, T.; Viana, M.M.; Castillo, S.; Pey, J.; Rodríguez, S.; Artiñano, B.; Salvador, P.; Sánchez, M.; Garcia Dos Santos, S.; Herce Garraleta M.D.; Fernandez-Patier, R.; Moreno-Grau, S.; Negral, L.; Minguillón M.C.; Monfort, E.; Sanz, M.J.; Palomo-Marín, R.; Pinilla-Gil, E.; Cuevas, E.; de la Rosa, J. & Sánchez de la Campa, A. (2008) *Spatial and temporal variations in airborne particulate matter (PM10 and PM2.5) across Spain 1999–2005.* Atmospheric Environment 42, 3964-3979.

R&P-Co. (2001). Radiello® Model 3310 passive sampling system. Passive gas sampling system for industrial indoor/Radiello sampler, version 1/2003. http://www.radiello.com.

Saiz-Lopez, A.; Notario, A.; Martínez, E.; & Albadalejo, J. (2006). *Seasonal evolution of levels of gaseous pollutants in an urban area (Ciudad Real) in central-southern Spain: A DOAS Study.* Water, Air & Soil Pollution 171, 153-167.

Sillman, S. & Samson, P. J. (1995). *Impact of temperature on oxidant photochemistry in urban, polluted rural and remote environment.* Journal of Geophysical Research, 100, 11497–11508.

Tang, H. & Lau, T. (1999). *A new all-season passive sampling system for monitoring NO2 in air.* Field Analytical Chemistry and Technology, 3(6), 338–345.

Toll, I. & Baldasano, J. M. (2000). *Modeling of photochemical air pollution in the Barcelona area with highly disaggregated anthropogenic and biogenic emissions,* Atmospheric Environment. 34, 3069–3084.

Vandaele, A. C.; Tsouli, A.; Carleer, M. & Colin, R. (2002). *UV Fourier transform measurements of tropospheric O3, NO2, SO2, benzene and toluene,* Environmental. Pollution. 116, 193-201.

Varshney, C. K. & Singh, A. P. (2003). *Passive samplers for NOx monitoring: a critical review.* The Environmentalist, 23(2), 127–136.

Villanueva, F.; Notario A.; Albaladejo, J; Millán M.C. & Mabilia R; (2010). *Ambient air quality in an urban area (Ciudad Real) in central-southern* Spain. Fresenius Environmental Bulletin., 19, 2064-2070.

WHO (2000a). *Chapter 7.2 Ozone and other photochemical oxidants.* Air Quality Guidelines-Second Edition. Copenhagen: WHO Regional Office for Europe.

WHO (2000b). *Air quality guidelines for Europe, European series 91.* Copenhagen, Denmark: World Health authority Regional Publications, 1-198.

Zhang, R.W.; Lei, X. & TieHess, P. (2004) *Industrial emissions cause extreme diurnal urban ozone variability.* Proceedings of the National Academy of Sciences, USA, 101, 6346-6350.

5

Analytical Model for Air Pollution in the Atmospheric Boundary Layer

Daniela Buske[1], Marco Tullio Vilhena[2], Bardo Bodmann[2]
and Tiziano Tirabassi[3]
[1]*Federal University of Pelotas, Pelotas, RS*
[2]*Federal University of Rio Grande do Sul, Porto Alegre, RS*
[3]*Institute ISAC, National Research Council, Bologna*
[1,2]*Brazil*
[3]*Italy*

1. Introduction

The worldwide concern due to the increasing frequency of ecological disasters on our planet urged the scientific community to take action. One of the possible measures are analytical descriptions of pollution related phenomena, or simulations by effective and operational models with quantitative predictive power. The atmosphere is considered the principal vehicle by which pollutant materials are dispersed in the environment, that are either released from the productive and private sector or eventually in accidental events and thus may result in contamination of plants, animals and humans. Therefore, the evaluation of airborne material transport in the Atmospheric Boundary Layer (ABL) is one of the requirements for maintenance, protection and recovery of the ecological system. In order to analyse the consequences of pollutant discharge, atmospheric dispersion models are of need, which have to be tuned using specific meteorological parameters and conditions for the considered region. Moreover, they shall be subject to the local orography and shall supply with realistic information concerning environmental consequences and further help reduce impact from potential accidents such as fire events and others. Moreover, case studies by model simulations may be used to establish limits for the escape of pollutant material from specific sites into the atmosphere.

The theoretical approach to the problem essentially assumes two basic forms. In the Eulerian approach, diffusion is considered, at a fixed point in space, proportional to the local gradient of the concentration of the diffused material. Consequently, it considers the motion of fluid within a spatially fixed system of reference. These kind of approaches are based on the resolution, on a fixed space-time grid, of the equation of mass conservation of the pollutant chemical species. Lagrangian models are the second approach and they differ from Eulerian ones in adopting a system of reference that follows atmospheric motions. Initially, the term Lagrangian was used only to refer to moving box models that followed the mean wind trajectory. Currently, this class includes all models that decompose the pollutant cloud into discrete "elements", such as segments, puffs or pseudo-particles (computer particles). In Lagrangian particle models pollutant dispersion is simulated through the motion of computer particles whose trajectories allow the calculation of the concentration field of the emitted substance. The underlying hypothesis is that the combination of the trajectories of such particles simulates the paths of the air particles situated, at the initial instant, in the same position. The motion of the particles can be reproduced both in a deterministic way and in a stochastic way.

In the present discussion we adopt an Eulerian approach, more specifically to the K model, where the flow of a given field is assumed to be proportional to the gradient of a mean variable. K-theory has its own limits, but its simplicity has led to a widespread use as the mathematical basis for simulating pollution dispersion. Most of of Eulerian models are based on the numerical resolution of the equation of mass conservation of the pollutant chemical species. Such models are most suited to confronting complex problems, for example, the dispersion of pollutants over complex terrain or the diffusion of non-inert pollutants.

In the recent literature progressive and continuous effort on analytical solution of the advection-diffusion equation (ADE) are reported. In fact analytical solutions of equations are of fundamental importance in understanding and describing the physical phenomena. Analytical solutions explicitly take into account all the parameters of a problem, so that their influence can be reliably investigated and it is easy to obtain the asymptotic behaviour of the solution, which is usually more tedious and time consuming when generated by numerical calculations. Moreover, like the Gaussian solution, which was the first solution of ADE with the wind and the K diffusion coefficients assumed constant in space and time, they open pathways for constructing more sophisticated operative analytical models. Gaussian models, so named because they are based on the Gaussian solution, are forced to represent real situations by means of empirical parameters, referred to as "sigmas". In general Gaussian models are fast, simple, do not require complex meteorological input, and describe the diffusive transport in an Eulerian framework exploring the use of the Eulerian nature of measurements. For these reasons they are widely employed for regulatory applications by environmental agencies all over the world although their well known intrinsic limits. For the further discussion we restrict our considerations to scenarios in micro-meteorology were the Eulerian approach has been validated against experimental data and has been found useful for pollution dispersion problems in the ABL.

A significant number of works regarding ADE analytical solution (mostly two-dimensional solutions) is available in the literature. By analytical we mean that starting from the advection-diffusion equation, which we consider adequate for the problem, the solution is derived without introducing approximations that might simplify the solution procedure. A selection of references considered relevant by the authors are Rounds (1955), Smith (1957), Scriven & Fisher (1975), Demuth (1978), van Ulden (1978), Nieuwstadt & Haan (1981), Tagliazucca et al. (1985), Tirabassi (1989), Tirabassi & Rizza (1994), Sharan et al. (1996), Lin & Hildemann (1997), Tirabassi (2003). It is noteworthy that solutions from citations above are valid for specialized problems. Only ground level sources or infinite height of the ABL or specific wind and vertical eddy diffusivity profiles are admitted. A more general approach that may be found in the literature is based on the Advection Diffusion Multilayer Method (ADMM), which solves the two-dimensional ADE with variable wind profile and eddy diffusivity coefficient (Moreira et al. (2006a)). The main idea here relies on the discretisation of the Atmospheric Boundary Layer (ABL) in a multilayer domain, assuming in each layer that the eddy diffusivity and wind profile take averaged values. The resulting advection-diffusion equation in each layer is then solved by the Laplace Transform technique. For more details about this methodology see reference Moreira et al. (2006a).

A further mile-stone in the direction of the present approach is given in references Costa et al. (2006; 2011) by the Generalized Integral Advection Diffusion Multilayer Technique (GIADMT), that presents a general solution for the time-dependent three-dimensional ADE again assuming the stepwise approximation for the eddy diffusivity coefficient and wind profile and proceeding further in a similar way as the multi-layer approach (Moreira et al. (2006a)), however in a semi-analytical fashion. In this chapter we report on a generalisation of the afore mentioned models resulting in a general space-time ($3 \oplus 1$) solution for this problem, given in closed form.

To this end, we start from the three-dimensional and time dependent ADE for the ABL as the underlying physical model. In the subsequent sections we show how the model is solved analytically using integral transform and a spectral theory based methods which then may provide short, intermediate and long term (normalized) concentrations and permit to assess the probability of occurrence of high contamination level case studies of accidental scenarios. The solution of the $3 \oplus 1$ space-time solution is compared to other dispersion process approaches and validated against experiments. Comparison between different approaches may help to pin down computational errors and finally allows for model validation. In this line we show with the present discussion, that our analytical approach does not only yield an acceptable solution for the time dependent three dimensional ADE, but further predicts tracer concentrations closer to observed values compared to other approaches from the literature. This quality is also manifest in better statistical coefficients including model validation.

Differently than in most of the approaches known from the literature, where solutions are typically valid for scenarios with strong restrictions with respect to their specific wind and vertical eddy diffusivity profiles, the discussed method is general in the sense that once the model parameter functions are determined or known from other sources, the hierarchical procedure to obtain a closed form solution is unique. More specifically, the general analytical solution for the advection-diffusion problem works for eddy diffusivity and wind profiles, that are arbitrary functions with a continuous dependence on the vertical and longitudinal spatial variables, as well as time.

A few technical details are given in the following, that characterise the principal steps of the procedure. In a formal step without physical equivalence we expand the contaminant concentration in a series in terms of a set of orthogonal eigenfunctions. These eigenfunctions are the solution of a simpler but similar problem to the existing one and are detailed in the next section. Replacing this expansion in the time-dependent, three-dimensional ADE in Cartesian geometry, we project out orthogonal components of the equation, thus inflating the tree-dimensional advection-diffusion equation into an equation system. Note, that the projection integrates out one spatial dimension, so that the resulting problem reduces the problem effectively to $2 \oplus 1$ space-time dimensions. Such a problem was already solved by the Laplace transform technique as shown in Moreira et al. (2009a), Moreira et al. (2009b), Buske et al. (2010) and references therein. In the next sections we present the prescription for constructing the solution for the general problem, but for convenience and without restricting generality we resort to simplified models. Further, we present numerical simulations and indicate future perspectives of this kind of modelling.

2. The advection-diffusion equation

The advection-diffusion equation of air pollution in the atmosphere is essentially a statement of conservation of the suspended material and it can be written as (Blackadar (1997))

$$\frac{D\bar{c}}{Dt} = \partial_t \bar{c} + \bar{\mathbf{U}}^T \nabla \bar{c} = -\nabla^T \overline{\mathbf{U}'c'} + S \tag{1}$$

where $\bar{c}$ denotes the mean concentration of a passive contaminant (in units of $\frac{g}{m^3}$), $\frac{D}{Dt}$ is the substantial derivative, ∂_t is a shorthand for the partial time derivative and $\bar{\mathbf{U}} = (\bar{u}, \bar{v}, \bar{w})^T$ is the mean wind (in units of $\frac{m}{s}$) with Cartesian components in the directions x, y and z, respectively, $\nabla = (\partial_x, \partial_y, \partial_z)^T$ is the usual Nabla symbol and S is the source term. The terms $\overline{\mathbf{U}'c'}$ represent the turbulent flux of contaminants (in units of $\frac{g}{s\,m^2}$), with its longitudinal, crosswind and vertical components.

Observe that equation (1) has four unknown variables (the concentration and turbulent fluxes) which lead us to the well known turbulence closure problem. One of the most widely used closures for equation (1), is based on the gradient transport hypothesis (also called K-theory) which, in analogy with Fick's law of molecular diffusion, assumes that turbulence causes a net movement of material following the negative gradient of material concentration at a rate which is proportional to the magnitude of the gradient (Seinfeld & Pandis (1998)).

$$\overline{\mathbf{U}'c'} = -\mathbf{K}\nabla \bar{c} \tag{2}$$

Here, the eddy diffusivity matrix $\mathbf{K} = \mathrm{diag}(K_x, K_y, K_z)$ (in units of $\frac{m^2}{s}$) is a diagonal matrix with the Cartesian components in the x, y and z directions, respectively. In the first order closure all the information on the turbulence complexity is contained in the eddy diffusivity.

Equation (2), combined with the continuity equation of mass, leads to the advection-diffusion equation (see reference Blackadar (1997)).

$$\partial_t \bar{c} + \bar{\mathbf{U}}^T \nabla \bar{c} = \nabla^T (\mathbf{K}\nabla \bar{c}) + S \tag{3}$$

The simplicity of the K-theory of turbulent diffusion has led to the widespread use of this theory as a mathematical basis for simulating pollutant dispersion (open country, urban, photochemical pollution, etc.), but K-closure has its known limits. In contrast to molecular diffusion, turbulent diffusion is scale-dependent. This means that the rate of diffusion of a cloud of material generally depends on the cloud dimension and the intensity of turbulence. As the cloud grows, larger eddies are incorporated in the expansion process, so that a progressively larger fraction of turbulent kinetic energy is available for the cloud expansion.

Equation (3) is considered valid in the domain $(x, y, z) \in \Gamma$ bounded by $0 < x < L_x, 0 < y < L_y$ and $0 < z < h$ and subject to the following boundary and initial conditions,

$$\mathbf{K}\nabla \bar{c}\big|_{(0,0,0)} = \mathbf{K}\nabla \bar{c}\big|_{(L_x, L_y, h)} = \mathbf{0} \tag{4}$$

$$\bar{c}(x, y, z, 0) = 0 . \tag{5}$$

Instead of specifying the source term as an inhomogeneity of the partial differential equation, we consider a point source located at an edge of the domain, so that the source position $\mathbf{r}_S =$

$(0, y_0, H_S)$ is located at the boundary of the domain $\mathbf{r}_S \in \delta\Gamma$. Note, that in cases where the source is located in the domain, one still may divide the whole domain in sub-domains, where the source lies on the boundary of the sub-domains which can be solved for each sub-domain separately. Moreover, a set of different sources may be implemented as a superposition of independent problems. Since the source term location is on the boundary, in the domain this term is zero everywhere ($S(\mathbf{r}) \equiv 0$ for $\mathbf{r} \in \Gamma \backslash \delta\Gamma$), so that the source influence may be cast in form of a condition, where we assume that our coordinate system is oriented such that the x-axis is aligned with the mean wind direction. Since the flow crosses the plane perpendicular to the propagation (here the y-z-plane) the source condition reads

$$\bar{u}\bar{c}(0, y, z, t) = Q\delta(y - y_0)\delta(z - H_S) , \tag{6}$$

where Q is the emission rate (in units of $\frac{g}{s}$), h the height of the ABL (in units of m), H_S the height of the source (in units of m), L_x and L_y are the horizontal domain limits (in units of m) and $\delta(x)$ represents the Cartesian Dirac delta functional.

3. A closed form solution

In this section we first introduce the general formalism to solve a general problem and subsequently reduce the problem to a more specific one, that is solved and compared to experimental findings.

3.1 The general procedure

In order to solve the problem (3) we reduce the dimensionality by one and thus cast the problem into a form already solved in reference Moreira et al. (2009a). To this end we apply the integral transform technique in the y variable, and expand the pollutant concentration as

$$\bar{c}(x, y, z, t) = \mathbf{R}^T(x, z, t)\mathbf{Y}(y), \tag{7}$$

where $\mathbf{R} = (R_1, R_2, \ldots)^T$ and $\mathbf{Y} = (Y_1, Y_2, \ldots)^T$ is a vector in the space of orthogonal eigenfunctions, given by $Y_m(y) = \cos(\lambda_m y)$ with eigenvalues $\lambda_m = m\frac{\pi}{L_y}$ for $m = 0, 1, 2, \ldots$. For convenience we introduce some shorthand notations, $\nabla_2 = (\partial_x, 0, \partial_y)^T$ and $\hat{\partial}_y = (0, \partial_y, 0)^T$, so that equation (3) reads now,

$$(\partial_t \mathbf{R}^T)\mathbf{Y} + \bar{\mathbf{U}}\left(\nabla_2 \mathbf{R}^T \mathbf{Y} + \mathbf{R}^T \hat{\partial}_y \mathbf{Y}\right) = \left(\nabla^T \mathbf{K} + (\mathbf{K}\nabla)^T\right)\left(\nabla_2 \mathbf{R}^T \mathbf{Y} + \mathbf{R}^T \hat{\partial}_y \mathbf{Y}\right)$$
$$= \left(\nabla_2^T \mathbf{K} + (\mathbf{K}\nabla_2)^T\right)(\nabla_2 \mathbf{R}^T \mathbf{Y}) + \left(\hat{\partial}_y^T \mathbf{K} + (\mathbf{K}\hat{\partial}_y)^T\right)(\mathbf{R}^T \hat{\partial}_y \mathbf{Y}) . \tag{8}$$

Upon application of the integral operator

$$\int_0^{L_y} dy \mathbf{Y}[\mathbf{F}] = \int_0^{L_y} \mathbf{F}^T \wedge \mathbf{Y} \, dy \tag{9}$$

here $\mathbf{F}$ is an arbitrary function and $\wedge$ signifies the dyadic product operator, and making use of orthogonality renders equation (8) a matrix equation. The appearing integral terms are

$$\mathbf{B}_0 = \int_0^{L_y} dy\, \mathbf{Y}[\mathbf{Y}] = \int_0^{L_y} \mathbf{Y}^T \wedge \mathbf{Y}\, dy\,,$$

$$\mathbf{Z} = \int_0^{L_y} dy\, \mathbf{Y}[\hat{\partial}_y \mathbf{Y}] = \int_0^{L_y} \hat{\partial}_y \mathbf{Y}^T \wedge \mathbf{Y}\, dy\,,$$

$$\boldsymbol{\Omega}_1 = \int_0^{L_y} dy\, \mathbf{Y}[(\nabla_2^T \mathbf{K})(\nabla_2 \mathbf{R}^T \mathbf{Y})] = \int_0^{L_y} \left((\nabla_2^T \mathbf{K})(\nabla_2 \mathbf{R}^T \mathbf{Y})\right)^T \wedge \mathbf{Y}\, dy\,, \tag{10}$$

$$\boldsymbol{\Omega}_2 = \int_0^{L_y} dy\, \mathbf{Y}[(\mathbf{K}\nabla_2)^T (\nabla_2 \mathbf{R}^T \mathbf{Y})] = \int_0^{L_y} \left((\mathbf{K}\nabla_2)^T (\nabla_2 \mathbf{R}^T \mathbf{Y})\right) \wedge \mathbf{Y}\, dy\,,$$

$$\mathbf{T}_1 = \int_0^{L_y} dy\, \mathbf{Y}[((\hat{\partial}_y^T \mathbf{K})(\hat{\partial}_y \mathbf{Y})] = \int_0^{L_y} \left(((\hat{\partial}_y^T \mathbf{K})(\hat{\partial}_y \mathbf{Y}))\right)^T \wedge \mathbf{Y}\, dy\,,$$

$$\mathbf{T}_2 = \int_0^{L_y} dy\, \mathbf{Y}[(\mathbf{K}\hat{\partial}_y)^T (\hat{\partial}_y \mathbf{Y})] = \int_0^{L_y} \left((\mathbf{K}\hat{\partial}_y)^T (\hat{\partial}_y \mathbf{Y})\right)^T \wedge \mathbf{Y}\, dy\,.$$

Here, $\mathbf{B}_0 = \frac{L_y}{2}\mathbf{I}$, where $\mathbf{I}$ is the identity, the elements $(\mathbf{Z})_{mn} = \frac{2}{1-n^2/m^2}\delta_{1,j}$ with $\delta_{i,j}$ the Kronecker symbol and $j = (m+n)\mathrm{mod}2$ is the remainder of an integer division (i.e. is one for $m+n$ odd and zero else). Note, that the integrals $\boldsymbol{\Omega}_i$ and $\mathbf{T}_i$ depend on the specific form of the eddy diffusivity $\mathbf{K}$. The integrals (10) are general, but for practical purposes and for application to a case study we truncate the eigenfunction space and consider M components in $\mathbf{R}$ and $\mathbf{Y}$ only, though continue using the general nomenclature that remains valid. The obtained matrix equation determines now together with initial and boundary condition uniquely the components R_i for $i = 1,\ldots,M$ following the procedure introduced in reference Moreira et al. (2009a).

$$(\partial_t \mathbf{R}^T)\mathbf{B} + \bar{\mathbf{U}}\left(\nabla_2 \mathbf{R}^T \mathbf{B} + \mathbf{R}^T \mathbf{Z}\right) = \boldsymbol{\Omega}_1(\mathbf{R}) + \boldsymbol{\Omega}_2(\mathbf{R}) + \mathbf{R}^T(\mathbf{T}_1 + \mathbf{T}_2) \tag{11}$$

3.2 A specific case for application

In order to discuss a specific case we recall the convention already adopted above, that considered the average wind velocity aligned with the x-axis. Since we consider length scales L_x, L_Y, h typical for micro-meteorological scenarios we simplify $\bar{\mathbf{U}} = (\bar{u}, 0, 0)^T$. By comparison of physically meaningful cases, one finds for the operator norm $||\partial_x K_x \partial_x|| << |\bar{u}|$, which can be understood intuitively because eddy diffusion is observable predominantly perpendicular to the mean wind propagation. As a consequence we neglect the terms with K_x and $\partial_x K_x$.

The principal aspect of interest in pollution dispersion is the vertical concentration profile, that responds strongly to the atmospheric boundary layer stratification, so that the simplified eddy diffusivity $\mathbf{K} \to \mathbf{K}_1 = \mathrm{diag}(0, K_y, K_z)$ depends in leading order approximation only on the vertical coordinate $\mathbf{K}_1 = \mathbf{K}_1(z)$. For this specific case the integrals $\boldsymbol{\Omega}_i$ reduce to

$$\boldsymbol{\Omega}_1 \to (\partial_z K_z)(\partial_z \mathbf{R}^T)\mathbf{B}\,,$$
$$\boldsymbol{\Omega}_2 \to K_z(\partial_z^2 \mathbf{R}^T)\mathbf{B}\,, \tag{12}$$
$$\mathbf{T}_1 \to 0\,,$$
$$\mathbf{T}_2 \to -K_y \Lambda \mathbf{B}\,, \tag{13}$$

where $\Lambda = \mathrm{diag}(\lambda_1^2, \lambda_2^2, \ldots)$. The simplified equation system to be solved is then,

$$\partial_t \mathbf{R}^T \mathbf{B} + \bar{u}\partial_x \mathbf{R}^T \mathbf{B} = (\partial_z K_z)\partial_z \mathbf{R}^T \mathbf{B} + K_z\partial_z^2 \mathbf{R}^T \mathbf{B} - K_y \mathbf{R}^T \Lambda \mathbf{B} \tag{14}$$

which is equivalent to the problem

$$\partial_t \mathbf{R} + \bar{u}\partial_x \mathbf{R} = (\partial_z K_z)\partial_z \mathbf{R} + K_z\partial_z^2 \mathbf{R} - K_y \Lambda \mathbf{R} \tag{15}$$

by virtue of $\mathbf{B}$ being a diagonal matrix.

The specific form of the eddy diffusivity determines now whether the problem is a linear or non-linear one. In the linear case the $\mathbf{K}$ is assumed to be independent of $\bar{c}$, whereas in more realistic cases, even if stationary, $\mathbf{K}$ may depend on the contaminant concentration and thus renders the problem non-linear. However, until now no specific law is known that links the eddy diffusivity to the concentration so that we hide this dependence using a phenomenologically motivated expression for $\mathbf{K}$ which leaves us with a partial differential equation system in linear form, although the original phenomenon is non-linear. In the example below we demonstrate the closed form procedure for a problem with explicit time dependence, which is novel in the literature.

The solution is generated making use of the decomposition method (Adomian (1984; 1988; 1994)) which was originally proposed to solve non-linear partial differential equations, followed by the Laplace transform that renders the problem a pseudo-stationary one. Further we rewrite the vertical diffusivity as a time average term $\bar{K}_z(z)$ plus a term representing the variations $\kappa_z(z, t)$ around the average for the time interval of the measurement $K_z(x, z, t) = \bar{K}_z(z) + \kappa_z(z, t)$ and use the asymptotic form of K_y, which is then explored to set-up the structure of the equation that defines the recursive decomposition scheme.

$$\partial_t \mathbf{R} + \bar{u}\partial_x \mathbf{R} - \partial_z (\bar{K}_z\partial_z \mathbf{R}) + K_y \Lambda \mathbf{R} = \partial_z (\kappa_z\partial_z \mathbf{R}) \tag{16}$$

The function $\mathbf{R} = \sum_j \mathbf{R}_j = \mathbf{1}^T \mathbf{R}^{(c)}$ is now decomposed into contributions to be determined by recursion. For convenience we introduced the one-vector $\mathbf{1} = (1, 1, \ldots)^T$ and inflate the vector $\mathbf{R}$ to a vector with each element being itself a vector $\mathbf{R}_j$. Upon inserting the expansion in equation (16) one may regroup terms that obey the recursive equations and starts with the time averaged solution for K_z.

$$\partial_t \mathbf{R}_0 + \bar{u}\partial_x \mathbf{R}_0 - \partial_z (\bar{K}_z\partial_z \mathbf{R}_0) + K_y \Lambda \mathbf{R}_0 = 0 \tag{17}$$

The extension to the closed form recursion is then given by

$$\partial_t \mathbf{R}_j + \bar{u}\partial_x \mathbf{R}_j - \partial_z \left(\bar{K}_z\partial_z \mathbf{R}_j\right) + K_y \Lambda \mathbf{R}_j = \partial_z \left(\kappa_z\partial_z \mathbf{R}_{j-1}\right). \tag{18}$$

From the construction of the recursion equation system it is evident that other schemes are possible. The specific choice made here allows us to solve the recursion initialisation using the procedure described in references Moreira et al. (2006a; 2009a), where a stationary $\mathbf{K}$ was assumed. For this reason the time dependence enters as a known source term from the first recursion step on.

3.3 Recursion initialisation

The boundary conditions are now used to uniquely determine the solution. In our scheme the initialisation solution that contains $\mathbf{R}_0$ satisfies the boundary conditions (equations (4)-(6)) while the remaining equations satisfy homogeneous boundary conditions. Once the set of problems (18) is solved by the GILTT method, the solution of problem (3) is well determined. It is important to consider that we may control the accuracy of the results by a proper choice of the number of terms in the solution series.

In references Moreira et al. (2009a;b), a two dimensional problem with advection in the x direction in stationary regime was solved which has the same formal structure than (18) except for the time dependence. Upon rendering the recursion scheme in a pseudo-stationary form problem and thus matching the recursive structure of Moreira et al. (2009a;b), we apply the Laplace Transform in the t variable, $(t \rightarrow r)$ obtaining the following pseudo-steady-state problem.

$$r\tilde{\mathbf{R}}_0 + \bar{u}\partial_x\tilde{\mathbf{R}}_0 = \partial_z\left(K_z\partial_z\tilde{\mathbf{R}}_0\right) - \Lambda K_y\tilde{\mathbf{R}}_0 \qquad (19)$$

The x and z dependence may be separated using the same reasoning as already introduced in (7). To this end we pose the solution of problem (19) in the form:

$$\tilde{\mathbf{R}}_0 = \mathbf{PC} \qquad (20)$$

where $\mathbf{C} = (\zeta_1(z), \zeta_2(z), \ldots)^T$ are a set of orthogonal eigenfunctions, given by $\zeta_i(z) = \cos(\gamma_i z)$, and $\gamma_i = i\pi/h$ (for $i = 0, 1, 2, \ldots$) are the set of eigenvalues.

Replacing equation (20) in equation (19) and using the afore introduced projector (9) now for the z dependent degrees of freedom $\int_0^h dz\mathbf{C}[\mathbf{F}] = \int_0^h \mathbf{F}^T \wedge \mathbf{C}\, dz$ yields a first order differential equation system.

$$\partial_x\mathbf{P} + \mathbf{B}_1^{-1}\mathbf{B}_2\mathbf{P} = 0, \qquad (21)$$

where $\mathbf{P}_0 = \mathbf{P}_0(x, r)$ and $\mathbf{B}_1^{-1}\mathbf{B}_2$ come from the diagonalisation of the problem. The entries of matrices $\mathbf{B}_1$ and $\mathbf{B}_2$ are

$$(\mathbf{B}_1)_{i,j} = -\int_0^h \bar{u}\zeta_i(z)\zeta_j(z)\, dz$$

$$(\mathbf{B}_2)_{i,j} = \int_0^h \partial_z K_z\partial_z\zeta_i(z)\zeta_j(z)\, dz - \gamma_i^2\int_0^h K_z\zeta_i(z)\zeta_j(z)\, dz$$

$$-r\int_0^h \zeta_i(z)\zeta_j(z)\, dz - \lambda_i^2 K_y\int_0^h \zeta_i(z)\zeta_j(z)\, dz.$$

A similar procedure leads to the source condition for (21).

$$\mathbf{P}(0, r) = Q\mathbf{B}_1^{-1}\int dz\mathbf{C}[\delta(z - H_S)]\int dy\mathbf{Y}[\delta(y - y_0)] = Q\mathbf{B}_1^{-1}\left(\mathbf{C}(H_S) \wedge \mathbf{1}\right)\left(\mathbf{1} \wedge \mathbf{Y}(y_0)\right) \quad (22)$$

Following the reasoning of Moreira et al. (2009b) we solve (21) applying Laplace transform and diagonalisation of the matrix $\mathbf{B}_1^{-1}\mathbf{B}_2 = \mathbf{XDX}^{-1}$ which results in

$$\tilde{\mathbf{P}}(s, r) = \mathbf{X}(s\mathbf{I} + \mathbf{D})^{-1}\mathbf{X}^{-1}\mathbf{P}(0, r) \qquad (23)$$

where $\tilde{\mathbf{P}}(s,r)$ denotes the Laplace Transform of $\mathbf{P}(x,r)$. Here $\mathbf{X}^{(-1)}$ is the (inverse) matrix of the eigenvectors of matrix $\mathbf{B}_1^{-1}\mathbf{B}_2$ with diagonal eigenvalue matrix $\mathbf{D}$ and the entries of matrix $(s\mathbf{I} + \mathbf{D})_{ii} = s + d_i$. After performing the Laplace transform inversion of equation (23), we come out with

$$\mathbf{P}(x,r) = \mathbf{X}\mathbf{G}(x,r)\mathbf{X}^{-1}\mathbf{\Omega}, \tag{24}$$

where $\mathbf{G}(x,r)$ is the diagonal matrix with components $(\mathbf{G})_{ii} = e^{-d_i x}$. Further the still unknown arbitrary constant matrix is given by $\mathbf{\Omega} = \mathbf{X}^{-1}\mathbf{P}(0,r)$.

The analytical time dependence for the recursion initialisation (19) is obtained upon applying the inverse Laplace transform definition.

$$\mathbf{R}_0(x,z,t) = \frac{1}{2\pi i}\int_{\gamma-i\infty}^{\gamma+i\infty} \mathbf{P}(x,r)\mathbf{C}(z)e^{rt}\,dr. \tag{25}$$

To overcome the drawback of evaluating the line integral appearing in the above solution, we perform the calculation of this integral by the Gaussian quadrature scheme, which is exact if the integrand is a polynomial of degree $2M - 1$ in the $\frac{1}{r}$ variable.

$$\mathbf{R}_0(x,z,t) = \frac{1}{t}\mathbf{a}^T\left(\mathbf{p}\mathbf{R}_0(x,z,\frac{\mathbf{p}}{t})\right), \tag{26}$$

where $\mathbf{a}$ and $\mathbf{p}$ are respectively vectors with the weights and roots of the Gaussian quadrature scheme (Stroud & Secrest (1966)), and the argument $(x,z,\frac{\mathbf{p}}{t})$ signifies the k-th component of $\mathbf{p}$ in the k-th row of $\mathbf{p}\mathbf{R}_0$. Note, k is a component from contraction with $\mathbf{a}$. Concerning the issue of possible Laplace inversion schemes, it is worth mentioning that this approach is exact, however we are aware of the existence of other methods in the literature to invert the Laplace transformed functions (Valkó & Abate (2004), Abate & Valkó (2004)), but we restrict our attention in the considered problem to the Gaussian quadrature scheme, which is sufficient in the present case. Finally, the solution of the remaining equations of the recursive system (18) are attained in an analogue fashion expanding the source term in series and solving the resulting linear first order differential matrix equation with known and integrable source by the Laplace transform technique as shown above.

3.4 Reduction to solutions of simpler models

To establish the solution of the stationary, three and two-dimensional, ADE under Fickian flow regime, with and without longitudinal diffusion, one only need to take the limit $t \to \infty$ of the previously derived solutions, which is equivalent to take $r \to 0$ (Buske et al. (2007a)). Further models that make use of simplifications are not discussed in this chapter because they can be determined with the present developments and applying them to the models of references Buske et al. (2007b); Moreira et al. (2005; 2006b; 2009b; 2010); Tirabassi et al. (2008; 2009; 2011); Wortmann et al. (2005) and others.

4. Solution validation against experimental data

The solution procedure discussed in the previous section was coded in a computer program and will be available in future as a program library add-on. In order to illustrate the suitability of the discussed formulation to simulate contaminant dispersion in the atmospheric boundary layer, we evaluate the performance of the new solution against experimental ground-level concentrations.

4.1 Time dependent vertical eddy diffusivity

As already mentioned in the derivation of the solution, one has to make the specific choice for the turbulence parametrisation. From the physical point of view the turbulence parametrisation is an approximation to nature in the sense that we use a mathematical model as an approximated (or phenomenological) relation that can be used as a surrogate for the natural true unknown term, which might enter into the equation as a non-linear contribution. The reliability of each model strongly depends on the way turbulent parameters are calculated and related to the current understanding of the ABL (Degrazia (2005)).

The present parametrisation is based on the Taylor statistical diffusion theory and a kinetic energy spectral model. This methodology, derived for convective and moderately unstable conditions, provides eddy diffusivities described in terms of the characteristic velocity and length scales of energy-containing eddies. The time dependent eddy diffusivity has been derived by Degrazia et al. (2002) and can be expressed as the following formula.

$$\frac{K_\alpha}{w_* h} = \frac{0.583 c_i \psi^{\frac{2}{3}} \left(\frac{z}{h}\right)^{\frac{4}{3}} X^* \left[0.55 \left(\frac{z}{h}\right)^{\frac{2}{3}} + 1.03 c_i^{\frac{1}{2}} \psi^{\frac{1}{3}} (f_m^*)_i^{\frac{2}{3}} X^*\right]}{\left[0.55 \left(\frac{z}{h}\right)^{\frac{2}{3}} (f_m^*)_i^{\frac{1}{3}} + 2.06 c_i^{\frac{1}{2}} \psi^{\frac{1}{3}} (f_m^*)_i X^*\right]^2} \tag{27}$$

Here $\alpha = x, y, z$ are the Cartesian components, $i = u, v, w$ are the longitudinal and transverse wind directions,

$$c_i = \alpha_i (0.5 \pm 0.05)(2\pi \kappa)^{-\frac{2}{3}},$$

$\alpha_i = 1$ and $4/3$ for u, v and w components, respectively (Champagne et al. (1977)), $\kappa = 0.4$ is the von Karman constant, $(f_m^*)_i$ is the normalized frequency of the spectral peak, h is the top of the convective boundary layer height, w_* is the convective velocity scale, ψ is the non-dimensional molecular dissipation rate function and

$$X^* = \frac{x w_*}{\bar{u} h}$$

is the non-dimensional travel time, where $\bar{u}$ is the horizontal mean wind speed.

For horizontal homogeneity the convective boundary layer evolution is driven mainly by the vertical transport of heat. As a consequence of this buoyancy driven ABL, the vertical dispersion process of scalars is dominant when compared to the horizontal ones. Therefore, the present analysis will focus on the vertical time dependent eddy diffusivity only. This vertical eddy diffusivity can be derived from equation (27) by assuming that $c_w = 0.36$ and

$$(f_m^*)_w = 0.55 \frac{\frac{z}{h}}{1 - \exp(-\frac{4z}{h}) - 0.0003 \exp(\frac{8z}{h})}$$

for the vertical component (Degrazia et al. (2000)). Furthermore, considering the horizontal plane, we can idealize the turbulent structure as a homogeneous one with the length scale of energy-containing eddies being proportional to the convective boundary layer height h.

	$\bar{u}$ (115 m)	$\bar{u}$ (10 m)	u_*	L	w_*	h
Run	(ms^{-1})	(ms^{-1})	(ms^{-1})	(m)	(ms^{-1})	(m)
1	3.4	2.1	0.36	-37	1.8	1980
2	10.6	4.9	0.73	-292	1.8	1920
3	5.0	2.4	0.38	-71	1.3	1120
4	4.6	2.5	0.38	-133	0.7	390
5	6.7	3.1	0.45	-444	0.7	820
6	13.2	7.2	1.05	-432	2.0	1300
7	7.6	4.1	0.64	-104	2.2	1850
8	9.4	4.2	0.69	-56	2.2	810
9	10.5	5.1	0.75	-289	1.9	2090

Table 1. Meteorological conditions of the Copenhagen experiment Gryning & Lyck (1984).

Moreover, for the lateral eddy diffusivity we used the asymptotic behaviour of equation (27) for large diffusion travel times with $c_v = 0.36$ and $(f_m^*)_v = 0.66\frac{z}{h}$ (Degrazia et al. (1997)). The dissipation function ψ according to Højstrup (1982) and Degrazia et al. (1998) has the form

$$\psi^{\frac{1}{3}} = [(1 - \frac{z}{h})^2(-\frac{z}{L})^{-\frac{2}{3}} + 0.75]^{\frac{1}{2}}$$

where L is the Obukhov length in the surface layer.

4.2 Numerical results

The measurements of the contaminant dispersion in the atmospheric boundary layer consist typically from a sequence of samples over a time period. The experiment used to validate the previously introduced solution was carried out in the northern part of Copenhagen and is described in detail by Gryning & Lyck (1984). Several runs of the experiment with changing meteorological conditions were considered as reference in order to simulate time dependent contaminant dispersion in the boundary layer and to evaluate the performance of the discussed solutions against the experimental centreline concentrations.

The essential data of the experiment are reported in the following. This experiment consisted of a tracer released without buoyancy from a tower at a height of 115m, and a collection of tracer sampling units were located at the ground-level positions up to the maximum of three crosswind arcs. The sampling unit distances varied between two to six kilometers from the point of release. The site was mainly residential with a roughness length of the 0.6m. Table 1 summarizes the meteorological conditions of the Copenhagen experiment where L is the Obukhov length, h is the height of the convective boundary layer, w_* is the convective velocity scale and u_* is the friction velocity.

The wind speed profile used in the simulations is described by a power law expressed following the findings of reference Panofsky & Dutton (1988).

$$\frac{\bar{u}_z}{\bar{u}_1} = \left(\frac{z}{z_1}\right)^n \tag{28}$$

Here $\bar{u}_z$ and $\bar{u}_1$ are the horizontal mean wind speed at heights z and z_1 and n is an exponent that is related to the intensity of turbulence (Irwin (1989)). As is possible to see in Irwin (1989)

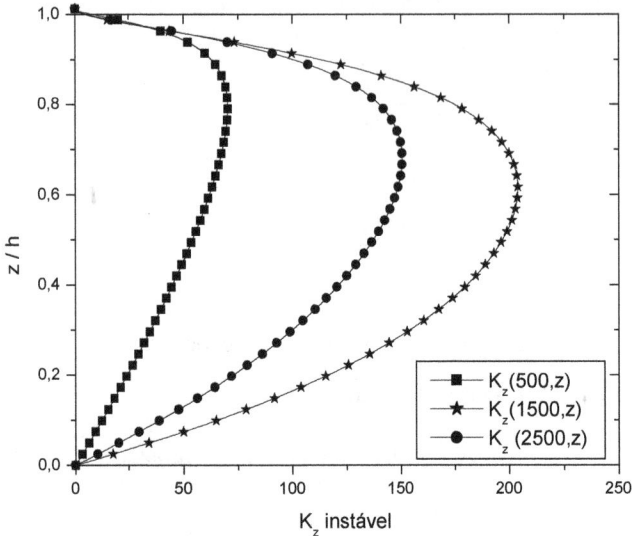

Fig. 1. The vertical eddy diffusivity $K_z(z,t)$ for three different travel times ($t = 500s, 1500s, 2500s$) using run 8 of the Copenhagen experiment.

$n = 0.1$ is valid for a power low wind profile in unstable condition. Moreover, U.S.EPA suggests for rural terrain (as default values used in regulatory models) to use $n = 0.15$ for neutral condition (class D) and $n = 0.1$ for stability class C (moderately unstable condition). In Figure 1 we present a plot of $K_z(z,t)$ for three different travel times ($t = 500s, 1500s, 2500s$) using run 8 of the Copenhagen experiment.

In order to exclude differences due to numerical uncertainties we define the numerical accuracy 10^{-4} of our simulations determining the suitable number of terms of the solution series. As an eye-guide we report in table 2 on the numerical convergence of the results, considering successively one, two, three and four terms in the solution series. One observes that the desired accuracy, for the solved problem solved is attained including only four terms in the truncated series, which is valid for all distances considered. Once the number of terms in the series solution is determined numerical comparisons of the 3D-GILTT results against experimental data may be performed and are presented in table 3.

Figure 2 shows the scatter plot of the centreline ground-level observed concentrations versus the simulated the 3D-GILTT model predictions, normalized by the emission rate and using two points in the time Gaussian Quadrature inversion (Moreira et al. (2006a), Stroud & Secrest (1966)). In the scatter diagram analysis, the closer the data are from the bisector line, the better are the results. The lateral lines indicate a factor of two (FA2), meaning if all the obtained data are between these lines FA2 equals to 1 (the maximum value for stochastically distributed data). From the scatter diagram (figure 2 one observes the fairly good simulation of dispersion data by the 3D-GILTT model, even for the two-point Gaussian quadrature scheme considered here.

Run	Recursion depth	$\bar{c}(x,y,z,t)$ $(10^{-7}\frac{g}{m^3})$	Run	Recursion	$\bar{c}(x,y,z,t)$
	0	8.68 4.49		0	3.65 2.10 1.31
	1	10.45 3.66		1	7.28 2.23 1.47
1	2	10.21 3.65	6	2	7.78 2.20 1.46
	3	10.21 3.65		3	7.78 2.20 1.46
	4	10.21 3.65		4	7.78 2.20 1.46
	5	10.21 3.65		5	7.78 2.20 1.46
	0	8.62 2.01		0	5.77 3.07 2.27
	1	6.60 2.30		1	6.73 3.33 1.65
2	2	7.09 2.28	7	2	6.61 3.32 1.65
	3	7.09 2.28		3	6.60 3.32 1.65
	4	7.09 2.28		4	6.60 3.32 1.65
	5	7.09 2.28		5	6.60 3.32 1.65
	0	5.58 6.38 3.72		0	6.38 3.94 2.76
	1	10.76 7.97 4.51		1	6.40 4.91 1.91
3	2	9.84 7.88 4.50	8	2	5.99 4.87 1.91
	3	9.79 7.88 4.50		3	5.97 4.87 1.91
	4	9.79 7.88 4.50		4	5.97 4.87 1.91
	5	9.80 7.88 4.50		5	5.97 4.87 1.91
	0	10.73		0	5.01 2.83 2.08
	1	15.32		1	5.73 3.05 2.20
4	2	15.23	9	2	5.65 3.04 2.19
	3	15.24		3	5.64 3.04 2.19
	4	15.24		4	5.64 3.04 2.19
	5	15.24		5	5.64 3.04 2.19
	0	7.14 4.11 2.47			
	1	9.11 4.77 3.78			
5	2	6.96 4.46 3.73			
	3	6.65 4.47 3.73			
	4	6.65 4.48 3.73			
	5	6.68 4.48 3.73			

Table 2. Pollutant concentrations for nine runs at various positions of the Copenhagen experiment and model prediction by the 3D-GILTT approach with time dependent eddy diffusivity.

To perform statistical comparisons between GILTT results against Copenhagen experimental data we consider the set of statistical indices described by Hanna (1989) and defined in by

- NMSE (normalized mean square error) = $\frac{\overline{(C_o-C_p)^2}}{\overline{C_o}\,\overline{C_p}}$,

- COR (correlation coefficient) = $\frac{\overline{(C_o-\overline{C_o})(C_p-\overline{C_p})}}{\sigma_o\sigma_p}$,

- FA2 = fraction of data (%, normalized to 1) for $0,5 \le \frac{C_p}{C_o} \le 2$,

- FB (fractional bias) = $\frac{\overline{C_o}-\overline{C_p}}{0,5(\overline{C_o}+\overline{C_p})}$,

- FS (fractional standard deviations) = $\frac{\sigma_0-\sigma_p}{0,5(\sigma_0+\sigma_p)}$,

Run	Distance (m)	Observed (Co) $(10^{-7}sm^{-3})$	Predictions (Cp) $(10^{-7}sm^{-3})$
1	1900	10.5	10.21
	3700	2.14	3.65
2	2100	9.85	7.09
	4200	2.83	2.28
3	1900	16.33	9.80
	3700	7.95	7.88
	5400	3.76	4.50
4	4000	15.71	15.24
5	2100	12.11	6.68
	4200	7.24	4.48
	6100	4.75	3.73
6	2000	7.44	7.78
	4200	3.47	2.20
	5900	1.74	1.46
7	2000	9.48	6.60
	4100	2.62	3.32
	5300	1.15	1.65
8	1900	9.76	5.97
	3600	2.64	4.87
	5300	0.98	1.91
9	2100	8.52	5.64
	4200	2.66	3.04
	6000	1.98	2.19

Table 3. Numerical convergence of the 3D-GILTT model with time dependent eddy diffusivity for the 9 runs of the Copenhagen experiment.

Recursion depth	NMSE	COR	FA2	FB	FS
0	0.38	0.83	0.83	0.32	0.59
1	0.16	0.90	1.00	0.11	-0.13
2	0.14	0.91	1.00	0.15	-0.07
3	0.14	0.91	1.00	0.15	-0.07
4	0.14	0.91	1.00	0.15	-0.07

Table 4. Statistical comparison between 3D-GILTT model results and the Copenhagen data set, changing the number of terms in equation (18).

where the subscripts o and p refer to observed and predicted quantities, respectively, and the bar indicates an averaged value. The best results are expected to have values near zero for the indices NMSE, FB and FS, and near 1 in the indices COR and FA2. Table 4 shows the findings of the statistical indices that show a fairly good agreement between the 3D-GILTT predictions and the experimental data. Moreover, the splitting proposed for the eddy diffusivity coefficient as a sum of the averaged eddy diffusivity coefficient plus time variation, appears to be a valid assumption, since we got compact convergence of the solution, in the sense that we attained results with accuracy of 10^{-4} with only a few terms in the solution series for all the distances considered.

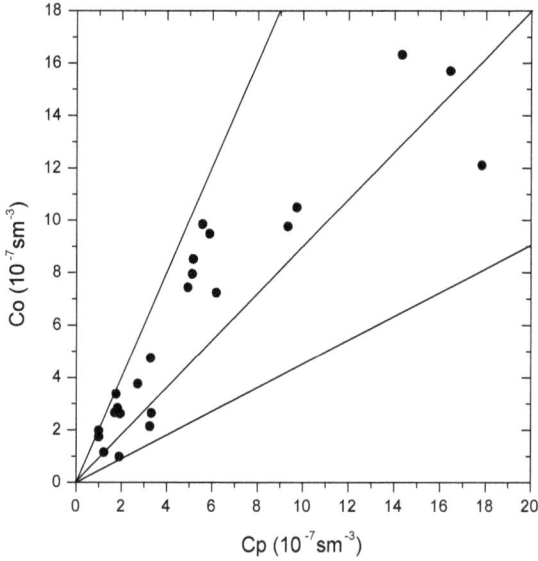

Fig. 2. Observed (Co) and predicted (Cp) scatter plot of centreline concentration using the Copenhagen dataset. Data between dotted lines correspond to ratio $Cp/Co \in [0.5, 2]$.

5. Conclusion

In the present contribution we focused on an analytical description of pollution related phenomena in a micro-scale that allows to simulate dispersion in an computationally efficient procedure. The reason why adopting an analytical procedure instead of using the nowadays available computing power resides in the fact that once an analytical solution to a mathematical model is found one can claim that the problem has been solved. We provided a closed form solution that may be tailored for numerical applications such as to reproduce the solution within a prescribed precision. As a consequence the error analysis reduces to model validation only, in comparison to numerical approaches where in general it is not straight forward to disentangle model errors from numerical ones.

Our starting point, i.e. the mathematical model, is the advection-diffusion equation, which we solved in $3 \oplus 1$ space-time dimensions for a general eddy diffusivity. The closed form solution is obtained using a hybrid approach by spectral theory together with integral transforms (in the present case the Laplace transform) and the decomposition method. The general formalism was simplified in order to attend the meteorological situation of the Copenhagen experiment. By comparison the present approach was found to yield an acceptable solution for the time dependent three dimensional advection-diffusion equation and moreover predicted tracer concentrations closer to observed values compared to other approaches from the literature. Although K-closure is known to have its limitations comparison of measurements and theoretical predictions corresponded on a satisfactory level and thus supported the usage of such an approach for micro-scale dispersion phenomena. Note, that the dispersion of the experimental data in comparison to the predictions might suggest a considerable discrepancy between theory and experiment, but it is worth mentioning that the measurements are a unique sample of a distribution around an average value, whereas the prediction of an average value is evaluated from a deterministic equation, where the stochastic character is hidden in the turbulence closure hypothesis, so that a spread of data along the bisector is to be expected.

A data dispersion due to numerical uncertainties may be excluded using convergence criteria to control the numerical precision. The quality of the solution is controlled by a genuine mathematical convergence criterion. Note, that for the t and x coordinate the Laplace inversion considers only bi-Lipschitz functions, which defines then a unique relation between the original function and its Laplace-transform. This makes the transform procedure manifest exact and the only numerical error comes from truncation, which is determined from the Sturm-Liouville problem. In order to determine the truncation index of the solution series we introduced a carbon-copy of the Cardinal theorem of interpolation theory.

Recalling, that the structure of the pollutant concentration is essentially determined by the mean wind velocity $\bar{\mathbf{U}}$ and the eddy diffusivity $\mathbf{K}$, means that the quotient of norms $\omega = \frac{\|\mathbf{K}\|}{\|\bar{\mathbf{U}}\|}$ defines a length scale for which the pollutant concentration is almost homogeneous. Thus one may conclude that with decreasing length ($\frac{\omega}{m}$ and m an increasing integer number) variations in the solution become spurious. Upon interpreting ω^{-1} as a sampling density, one may now employ the Cardinal Theorem of Interpolation Theory (Torres (1991)) in order to find the truncation that leaves the analytical solution almost exact, i.e. introduces only functions that vary significantly in length scales beyond the mentioned limit.

The square integrable function $\chi = \int_r \bar{c} \, dt \, dx \, d\eta \in L^2$ ($\eta = y$ or z) with spectrum $\{\lambda_i\}$ which is bounded by $m\omega^{-1}$ has an exact solution for a finite expansion. This statement expresses the *Cardinal Theorem of Interpolation Theory* for our problem. Since the cut-off defines some sort of sampling density, its introduction is an approximation and is related to convergence of the approach and Parseval's theorem may be used to estimate the error. In order to keep the solution error within a prescribed error, the expansion in the region of interest has to contain $n + 1$ terms, with $n = \text{int} \left\{ \frac{mL_{y,z}}{2\pi\omega} + \frac{1}{2} \right\}$. For the bounded spectrum and according to the theorem the solution is then exact. In our approximation, if m is properly chosen such that the cut-off part of the spectrum is negligible, then the found solution is almost exact.

Further, the Cauchy-Kowalewski theorem (Courant & Hilbert (1989)) guarantees that the proposed solution is a valid solution of the discussed problem, since this problem is a special case of the afore mentioned theorem, so that existence and uniqueness are guaranteed. It remains to justify convergence of the decomposition method. In general convergence by the decomposition method is not guaranteed, so that the solution shall be tested by an appropriate criterion. Since standard convergence criteria do not apply in a straight forward manner for the present case, we resort to a method which is based on the reasoning of Lyapunov (Boichenko et al. (2005)). While Lyapunov introduced this conception in order to test the influence of variations of the initial condition on the solution, we use a similar procedure to test the stability of convergence while starting from an approximate (initial) solution $\mathbf{R}_0$ (the seed of the recursive scheme). Let $|\delta Z_n| = \left\| \sum_{i=n+1}^{\infty} \mathbf{R}_i \right\|$ be the maximum deviation of the correct from the approximate solution $\Gamma_n = \sum_{i=0}^{n} \mathbf{R}_i$, where $\| \cdot \|$ signifies the maximum norm. Then strong convergence occurs if there exists an n_0 such that the sign of λ is negative for all $n \geq n_0$. Here, $\lambda = \frac{1}{\|\Gamma_n\|} \log \left(\frac{|\delta Z_n|}{|\delta Z_0|} \right)$.

Concluding, analytical solutions of equations are of fundamental importance in understanding and describing physical phenomena, since they might take into account all the parameters of a problem, and investigate their influence. Moreover, when using models, while they are rather sophisticated instruments that ultimately reflect the current state of knowledge on turbulent transport in the atmosphere, the results they provide are subject to a considerable margin of error. This is due to various factors, including in particular the uncertainty of the intrinsic variability of the atmosphere. Models, in fact, provide values expressed as an average, i.e., a mean value obtained by the repeated performance of many experiments, while the measured concentrations are a single value of the sample to which the ensemble average provided by models refer. This is a general characteristic of the theory of atmospheric turbulence and is a consequence of the statistical approach used in attempting to parametrise the chaotic character of the measured data. An analytical solution can be useful in evaluating the performances of numerical models (that solve numerically the advection-diffusion equation) that could compare their results, not only against experimental data but, in an easier way, with the solution itself in order to check numerical errors without the uncertainties presented above. Finally, the program of providing analytical solutions for realistic physical problems, leads us to future problems with different closure hypothesis considering full space-time dependence in the resulting dynamical equation, which we will also approach by the proposed methodology.

6. Acknowledgements

The authors thank to CNPq (Conselho Nacional de Desenvolvimento Científico e Tecnológico) for the partial financial support of this work.

7. References

Abate, J. & Valkó, P.P. (2004). Multi-precision Laplace transform inversion. *Int. J. for Num. Methods in Engineering*, Vol. 60, page numbers (979-993).

Adomian, G. (1984). A New Approach to Nonlinear Partial Differential Equations. *J. Math. Anal. Appl.*, Vol. 102, page numbers (420-434).

Adomian, G. (1988). A Review of the Decomposition Method in Applied Mathematics. *J. Math. Anal. Appl.*, Vol. 135, page numbers (501-544) .

Adomian, G. (1994). *Solving Frontier Problems of Physics: The Decomposition Method*, Kluwer, Boston, MA. .

Blackadar, A.K. (1997). *Turbulence and diffusion in the atmosphere: lectures in Environmental Sciences*, Springer-Verlag.

Bodmann, B.; Vilhena, M.T.; Ferreira, L.S. & Bardaji, J.B. (2010). An analytical solver for the multi-group two dimensional neutron-diffusion equation by integral transform techniques. *Il Nuovo Cimento C*, Vol. 33, page numbers (199-206).

Boichenko, V.A.; Leonov, G.A. & Reitmann, V. (2005). *NDimension theory for ordinary equations*, Teubner, Stuttgart.

Buske, D.; Vilhena, M.T.; Moreira, D.M. & Tirabassi, T. (2007a). An analytical solution of the advection-diffusion equation considering non-local turbulence closure. *Environ. Fluid Mechanics*, Vol. 7, page numbers (43-54).

Buske, D.; Vilhena, M.T.; Moreira, D.M. & Tirabassi, T. (2007b). Simulation of pollutant dispersion for low wind conditions in stable and convective planetary boundary layer. *Atmos. Environ.*, Vol. 41, page numbers (5496-5501).

Buske, D.; Vilhena, M.T.; Moreira, D.M. & Tirabassi, T. (2010). An Analytical Solution for the Transient Two-Dimensional Advection-Diffusion Equation with Non-Fickian Closure in Cartesian Geometry by Integral Transform Technique. In: *Integral Methods in Science and Engineering: Computational methods*, C. Constanda & M.E. Pérez,page numbers (33-40), Birkhauser, Boston.

Champagne, F.H.; Friche, C.A.; Larve, J.C. & Wyngaard, J.C. (1977). Flux measurements flux estimation techniques, and fine scale turbulence measurements in the unstable surface layer over land. *J. Atmos. Sci.*, Vol. 34, page numbers (515-520).

Costa, C.P.; Vilhena, M.T.; Moreira, D.M. & Tirabassi, T. (2006). Semi-analytical solution of the steady three-dimensional advection-diffusion equation in the planetary boundary layer. *Atmos. Environ.*, Vol. 40, No. 29, page numbers (5659-5669).

Costa, C.P.; Tirabassi, T.; Vilhena, M.T. & Moreira, D.M. (2011). A general formulation for pollutant dispersion in the atmosphere. *J. Eng. Math.*, Published online. Doi 10.1007/s10665-011-9495-z.

Courant, R. & Hilbert, D. (1989). *Methods of Mathematical Physics*. John Wiley & Sons, New York .

Degrazia, G.A.; Campos Velho, H.F. & Carvalho, J.C. (1997). Nonlocal exchange coefficients for the convective boundary layer derived from spectral properties. *Contr. Atmos. Phys.*, Vol. 70, page numbers (57-64).

Degrazia,G.A.; Mangia, C. & Rizza, U. (1998). A comparison between different methods to estimate the lateral dispersion parameter under convective conditions. *J. Appl. Meteor.*, Vol. 37, page numbers (227-231).

Degrazia,G.A.; Anfossi, D.; Carvalho, J.C.; Mangia, C.; Tirabassi, T. & Campos Velho, H.F. (2000). Turbulence parameterization for PBL dispersion models in all stability conditions. *Atmos. Environ.*, Vol. 33, page numbers (2007-2021).

Degrazia, G.A.; Moreira, D.M.; Campos, C.R.J.; Carvalho, J.C. & Vilhena, M.T. (2002). Comparison between an integral and algebraic formulation for the eddy diffusivity using the Copenhagen experimental dataset. *Il Nuovo Cimento*, Vol. 25C, page numbers (207-218).

Degrazia, G.A. (2005). Lagrangian Particle Models, In: *Air Quality Modeling: Theories, Methodologies, Computational Techniques and Avaiable Databases and Software, vol II - Advanced Topics*, D. Anfossi & W. Physick, page numbers (93-162), EnviroComp Institute, Fremont, California, USA.

Demuth, C. (1978). A contribution to the analytical steady solution of the diffusion equation for line sources. *Atmos. Environ.*, Vol. 12, page numbers (1255-1258).

Gryning, S.E. & Lyck, E. (1984). Atmospheric dispersion from elevated source in an urban area: comparison between tracer experiments and model calculations. *J. Appl. Meteor.*, Vol. 23, page numbers (651-654).

Hanna, S.R. (1989). Confidence limit for air quality models as estimated by bootstrap and jacknife resampling methods. *Atmos. Environ.*, Vol. 23, page numbers (1385-1395).

Højstrup, J.H. (1982). Velocity spectra in the unstable boundary layer. *J. Atmos. Sci.*, Vol. 39, page numbers (2239-2248).

Irwin, J.S. (1979). A theoretical variation of the wind profile power-low exponent as a function of surface roughness and stability. *Atmos. Environ.*, Vol. 13, page numbers (191-194).

Lin, J.S. & Hildemann, L.M. (1997). A generalised mathematical scheme to analytically solve the atmospheric diffusion equation with dry deposition. *Atmos. Environ.*, Vol. 31, page numbers (59-71).

Moreira, D.M.; Vilhena, M.T.; Tirabassi, T.; Buske, D. & Cotta, R.M. (2005). Near source atmospheric pollutant dispersion using the new GILTT method. *Atmos. Environ.*, Vol. 39, No.34, page numbers (6290-6295).

Moreira, D.M.; Vilhena, M.T.; Tirabassi, T.; Costa, C. & Bodmann, B. (2006a). Simulation of pollutant dispersion in atmosphere by the Laplace transform: the ADMM approach. *Water, Air and Soil Pollution*, Vol. 177, page numbers (411-439).

Moreira, D.M.; Vilhena, M.T.; Buske, D. & Tirabassi, T. (2006b). The GILTT solution of the advection-diffusion equation for an inhomogeneous and nonstationary PBL. *Atmos. Environ.*, Vol. 40, page numbers (3186-3194).

Moreira, D.M.; Vilhena, M.T.; Buske, D. & Tirabassi, T. (2009). The state-of-art of the GILTT method to simulate pollutant dispersion in the atmosphere. *Atmos. Research*, Vol. 92, page numbers (1-17).

Moreira, D.M.; Vilhena, M.T. & Buske, D. (2009). On the GILTT Formulation for Pollutant Dispersion Simulation in the Atmospheric Boundary Layer. *Air Pollution and Turbulence: Modeling and Applications*, D.M. Moreira & M.T. Vilhena, page numbers (179-202), CRC Press, Boca Raton - Flórida (USA).

Moreira, D.M.; Vilhena, M.T.; Tirabassi, T.; Buske, D.; Costa, C.P. (2010). Comparison between analytical models to simulate pollutant dispersion in the atmosphere. *Int. J. Env. and Waste Management*, Vol. 6, page numbers (327-344).

Nieuwstadt F.T.M. & de Haan B.J. (1981). An analytical solution of one-dimensional diffusion equation in a nonstationary boundary layer with an application to inversion rise fumigation. *Atmos. Environ.*, Vol. 15, page numbers (845-851).

Panofsky, A.H. & Dutton, J.A. (1988). *Atmospheric Turbulence*. John Wiley & Sons, New York .

Scriven R.A. & Fisher B.A. (1975). The long range transport of airborne material and its removal by deposition and washout-II. The effect of turbulent diffusion.. *Atmos. Environ.*, Vol. 9, page numbers (59-69).

Rounds W. (1955). Solutions of the two-dimensional diffusion equation. *Trans. Am. Geophys. Union*, Vol. 36, page numbers (395-405).

Seinfeld J.H. & Pandis S.N. (1998). *Atmospheric chemistry and physics*. John Wiley & Sons, New York, 1326 pp. .

Sharan, M.; Singh, M.P. & Yadav, A.K. (1996). A mathematical model for the atmospheric dispersion in low winds with eddy diffusivities as linear functions of downwind distance. *Atmos. Environ.*, Vol. 30, No.7, page numbers (1137-1145).

Smith F.B. (1957). The diffusion of smoke from a continuous elevated poinr source into a turbulent atmosphere. *J. Fluid Mech.*, Vol. 2, page numbers (49-76).

Stroud, A.H. & Secrest, D. (1966). *Gaussian quadrature formulas*. Prentice Hall Inc., Englewood Cliffs, N.J..

Tagliazucca, M.; Nanni, T. & Tirabassi, T. (1985). An analytical dispersion model for sources in the surface layer. *Nuovo Cimento*, Vol. 8C, page numbers (771-781).

Tirabassi, T. (1989). Analytical air pollution and diffusion models. *Water, Air and Soil Pollution*, Vol. 47, page numbers (19-24).

Tirabassi T. & Rizza U. (1994). Applied dispersion modelling for ground-level concentrations from elevated sources. *Atmos. Environ.*, Vol. 28, page numbers (611-615).

Tirabassi T. (2003). Operational advanced air pollution modeling. *PAGEOPH*, Vol. 160, No. 1-2, page numbers (05-16).

Tirabassi, T.; Buske, D.; Moreira, D.M. & Vilhena, M.T. (2008). A two-dimensional solution of the advection-diffusion equation with dry deposition to the ground. *J. Appl. Meteor. and Climatology*, Vol. 47, page numbers (2096-2104).

Tirabassi, T.; Tiesi, A.; Buske, D.; Moreira, D.M. & Vilhena, M.T. (2009). Some characteristics of a plume from a point source based on analytical solution of the two-dimensional advection-diffusion equation. *Atmos. Environ.*, Vol. 43, page numbers (2221-2227).

Tirabassi, T.; Tiesi, A.; Vilhena, M.T.; Bodmann, B.E.J. & Buske, D. (2011). An analytical simple formula for the ground level concentration from a point source. *Atmosphere*, Vol. 2, page numbers (21-35).

Torres, R.H. (1991). Spaces of sequences, sampling theorem, and functions of exponential type. *Studia Mathematica*, Vol. 100, No. 1, page numbers (51-74).

Valkó, P.P. & Abate, J. (2004). Comparison of sequence accelerators for the Gaver method of numerical Laplace transform inversion. *Computers and Mathematics with Application*, Vol. 48, page numbers (629-636).

van Ulden, A.P. (1978). Simple estimates for vertical diffusion from sources near the ground. *Atmos. Environ.*, Vol. 12, page numbers (2125-2129).

Wortmann, S.; Vilhena, M.T.; Moreira, D.M. & Buske, D. (2005). A new analytical approach to simulate the pollutant dispersion in the PBL. *Atmos. Environ.*, Vol. 39, page numbers (2171-2178).

6

Critical Episodes of PM10 Particulate Matter Pollution in Santiago of Chile, an Approximation Using Two Prediction Methods: MARS Models and Gamma Models

Sergio A. Alvarado O.[1,2,*], Claudio Z. Silva[1] and Dante L. Cáceres[1,2]

[1]Division of Epidemiology, School of Public Health,
Faculty of Medicine, University of Chile,
[2]Grups de Recerca d'Amèrica i Àfrica Llatines (GRAAL), Unitat de Bioestadística,
Facultat de Medicina, Universitat Autònoma de Barcelona, Barcelona,
[1]Chile
[2]Spain

1. Introduction

Several studies worldwide have reported the effects of air pollution on human health, especially those related with exposure to particulate matter [1,8,17,20,21,25], research it is currently focussed on studying the acute and short-term effects, especially the mortality and morbidity impact due to cardiovascular and respiratory reasons [1,8,17,21,25].

This has meant that countries take a series of environmental management measures to control particulate material emissions and also to generate early prediction of high air pollution episodes. Some measures are: a systematic shift to cleaner fuels, restriction of daily circulation of a certain percentage of motor vehicles, temporary shutdown of some industries, and so on. Air pollution causes are diverse, being anthropogenic activities the major contributor to the problem. But the air pollution level is also influenced by other factors such as climate and topography. Climate has a decisive influence on the persistence of air pollutants, and the winds, temperature and solar radiation drastically alter the dispersion and the type of contaminants present at one time. Topography influences the movement of air masses and hence the persistence of pollution levels in a given geographical area. The combination of these ultimately determines the quality of air [20].

Prediction of critical episodes of air pollution in large cities has become an environmental management tool aimed to protect the health of the population, allowing health authorities to know with some certainty the likely levels of air pollution border within a certain time interval. This prediction has been addressed through different models combining deterministic and probabilistic approaches using various types of information [6,9,13,14,22,28]. The official methodology in use by the administrative authority of the Metropolitan Region of Santiago de Chile to forecast PM10 concentrations is the Cassmassi Model, proposed in 1999

---

*Corresponding Author

by Joseph Cassmassi [2]. It uses multiple linear regression to predict the maximum concentration of 24-hour average PM10 for the next day (00 to 24 hours). This model includes observed meteorological variables, observed and forecasted weather conditions rates, pollutants expected concentrations rates of expected variations in emissions and others predictors.

Different statistical methods have been used in Chile to model airborne PM10 concentrations of air pollutants including time series [20], neural networks [13,22] and regression models based on multivariate adaptive smoothing functions (splines) MARS [23] .The predictive efficiency of these models is variable and is closely associated with the behavior and evolution of environmental characteristics. Models that use extreme value theory are widely used for this purpose, especially for episodes that occur over short periods of time and present extreme values or exceedances of emergency limits established by the authority [3].

The aim of this study is to compare the predictive efficiency of multivariate predictive models Gamma vs MARS to predict "tomorrow" maximum concentration of PM10 in Santiago de Chile in the period between April 1 and August 31 of the years 2001, 2002, 2003 and 2004.

2. Methods

2.1 Information sources

We used the databases of PM10 collected at the Pudahuel monitoring station, that it is element of the MACAM2-RM monitoring network, for the years 2001, 2002, 2003 and 2004. For each year measurements were selected from April 1 to August 31, which correspond to the time of year with less ventilation in the Santiago basin. We worked with the moving average of 24 hours. For missing data, imputed values were generated by a double exponential smoothing with a smoothing coefficient $\alpha = 0.7$ (see Annex 1). We selected this monitoring station because most of the year presents the highest levels of PM10 concentration. It is also the most influential in taking administrative decisions about forecasting critical episodes for the next day. Moreover, because of environmental management measures implemented when declaring a critical episode of PM10 pollution, the behavior of the time series, is affected generating lower levels of concentrations that do not reflect actual observed concentration that would have occurred without environmental management measures , so we penalize this effect by introducing a correction constant given by $\Delta C_I = mean[CPM_{24_{I-1}} - CPM_{24_I}]$ where $CPM_{24_{I-1}}$ and CPM_{24_I} are the average of PM10 concentration on the day before and the day of intervention, respectively, for each month of the study period; the number of episodes in excess of 240 ($\mu g/m^3$) for 2001, 2002, 2003 and 2004 corresponds to 4 (2.6%), 11 (7.2%) 5 (3.3%) and 2 (1.3%) respectively. For the construction of the Gamma models use the statistical program Stata 11.0 and MARS models use a demo of the program obtained from the web page of Salford Systems.

2.2 Procedures

152 multi-dimensional observations were used; they consisted of a response variable PM10 and 13 predictors. The modeling includes delays of 1 and 2 days for the variables of interest, corresponding (N+1) to tomorrow, i.e. the day to be modeled. The predictor variables for today (delay 1) were defined as **pm0** the hourly average PM10 concentration at 0:00 hrs of day N, **pm6** the average hourly PM10 concentration at 6:00 hrs on day N, **pm12** the average hourly PM10 concentration at 12:00 hrs on day N, **pm18** the hourly average PM10 concentration at 18:00 hrs on day N.

Some predictor variables incorporate delays of 2 days; they are: **pm10h** maximum PM10 concentration of 24-hour moving average between19 hrs of day N-1 and 18 hrs day N, **mth** maximum temperature between 19 hrs of day N-1 and 18 hrs o day N, **mhrh** minimum relative humidity between 19 hrs of day N-1 and 18 hrs of day N, **dth** difference between maximum and minimum temperature between 19 hrs of day N-1 and 18 hrs of day N, **vvh** , average wind velocity 19 hrs of day N-1 and 18 hrs of day N. The predictors of tomorrow (N+1) correspond to: **mtm** maximum temperature of day N+1, **mhrm** minimum relative humidity of day N+1, **dtm** difference between the maximum and minimum temperature of day N+1 and **vvm** average speed wind of day. The response in this study (**pm10m)** is the maximum concentration of the 24 hs moving average of PM10 of a day N+1. The values of the variables of tomorrow are forecasts validated and reported by the Chilean Meteorological Office using models Mesoscale Modeling System (MM5), which is a numerical model that uses the equations of physics of the atmosphere for weather forecasting in limited areas at the regional level [15],

The authorities of the National Environment Commission (CONAMA), have defined four levels of PM10, in order to make management decisions when critical events occur: **good** 0 - 193 ($\mu g/m^3$); **alert** 194 - 239 ($\mu g/m^3$), **pre-emergency** 240 - 329 ($\mu g/m^3$) and **emergency** PM10 > 330 ($\mu g/m^3$) [3]. For our study we dichotomized the response into two classes **I:** pm10m < 240 ($\mu g/m^3$) and **II:** pm10m > 240 ($\mu g/m^3$), ie, "good or alert" versus "pre-emergency or emergency." The aim of the dichotomy is to generate 2 x 2 cross-tabulations between the observed and predicted values of the response for each proposed model.

2.3 Construction of Gamma models and MARS models

The fit of the models of a given year was validated with the information of the following year, thus ensuring the independence of the data used for validation with respect to those used in its construction. Therefore no predictions are given for the model year built with year 2004 information since there is no information for 2005. Each model was estimated with data from the period between April 1 and August 31 of a year and applied to the following year's data for the same period, evaluating the fit of these estimates compared with actual observations for that second year.

Gamma regression

Gamma models are used in situations where the variable has non negative values; were originally used for continuous data, but now the family of Gamma generalized linear models is used with count data [7]. In general, these models consider different ways of how to work the response variable, such as exponentiation of the response using the log-gamma transformation [12].

The probability density function for the generalized gamma function is given by

$$f(y;\kappa,\mu,\sigma) = \frac{\gamma^\gamma}{\sigma y \sqrt{\gamma}\Gamma(\gamma)} e^{(z\sqrt{\gamma}-u)} \; ; y \geq 0 \text{ , where } \gamma = |\kappa|^{-2} \text{ , } z = sign(\kappa) \times \left\{ \frac{\ln(y)-\mu}{\sigma} \right\}$$

and $u = \gamma \exp(|\kappa|z)$. The parameter μ is equal to $x^t\beta$ where x is the matrix of predictors including the intercept and β is the vector of coefficients. For generalized Gamma distribution the expected value conditional on x is given by:

$$E(y|x) = \exp[x\hat{\beta} + (\hat{\sigma}/\hat{\kappa})\ln(\hat{\kappa}^2) + \ln(\Gamma\{(1/\hat{\kappa}^2) + (\hat{\sigma}/\hat{\kappa})\} - \ln(\Gamma(1/\hat{\kappa}^2))]$$

where $\hat{\sigma} = (1/n)\sum_i \exp(\alpha_0 + \alpha_1 \ln(f(x_i)))$ and $\ln(\sigma)$ is parameterized as $\alpha_0 + \alpha_1 \ln(f(x))$.

Since it is required to report estimates and results in the original measurement scale, we work with the exponentiation of the model using the log-gamma transformation $\mu = E(Y) = \exp(x^t\beta)$ and ensure that the transformation does not affect the interpretations which refers directly to the original scale [12].

Multivariate Adaptive Regression Splines (MARS)

MARS is a methodology proposed by Jerome Friedman in August 1991, it tries to build a model of non-linear regression that is based on a product of functions called base smoothing functions (splines). These functions incorporated into its structure predictors entering the model as part of a function, not directly as in classical regression, produces a model for the response in study that may be continuous or binary that automatically selects the predictors present in the final equation, they are incorporated in the smoothing basal functions [5,10,11].

Model for a predictor

The methodology MARS, proposed by Friedman [5], selects K nodes of the predictor variable x, denoted by t_k, $k = 1,\ldots\ldots,K$, which could correspond to each of the observations of the variable; then $K + 1$ defined regions on the range of x, where it is associated to each node the linear smoothing function, generating a family of basis functions:

$$B_K^{(q)}(x) = \begin{cases} x^j & j = 0,\ldots\ldots,q \\ (x - t_k)_+^q & k = 1,\ldots\ldots,K \end{cases}$$

Where $(x - t_k)_+^q$ it is known as a truncation function . For the approximation of order q, we estimate the function $\hat{f}_q(x) = \sum_{K=0}^{K+q} a_k B_K^{(q)}(x)$, usually the order of smoothing to be taken must be less than or equal to three, so the function and its $q - 1$ first derivatives are continuous. This restriction and the use of polynomial functions in each subregion produce smooth and tight functions.

Generalization to p predictors

For the vector of predictors $x = (x_1, x_2, \ldots, x_p)$ the smoothing function is defined analogously to the univariate case. In this case the space R^p is divided into a set of disjoint regions and within each region a polynomial of p variables is fitted.

For $p > 2$ disjoint regions are considered to define the smoothing approximation as tensor products of disjoint intervals in each of the variables associated to the node location. So placing K_j nodes in each variable produces a product of $K_j + 1$ regions, $j = 1,\ldots,p$. A set of basis functions generating the space of smoothing functions for the entire set of regions, is the tensor product of the corresponding basal one-dimensional smoothing functions associated with the location of the nodes in each variable given by:

$$\hat{f}_q(x) = \sum_{k_1=0}^{K_1+q} \cdots \sum_{k_p=0}^{K_p+q} a_{k_1},\ldots\ldots, a_{k_p} \prod_{j=1}^{p} B_{k_j}^{(q)}(x_j)$$

The selection of basis functions looks for a good set of regions to define a smoothing approach adequated to the problem; MARS generates the basis functions by a stepwise process. It starts with a constant in the model and then begins the search for a variable-node combination that improves the model. The improvement is measured in part by the change in the sum of squared errors (MSE) , adding basis functions is done whenever it reduces reduce the MSE.

To evaluate this model, Friedman proposes using the *Generalized Cross Validation* statistic

$$GCV = \frac{\dfrac{\sum_{i=1}^{N}\left(y_i - \hat{f}_q(x_i)\right)^2}{N}}{\left(1 - \dfrac{C(M)}{N}\right)^2} \quad \text{where } C(M) = 1 + trace(B(B^tB)^{-1}B^t),$$

B is the design matrix, the numerator is the lack of fit on the training data set and the denominator is a penalty term that reflects the complexity of the model.

To compare the models we considered the following statistics: (a) Pearson's linear correlation between the observed and the predicted value and (b) mean absolute error ratio (mpab) between the observed (obs) and the predicted value (pred) given by

$$mpab = \sum_{i=1}^{n}\left[\frac{\left|PM10_{obs} - PM10_{pred}\right|}{PM10_{obs}}\right] / n$$

that is equivalent to evaluate the average errors committed by both models in the predictions. We also considered the concordance proportion in each class.

In order to compare the settings of the models and the regions selected for predicting the response variable two MARS models were constructed to each year, one based on 20 and another using 40 base functions, allowing us to choose that pattern which best fits the expected response based on the partitions of the predictor variables [11]. In the case of Gamma regression we used a logarithm link function, with three sets of predictors: the first corresponds to all variables, the second set involved yesterday and today variables and the third included the variables: pm0, pm6 , pm18, dtm, dth and vvm. This last set of variables would better describe the behavior of the pm10 concentrations, as has been described by other authors [18].

3. Results

Table 1 shows the descriptive statistics of the variables incorporated in the final modeling of PM10 concentrations. We can see that the maxima for the years 2001, 2002 and 2003 exceed the value of 240 ($\mu g/m^3$), such behavior is not seen for 2004.

	2001			2002		
	Mean (sd)	**Minimum**	**Maximun**	**Mean (sd)**	**Minimum**	**Maximun**
pm10m	122,83 (58,4)	18,50	307,80	123,90 (67,2)	10,40	298,00
pm0	123,85 (106)	1,00	492,00	133,00 (126,6)	1,00	625,00
pm6	78,06 (57,9)	1,00	250,00	80,10 (56,9)	1,00	222,00
pm18	117,50 (83,5)	2,00	467,00	116,70 (83,6)	4,00	412,00
dtm	10,37 (5,3)	2,50	23,00	11,00 (5,1)	0,00	24,00
vvm	1,54 (0,48)	0,53	3,40	1,40 (0,4)	0,00	2,49

	2003			2004		
	Mean (sd)	**Minimum**	**Maximun**	**Mean (sd)**	**Minimum**	**Maximun**
pm10m	127,50 (50,50)	28,38	260,30	101,00 (43,10)	20,60	230,00
pm0	133,60 (101,40)	1,00	450,00	105,50 (88,30)	1,00	519,00
pm6	86,50 (55,20)	1,00	263,00	68,60 (42,30)	1,00	202,00
pm18	116,10 (72,10)	6,00	385,00	91,40 (59,20)	1,00	345,00
dtm	11,90 (5,03)	2,11	22,90	11,20 (5,02)	1,53	24,00
vvm	1,42 (0,40)	0,56	2,72	missing	missing	missing

sd: standard deviation;
pm0, pm6, pm18: PM10 concentration at 00, 06 and 18 previous day, respectively;
dtm: temperature difference of tomorrow; vvm: wind velocity of tomorrow.

Table 1. Mean and standard deviation of predictor variables used to model PM10. Years 2001 to 2004.

Table No. 2 shows the successes of the class I and II by the Gamma models and the MARS models. The correlations are significant for the three models per year, the mpab remain high for all models, except the 40 fb MARS 2002 model where we can see a 19% mean absolute error similar to the Gamma Model. In general for all three years the Gamma models have better performance than MARS for the PM10 concentrations < 240 ($\mu g/m^3$); MARS models have better performance for PM10 exceeding 240 ($\mu g/m^3$).

	2001 to 2002			2002 to 2003			2003 to 2004		
	Gamma	MARS 20 bf	MARS 40 bf	Gamma	MARS 20 bf	MARS 40 bf	Gamma	MARS 20bf	MARS 40bf
mpab %	38,00	44,00	32,00	19,00	29,00	19,00	28,00	14,00	25,00
$r_{pearson}$	0,83	0,86	0,74	0,82	0,81	0,86	0,78	0,92	0,82
Class I %	98,00	67,00	44,40	99,00	0,00	100,00	99,00	99,00	99,00
Class II %	50,00	97,00	95,40	12,50	98,00	97,00	0,00	100,00	0,00

mpab: mean abolute proportion
bf: basis function.

Table 2. Results of MARS and Gamma modeling.

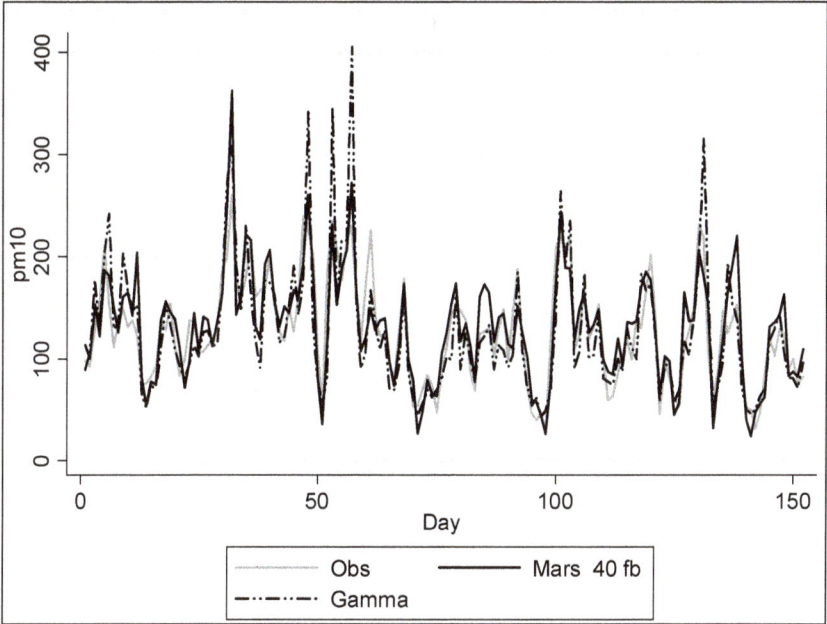

Fig. 1. Shows that the Gamma model predictions are further away from observed PM10 than
MARS predictions for high PM10 concentrations for year 2003 using model of 2002, but
Gamma model gives better predictions for values below 200 ($\mu g/m^3$).

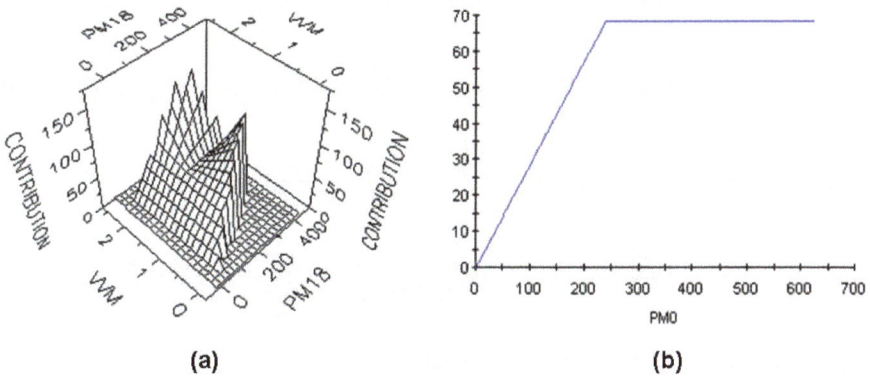

(a)   (b)

Fig. 2. Corresponds to the MARS model with 40 basis functions for 2002 and illustrates in
part (a) the interaction of the basis functions max (0; pm18-137) and max (0; vvm-1, 522) that
generate the basis function BF8 which represents an interaction surface (Table No. 3) with
maximum value 165 ($\mu g/m^3$). This implies that for PM10 concentrations at 18 pm (pm18)
above 137 ($\mu g/m^3$) and wind speed for tomorrow of 1,522 m/s, the contribution to the PM10
tomorrow concentration due to the interaction has a maximun of 165 ($\mu g/m^3$). On the other
hand, figure (b) shows that for concentrations above 240 ($\mu g/m^3$), the variable pm0 has a
maximum contribution to the response of 70 ($\mu g/m^3$), which represents the value with
which the base function BF4 contributes to the predicted value.

Table No. 3 shows the explicit model for the MARS models with 20 and 40 basis functions and the Gamma models for the years 2001, 2002 and 2003. The complexity of the model is seen in the number of basis functions incorporated into the explicit model, the model for 2002 has nine basis functions of which five are interactions of univariate basis functions, the BF9 and BF8 functions correspond to interactions of the mirror variable for vvm and the corresponding basis function associated with pm18.

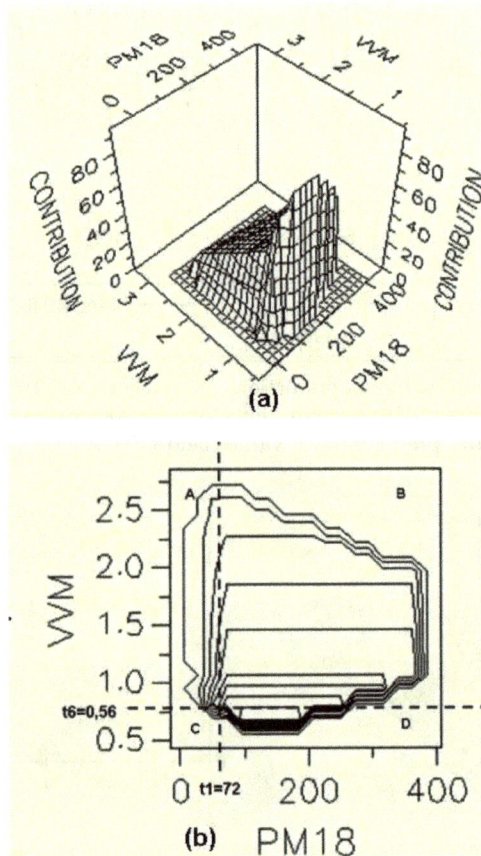

Fig. 3. Shows MARS model with 20 basis functions for 2003 and illustrates in part (a) the interaction of the basis functions max (0;72-pm18) and max (0; vvm-0,56) that generate the basis function BF10 which represents an interaction surface (Table N . 3). In the other figure (b) shows the surface contour are four regions where the letters A, B, C and D, D represents the area for values less than 0,56 meters per second (vvm) and particulate matter concentrations at 18 hours greater than 72 micrograms per cubic meter that generated by the interaction of both base functions with the maximum contribution values to the response variable of interest.

Critical Episodes of PM10 Particulate Matter Pollution in Santiago of Chile, an Approximation Using Two Prediction
Methods: MARS Models and Gamma Models

149

Year 2001

Mars		
20 basis function	BF1 = max(0;pm18-180)	BF7 = max(0; 1,035-vvm)
	BF2 = max(0; 180-pm18)	BF9 = max(0;1,56-vvm)*BF1
	BF3 = max(0; pm6-13)	BF10 = max(0; pm0-241)
	BF5 = max(0; dtm-2,55)	BF15 = max(0;87-pm6)*BF1
	BF6 = max(0;vvm-1,03)	BF16 = max(0;pm18-183)*BF6

pm10m=109,5-0,48*BF2+0,24*BF3+2,98*BF5+204,5*BF7+0,61*BF9+0,26*BF10-0,008*BF15+0,68*BF16

Gamma

pm10m=exp(4,15+0,00076*pm0+0,00223*pm6+0,00292*pm18+0,02504*dtm-0,20295*vvm)

Year 2002

Mars		
40 basis function	BF1 = max(0; pm18-137)	BF7 = max(0; 1,307-vvm)*BF5
	BF2 = max(0; 137-pm18)	BF8 = max(0; vvm-1,522)*BF1
	BF4 = max(0; 240-pm0)	BF9 = max(0; 1,522-vvm)*BF1
	BF5 = max(0; dtm+ 0,00000512)	BF12 = max(0; vvm-0,998)*BF5
		BF24 = max(0; vvm-0,867)*BF5

pm10m=146,343-0,553*BF2-0,286*BF4+7,072*BF7+2,093*BF8+0,537*BF9-22,888*BF12+21,844*BF24

Gamma

pm10m=exp(3,818+0,00118*pm0+0,002267*pm6+0,00298*pm18+0,035488*dtm-0,149482*vvm)

Year 2003

Mars		
20 bases function	BF1 = max(0; pm18-72)	BF8 = max(0;212-pm0)*BF1
	BF2 = max(0; 72-pm18)	BF9 = max(0;pm0-1)*BF3
	BF3 = max(0;pm6-16)	BF10 = max(0;vvm-0,56)*BF2
	BF5 = max(0;vvm-1,07)	BF11 = max(0;dtm-8,17)*BF1
	BF6 = max(0;1,07-vvm)	BF16 = max(0;dtm-2,11)
	BF7 = max(0;pm0-212)*BF1	BF17 = max(0;pm18-79)*BF16

pm10m=94,2-1,58*BF2-25,99*BF5+111,24*BF6-0,0022*BF7-0,002*BF8+ 0,0009*BF9+0.75*BF10-0,1*BF11+2,02*BF16+0,1*BF17

Gamma

pm10m=exp(3,885+0,00108*pm0+0,00107*pm6+0,0030047*pm18+0,02658*dtm)

bf: basis functions;
pm0, pm6, pm18: PM10 concentration at 00, 06 and 18 previous day, respectively;
dtm: temperature difference of tomorrow; vvm: wind velocity of tomorrow.

Table 3. Explicit MARS models for 20 and 40 basis functions and Gamma models for years
2001, 2002 and 2003.

4. Discussion

The MARS modeling selects those significant predictors and detects possible interactions
between them generating more flexible models from the point of view of interpretation;
since interactions are always restricted to a subregion, they are expressed algebraically
through the basis functions, generating a parsimonious model that represents without any
further transformation the nature of the working variables. This procedure creates nodes or

cutting points that act as threshold values for each predictor variable selected, indicating the change in the contribution generated by the basis function to the response under study.

After selecting the optimal model, MARS fits the model removing one variable, in order to determine the impact on the quality of the model due to eliminate that variable and assigns a relative ranking from the variable most important to the least important. Thus, replacement or competing variables are defined in the MARS methodology allowing us to treat missing values generating a special basis function for those variables that have no information, ie, generate an allocation basis function whose purpose is to impute the average value of the predictor without information.

As could be determined in this study, the MARS models performed similarly in their predictive efficiency when changing the number of basis functions, regardless of the year for which the model was built and the year used for validation. Partitioning the range of the predictor variables did not improve the quality of predictions, showing that sometimes a less fine partition (minus subregions) methodology was more robust than a thin partition. This could be explained as the time series of PM10 over time show a downward trend and variations in lower concentrations since the intervention measures applied [3].

The Pudahuel monitoring station registered changes in annual and monthly concentrations of PM10 between 1998 and 2004. These changes were due mainly to global decontamination measures, extraordinary administrative actions on critical air pollution events, use the forecast model of critical events to reduce their negative impact and meteorological factors. Such changes affect the performance of these models allowing a less complex structure without loss of efficiency because the relationship established between the predictors is not as complex, this is reflected in the basis functions constructed. In turn there are differences from one year to another, which could be influenced by weather conditions of each year, such as dry year, El Niño and other climate changes. Another factor to evaluate and that could be relevant in this behavior is the increase in car ownership in recent years.

Additionally, different methodologies have been implemented to model the concentration of particulate matter in the metropolitan region of Santiago de Chile. For example, Silva and colleagues applying time series using transfer functions involving meteorological variables, reported a 40% average ratio of absolute error in the prediction of critical events[24]. Moreover Perez and colleagues, using neural networks with pre-smoothing, reported approximately 30% mean absolute error in predicting the critical episodes [19]. Subsequent work by the same author using a neural network shows results superior to classical linear regression [18]. Furthermore, Silva and colleagues assessed two methodological approaches to the problem of predicting air pollution by particulate matter, reporting that MARS generated better predictions than the discriminant analysis [23].

Application of these models to the monitoring station Pudahuel shows that the predictor variables that best predict the answer would be (i) some PM10 concentrations: pm0, pm6 and pm18 and (ii) meteorological variables: vvm dtm; this is consistent since MARS selected variables related to persistence of ventilation conditions, which are related to meteorology [23.]

Prediction methodologies give us adequate models to study particulate matter pollution. Gamma regression was generally lower in Class II hits than MARS models, except for predicting from 2003 to 2004, so MARS appears as a better prediction tool of pollution episodes above the 240 (μg/m³). Additionally, an advantage of the Gamma model is that generally makes better predictions for the class I (PM10 < 240 μg/m³), ie, is sensitive to values of concentrations of particulate matter associated to air quality good or warning; this would be given by the behavior of the variable of interest. For example, for 2001 the intervention measures explain the best fit of the Gamma modeling in that year, whereas later on these measures had a moderate impact on the values of the series of PM10, making these models less efficient compared to MARS, as happens for the years 2002, 2003 and 2004.

This study basically aims to proposed models being able to detect concentrations above the threshold of 240 μg/m³ that mandate periods of epidemiological alert in Santiago de Chile; this consideration would place the MARS modeling as a tool statistically more powerful than Gamma modeling. This last point is consistent with previous findings showing that MARS is more efficient than other techniques [4,23,27]. This could be explained by the smoothing approximation that uses this methodology generating breakdowns in the predictors time series and locally adjusting the basis functions in function on such nodes.

5. Appendix 1. – Exponential smoothing

This technique uses a smoothing constant; if the constant is close to 1 it affects very much the new forecast and conversely when this constant is close to 0, the new forecast will be very similar to the old observation . If you want a sharp response to changes in the predictor variable you must choose a large smoothing constant. The formula that relates the coefficient and the time series is given by $S_t^{[2]} = \alpha S_t + (1-\alpha)S_{t-1}^{[2]}$ where $S_t = \alpha x_t + (1-\alpha)S_{t-1}$; to generate the smoothing is necessary to have the values S_0 and S_t, where x_t corresponds to the values of the original series [16,26].

6. References

[1] Biggeri A, Bellini P, Terracini B. [Meta-analysis of the Italian studies on short-term effects of air pollution--MISA 1996-2002]. Epidemiol Prev 2004;28(4-5 Suppl):4-100.

[2] Cassmassi J. Improvement of the forecast of air quality and of the knowledge of the local meteorological conditions in the Metropolitan Region; 1999.

[3] CONAMA. Comisión Nacional del Medio Ambiente. Decreto Ley N° 59, Diario Oficial de la Republica de Chile. Lunes 25 de mayo de 1998". 1998.

[4] De Veaux R, Psichogios D, Ungar L. A comparison of two nonparametric estimation Schemes: MARS and Neural Networks. *Computers Chemical Engineering*, 1993: 17(8): 819-837.

[5] Friedman J. Multivariate Adaptive Regression Splines *The Annals of Statistics* 1991: 19(1): 1-141.

[6] Gokhale S, Khare M. A theoretical framework for episodic-urban air quality management plan (e-UAQMP). *Atmospheric Environment* 2007: 41(2007): 7887-7894.

[7] Hardin J, Hilbe J. Generalized linear models and extensions. USA: College Station Texas; 2001.

[8] Janke K, Propper C, Henderson J. Do current levels of air pollution kill? The impact of air pollution on population mortality in England. Health Econ 2009;18(9):1031-55.

[9] Kurt A, Gulbagci B, Karaca F, Alagha O. An online air pollution forecasting system using neural networks. *Environ Int* 2008: 34(5): 592-598.

[10] Lewis PAW, Stevens JG. Nonlinear modeling of time series using multivariateadaptive regression splines (MARS). J Am Statist Assoc. 1991;86:416.

[11] MARS. User guide. Salfords-Systems 2001. Multivariate Adaptive RegressionSplines Vol. 2010; 2001.

[12] McCullagh P, Nelder J. Generalized Linear Models. Londin, UK: Chapman Hall, 1989.

[13] McKendry IG. Evaluation of artificial neural networks for fine particulate pollution (PM10 and PM2.5) forecasting. J Air Waste Manag Assoc 2002: 52(9): 1096-1101.

[14] Meenakshi P, Saseetharan MK. Urban air pollution forecasting with respect to SPM using time series neural networks modelling approach--a case study in Coimbatore City. J Environ Sci Eng 2004: 46(2): 92-101.

[15] MM5 Community Model. Vol. 2010. Pennsylvania State University/National Center for Atmospheric Research Numerical Model 2008.

[16] Montgomery DC, L. A. Johnson, Gardiner. JS. Forecasting and Time Series Analysis. , 1990.

[17] Pascal L. [Short-term health effects of air pollution on mortality]. Rev Mal Respir 2009;26(2):207-19.

[18] Perez P, Reyes J. Prediction of maximum of 24-h average of PM10 concentrations 30 h in advance in Santiago, Chile. Atmospheric Environment 36: 2002: 36: 4555-4561.

[19] Pérez P, Trier A, Silva C, Montaño RM. Prediction of atmospheric pollution by particulate matter using a neural network. In: Conf on Neural Inf Proc; 1998 Proc. of the 1997; Dunedin, New Zealand: Springer-Verlag, Vol 2; 1998.

[20] Sanhueza P, Vargas C, Jimenez J. [Daily mortality in Santiago and its relationship with air pollution]. Rev Med Chil 1999: 127(2): 235-242.

[21] Sanchez-Carrillo CI, Ceron-Mireles P, Rojas-Martinez MR, Mendoza-Alvarado L, Olaiz-Fernandez G, Borja-Aburto VH. Surveillance of acute health effects of air pollution in Mexico City. Epidemiology 2003;14(5):536-44.

[22] Scott GM, Diab RD. Forecasting air pollution potential: a synoptic climatological approach. J Air Waste Manag Assoc 2000: 50(10): 1831-1842.

[23] Silva C, Pérez P, Trier A. Statistical modelling and prediction of atmospheric pollution by particulate material: two nonparametric approaches. Environmetrics, 2001: 12: 147-159.

[24] Silva C, Firinguetti L, Trier A. Contaminación ambiental por partículas en suspensión: Modelamiento Estadístico. In; 1994; Actas XXI Jornadas Nacionales de Estadística, Concepción, Noviembre; 1994.

[25] Staniswalis JG, Yang H, Li WW, Kelly KE. Using a continuous time lag to determine the associations between ambient PM2.5 hourly levels and daily mortality. J Air Waste Manag Assoc 2009;59(10):1173-85.

[26] Stata. Stata: Release 11. Statistical Software. In: College Station TSLTS, ed., 2009.

[27] Steinberg D. An Alternative to Neural Nets: Multivariate Adaptive Regression Splines (MARS). PC AI's 2001: 15(1): 38.

[28] Tobias A, Scotto MG. Prediction of extreme ozone levels in Barcelona, Spain. Environ Monit Assess 2005: 100(1-3): 23-32.

7

Train-Based Platform for Observations of the Atmosphere Composition (TROICA Project)

N.F. Elansky[1], I.B. Belikov[1], O.V. Lavrova[1], A.I. Skorokhod[1],
R.A. Shumsky[1], C.A.M. Brenninkmeijer[2] and O.A. Tarasova[3]

[1]*A.M. Obukhov Institute of Atmospheric Physics RAS*
[2]*Max Planck Institute for Chemistry*
[3]*World Meteorological Organization*
[1]*Russia*
[2]*Germany*
[3]*Switzerland*

1. Introduction

Composition of the Earth's atmosphere is evolving in time in continuous interaction with the land, ocean, and biosphere. Growth of the world's population leads to increased human impact on the nature, causing substantial changes in atmospheric composition due to additional emissions of gases and particles. The concerns about the consequences of the human induced changes have resulted in the ratification of several important international agreements regulating anthropogenic emissions of different substances. To understand the reasons and consequences of the atmospheric composition change complex multi-component global observations are essential. In spite of the long enough history of the observations several parts of the world still remain sparsely covered by observations, and Russia is among such regions.

Due to a huge territory and diverse climatic conditions it is extremely difficult to cover the territory of the world's biggest country, Russia, by the observational network of the proper spatial resolution. Moreover, disproportions of the industrial development of different regions, the dominance of natural resource-based industry, and varied application of imperfect technologies, have produced non-uniformities in pollution level in the different regions of the country. At the same time, over 65% of the territory contains almost no industrial activity, and the ecosystem state is close to the background conditions.

Ecosystems are very diverse in Russia as well. The Russian boreal forests (73% of the world area of such forests), wetlands and peatland ecosystems are the most capacious reservoir of carbon and powerful sources/sinks of the greenhouse and reactive gases. These systems are climate sensitive and need careful assessment under changing conditions.

Due to degradation of the efficiency of the Russian meteorological network in the beginning of 1990s and termination of the background monitoring of O_3, NOx, SO_2, and aerosol by many stations the question arose on how to continue with the observations of the key pollutants in the most efficient way. In this crucial period (February, 1995) Dr. Prof. P.J. Crutzen proposed to the director of the A.M. Obukhov Institute of Atmospheric Physics (OIAP), academician G.S. Golitsyn, carrying out observations of surface ozone and its precursors from a passenger train moving along the Trans-Siberian Railroad. In November–December 1995 the first experiment of such kind was performed utilizing a specialized car-laboratory. Experience of the first campaign demonstrated that reliable information on the regional boundary layer composition can be obtained if the car-laboratory is coupled at the head of a train moving along electrified railroads (Crutzen et al., 1996). Since then the experiments are performed on a regular basis and comprise the Trans-Continental (or Trans-Siberian) Observations Into the Chemistry of the Atmosphere (TROICA) project.

The routes and the dates of expeditions are summarized in Table 1. TROICA expeditions were mainly performed along the Trans-Siberian Railroad on the Moscow-Khabarovsk or Moscow-Vladivostok routes. The most complicated program was performed in 2000; it consisted of monitoring along the railroad from Murmansk to Kislovodsk and monitoring based on stationary scientific stations located on the Kola Peninsula, in central regions of Russia, in the town of Kislovodsk, and in the mountainous North-Caucasian region. The measurements were performed everywhere in "local spring" allowing for study of the transition from winter to summer conditions.

Experiment	Date	Route
TROICA-1	17/11 - 2/12/1995	N.Novgorod-Khabarovsk-Moscow
TROICA-2	26/07 – 13/08/1996	N.Novgorod-Vladivostok-Moscow
TROICA-3	1/04 – 14/04/1997	N.Novgorod-Khabarovsk-Moscow
TROICA-4	17/02 – 7/03/1998	N.Novgorod-Khabarovsk-N.Novgorod
TROICA-5	26/06 - 13/07/1999	N.Novgorod-Khabarovsk-Moscow Ob river beating
TROICA-6	6/04 – 25/06/2000	Moscow-Murmansk-Kislovodsk-Kislovodsk-Murmansk-Moscow Observations at 4 stations
TROICA-7	27/06 – 10/07/2001	Moscow-Khabarovsk-Moscow
TROICA-8	19/03 – 1/04/2004	Moscow-Khabarovsk-Moscow
TROICA-9	4/10 – 18/10/2005	Moscow-Vladivostok-Moscow
TROICA-10	4/10 - 7/10/2006	Around and cross-section of Moscow megacity
TROICA-11	22/07 - 5/08/2007	Moscow-Vladivostok-Moscow
TROICA-12	21/07 - 4/08/2008	Moscow-Vladivostok-Moscow
TROICA-13	9/10 – 23/10/2009	Moscow-Vladivostok-Moscow
TROICA-14	26/05 – 24/06/2010	Moscow-Murmansk-Sochi-Murmansk-Moscow
TROICA-15	7/12 – 10/12/2010	Moscow-S.-Petersburg- N.Novgorod-Moscow

Table 1. TROICA experiments: dates and routes.

In October 2006, the Moscow megacity was circumnavigated (mean radius ~70km) three times via the electrified circuit railroad and Moscow city was traversed twice (TROICA-10). In the course of the TROICA-5 experiment, shipboard atmospheric monitoring was performed while traveling down the 2000 km Ob River starting from the city of Novosibirsk.

2. TROICA observing system

In 2003 OIAP together with the Russian Research Institute of Railway Transport (RRIRT) constructed and equipped a new mobile laboratory (Fig.1). It consists of two cars intended for continuous measurements of gas and aerosol concentrations and of radiative and meteorological parameters. It also housed equipment for chemical analysis of air, water, soil, and vegetation samples (Elansky, 2007; Elansky et al., 2009). Interior of the car is shown in Fig.2.

Fig. 1. Mobile laboratory for TROICA experiments (from TROICA-8 on) and its set up in the passenger train.

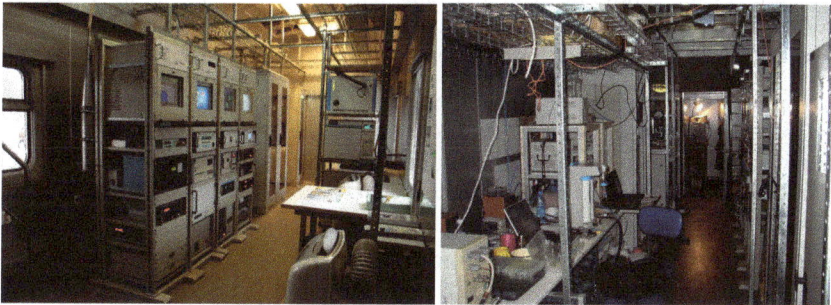

Fig. 2. Interior of the laboratory for atmospheric monitoring.

Since January 2009, the scientific instrumentation included the:

- instruments for continuous measurement of the gas mixing ratios: gas analyzers, chromatographs, PTR-MS;
- instruments for continuous measurements of concentration and microphysical characteristics of aerosols: particle counters, nephelometers, aerosol spectrometer, aetholometer, trap for bio-aerosol and radiometer-spectrometer analyzing gamma-radiation;
- instruments for remote sounding of the composition of the troposphere and the middle atmosphere: photometer, spectrophotometers operating in the UV-VIS and IR spectral regions (including MAX-DOAS and microwave spectro-radiometer);
- instruments for measurements of the solar radiation, optical and meteorological characteristics of the atmosphere: spectrometer, fluxmeters operating in the UV, VIS and IR spectral regions, photometers for determining the rate of NO_2 photodissociation, a temperature profiler, sonic anemometer and meteorological sensors;

- systems for collection of gas and aerosol samples and instruments for express chemical analysis of some samples (gas chromatographs, mass-spectrometer, x-ray fluorimeter);
- integrated (combined with GIS) PC-system for data collection;
- communication TV and audio systems

For some special experiments various instruments belonging to the partner institutions were deployed, namely, multi-channel chromatograph ACATS-IV (NOAA-ESRL, Hurst et al., 2004); proton-transfer-reaction mass spectrometer (MPIC); different mobility particle sizer, air ion spectrometer, humidity – tandem differential mobility analyzer (UH); particle into liquid sampler with ion chromatograph, aethalometer (FMI); and some others.

A computerized integrated instrumentation set was developed for continuous atmospheric observations on the basis of the mobile laboratory. Detailed description of the used instruments can be found in the TROICA related publications.

It is important to estimate the degree of influence of the leading train and of oncoming trains in the area where the measurements are performed for adequate interpretation of the results obtained from the TROICA mobile observatory. Detailed analysis of the collected dataset showed that measurements performed on board of the laboratory moving along electrified railroads (air samples are taken above the roof just after locomotive) is capable of reflecting the principal peculiarities of the background state of the atmosphere over the continent. However, under calm conditions with nighttime temperature inversions, air pollution along the route can be accumulated within the boundary layer and in these cases a railroad can be considered as an enterprise polluting the atmosphere (Panin et al., 2001).

Therefore prior to the analysis of the phenomena of different scale all data were separated into several groups representing different pollution conditions, namely polluted (urban and industrial) and unpolluted (rural) route sections. The special and temporal features of trace gases variations were analyzed for each group separately.

3. Gases and aerosol over the continent

3.1 Surface ozone variations

Most of the observations were performed along the Trans-Siberian railroad covering a belt of the North-Eurasia between 48o and 58oN and 37o and 135oE (exclusive of the Khabarovsk-Vladivostok zonal route section). The entire data file was divided into two groups of data corresponding to polluted areas (towns/villages and industrial regions) and unpolluted areas. The regions with the NO and CO mixing rations lower than 0.4 ppb and 0.2 ppm, respectively, were taken as the unpolluted areas; the rest regions were considered as the polluted ones.

Surface ozone mixing ratio was measured with 1008RS and 1008AH gas analyzers Dasibi. These instruments are based on the photometric method. They are capable of measuring mixing ratios in the range from 1 to 1000 ppb with a total uncertainty of ±1 ppb.

Analysis of the data of all expeditions shows that minimum ozone mixing ratios were observed in the boundary layer under nighttime temperature inversions. Maximum mixing ratios were always associated with daytime conditions which are favorable for ozone generation. The spatial distribution of ozone is characterized by slight zonal gradient to the

East with a magnitude of 0.47±0.02 ppb per 10 degrees longitude. Such a spatial gradient is probably due to increasing availability of precursors toward the eastern part of the country (biogenic CH_4 and VOC emissions in Siberia, forest fires, transport of ozone precursors from China), intensive air exchange between stratosphere and troposphere over east continental regions, and some other factors (Crutzen et al., 1998; Elansky et al., 2001b; Golitsyn et al., 2002; Markova & Elansky, 2002).

Analysis of the campaign averaged ozone mixing ratios under unpolluted conditions showed that spring campaigns are characterized by the highest ozone mixing rations. This is likely to be connected with intensive stratosphere–troposphere exchange (Shakina et al., 2001), bringing ozone from the stratosphere, especially at the eastern part of the continent. The absence of the pronounced summer maximum, which is often reported for the stations in the Central Europe, indicates that photochemical ozone production in temperate latitudinal belt of the substantial part if Eurasia is weak and that ozone destruction under powerful continental temperature inversions plays very important role.

Diurnal variations of the surface ozone mixing ratio averaged over several expeditions and over the whole expedition route (Fig.3) confirm an important role of vertical mixing and ozone destruction on the surface by pronounced morning minimum and daily maximum (re-build ozone levels due to the developed mixing in the boundary layer). Comparison of the diurnal variations for different seasons allowed estimation of the average rate of ozone dry deposition over extended continental regions: 0.08 and 0.65 cm/s on the surfaces covered and not covered with snow, respectively (Elansky et al., 2001b).

3.2 Nitrogen oxides variations

Nitrogen oxides distribution along the route is characterized by high and quite localized peaks, associated with main anthropogenic emission sources (Markova et al, 2004). The city plumes stretch over several tens of kilometers and, under some conditions, can reach even several hundreds of kilometers (e.g. under stable atmospheric stratification with slow mixing in the boundary layer).

The NO and NO_2 mixing ratios were measured in different expeditions either with TE42C-TL instrument (Thermo Electron Corp., USA) or with M200AU instrument produced by Teledyne Corp. (USA). Both instruments are based on the chemiluminescent method. The minimum NO and NO_2 mixing ratios detectable with these instruments are 0.05 ppb allowing for measurements under unpolluted conditions.

Numerous anthropogenic emission sources in the latitudinal belt 48-58º N lead to the excesses in the NO and NO_2 mixing ratios averaged over this part of the route (all data) above NO and NO_2 mixing ratios measured at background (unpolluted) conditions. The excess in NO_2 mixing ratio is less pronounced than the one in NO mixing ratio (Markova et al, 2004) due to the fact that NO is a primary form for nitrogen oxides emission while NO_2 is already oxidized and diluted. There was no strong seasonal cycle detected in the NOx mixing rations. Diurnal cycle of NOx is also weak in the considered latitudinal belt, except for the NO mixing ratio which is enhanced during the daytime over small and large cities as a result of household heating. Burning of the dry grass in spring and autumn can have some impact on NOx levels as well.

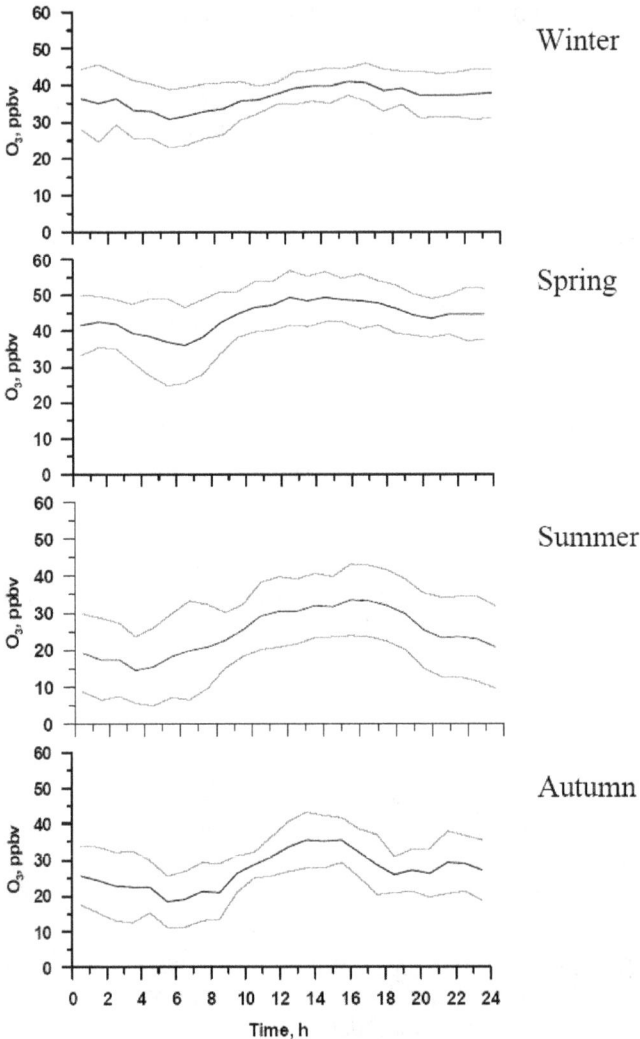

Fig. 3. Mean daily variations of ozone mixing ratio along Moscow – Vladivostok route.

3.3 Carbon compounds CO, CO₂ and CH₄

Observations of carbon dioxide and methane were used to study natural and anthropogenic sources of these important long-lived greenhouse gases. Taking into consideration diversity of ecosystems across the Eurasian continent and very limited availability of the stationary observations, train based observations can be used to validate emission inventories used for atmospheric greenhouse gas modeling.

Carbon monoxide (CO) mixing ratios were detected continuously using an automated infrared absorption gas filter correlation instrument TE48S instrument (Thermo Electron

Corp). It allows for measuring background CO mixing ratios at a level of less than 100 ppb with the total uncertainty of ±10 ppb.

Carbon dioxide (CO_2) measurements were carried out in situ using a differential non-dispersive infrared analyzer LI-6262 (LiCor, USA). The measurement range is 0–3000 ppm. The outputs from the analyzer were averaged over 10-min periods and reported in ppm on a dry air basis. The accuracy of the measurements is ±1 ppm.

Mixing ratios of CH_4 and non-methane hydrocarbons (NMHC) were measured with APHA-360 instrument (Horiba Company, Japan). This gas analyzer separates CH_4 and NMHC using selective catalytic absorbers and measures gas mixing ratios with a flame-ionization detector. The total uncertainty of CH_4 and NMHC mixing ratio measurements doesn't exceed ±5 ppb. To supply the flame-ionization detector with hydrogen necessary for its operation, the instrumentation set included hydrogen generators of different types.

Observation of CO, CH_4 and CO_2 averaged over 10 degrees latitude for different seasons are presented in Fig. 4. The plot shows no systematic large scale gradient either Eastward or Westward in CO_2 spatial distribution. Analysis of CO_2 diurnal variations (Belikov et al., 2006) demonstrated that high CO_2 mixing ratios are observed during night-time, when vegetation and soil respiration fluxes are intensive and mixing is weak, so that CO_2 is accumulated in the boundary layer under temperature inversions. In industrial regions CO_2 from anthropogenic sources is accumulated as well. CO_2 peaks in many cases are accompanied by the accumulation of anthropogenic CH_4.

Important findings on the key greenhouse gases and CO distribution and variability are summarized in the number of papers (Oberlander et al., 2002; Belikov et al., 2006, Tarasova et al., 2005a and Tarasova et al., 2005b). Below the results are presented for rural areas, where anthropogenic emissions do not influence observed mixing ratios.

Mean CO mixing ratio equals to 0.21±0.03 ppm for the cold periods. For warm periods, the mean CO mixing ratio is significantly lower, 0.12±0.03 ppm. This seasonality is controlled by CO sink in the reaction with OH in summer, which is rather typical for unpolluted stations in the mid latitudes (Tarasova et al., 2005b). CO distribution function is a bit broader for winter, showing bigger variability around the average than in summer months. Such variability is connected with a higher stability of the PBL in winter than in summer, which leads to irregular accumulation of CO from local sources along the train track.

Different shape of the distribution function is observed for CO_2 in different seasons (Belikov et al., 2006). If in winter CO_2 distribution is quite narrow and most of values are in the range 380-400 ppm, then in summer period, active biosphere causes substantial variations in CO_2 mixing ratio, especially under the influence of strong surface temperature inversions. The mean summertime CO_2 mixing ratio is 376±34 ppm. Excess of the winter CO_2 levels over the summer ones demonstrates that biosphere in the studied latitudinal belt of Russia servers as a sink for CO_2. Nevertheless, summer peaks in CO_2 (e.g. under night temperature inversions or in the zones of biomass burning) exceed winter CO_2 levels which are primarily controlled by anthropogenic sources.

The frequency distribution of CH_4 mixing ratios is similar to that of CO_2 (Tarasova et al., 2005a; Belikov et al., 2006); however, the CH_4 mixing ratio depends on the surface temperature inversions to a lesser degree than CO_2 mixing ratio does.

Figure 4 shows spatial gradients of CO, CO_2 and CH_4 over the continent during cold and warm seasons (averaged over two expeditions each time). As it can be seen in the figure, CO distribution over the continent in cold periods is almost uniform. Rather small elevations of the longitudal zone averages are observed for the parts of the route which include big cities and their plumes. Longitudinal gradient of CO mixing ratio of 0.6 ppb/degree in the East direction, is clearly pronounced in Siberia in warm periods. This gradient is connected with intensive biomass burning in Eastern Siberia and Far East, which is known to be a substantial CO emission source.

CO_2 spatial distribution (Fig. 4) in cold periods is also almost uniform. The non-uniformities observed in warm periods are caused by variations in the landscape and vegetation. These variations are unlikely to be connected with anthropogenic sources as they are either seasonally independent or would be stronger in winter (heating season).

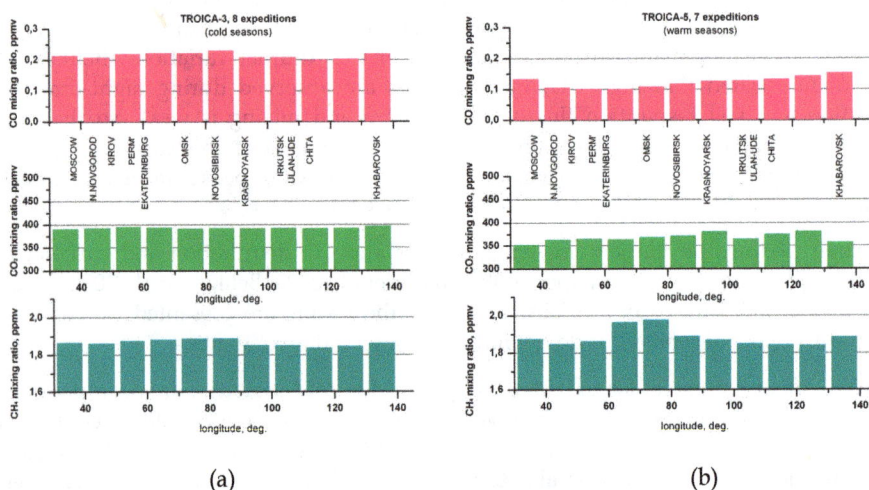

(a)   (b)

Fig. 4. Spatial distributions of the CO, CO_2, and CH_4 mixing ratio over the continent in (a) cold and (b) warm periods (averaged over 10 degrees longitude).

CH_4 mixing ratio distribution non-uniformities (Fig. 4) observed in cold periods are caused by anthropogenic sources (similar to the case of CO). The enhance mixing ratio near 70-90 E is due to the plume of the largest in Russia Kuznetsky coil basin which is one of the important sources of atmospheric CH_4 in any season. This conclusion was further confirmed using isotopic analysis of methane (Tarasova et al., 2006). In warm periods, a well-pronounced increase in the CH_4 mixing ratios is observed over West Siberia. This effect is associated with intensive CH_4 biogenic production by bogs.

3.4 Isotopic composition studies

Analysis of the carbon, oxygen, and hydrogen isotopologues of the atmospheric methane and carbon monoxide in the air samples collected from the mobile observatory was used to identify the sources of these compounds and to estimate their contribution to the global and regional carbon budgets. The ^{13}C, ^{14}C, ^{18}O, and D isotopes analysis of methane and carbon monoxide in the air samples was done in Max-Planck Institute for Chemistry. CH_4 isotopes measurements in the air samples were performed in the campaigns TROICA-2, 5, 7, 8 along Trans-Siberian railroad. The TROICA-5 campaign included also an exploratory expedition along the river Ob using a boat as a measurements platform (Oberlander et al., 2002). To identify primary processes contributing to the measured methane levels a two component mixing model was used.

Analysis of methane isotopic composition (^{13}C and D) showed that a predominant source of atmospheric methane observed in the boundary layer has a biogenic origin and is likely emitted to the atmosphere from bogs (Tarasova et al., 2006). Natural-gas leakages during the production, processing, and transportation of natural gas also contribute to the atmospheric methane levels. Shipboard measurements showed substantial increase of methane mixing ratio of up to 50% above the background level at a distance of 0.5 km from natural gas-extraction enterprise. Methane accumulation took place under temperature inversion, while its isotopic composition has not been analyzed.

In majority of expeditions an elevated mixing ratio of CH_4 was observed in the Perm region where train crosses several gas pipe lines connecting Polar Ural and Western Siberia gas fields with Central Russia. For example in the TROICA-5 the excess of CH_4 reached about 200 ppb above background mixing ratio and identified source isotopic signature $\delta^{13}C_{source}$=-52.42% showed the presence of the natural gas in the collected air sample.

The performed analysis (Tarasova et al., 2006) showed that biogenic methane sources play the principal role not only in Western Siberia but in other parts of Russia. At the same time for a few samples strong contributions of natural gas can be observed. In most of cases such samples were taken in the vicinity of the objects of gas industry or of the large cities (like Novosibirsk). Several examples of the city footprinted were observed in the spring campaign TROICA-8, when the strong local peaks were associated with the gas leakages from the low pressure gas distribution networks.

Analysis of the ^{14}C and [18]O in CO (Bergamaschi et al., 1998; Rochmann et al., 1999; Tarasova et al., 2007) in combination with trajectory analysis allowed identification of CO source in the summer campaign TROICA-2 as biomass burning, in particular, forest fires and combustion of agriculture waste in China. Identified sources of CO along the Ob river in the TROICA-5 campaign is likely to be connected with methane oxidation based on an inferred $\delta^{13}C_{source}$=-36.8±0.6‰, while the value for $\delta^{18}O_{source}$=9.0±1.6‰ identifies it as burning (Tarasova et al., 2007). Thus flaring in the oil and gas production was supposed to be a source. The extreme ^{13}C depletion and concomitant [18]O enrichment for two of the boat samples unambiguously indicates contamination by CO from combustion of natural gas (inferred values $\delta^{13}C_{source}$= -40.3‰ and $\delta^{18}O_{source}$=17.5‰). The impact of industrial burning was discernable in the vicinity of Perm-Kungur.

3.5 Carbon dioxide $\Delta^{14}CO_2$

The unique spatial distribution of $^{14}CO_2$ over Russia was obtained as part of TROICA-8 expedition. The details of the method of $^{14}CO_2$ measurement in the collected flasks are described by Turnbull et al. (2009). It was shown that the addition of fossil fuel, ^{14}C-free CO_2 had the greatest influence on the spatial distribution of $\Delta\,^{14}CO_2$. In addition, $^{14}CO_2$ produced by nuclear reactors caused local enhancement in $\Delta\,^{14}CO_2$ in some samples.

$\Delta^{14}CO_2$ increases by $5\pm 1.0\%$ across the transect from 40^0E to 120^0E (Fig. 5) and this difference is significant at the 99% confidence level. The magnitude of the $\Delta^{14}CO_2$ gradient is consistent with the dispersion of fossil fuel CO_2 emissions produced in Europe and atmospheric transport across northern Asia dispersing and diluting the fossil fuel plume. The observed isotopic change implies a gradient of 1.8 ppm of fossil fuel derived CO_2 (assuming a $-2.8‰$ change in $\Delta\,^{14}CO_2$ per ppm of fossil fuel CO_2 added). $\Delta^{14}CO_2$ measurements from Niwot Ridge, Colorado, USA (NWR, 40.05° N, 105.58° W, 3475 m.a.s.l.) representing relatively clean free troposphere, give a mean $\Delta^{14}CO_2$ value of $66.8\pm1.3‰$ for 3 sampling dates during the time of the TROICA-8 campaign (grey bar in Fig. 5). The Eastern Siberian part of the TROICA transect shows values ($62.8\pm0.5‰$) most similar to the NWR value, consistent with an easterly dilution of the fossil fuel content in boundary layer air away from the primary source, but suggesting that some influence from the European fossil fuel CO_2 source remains in the Eastern Siberian air mass.

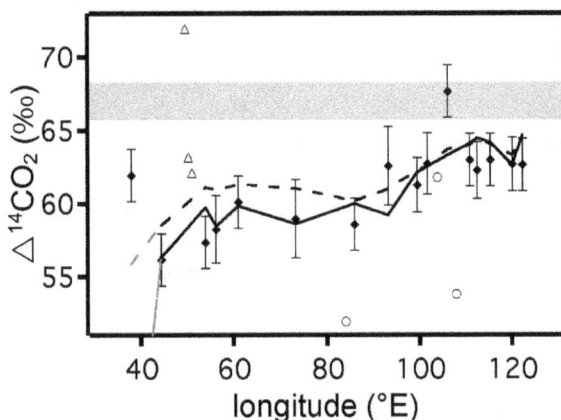

Fig. 5. $\Delta^{14}CO_2$ as a function of longitude. Closed diamonds are the clean air $\Delta^{14}CO_2$ dataset. Open symbols indicate samples that may be influenced by either nuclear reactor effluent (triangles) or local city pollution (circles). The shaded bar indicates the Niwot Ridge $\Delta^{14}CO_2$ value measured over the same time period and its 1-sigma error envelope. Modeled estimates for each sampling time and location have been done on the basis standard (solid line) and fast (dashed line) mixing (see Turnbull et al., 2009).

3.6 Volatile organic compounds

Volatile organic compounds (VOCs) were measured in a number of expeditions (Elansky et al., 2000; Elansky et al., 2001a). VOCs were retrieved analyzing sorbent samples or in air samples pumped into stainless-steel canisters. These measurements can be performed in mobile or stationary chemical laboratories.

Among all campaigns, the highest VOC mixing ratios were observed in summer 1999. Extremely high levels and unusual distribution of pollutants during TROICA-5 were caused by the abnormally torrid weather that led to intense evaporation of organic substances and a high atmospheric oxidative capacity.

In the TROICA-6 experiment the mixing ratios of a number of VOCs which are important for understand of reactivity of many species were measured using a proton mass-spectrometer (Elansky et al., 2001a). The highest VOC mixing ratios in this campaign along the Murmansk-Kislovodsk railroad, were seen in vicinities of cities of Novocherkassk, Ryazan', Moscow and St. Petersburg. A spring zonal VOC gradient associated with a northward decrease in biogenic emissions was revealed (Table 2) over the unpolluted rural areas.

Substance	Gradient, ppb / 1000 km
Methanol	2.37 ± 0.11
Acetonitrile	0.08 ± 0.01
Acetaldehyde	0.23 ± 0.03
Acetone	0.89 ± 0.04
Isoprene, vegetable alcohols	0.06 ± 0.01
C-pentane	0.03 ± 0.01
Benzene	0.07 ± 0.01
Toluene	0.05 ± 0.01
C8-benzenes	0.03 ± 0.01
C9-benzenes	0.07 ± 0.01

Table 2. Zonal gradient of VOC mixing ratios for TROICA-6 Kislovodsk-Murmansk route (May 27-29, 2000).

Detailed Trans-Siberian VOC measurements were carried out using PTR-MS during summer experiment TROICA-12 (Timkovsky et al., 2010). Figure 6 shows a spatial distribution of isoprene from Moscow to Vladivostok. This plot in particular demonstrates an influence of meteorological conditions on isoprene levels. If for the Eastern transect the weather over most of the Trans-Siberian railroad was warm and sunny, the weather was colder and less sunny of the back route (Westward). Due to these weather difference biogenic emissions in boreal and broad-leaved forests at the East of the continent on Eastern transect were more active compared with western transect which can be clearly seen in the Fig. 6. Anthropogenic emissions are mostly 'weather independent'. Observations of benzene and toluene which are connected with industrial activities or transport do not show difference between two transects of campaign. Most of peaks in the mixing ration of these compounds are associated with cities and industrial centers.

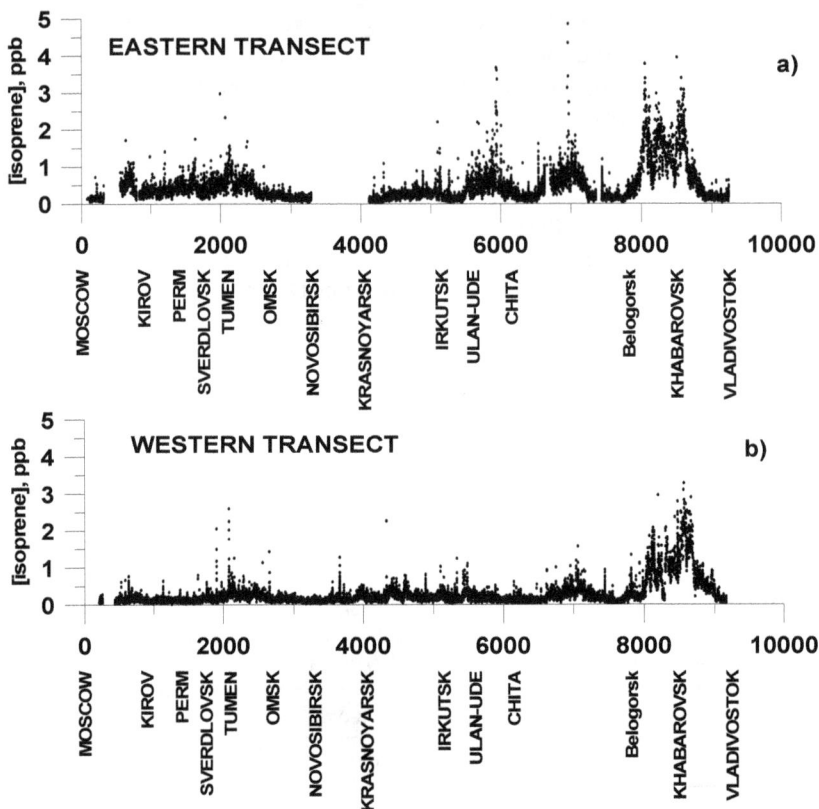

Fig. 6. Spatial distribution of isoprene mixing ratio along eastern (a) and western (b) transects in the summer campaign of 2008 (TROICA-12).

3.7 Aerosols

Concentration and microphysical and chemical properties of aerosol have been measured in most of the TROICA experiments. A big archive of data on aerosol concentration, size distribution and chemical composition was collected. Up-to-now the complete analysis of these data has not been finished yet. Some results were published by Andronova et al. (2003).

Distribution of the aerosol mass concentration over the route of the TROICA campaigns is shown in Figure 7. Aerosol size distribution for unpolluted region is presented in Figure 8.

Aerosol mass was compared between rural and industrial regions. In the plumes of industrial zones and cities, the mean aerosol mass concentrations were 100-120 μg/m³. A pronounces weekly cycle was observed in these regions as well. For comparison, the rural aerosol mass concentrations varied between 10 and 20 μg/m³. In the plumes of forest fires observed by the TROICA expeditions, maximum identified aerosol mass concentrations exceeded 800 μg/m³.

The soot aerosol concentration was measured using atmospheric aerosol sampling with quartz fiber filters and subsequent measurements of the light absorption by the aerosol samples (Kopeikin, 2007 and Kopeikin, 2008). Along the Murmansk-Kislovodsk transect, the mean level of atmospheric pollution by soot is about 1-2 µg/m³. However, the atmospheric pollution by soot is higher by an order of magnitude at the parts of the railroad where diesel locomotives are used. Large-scale (extended over 500-1000 km) non-uniformities in the soot aerosol distribution in the atmospheric boundary layer are seen during winter. In the spring of 1997, large-scale polluted zones extending over about 1000 km resulted from grass fires. In the winter-spring period, the atmospheric soot concentration over South Siberia and Far East were twice as high as that over European Russia. In summer, the atmosphere is weakly polluted by soot nearly along the entire railroad.

More detailed information on aerosol physical and chemical properties have been obtained by groups of specialists leaded by M.Kulmala and V.-M. Kerminen (Kuokka et. al., 2007; Vartiainen et al., 2007). During the TROICA-9 expedition the equipment for continuous monitoring of aerosol parameters with high time resolution was mounted in the moving laboratory. During this expedition the total particle concentration was typically of the order of few thousand particles /cm³ varying between 300 and 40 000 particles /cm³. The concentrations were the lowest in the rural area between Chita and Khabarovsk and the highest near larger villages and towns. Particle concentration levels measured on the way to Vladivostok and back were similar at both ends of the route but differed in the middle, between 4000 and 7000 km, were the concentration was notably lower on the westward route.

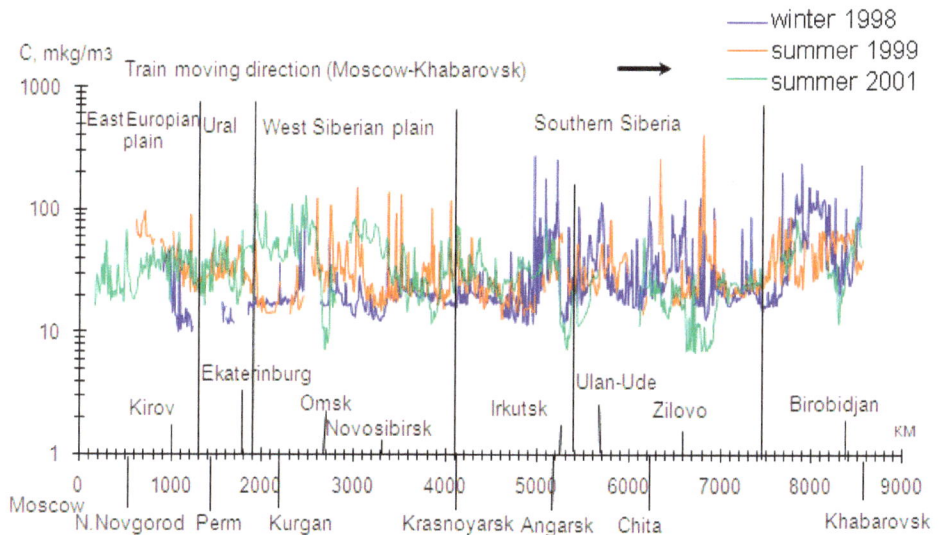

Fig. 7. Mass aerosol concentration variations along the Moscow-Khabarovsk railroad; the TROICA expeditions of 1998, 1999, and 2001.

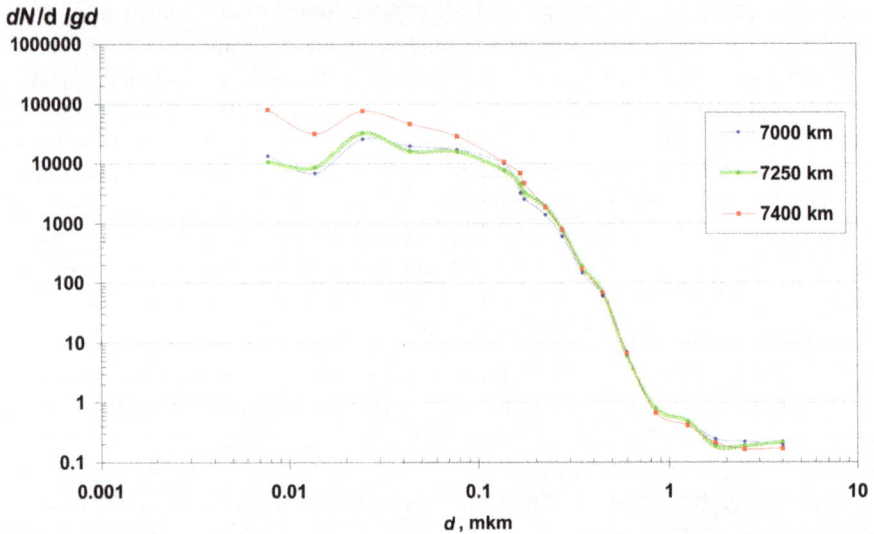

Fig. 8. Aerosol particle size distribution over the rural region covering the 7000 – 7400 km (from Moscow) railroad section (TROICA-5).

Concentrations of all ions and black carbon were quite low between the 3500 and 6500 km from Moscow, whereas high concentrations were observed between Moscow and Novosibirsk and in Asia before and after Khabarovsk on both (going to Vladivostok and coming back to Moscow) routes. On the way back to Moscow, the concentrations of all ions and black carbon were at extremely high levels near Khabarovsk. It was shown using trajectory analysis that high concentration events were mostly associated with long-range transport of aerosol particles from North-East China. The increased potassium and oxalate concentrations in this area are indicative of biomass burning.

Detailed analysis of the TROICA-9 aerosol measurements showed that fine particles consisted mainly of soot (BC, 15.8–48.7%, average 27.6%), SO_4^{2-} (2.7–33.5%, 13.0%), NH_4^+ (1.2–10.5%, 4.1%), and NO_3^- (0.5–2.4%, 1.4%). Trace metals in total accounted for 0.4–9.8% of the fine particulate mass. The fraction of monosaccharide anhydrides was in the range 0.4–1.6%, except for one sample (5926-7064 km from Moscow) on the way to Vladivostok where it was as high as 4%. The measured chemical components accounted for 27.7–78.5% of the PM2.5 mass. The unidentified fraction is expected to include organic particulate matter, water-insoluble material and water. The contribution of BC to PM2.5 was much higher during the whole TROICA-9 expedition than the fraction observed in Europe, where the BC contribution was estimated to be in the range 5–10% (Kuokka et al., 2007).

4. Air pollution in cities

The mobile laboratory provides an opportunity to study air pollution in the cities along the expedition route. The cross-sections of the cities and individual pollution plumes from different sources under different conditions can help to estimate the intensity of different anthropogenic sources. Topography, vegetation structure and state and a number of other factors are also important for the formation of city plumes in addition to meteorological factors. Figure 9 gives an example of the atmospheric pollutants distribution measured under different wind directions along the railroad traversing the city of Tyumen'. The structure of the spatial NO distribution demonstrates that several distinct sources are responsible for the formation of the observed NO levels in the city. These sources were the same for the movement of expedition in both directions. Observed differences of the plume on the ways forward and back are caused by the plume shift resulted from the wind-direction variation (different remoteness of the sources from the railroad).

During the TROICA expeditions, the train travelling along the Trans-Siberian Railroad crossed repeatedly about 110 cities with a population from 20 to 1500 thousand people (except Moscow). In addition, several tens of cities were crossed by the train in the course of the Kislovodsk–Murmansk expedition (TROICA-6). One particular expedition (TROICA-10) was aimed at the studying of the atmosphere composition around Moscow megacity.

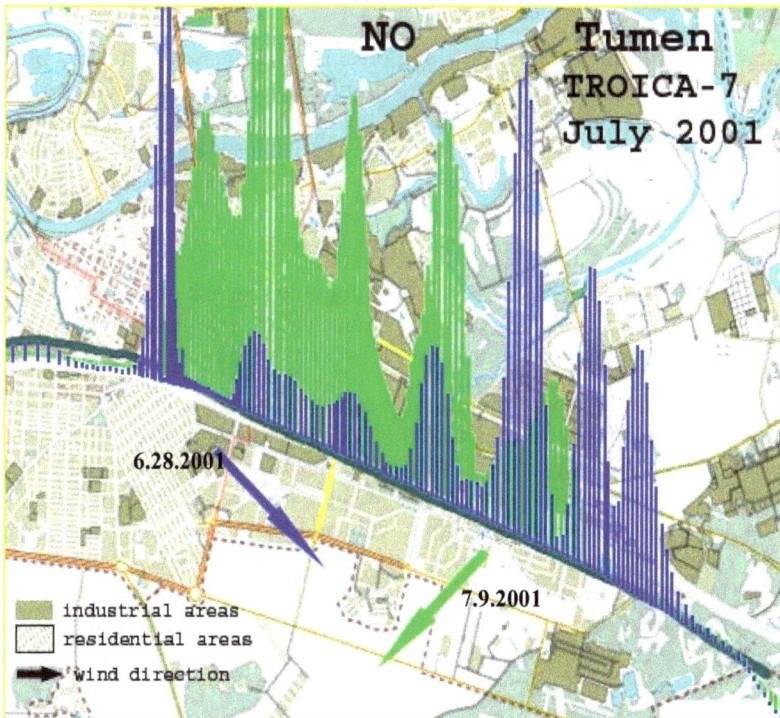

Fig. 9. Transects of Tumen' city on the routes eastward and westward. The structure of the plume profile from city's air pollution sources at the railroad depends on the wind direction.

Comparison of the different pollutants profile across the cities of different size (Table 3) is presented on Fig. 10. Observations show that ozone mixing rations in the cities are slightly lower than the ozone mixing ratios in rural areas due to ozone destruction by nitrogen oxides in urban air. Ozone mixing ratio in the cities of all sizes is the highest in spring. Spring maximum of ozone mixing ratio is typical for unpolluted sites of midlatitudes, while most of big cities are characterized by the broad spring-summer maximum with a clear dominance of the summer levels. Ozone destruction intensity in polluted air is significantly higher in Moscow than in the other Russian cities located along the Trans-Siberian railroad, where ozone behavior corresponds, on average, to the one in slightly polluted areas.

City groups	Population	Number of cities
Large	> 500 000	11
Middle	50 000÷ 500 000	34
Small	< 50 000	62

Table 3. Cities statistics according to population.

In all seasons (Fig. 10) mixing ratio of ozone precursors decreases in the city plumes. The daytime ozone mixing ratios exceed the background level in the medium and small size cities. Ozone destruction by nitrogen oxides prevails over ozone generation within large cities. During the warm period ozone mixing ratios in the cities' plume are comparable with the background (some generation takes place). More active ozone generation leading to the small exceed over background levels is observed in the plumes of the medium size and small size cities.

The level of atmospheric pollution characteristic for cold seasons exceeds significantly that characteristic for warm seasons due to the contribution of urban heating systems and coal heating systems functioning in passenger cars at railroad stations.

The daily variations of the primary pollutants in the cities are determined by the primary emissions, including traffic, and they are in less degree dependent on the functioning of industrial enterprises (Elansky et al., 2007). Pollution levels are influenced by the set up and destruction of temperature inversions, which are rather intensive in Siberia. If morning and evening rush hours often coincide with the inversion period, fast accumulation of pollutants in the boundary layer occurs. Such situations are quite common during cold seasons. In summer the duration and magnitude of the night temperature inversions is less than in cold period, diurnal peak of emissions occurs during the hours with substantial air mixing processes, which prevents accumulation of primary pollutants in the boundary layer.

Change of the pollution level in the cities gets clear when comparing observations of the primary pollutants level during different years. This difference is clearly related to city traffic grows and intensification of industrial activities after the economic crisis of 1991-1995. This correlation is illustrated by the Figures 11 (a,b), where NOx and CO mixing ratio trends calculated on the basis of the measurements performed in the course of 1996-2007 expeditions are shown. Fig. 11 shows a trend in the number of motor vehicles in Russian cities, which correlates with the observed changes in the primary pollutants levels.

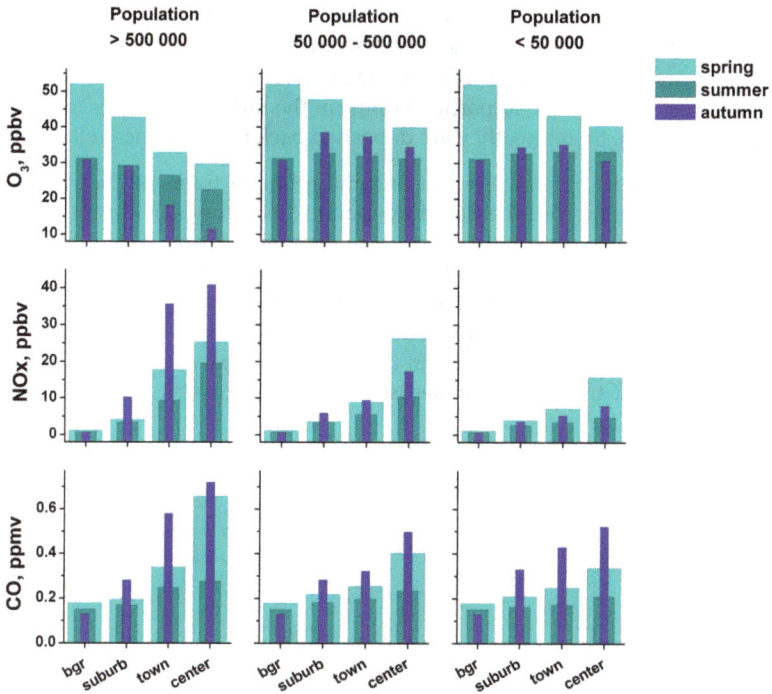

Fig. 10. Daytime mean (11:00-18:00 h. local time) O_3, NOx, CO mixing ratios averaged over the background (brg) and different parts of cities territories for spring, summer and autumn and for three groups of cities (TROICA-3 – TROICA-11).

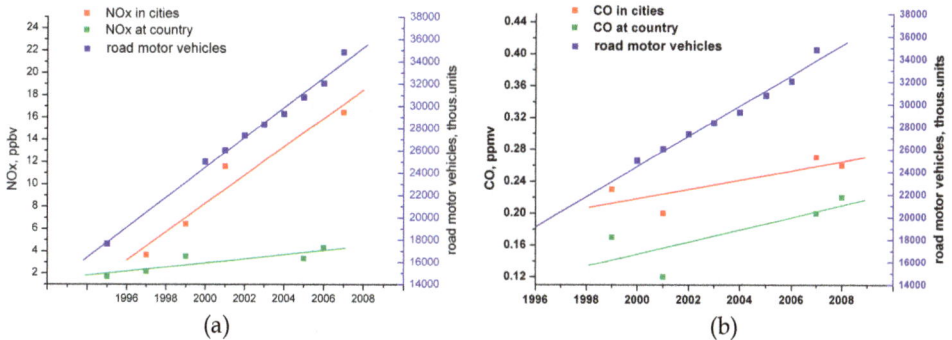

Fig. 11. The growth of the number of road motor vehicles in Russia and mean summer (a) NOx and (b) CO mixing ratios in cities and outside cities (TROICA-2,5,7,11 experiments).

5. Local impacts of railway transport on the atmospheric composition

The adverse effects of railroad systems on the environment are due to effects/emissions of stationary industrial objects (freight terminals, repair plants and depots) and freight and passenger trains themselves. The latter one is due abrasion of solid surfaces of car and locomotive components, pouring, leaking, and dusting out of transported loads, emissions from locomotive diesels and coal car-heating boilers. Large amounts of solids of different chemical species can be blown away from the trains transporting ores, coal, and other loose goods. Upward airflows blow solid particles up from the railroad bed. The majority of these particles are coarse, and, therefore, they are deposited back to the surface rather quickly.

Figures 12 and 13 demonstrate an effect of oncoming freight trains on the size distribution and aerosol mass concentrations measured by mobile laboratory. An abrupt increase of the portion of coarse particles after the passage of oncoming trains is caused by the lifting of solid particles from railroad beds and from freight cars. In contrast, oncoming trains have almost no influence on the atmospheric concentration of fine particles. An abrupt increase in the aerosol concentration is observed immediately after the passage of oncoming trains. The height of the concentration peak depends on the length and speed of the oncoming train and on its type: long freight trains increase aerosol concentrations to a greater extent than the passenger trains do.

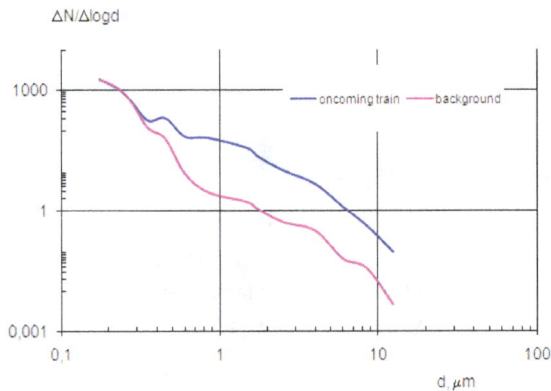

Fig. 12. Effect of oncoming trains on the aerosol size distribution.

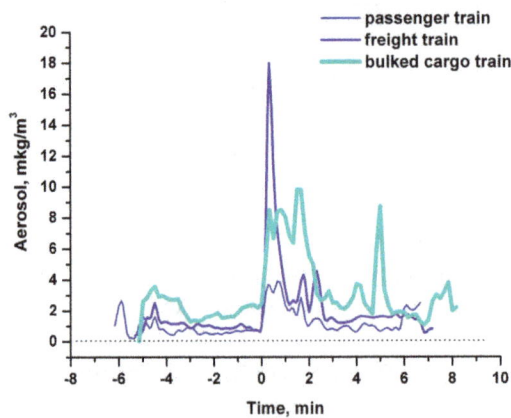

Fig. 13. Aerosol concentration peaks caused by the passage of oncoming trains of different types.

Enhanced atmospheric mixing ratios of unsaturated hydrocarbons (UHCs) were measured when trains containing oil-gasoline tanks passed by and also in vicinities of stations and siding lines where similar tanks were concentrated (Fig. 14). Emissions caused by trains consisting of tanks only and of different cars, including tanks, were estimated. The degree of UHC enhancement depends on the tank quantity in the train, the type of the transported good, hermeticity of the tank hatches, wind speed and direction, air temperature, and landscape.

It is important that hatches of empty railroad oil-gasoline tanks are open to avoid explosive vapor concentrations within tanks containing hydrocarbon residuals. As it can be seen in Fig. 14 the highest peaks are associated with the empty oil tank wagons. It should be noted that the effects of the train on the observed changes of the atmosphere composition are very local and can be seen only in the data of the highest spatial resolution. Other examples of the local-scale effects (power lines and industrial plumes) can be found in the relevant publications (Elansky & Nevraev, 1999; Elansky et al., 2001c).

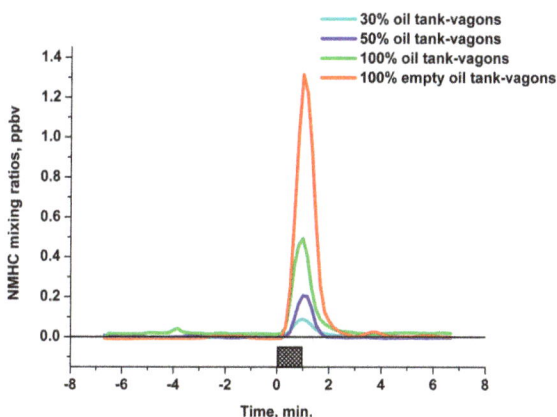

Fig. 14. Atmospheric UHC mixing ratios enhancement associated with oncoming trains with oil-gasoline tanks.

6. Conclusions

Researchers and engineers from different countries participated in the TROICA experiments. Due to close cooperation and unification of technical capabilities new and important results which increased our knowledge about atmospheric conditions above the North Eurasia were obtained. The mobile laboratory was a basis for this work. The laboratory was specially developed and constructed to carry out observations on the network of electrified railroads in the countries of the former Soviet Union.

By January 2009, the TROICA laboratory was equipped with the up-to-date measuring system for many atmospheric parameters. Field testing of the laboratory performed in the framework of the TROICA International Experiments showed that this laboratory has unique characteristics and has a number of advantages over other Russian means of monitoring the atmospheric gaseous and aerosol constituents:

-     wide variety of the parameters can be measured, including most of the key gaseous and aerosol pollutants, the radiative and thermodynamic parameters characterizing

atmospheric photochemical activity, and the parameters describing transport and deposition of atmospheric components;
- wide range of detectable mixing ratios variability from small natural variations of potential pollutants in the non-polluted atmosphere to the extreme variations of pollutants under critical conditions;
- multi-functionality, namely, simultaneous observations of the different species in different media (in atmosphere, water, soils, and vegetation bodies) which allows for direct observations of pollution on the state of ecosystems and on the environment;
- conformance with the international standards, namely, the instruments applied at the TROICA laboratory are supplied with international certificates, and undergo regular calibrations and standard validations of their applicability;
- universality of the measuring methods used in the moving platform, namely, individual elements of the instrumentation can be transported into other mobile platform or any buildings and installed there as mobile or stationary observational stations, and, in addition, simplified variants of the measuring system and software can be replicated and used for equipping the future Russian network stations;
- on-line functioning, namely, the software allows on-line analysis of the measurements, numerical simulation of photochemical and dynamic processes, prediction of extreme ecological situations, and on-line danger warning on the basis of satellite communications.

The support of TROICA experiments by the International Science and Technology Center played a very important role. The center united many organizations and made collaboration very efficient.

Accumulated data have proven to have a high potential to be further used for models (Tarasova et al., 2009) and satellite observations validation as well as provided better understanding of the atmosphere composition features over substantial part of the Russian territory.

7. References

Andronova, A.V.; Granberg, I.G.; Iordansky, M.A.; Kopeikin, V.M.; Minashkin, M.A.; Nevsky, I.A. & Obvintsev, Yu.I. (2003). Studies of the Spatial and Temporal Distribution of Surface Aerosol along the Trans-Siberian Railroad. *Izvestiya, Atmospheric and Oceanic Physics*, Vol. 39, Suppl. 1, pp. S27-S34.

Belikov, I. B.; Brenninkmeijer, C. A. M.; Elansky, N. F. & Ral'ko, A. A. (2006). Methane, carbon monoxide, and carbon dioxide concentrations measured in the atmospheric surface layer over continental Russia in the TROICA experiments. *Izvestiya, Atmospheric and Oceanic Physics*, Vol.42, Issue 1, pp. 46-59.

Bergamaschi P.; Brenninkmeijer, C.A.M.; Hahn, M.; Rockmann, T.; Schaffe, D.; Crutzen, P.J.; Elansky, N.F.; Belikov, I.B.; Trivett, N.B.A. & Worthy, D.E.J. (1998). Isotope analysis based on source identification for atmospheric CH_4 and CO sampled across Russia using the Trans-Siberian railroad. *J.Geophys.Res.*, Vol. 103, No. D7, pp. 8227-8235.

Crutzen, P.J.; Golitsyn, G.S.; Elansky, N.F.; Brenninkmeijer, C.A.M.; Scharffe, D.; Belikov, I.B. & Elokhov, A.S. (1996). Monitoring of the atmospheric pollutants over the Russian territory on the basis of a railroad mobile laboratory. *Dokl. Akad. Nauk*, Vol. 350, No. 6, pp. 819-823.

Crutzen, P.J.; Elansky, N.F.; Hahn, M.; Golitsyn, G.S.; Brenninkmeijer, C.A.M.; Scharffe, D.; Belikov, I.B.; Maiss, M.; Bergamaschi, P.; Rockmann, T.; Grisenko, A.M. &

Sevastyanov, V.V. (1998). Trace gas measurements between Moscow and Vladivostok using the Trans-Siberian Railroad. *Journal of Atmospheric Chemistry*, No. 29, pp. 179-194.

Elansky, N.F. & Nevraev, A.N. (1999). The electric tension lines as a possible source of ozone in the troposphere. *Doklady Earth Sciences*, Vol. 365, No. 4, pp. 533-536.

Elansky, N.F.; Golitsyn, G.S.; Vlasenko, T.S. & Volokh, A.A. (2000). Volatile organic compounds observed in the atmospheric surface layer along the Trans-Siberian Railroad. *Dokl. Akad. Nauk*, Vol. 373, No. 6, pp. 816-821.

Elansky, N.F.; Golitsyn, G.S.; Vlasenko, T.S. & Volokh, A.A. (2001a). Concentrations of Volatile Organic Compounds in Surface Air along the Trans-Siberian Railroad. *Izvestiya, Atmospheric and Oceanic Physics*, Vol. 37, Suppl. 1, pp. S10-S23.

Elansky, N.F.; Markova, T.A.; Belikov, I.B. & Oberlander, E.A. (2001b). Transcontinental Observations of Surface Ozone Concentration in the TROICA Experiments: 1. Space and Time Variability. *Izvestiya, Atmospheric and Oceanic Physics*, Vol. 37, Suppl. 1, pp. S24 - S38.

Elansky, N.F.; Panin, L.V. & Belikov, I.B. (2001c). Influence of High-Voltage Transmission Lines on Surface Ozone Concentration. *Izvestiya, Atmospheric and Oceanic Physics*, Vol. 37, Suppl. 1, pp. S92-S101.

Elansky, N.F.; Lokoshchenko, M. A.; Belikov.; Skorokhod, A.I. & Shumskii, R.A. (2007). Variability of Trace Gases in the Atmospheric Surface Layer from Observations in the City of Moscow. *Izvestiya, Atmospheric and Oceanic Physics*, Vol. 43, No. 2, pp. 219-231.

Elansky, N. F. (2007). Observations of the atmospheric composition over Russia using a mobile laboratory: the TROICA experiments. *International Global Atmospheric Chemistry*. Newsletter, No. 37, pp. 31-36.

Elansky, N.F.; Belikov, I.B; Berezina, E.V; et al. (2009). Atmospheric composition observations over Northern Eurasia using the mobile laboratory: TROICA experiments. *ISTC*. Moscow. 72 p.

Golitsyn, G.S.; Elansky, N.F.; Markova, T.A. & Panin, L.V. (2002). Surface Ozone behavior over Continental Russia. *Izvestiya, Atmospheric and Oceanic Physics*, Vol. 38, Suppl. 1, pp. S116-S126.

Hurst, D.F.; Romashin, P.A.; Elkins, J.W.; Oberlander, E.A.; Elansky, N.F.; Belikov, I.B.; Granberg, I.G.; Golitsyn, G.S.; Grisenko, A.M.; Brenninkmeijer, C.A.M. & Crutzen, P.J. (2004). Emissions of ozone-depleting substances in Russia during 2001. *J. Geophys. Res.*, Vol. 109, No. D14303, doi: 10.1029/2004JD004633.

Kopeikin, V.M. (2007). Monitoring of the soot aerosol of the atmosphere over Russia the TROICA international experiments. *Atmospheric and oceanic optics*, Vol. 20, No. 7, pp. 641 – 646.

Kopeikin, V.M. (2008). Monitoring of the submicron aerosol content of the atmosphere over Russia the TROICA international experiments. *Atmospheric and oceanic optics*, Vol. 21, No. 11, pp. 970 – 976.

Kuokka, S.; Teinilä, K.; Saarnio, K..; Aurela, M.; Sillanpää, M.; Hillamo, R.; Kerminen, V.-M.; Pyy, K.; Vartiainen, E.; Kulmala, M.; Skorokhod, A. I.; Elansky, N.F. & Belikov, I.B. (2007). Using a moving measurement platform for determining the chemical composition of atmospheric aerosols between Moscow and Vladivostok. *Atmos. Chem. and Phys.*, V. 7, No 18, 4793–4805.

Markova, T.A. & Elansky., N.F. (2002). Transcontinental Observations of the Surface Ozone and Nitrogen Oxide concentrations by using the Carriage-Laboratory. Ed. I. Barnes, *Global Atmospheric Change and its Impact on Regional Air Quality*, Kluwer Academic Publishers, Netherlands, pp. 249-254.

Markova, T.A.; Elansky, N.F.; Belikov, I.B.; Grisenko, A.M. & Sevast'yanov, V.V. (2004). Distribution of nitrogen oxides in the atmospheric surface layer over continental Russia, *Izvestiya, Atmospheric and Oceanic Physics*, Vol. 40, No. 6, pp. 811-813.

Oberlander, E.A.; Brenninkmeijer, C.A.M.; Crutzen, P.J.; Elansky, N.F.; Golitsyn, G.S.; Granberg, I.G.; Scharffe, D.H.; Hofmann, R.; Belikov, I.B.; Paretzke, H.G. & van Velthoven, P.F.J. (2002). Trace gas measurements along the Trans-Siberian railroad: The TROICA 5 expedition. *J.Geophys.Res.*, Vol. 107, No. D14, doi: 10.1029/2001JD000953.

Panin, L.V.; Elansky, N.F.; Belikov, I.B.; Granberg, I.G.; Andronova, A.V.; Obvintsev, Yu.I.; Bogdanov, V.M.; Grisenko, A.M. & Mozgrin, V.S. (2001). Estimation of Reliability of the Data on Pollutant Content Measured in the Atmospheric Surface Layer in the TROICA Experiments. *Izvestiya, Atmospheric and Oceanic Physics*, Vol. 37, Suppl. 1, pp. S81-S91.

Rockmann, T.; Brenninkmeijer, C.A.M.; Hahn, M. & Elansky, N. (1999). CO mixing and isotope ratios across Russia; trans-Siberian railroad expedition TROICA 3, April 1997. *Chemosphere: Global Change Science*, No. 1, pp. 219-231.

Shakina, N.P.; Ivanova, A.R.; Elansky, N.F. & Markova, T.A. (2001). Transcontinental Observations of Surface Ozone Concentration in the TROICA Experiments: 2. The Effect of the Stratosphere-Troposphere Exchange. *Izvestiya, Atmospheric and Oceanic Physics*, Vol. 37, Suppl. 1, pp. S39 - S48.

Tarasova, O.A.; Brenninkmeijer, C.A.M.; Assonov, S.S.; Elansky, N.F & Hurst, D.F. (2005a). Methane variability measured across Russia during TROICA expeditions. *Environmental Sciences*, Vol. 2(2-3), pp. 241-251.

Tarasova, O.A.; Brenninkmeijer, C.A.M.; Elansky, N.F. & Kuznetsov, G.I. (2005b). Studies of variations in the carbon monoxide concentration over Russia from the data of TROICA expeditions. *Atmospheric and Oceanic Optics*, Vol. 18, No. 5-6, pp. 511-516.

Tarasova, O.A.; Brenninkmeijer, C. A. M.; Assonov, S.S.; Elansky, N. F.; Röckmann, T. & Brass, M. (2006). Atmospheric CH_4 along the Trans-Siberian Railroad (TROICA) and River Ob: Source Identification using Stable Isotope Analysis. *Atmospheric Environment*, Vol. 40. No. 29, pp. 5617-5628.

Tarasova, O.A.; Brenninkmeijer, C. A. M.; Assonov, S.S.; Elansky, N.F.; Röckmann, T. & Sofiev, M. A. (2007). Atmospheric CO along the Trans-Siberian Railroad and River Ob: Source Identification using Isotope Analysis. *J. Atmos. Chem.*, Vol. 57, No. 2, pp. 135-152.

Tarasova, O.A.; Houweling, S.; Elansky, N. & Brenninkmeijer, C.A.M. (2009). Application of stable isotope analysis for improved understanding of the methane budget: comparison of TROICA measurements with TM3 model simulations. *J. Atmos. Chem.*, Vol. 63, No. 1, pp. 49-71

Timkovsky, I. I.; Elanskii, N. F.; Skorokhod, A. I. & Shumskii, R. A. (2010). Studying of Biogenic Volatile Organic Compounds in the Atmosphere over Russia. *Izvestiya, Atmospheric and Oceanic Physics*, Vol. 46, No. 3, pp. 319–327.

Turnbull, J. C.; Miller, J. B.; Lehman, S. J.; Hurst, D. .; Peters, W.; Tans, P. P.; Southon, J.; Montzka, S.; Elkins, J.; Mondeel, D. J.; Romashkin, P. A.; Elansky, N. & Skorokhod, A. (2009). Spatial distribution of $\Delta^{14}CO_2$ across Eurasia: measurements from the TROICA-8 expedition. *Atmos. Chem. and Phys.*, Vol. 9, pp. 175-187.

Vartiainen, E.; Kulmala, M.; Ehn, M.; Hirsikko, A.; Junninen, H.; Petäjä, T.; Sogacheva, L.; Kuokka, S., Hillamo, R., Skorokhod, A., Belikov, I., Elansky, N. & Kerminen, V.-M. (2007). Ion and particle number concentrations and size distributions along the Trans-Siberian railroad. *Boreal. Env. Res.*, Vol. 12, No. 3, pp. 375–396.

8

Mapping the Spatial Distribution of Criteria Air Pollutants in Peninsular Malaysia Using Geographical Information System (GIS)

Mohd Zamri Ibrahim, Marzuki Ismail and Yong Kim Hwang
Department of Engineering Science,
Faculty of Science and Technology,
Universiti Malaysia Terengganu
Kuala Terengganu
Malaysia

1. Introduction

Air pollution is defined as the introduction by man, directly or indirectly, of substances into the air which results in harmful effects of such nature as to endanger human health, harm living resources and ecosystems, cause material damage, interfere with amenities and other legitimate uses of the environment (United Nations Environment Programme [UNEP], 1999). The air pollution sources is categorized according to form of emissions whether gaseous or particulates. Air pollution sources also can be distinguished by primary or secondary air pollutants. Primary air pollutants are in the atmosphere that exists in the same form as in source emissions, whereas, secondary air pollutants are pollutants formed in the atmosphere as a result of reactions such as hydrolysis, oxidation, and photochemical oxidation (David Liu & Lipták, 2000). World Health Organization (WHO) had been listed six "classic" air pollutants: carbon monoxide (CO), lead, nitrogen dioxide (NO_2), suspended particulate matter (SPM) sulphur dioxide (SO_2) and tropospheric ozone (O_3) (World Health Organization [WHO], 1999).

In Malaysia, the country's air qualities were monitored by Department of Environment (DOE). The five main air pollutants focus in Malaysia air quality monitoring are CO, O_3, NO_2, SO_2 and respirable SPM of less than 10 microns in size (PM_{10}).

CO is a colorless, odorless, tasteless gas. It produced by the incomplete combustion of carbon-based fuels and by some industrial and natural processes. NO_2 is colorless, slightly sweet, relatively nontoxic gas. It has a pungent, irritating odor and, because of high oxidation rate, is relatively toxic and corrosive (Godish, 2004). O_3 is formed in the air by the action of sunlight on mixtures of nitrogen oxides and VOCs. O_3 concentrations are higher in the suburbs and in rural areas downwind of large cities than in the city centre, due to ozone removal from the air by reactions with nitric oxide and other components. SO_2 is a colorless gas, which emitted from the combustion of fossil fuels and industrial refining of sulphur-containing ores (McGranahan & Murray, 2003). Respirable SPM is a suspension of solid and liquid particles in the air (Moussiopoulos, 2003). The aerosol particles with an aerodynamic diameter of less than 10 μm and 2.5 μm are referred to as PM_{10} and $PM_{2.5}$ respectively (Mkoma et al., 2010). The particles are also sites for accumulation of compounds of moderate volatility (Brimblecombe & Maynard, 2001). Many particulates in the air are metal compounds that can catalyze secondary reactions in the air or gas phase to produce aerosols as secondary products (David Liu & Lipták, 2000).

1.1 Monitoring stations in Malaysia

Department of Environment (DOE) monitors the country's ambient air quality through a network of 51 Continuous Air Quality Monitoring stations (CAQM) (Department of Environment [DOE], 2008). Location of the CAQM in Peninsular Malaysia and East Malaysia are shown in Figure 1 and Figure 2 respectively.

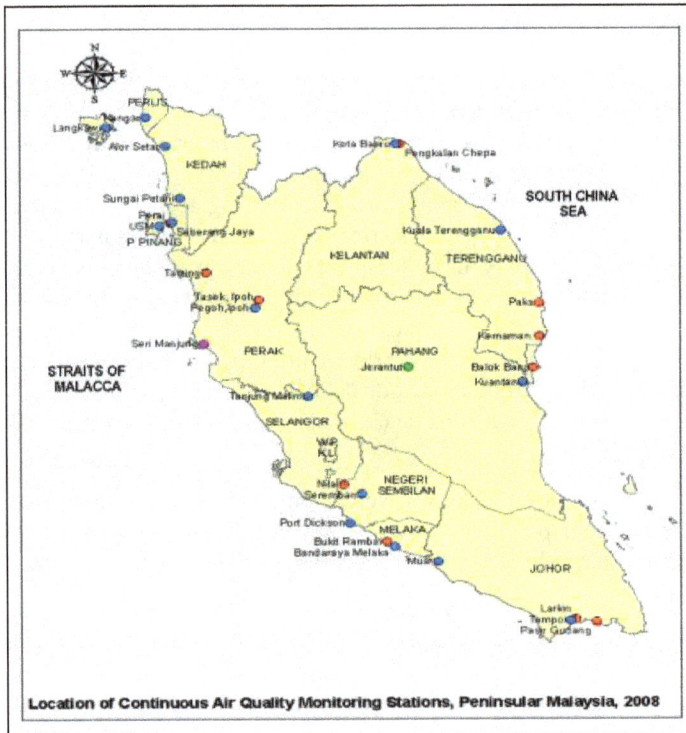

Fig. 1. Location of CAQM, Peninsular Malaysia, 2008 (Source: DOE, 2008).

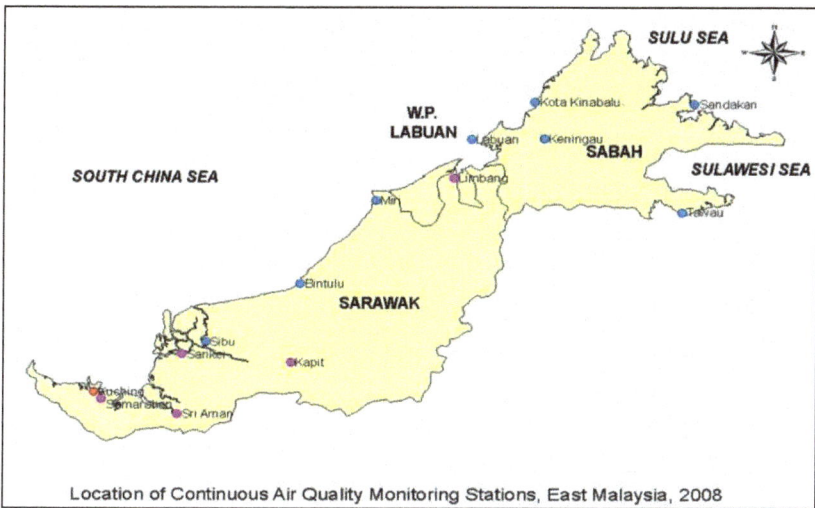

Fig. 2. Location CAQM, East Malaysia, 2008 (Source: DOE, 2008).

CAQM monitors a wide range of anthropogenic and natural emissions, there are SO_2, NO_2, O_3, CO, and PM_{10} (Parkland Airshed Management Zone [PAMZ], 2009).

CAQM (Figure 3) is designed to collect or measure data continuously during the monitoring period. CAQM typically include measurement instrumentation (for both pollutant gases and meteorological parameters); support instrumentation (support gases, calibration equipment); instrument shelters (temperature controlled enclosures); and data acquisition system (to collect and store data) (DOE, 2008).

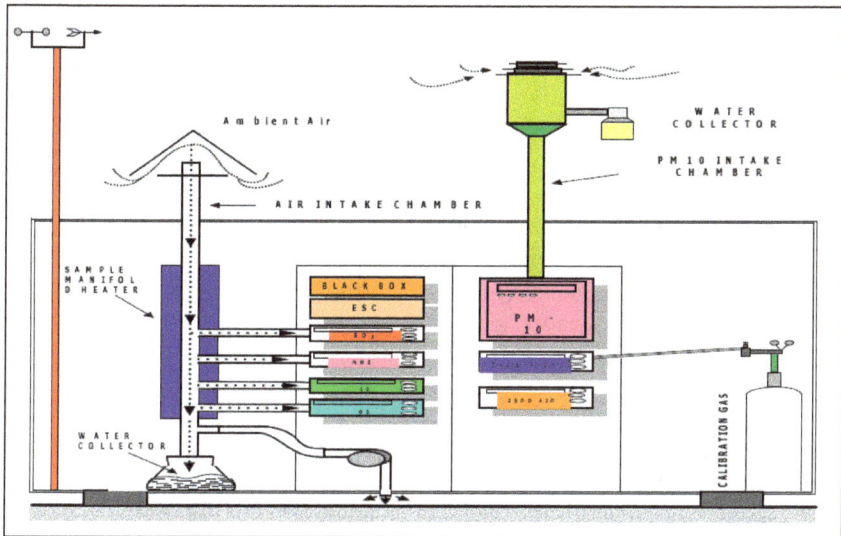

Fig. 3. Schematic Diagram of CAQM (Source: DOE, 2008).

DOE Malaysia publishes the air quality status to public using Air Pollutant Index (API) system on its website. The API system of Malaysia closely follows the Pollutant Standard Index (PSI) system of the United States.

An API system normally includes the major air pollutants which could cause potential harm to human health should they reach unsafe levels. The air pollutants included in Malaysia's API are O_3, CO, NO_2, SO_2 and PM_{10}.

The Table 1 show the category corresponds to a different level of health concern. The five levels of health concern and what they mean are (DOE, 1997):

- "Good" API is 0 - 50. Air quality is considered satisfactory, and air pollution poses little or no risk.
- "Moderate" API is 51 - 100. Air quality is acceptable; however, for some pollutants there may be a moderate health concern for a very small number of people.
- "Unhealthy" API is 101 - 200. People with lung disease, older adults and children are at a greater risk from exposure to ozone, whereas persons with heart and lung disease, older adults and children are at greater risk from the presence of particles in the air. Everyone may begin to experience some adverse health effects, and members of the sensitive groups may experience more serious effects.
- "Very Unhealthy" API is 201 - 300. This would trigger a health alert signifying that everyone may experience more serious health effects.
- "Hazardous" API greater than 300. This would trigger a health warning of emergency conditions. The entire population is more likely to be affected.

API	Descriptor
0-50	Good
51-100	Moderate
101-200	Unhealthy
201-300	Very unhealthy
>300	Hazardous

(Source: DOE, 1997)

Table 1. General human health effect and cautionary statements within each of the API categories.

1.2 Trend/status pollutions and air pollutants sources in Malaysia

An annual Environmental Quality Report (EQR) had been published in compliance with the Section 3(1)(i) of the Environmental Quality Act 1974. The EQR reported that air quality, noise monitoring, river water quality, groundwater quality, marine and island marine water quality, and pollution sources inventor in Malaysia.

In Malaysia EQR 2009, the air quality status reported in 5 regions of Malaysia. There are Selangor, northern region of west coast of Peninsular Malaysia (Perlis, Kedah, Pulau Pinang and Perak), southern region of west coast of Peninsular Malaysia (Negeri Sembilan, Melaka and Johor), east coast of Peninsular Malaysia (Pahang, Terengganu and Kelantan), as well as Sabah, Labuan and Sarawak. Meanwhile, the overall air quality for Malaysia in 2009 was between good to moderate levels most of the time. However, there was a slight decrease in the number of good air quality days recorded in 2009 (55.6 percent of the time) compared to that in 2008 (59 percent of the time) while remaining 43 percent at moderate level and only 1.4 percent at unhealthy level (DOE, 2009b).

The highest number of unhealthy air quality status days was recorded in Shah Alam (41 days) for the state of Selangor (Figure 4). Figure 4 also showed that, from year 2001 until 2009, Shah Alam has the highest number of unhealthy air quality than other in the states of Selangor. The air quality of the northern and southern region of west coast of Peninsular Malaysia was between good to moderate most of the time. Then, east coast of Peninsular Malaysia the air quality remained good most of the time and occasionally moderate. Last, Sabah, Labuan and Sarawak were generally good and moderate.

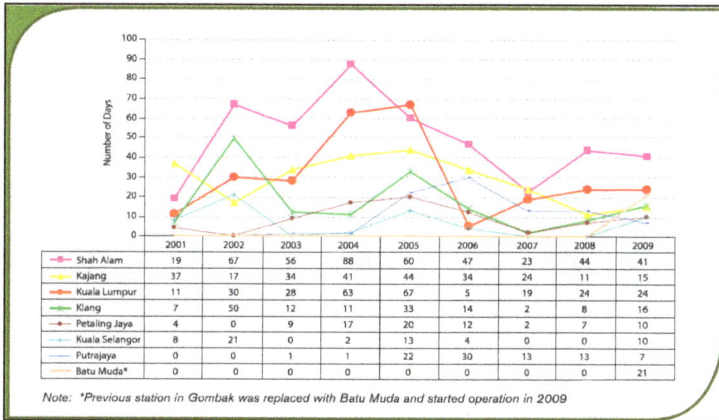

	2001	2002	2003	2004	2005	2006	2007	2008	2009
Shah Alam	19	67	56	88	60	47	23	44	41
Kajang	37	17	34	41	44	34	24	11	15
Kuala Lumpur	11	30	28	63	67	5	19	24	24
Klang	7	50	12	11	33	14	2	8	16
Petaling Jaya	4	0	9	17	20	12	2	7	10
Kuala Selangor	8	21	0	2	13	4	0	0	10
Putrajaya	0	0	1	1	22	30	13	13	7
Batu Muda*	0	0	0	0	0	0	0	0	21

Note: *Previous station in Gombak was replaced with Batu Muda and started operation in 2009

Fig. 4. Malaysia: Number of Unhealthy Days, Klang Valley, 2001-2009 (Source: DOE, 2009b).

The air quality trend of five air pollutants for the period of 1998 to 2009 was showed in the EQR 2009. In 2009 the annual average value of PM_{10} (Figure 5) was 44 μg/m^3 and no significant change compared to the annual average of PM_{10} (42 μg/m^3) in 2008. The higher level of PM_{10} recorded in several areas in Selangor and Sarawak from June to August 2009

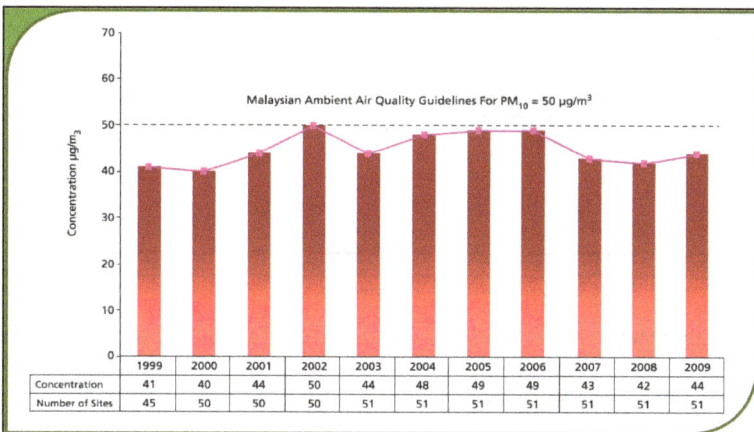

	1999	2000	2001	2002	2003	2004	2005	2006	2007	2008	2009
Concentration	41	40	44	50	44	48	49	49	43	42	44
Number of Sites	45	50	50	50	51	51	51	51	51	51	51

Fig. 5. Malaysia: Annual Average Concentration of PM_{10}, 1999-2009 (Source: DOE, 2009b).

was caused by the incidences of local peat land fires and trans-boundary haze (DOE, 2009b). However, the trend of the annual average levels of PM_{10} concentration in the ambient air from 1999 to 2009 complied with the Malaysian Ambient Air Quality Guidelines (PM_{10} concentration not exceed 50 $\mu g/m^3$).

In EQR 2009, it is estimated that, the combined air pollutant emission load was 1,621,264 metric tonnes of CO; 756,359 metric tonnes of NO_2; 171,916 metric tonnes of SO_2 and 27,727 metric tonnes of PM_{10} (DOE, 2009b). There was an increase in emission load for CO, NO_2 and SO_2 compared to 2008. Figure 6 (a-d) showed the emission by source for SO_2, PM_{10}, NO_2, and CO respectively. The results reveal that, power stations contributed the highest SO_2 and NO_2 emission load, 47% and 57% respectively. On the other hand, PM_{10} and CO the highest contributor were industries (49%) and motor vehicles (95%) respectively.

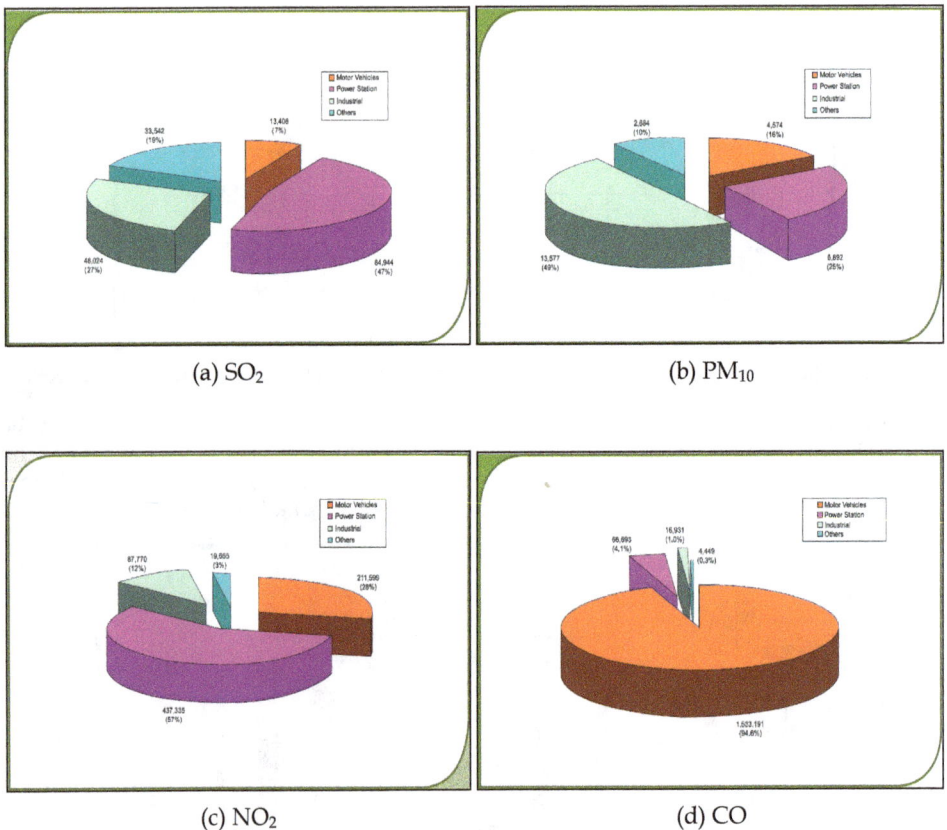

(a) SO_2 (b) PM_{10}

(c) NO_2 (d) CO

Fig. 6. (a-d): Malaysia: Emission by Sources (Metric Tonnes), 2009 (Source: DOE, 2009b).

2. Spatial air pollutants mapping in Malaysia

Presenting of the air pollutant concentration or API to public always is a challenge for the DOE. Air pollutants data from the monitoring station present in numerical or literal form are wearisome and cannot well present a large surrounding area. Meanwhile, actual situation of air pollution cannot be figured out to public. Moreover, the air pollutants data present in the numerical form have lack of geographical information. Beside, the way to get more actual pollution status in a huge area, have more monitoring stations is a high cost of the solvent method. Increase the monitoring station will increase the persistence of the pollution information. However, building new monitoring station is very costly. According to Alam Sekitar Malaysia Sdn. Bhd. (ASMA), an air monitoring station is cost about one million and the station only can present the area with the radius of 15 km. In other word, bigger areas for a States require more monitoring station and the higher cost of the air quality monitoring in the States.

A picture can describe thousand words, air pollutants statuses describe in an image (spatial map) can be more easily been visualized. In this century, presenting the data in a compact and full of information ways are preferred. There are many researcher nowadays attempt to study the dispersion of air pollutants with an image. In this study, a GIS-based approach of spatio-temporal analysis is attempted to use for presenting the air pollutions situation in an area.

Geographic information system (GIS) is an integrated assembly of computer hardware, software, geographic data and personnel designed to efficiently acquire, store, manipulate, retrieve, analyze, display and report all forms of geographically referenced information geared towards a particular set of purposes (Burrough, 1986; Kapetsky & Travaglia, 1995, as cited in Nath et al., 2000). The power of a GIS within the framework of spatio-temporal analysis depends on its ability to manage a wide range of data formats, which are represented by digital map layers extended by attributes with various observations, measurements and preprocessed data (Matějíček et al., 2006). The statistical data of the GIS can include area, perimeter and other quantitative estimates, including reports of variance and comparison among images (Nath et al., 2000). GIS is useful to produce the interpolated maps for visualization, and for raster GIS maps algebraic functions can calculate and visualize the spatial differences between the maps (Zhang & McGrath, 2003).

Nowadays, the applications of the GIS become wider. The increased use of GIS creates an apparently insatiable demand for new, high resolution visual information and spatial databases (Pundt & Brinkkötter-Runde, 2000). GIS able to do spatio-temporal analysis due to its ability to manage a wide range of data formats, which are represented by digital map layers extended by attributes with various observations, measurements and preprocessed data (Matějíček et al., 2006).

Concentration of pollutant present in spatio-temporal GIS-based image allow the reader more understand the real pollution level of the area. A GIS-based image with the coordination, geographical information, and the concentration of the air pollution can figure out or visualized the pollutant level of the study area. A GIS-based image of the spatio-temporal analysis only requires few set of data from different monitoring stations. Dispersion of the air pollutants can be produced by the spatio-temporal analysis with few point of the air monitoring data. Pollution level of Malaysia can be present with lesser monitoring station build.

Various technique of interpolating that GIS allow user to interpolate the variation of air pollutants. Inverse Distance Weighted (IDW), example, estimate of unknown value via a known value with the decrease of value through the increase of the distance as a simple interpolation method for air pollutants.

2.1 Inverse distance weighted of the PM_{10} spatial mapping federal territory Kuala Lumpur

Federal Territory Kuala Lumpur has the total areas of 243 km² with the population of 1,655,100 people in 2009. In year 2009, Kuala Lumpur has the highest population density in Malaysia with 6,811 people per km² (Department of Statistics, 2009). Federal Territory Kuala Lumpur has a rapid transformation and its wider urban region during the last decade of the twentieth century demands greater critical scrutiny than it has so far attracted (Bunnell et al., 2002). Kuala Lumpur is the social and economic driving force of a nation eager to better itself, a fact reflected in the growing number of designer bars and restaurants in the city, and in the booming manufacturing industries surrounding it (Ledesma et al., 2006). Figure 7 shows the average temperature and the rainfall in area of Kuala Lumpur, temperatures have not much different and humidity is high all year around.

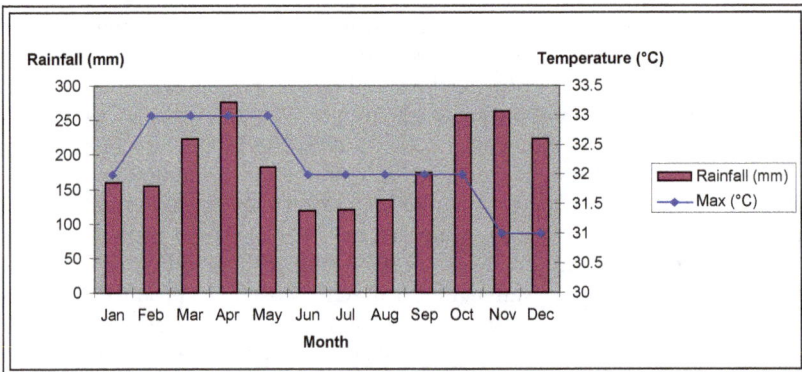

Fig. 7. Graph average daily temperature and rainfall (Source: Ledesma et al., 2006).

Air quality of a small area with a high population is the most concern issue for the government. It cannot be ignoring to study the dispersion of the pollutants. In year 1997, Kuala Lumpur and surrounding areas had been shrouded by haze with a pall of noxious fumes, smelling of ash and coal, caused by the fires in the forests to clear land during dry weather at neighbor country Indonesia's Sumatra Island (msnbc.com, 2010). In year 2005, the highest number, 67 days, of unhealthy day were recorded in Kuala Lumpur (DOE, 2006).

In section 2.1 and 2.2, IDW is the method used to interpolate the dispersion of PM_{10} concentrations in Kuala Lumpur. Changing of the study of pollutant with the air monitoring station had to change to more presentable of dispersion image form. The dispersion of the PM_{10} concentration in Kuala Lumpur using the interpolation of IDW is the first attempted to have spatial air pollutants mapping in Malaysia. Some more, API in Malaysia always show by the concentration of PM_{10} (DOE, 2009a), which mean that PM_{10} is the parameter most contribute to the air pollution in Malaysia. So that, PM_{10} was chose as study air pollutant. PM_{10} concentration data which collected from DOE are the main data were used.

There are three air monitoring stations located at the Federal Territory Kuala Lumpur. CAQM which build at Kuala Lumpur are station CA0012, station CA0016 and station CA0054 (Figure 8). Station CA0012 operated since December 1996 and ceased operation on February 2004. Whereas, Station CA0016 and CA0054 operated since December 1996 and February 2004 respectively until today both still well monitoring the air quality in Federal Territory Kuala Lumpur. However, to make the interpolation more persistent, all the PM_{10} data which the CAQM located in States of Selangor were obtained to do the interpolation of the dispersion in Kuala Lumpur. Figure 9 show the CAQM station location on Selangor map.

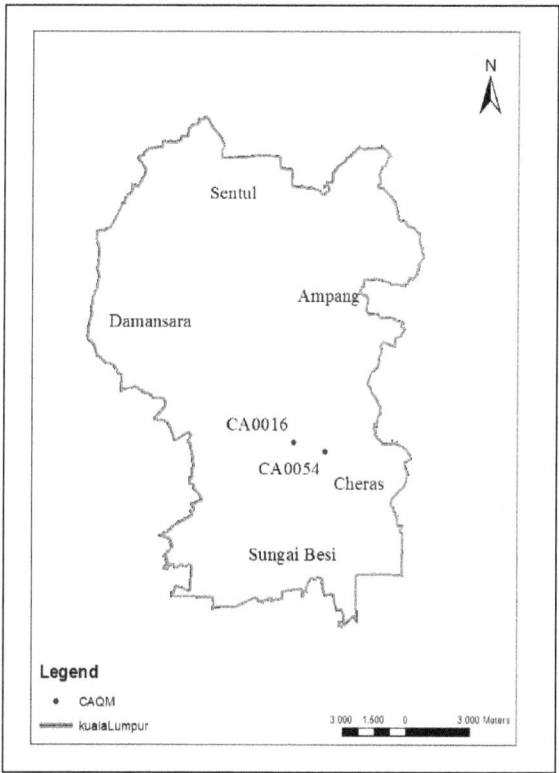

Fig. 8. Kuala Lumpur map with the located CAQM.

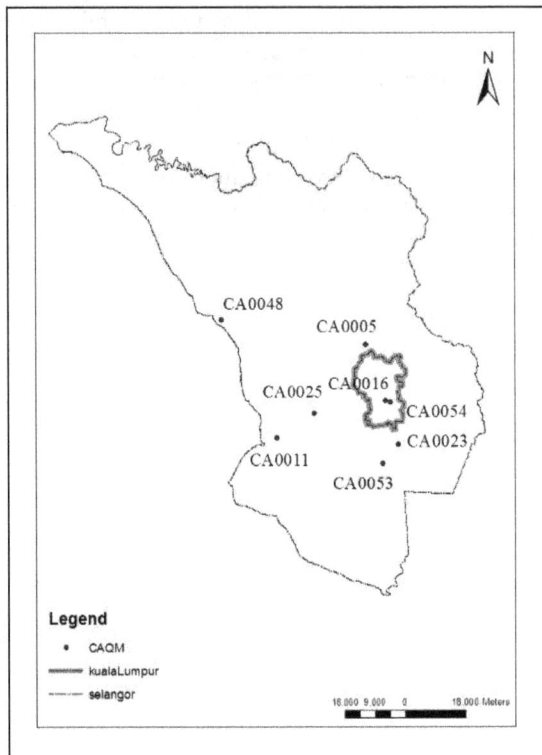

Fig. 9. Selangor map with the located CAQM.

Table 2 showed geographical information for the CAQM stations in Selangor, latitude and longitude had been converted to Rectified Skewered Orthomocphic (RSO) map projection, X-Y coordinate. RSO is a local map projection commonly used in Malaysia.

Station Code	Latitude	Longtide	x (meters)	y (meters)	Jan-04 ($\mu g/m^3$)	Feb-04 ($\mu g/m^3$)
CA0005	3°15.702'N	101°39.103'E	406223.4	360936.1	53	62
CA0011	3°00.620'N	101°24.484'E	379076.9	333208.9	65	73
CA0016	3°06.612'N	101°42.274'E	412048.4	344183.5	47	64
CA0023	2°59.645'N	101°44.417'E	416001.4	331336.1	33	48
CA0025	3°04.636'N	101°30.673'E	390551.5	340581.4	54	61
CA0048	3°19.592'N	101°15.532'E	362596.0	368240.6	57	68
CA0053	2°56'37.81N	101°41'54.25E	411326.1	325756.9	33	45
CA0054	3°06.376'N	101°43.072'E	413529.3	343750.0	44	77

Table 2. Geographical information for the CAQM stations in Selangor.

IDW interpolations for the dispersion PM_{10} concentration in Kuala Lumpur was computed by the GIS software validate with the in-situ monitoring (section 2.2.1). Then, the result will be further discussed in section 2.2. GIS software integrated collection of computer software and data used to view and manage information about geographic places, analyze spatial relationships, and model spatial processes. GIS provides a framework for gathering and organizing spatial data and related information so that it can be displayed and analyzed.

2.2 Decision making mapping

IDW method interpolates the pollutants concentration to a spatial air pollutants mapping. Spatial air pollutants mapping clearly show the dispersion of pollutant in a study area and leave a visual tool to the decision maker or public. Referring to the dispersion of PM_{10} concentrations in Kuala Lumpur, it is more easily been visualized the air pollution status in Kuala Lumpur.

Figure 10 show the IDW interpolation of the dispersion of PM_{10} concentration in Federal Territory Kuala Lumpur March 2004. Two points shown in the map are the located CAQM stations in Kuala Lumpur. Interpolation of the dispersion PM_{10} concentration show the PM_{10} concentrations decreasing when the distance from the CAQM station increasing. The nearest areas, Cheras, shows highest interpolate PM_{10} concentration which in the range 70 – 73 $\mu g/m^3$. The interpolation show areas of Cheras having a highest value this is due to the high

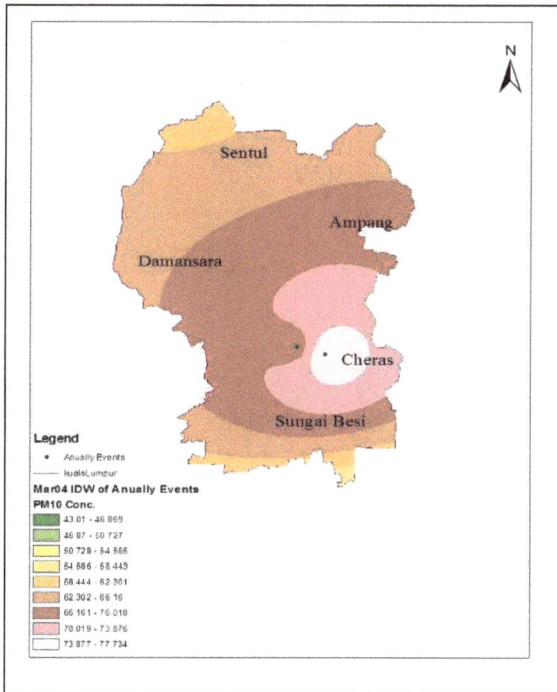

Fig. 10. Dispersion of PM_{10} Concentration at Kuala Lumpur March 2004.

PM_{10} concentration at the CA0054 obtained is high and the Cheras is the nearest to the CA0054. Then, areas of Ampang and Sungai Besi have the estimated PM_{10} concentrations about 66 – 70 $\mu g/m^3$ as well as areas of Damansara and Sentul having 62 – 66 $\mu g/m^3$ of estimation PM_{10} concentrations. The areas of Sungai Besi, Ampang, Damansara and Sentul respectively far from the CAQM areas, so that, the interpolation PM_{10} concentration decreasing across the areas.

One of the great powers of the GIS is the analysis of the values for every raster. Users can analysis the value of mean, maximum or minimum for a set of spatial air pollutants mapping. Figure 11 show the mean for the estimation dispersion of PM_{10} concentration from year 2004 until 2008. Averages of the dispersion of PM_{10} concentrations from year 2004 until 2008 are in the range of 44 – 54 $\mu g/m^3$. Areas of Cheras and Sungai Besi show the PM_{10} concentration between 52 – 54 $\mu g/m^3$. There are two factors effected the high average of the interpolation PM_{10} concentration at the middle areas of Kuala Lumpur. First, the values use for the interpolation, CAQM data, located at the middle areas of Kuala Lumpur. Second, the centering of human daily activities at the middle of Kuala Lumpur also effected the highest concentration of PM_{10} at the middle areas of Kuala Lumpur. Next, areas of Damansara and Ampang have about 49 – 52 $\mu g/m^3$ estimation of PM_{10} concentrations. As well as areas of Sentul show 47 – 49 $\mu g/m^3$ of PM_{10} concentrations. Areas far apart from the CAQM station, CA0016 and CA0054, have lower average PM_{10} concentration.

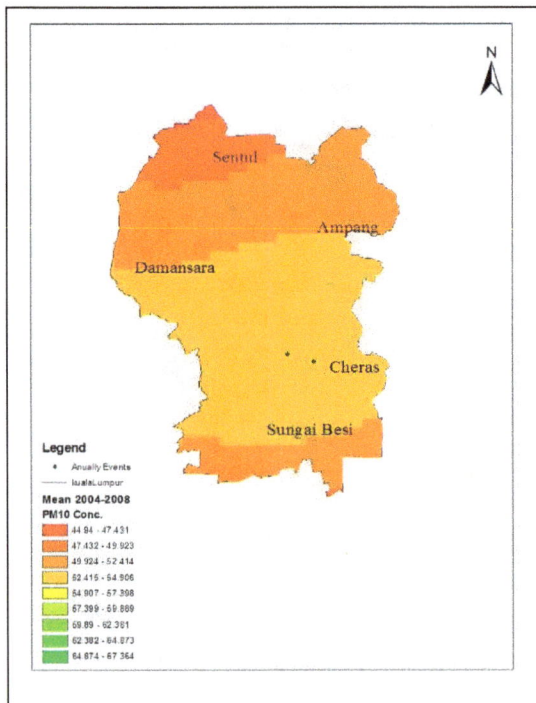

Fig. 11. Mean PM_{10} Concentration at Kuala Lumpur from 2004-2008.

Figure 12 shows the maximum for the estimation dispersion of PM_{10} concentration from year 2004 until 2008. Maximum of the dispersion of PM_{10} concentrations from year 2004 until 2008, at the middle areas of Federal Territory Kuala Lumpur, is in the range of 112 – 118 $\mu g/m^3$. The highest maximum values show are nearest to the CA0016 station, which mean that the maximum PM_{10} concentrations are obtained from the CA0016 station. The maximum values of 118 $\mu g/m^3$ have the API of 84 and this shows "Moderate" in the API status. Areas of Cheras, Sungai Besi, Damansara and Ampang show the PM_{10} concentration between 107 – 112 $\mu g/m^3$. Next, areas of Sentul show about 102 – 107 $\mu g/m^3$ of PM_{10} concentrations. All the areas in Kuala Lumpur show "Moderate" API status for the maximum PM_{10} concentration.

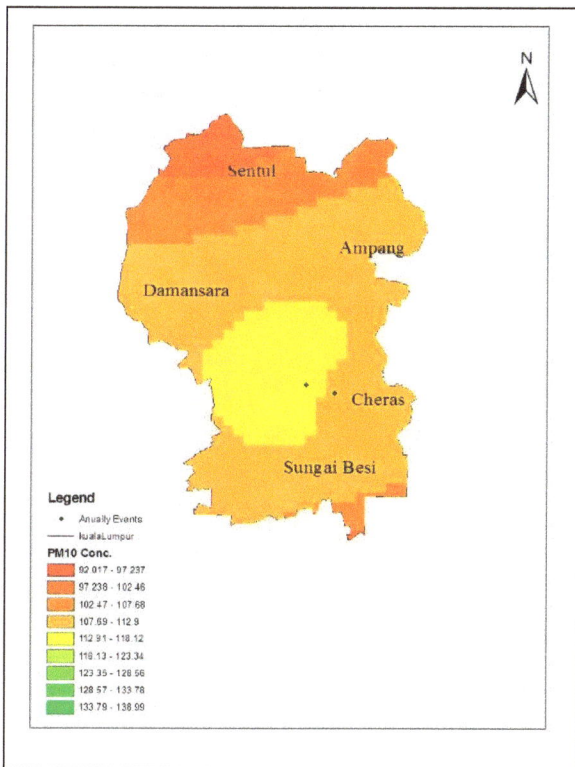

Fig. 12. Maximum PM_{10} Concentration at Kuala Lumpur from 2004-2008.

The weather in Malaysia is characterized by two monsoon regimes, namely, the Southwest Monsoon from late May to September, and the Northeast Monsoon from November to March (Weather Phenomena, 2009). The dispersion of PM_{10} concentration in Federal Territory Kuala Lumpur during Southwest Monsoon and Northeast Monsoon can be easily showed by the aid of analysis tool of GIS. The haze which shrouded the Kuala Lumpur and surrounding areas in year 1997 is one of the effects of the Southwest Monsoon. The wind blowing from the Southwest bring along the air pollutant from the neighbor Indonesia's Sumatra Island. Addition, during Southwest monsoon, Peninsular Malaysia will have less rain and dry. Otherwise, during Northeast monsoon, Peninsular Malaysia will have heavy rain and flooding (Ecographica, 2010).

The mean of the PM_{10} concentration dispersion in Kuala Lumpur during Northeast Monsoon year 2004 until 2008 is show in Figure 13 Averages of the dispersion of PM_{10} concentrations during Northeast Monsoon from year 2004 until 2008 are in the range of 39 – 50 $\mu g/m^3$. The higher average of PM_{10} concentration is about 47 – 50 $\mu g/m^3$, it shown in the middle areas of Kuala Lumpur. The highest mean PM_{10} concentrations shows are nearest to the station CA0016. Areas of Cheras, Sungai Besi, Damansara and Ampang show the PM_{10} concentration between 45 – 47 $\mu g/m^3$. Next, areas of Sentul show about 42 – 45 $\mu g/m^3$ of PM_{10} concentrations. The dispersion PM_{10} concentration of mean during Northeast monsoon is similar with the dispersion of maximum PM_{10} concentration along the year 2004 till 2008. Both image shown the highest value of PM_{10} concentrations are nearest to the CAQM station, CA0016.

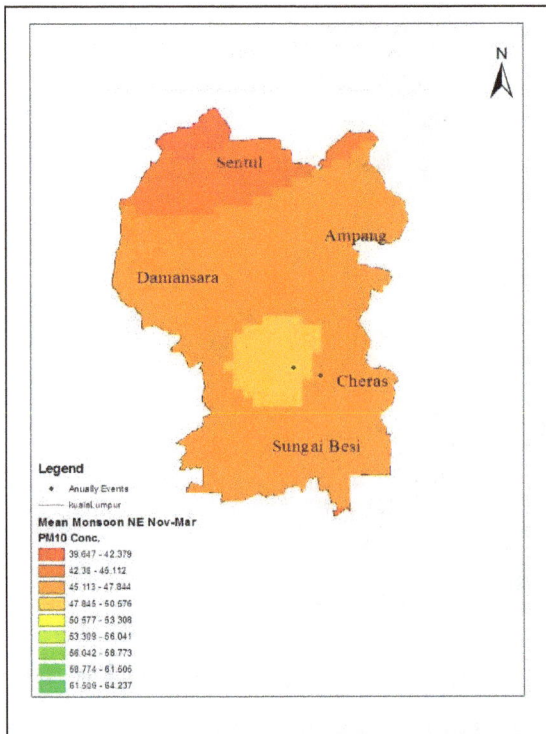

Fig. 13. Mean PM_{10} Concentration at Kuala Lumpur during Northeast Monsoon Nov-Mar 2004-2008.

Figure 14 shows the maximum for the dispersion of PM_{10} concentration during Northeast Monsoon from year 2004 until 2008. Maximum of the dispersion of PM_{10} concentrations during Northeast Monsoon from year 2004 until 2008, at the areas of Cheras and Damansara, is in the range of 54 – 58 $\mu g/m^3$. Areas of Sungai Besi and Ampang show the

PM_{10} concentration between 50 – 54 $\mu g/m^3$. Next, areas of Sentul show about 47 – 50 $\mu g/m^3$ of PM_{10} concentrations during the Northeast Monsoon.

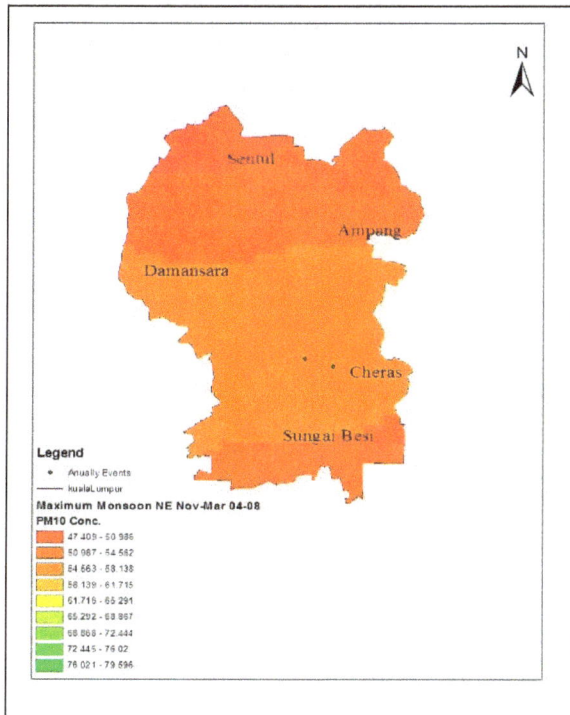

Fig. 14. Maximum PM_{10} Concentration at Kuala Lumpur during Northeast Monsoon Nov-Mar 2004-2008.

The dispersion PM_{10} concentration of mean during Northeast monsoon is similar with the dispersion of mean PM_{10} concentration along the year 2004 till 2008. During the wet season or Northeast monsoon, the maximum of PM_{10} concentration can be seen are the averaging of the PM_{10} concentration. Meanwhile, the mean of the PM_{10} concentration dispersion in Kuala Lumpur during Southwest Monsoon year 2004 until 2008 is shows in Figure 15. Averages of the dispersion of PM_{10} concentrations during Southwest Monsoon from year 2004 until 2008 are in the range of 52 – 60 $\mu g/m^3$. Areas of Cheras, Sungai Besi, and Damansara show the highest PM_{10} concentration between 57 – 60 $\mu g/m^3$. Next, areas of Ampang show about 55 – 57 $\mu g/m^3$ of PM_{10} concentrations. As well as areas of Sentul, the PM_{10} concentrations are in the range of 52 – 55 $\mu g/m^3$. The averaging of the PM_{10} concentration during Southwest monsoon or dry season is higher than the overall averaging PM_{10} concentration from year 2004 until 2008. This shows that Kuala Lumpur had a higher pollution during the dry season.

Figure 16 shows the maximum for the dispersion of PM_{10} concentration during Southwest Monsoon from year 2004 until 2008. Maximum of the dispersion of PM_{10} concentrations during Southwest Monsoon from year 2004 until 2008, at the middle areas of Federal

Territory Kuala Lumpur, is in the range of 72 – 75 $\mu g/m^3$. The highest maximum PM_{10} concentrations shows are around the areas of station CA0016. Areas of Cheras, Damansara, Sungai Besi and Ampang show the PM_{10} concentration between 69 – 72 $\mu g/m^3$. Next, areas of Sentul show about 63 – 66 $\mu g/m^3$ of PM_{10} concentrations during the Southwest Monsoon.

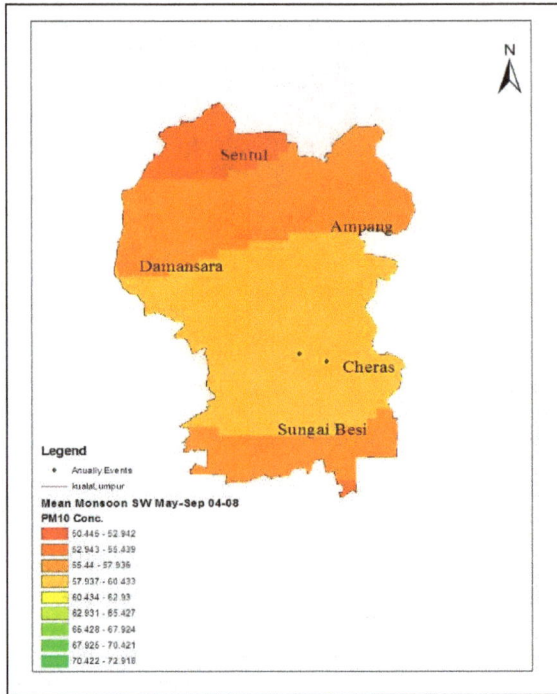

Fig. 15. Mean PM_{10} Concentration at Kuala Lumpur during Southwest Monsoon May-Sep 2004-2008.

The dispersion of PM_{10} concentration during Southwest monsoon or dry season is show similarity with the overall maximum PM_{10} concentration from year 2004 until 2008. During dry season or Southwest monsoon, the average of the PM_{10} concentration show higher value than wet season or Northeast monsoon. As well as the maximum PM_{10} concentration during Southwest monsoon also show higher than during Northeast monsoon. From Figure 15 and Figure 16, the highest PM_{10} concentration show during Southwest Monsoon. The maximum concentration of PM_{10} during Southwest Monsoon is 75 $\mu g/m^3$, whereas, during Northeast Monsoon the PM_{10} concentration is 58 $\mu g/m^3$. The overall maximum PM_{10} concentrations from year 2004 until 2008, it can be said similar with the mean PM_{10} concentration during Northeast monsoon and maximum PM_{10} concentration during Southwest monsoon. In other word, the overall maximum PM_{10} concentrations from year 2004 till 2008 most of it are contribute by the maximum PM_{10} concentrations during the dry season, Southwest monsoon. However, the maximum PM_{10} concentration shown in the Figure 12, Maximum PM_{10} concentration from year 2004 till 2008 is 118 $\mu g/m^3$. This value is not found in the PM_{10} concentration during monsoon, it can be said that it is in intermediate of the monsoon.

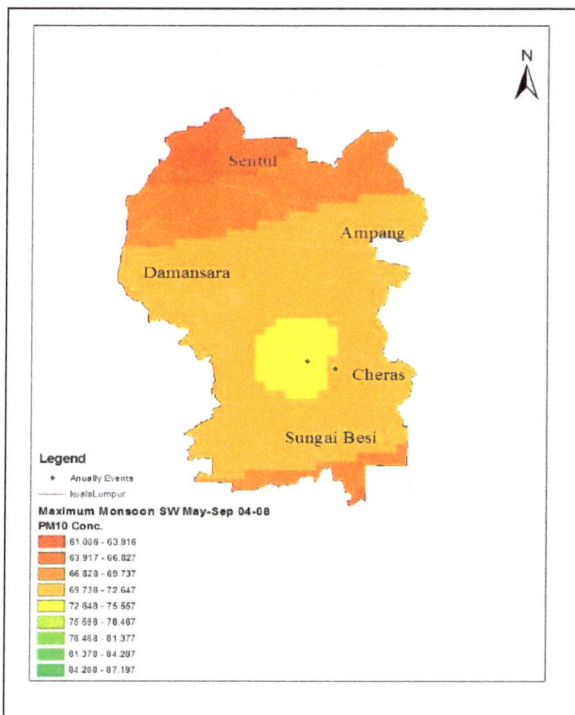

Fig. 16. Maximum PM_{10} Concentration at Kuala Lumpur during Southwest Monsoon May-Sep 2004-2008.

2.2.1 Validation of IDW interpolation

IDW interpolations for the dispersion PM_{10} concentration in Kuala Lumpur have to validate with the in-situ monitoring. Due to the differences device are using to detect the PM_{10} concentration, one relationship have to make between the data of both devices. The DOE PM_{10} concentration data is the main reference in this study. The compatible in-situ PM_{10} concentrations, after that, are used to validate the interpolation of the dispersion.

Equation 1, the regression between CAQM data and Casella Microdust Pro (PM_{10} detection device) data (Figure 17), had been formed. Equation 1 showed the linear relationship between CAQM data and Casella Microdust Pro data with the correlation coefficient of 0.148, weak relationship:

$$y = 46.173 + 0.098x \tag{1}$$

where, y is CAQM PM_{10} concentrations and x is Casella Microdust Pro PM_{10} concentrations respectively. However, equation 1 makes the Casella Micodust Pro data more compatible to the CAQM data.

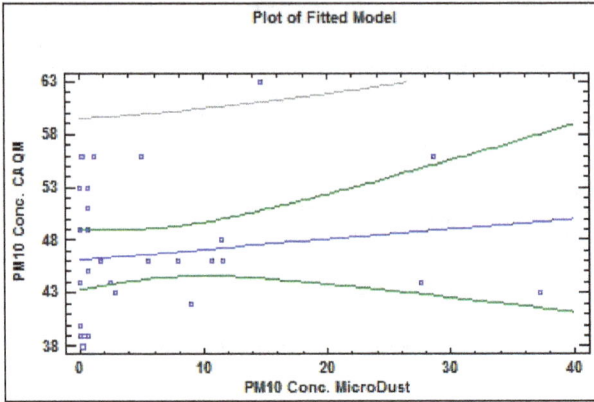

Fig. 17. Regression between CAQM data and Casella Mircodust Pro data.

Figure 18 showed the dispersion of PM_{10} concentration in Kuala Lumpur at 3pm 1/12/2009 with the selected in-situ monitoring point. The interpolated PM_{10} concentration for the selection monitoring point, Sekolah Jenis Kebangsaan (C) Chi Man, is extracted from Figure 18. The interpolated PM_{10} concentration the selected point is 30.178 $\mu g/m^3$. Then, the interpolated PM_{10} concentration, data on map, compare with the in-situ monitoring PM_{10} concentration.

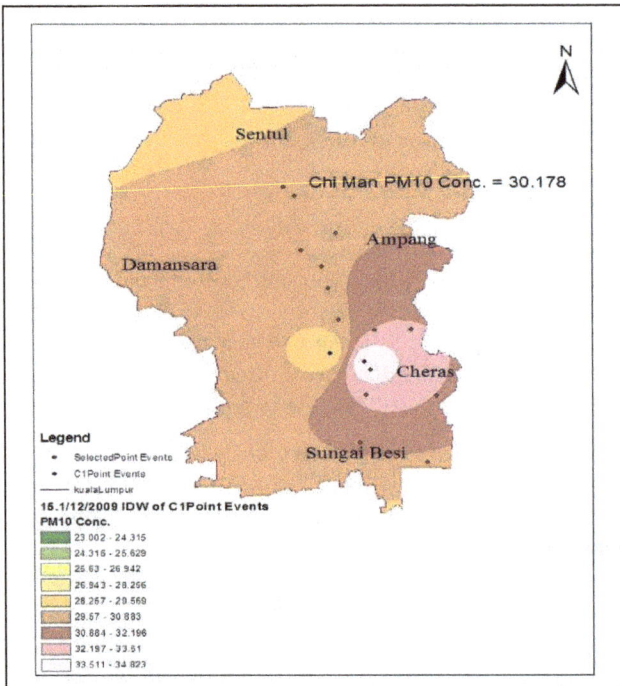

Fig. 18. Dispersion of PM_{10} concentration in Kuala Lumpur at 3pm 1/12/2009.

Equation 2, the regression between interpolated and in-situ monitoring PM_{10} concentrations, formed (Figure 19). The relationship between interpolated and in-situ monitoring PM_{10} concentrations is double reciprocal with the correlation coefficient 0.289:

$$\frac{1}{y} = 0.020 + \frac{0.042}{x} \tag{2}$$

where, y is the in-situ monitoring PM_{10} concentrations or the actual PM_{10} concentration; then, x is the interpolated or Spatio-temporal PM_{10} concentrations.

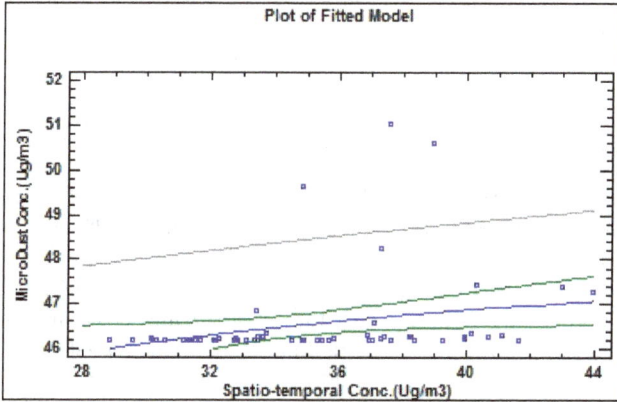

Fig. 19. Regression between interpolated and in-situ monitoring PM_{10} concentrations.

A model is considered validate if the calculated and measured values do not differ by more than approximately a factor of 2 (Pratt et al., 2004; Weber, 1982). Table 3 show a part of the interpolated (Spatio-temporal) and in-situ monitoring (Microdust) PM_{10} concentrations and have not differ by more than a factor of 2, thus, the IDW interpolation method can be used to describe the PM_{10} dispersion in Kuala Lumpur.

MicroDustConc.($\mu g/m^3$)	Spatio-temporal Conc.($\mu g/m^3$)
46.206	32.190
46.176	30.303
46.179	28.833
46.229	30.178
46.304	36.872
46.272	38.280
46.584	37.103
46.205	34.538
46.179	33.101
46.182	30.574

Table 3. A part of Microdust and Spatio-temporal PM_{10} concentration.

3. Conclusion

Malaysia air quality monitoring generally are focused on five air pollutants which are CO, O_3, NO_2, SO_2 and PM_{10}. Ambient air quality monitoring in Malaysia was installed, operated and maintained by ASMA under concession by the DOE through a network of 51 CAQM. CAQM monitors a wide range of anthropogenic and natural emissions; there are SO_2, NO_2, O_3, CO, and PM_{10}.

Similar to other ASEAN countries, the air quality in Malaysia is reported as the API. API system of Malaysia closely follows the PSI system of the United States. Four of the index's pollutant components (i.e., CO, O_3, NO_2, and SO_2) are reported in ppmv on the other hand PM_{10} is reported in $\mu g/m^3$. An individual score is assigned to the level of each pollutant and the final API is the highest of those 5 scores. To reflex the status of the air quality and its effects on human health, the ranges of index values is categorized as follows: good (0-50), moderate (51-100), unhealthy (101-200), very unhealthy (201-300) and hazardous (>300).Most of the time, API in Malaysia is always based on the concentration of PM_{10}.

EQR 2009 reported that, there was a slight decrease in the number of good air quality days recorded in 2009 (55.6 percent of the time) compared to that in 2008 (59 percent of the time) while remaining 43 percent at moderate level and only 1.4 percent at unhealthy level. It is estimated that, the combined air pollutant emission load was 1,621,264 metric tonnes of CO; 756,359 metric tonnes of NO_2; 171,916 metric tonnes of SO_2 and 27,727 metric tonnes of PM_{10}. There have a change of presenting air pollutants data in spatial from the wearisome numerical or literal form. The dispersion PM_{10} concentration, spatial PM_{10} mapping, was drawn by the IDW method. The results reveal that decreasing of estimation PM_{10} concentration with the increasing of the distance between the CAQM. The average of the dispersion of PM_{10} from year 2004 until 2008 is in the range of 44 – 54 $\mu g/m^3$. The maximum of the dispersion of PM_{10} concentrations from year 2004 until 2008 is about 112 – 118 $\mu g/m^3$. During the Northeast Monsoon and Southwest Monsoon, the mean are in the range 39 – 50 $\mu g/m^3$ and 52 – 60 $\mu g/m^3$ respectively. The maximum concentration of PM_{10} during Southwest Monsoon is 75 $\mu g/m^3$, whereas, during Northeast Monsoon the PM_{10} concentration is 58 $\mu g/m^3$. The intermediate monsoon period PM_{10} concentration is the highest value contribution of the PM_{10} concentrations to the maximum overall the year 2004 until 2008.

The different between the interpolated and in-situ monitoring PM_{10} concentration at the selected point have not differ by more than a factor of 2, thus, the IDW interpolation method can be used to describe the PM_{10} dispersion in Kuala Lumpur. Double reciprocal relationship formed between in-situ monitoring data and estimation IDW data with the correlation coefficient 0.289. Therefore, the IDW interpolation method is suitable for determining the air pollution status in areas which are not covered by the monitoring stations.

4. Acknowledgement

The authors would like to thank the Malaysian Department of Environment (DOE), for the air pollutions data provided. The authors also wish to express appreciation for the technical support provided by Dr. Razak Zakaria, Department of Marine Science, Faculty of Maritime Studies and Marine Science, Universiti Malaysia Terengganu.

5. References

Brimblecombe, P. & Maynard, R.L. (2001). *The urban atmosphere and its effects*, Imperial College Press, London.

Bunnell, T., Barter, P. A., & Morshidi, S. (2002). City Profile Kuala Lumpur metropolitan area A globalizing city-region. *Cities*, Vol 19, No. 5, p. 357-370.

Burrough, P.A. (1986). *Principles of Geographic Information Systems (1st ed)*, Oxford University Press, New York, 336pp.

David Liu, H.F & Lipták, B.G. (2000). *Air pollution*, Lewis, U.S.

Department of Statistics, Malaysia. (2009). *Basic population characteristics by administrative districts*, Department of Statistics, Malaysia.

DOE, (1997). *A guide to air pollutant index in Malaysia (API) (Third Edition)*, Department of Environment, Malaysia.

DOE, (2008). *Air Quality Monitoring*, Feb 27, 2010,
<http://www.doe.gov.my/en/content /air-quality-monitoring>.

DOE. (2006). *Malaysia environmental quality report 2005*, Department of Environment, Malaysia.

DOE. (2009a). *Air Pollutants Index Management System (APIMS)*, Sep 8, 2009,
<http://www.doe.gov.my/apims/>.

DOE. (2009b). *Malaysia environmental quality report 2009*, Department of Environment, Malaysia.

Ecographica Sdn. Bhd. (2010). *Tropical Malaysia weather "Be prepared for dry, wet and wild"*, March 26, 2010, <http://www.nature-escapes-kuala-lumpur.com/Malaysia-Weather.html>.

Godish, T. (2004). *Air quality (4th Edition)*, Lewis, U.S.

Kapetsky, J.M. & Travaglia, C. (1995). Geographical information system and remote sensing: an overview of their present and potential applications in aquaculture. In: Nambiar, K.P.P. & Singh, T. (Eds), *AquaTech '94: Aquaculture Towards the 21st Century*. INFOFISH, Kuala Lumpur, pp. 187-208.

Ledesma, C., Lewis, M., Lim, R. Martin, S., & Savage, P. (2006). *The rough guide to Malaysia, Singapore & Brunei*, Rough Guide, New York.

Matějíček, L., Engst, P., & Jaňour, Z. (2006). A GIS-based approach to spatio-temporal analysis of environmental pollution in urban areas: A case study of Prague's environment extended by LIDAR data. *Ecological Modelling*, 199 (2006), 261-277.

McGranahan, G. & Murray, F. (2003). *Air pollution & health in rapidly developing countries*, Earthscan, UK.

Mkoma, S.L., Wang, W., Maenhaut, W. & Tungaraza, C.T. (2010). Seasonal Variation of Atmospheric Composition of Water-Soluble Inorganic Species at Rural Background Site in Tanzania, East Africa. *Ethiopian Journal of Environmental Studies and Management*, Vol.3 (2).

Moussiopoulos, N. (2003). *Air Quality in Cities*, Springer, Germany.

Msnbc.com. (2010). *Hazardous haze shrouds Kuala Lumpur, port schools closed as government encourages residents to wear masks*, March 26, 2010,
<http://www.msnbc.msn.com/id/8908221/>.

Nath, S.S., Bolte, J.P., Ross, L.G., & Aguilar-Manjarrez, J. (2000). Applications of geographical information systems (GIS) for spatial decision support in aquaculture. *Aquacultural Engineering*, 23 (2000), 233-278.

Parkland Airshed Management Zone (PAMZ). (2009). *Continuous*, Jan 10, 2010, <http://www.pamz.org/air-quality/continuous/>.

Pratt, G.C., Chun, Y.W., Bock, D., Adgate, J.L., Ramachandran, G., Stock, T.H., Morandi, M., & Sexton, K. (2004). Comparing Air Dispersion Model Predictions with Measured Concentrations of VOCs in Urban Communities. *Environmental Science & Technolology*. 38 (7), 1949–1959.

Pundt, H. & Brinkkötter-Runde, K. (2000). Visualization of spatial data for field based GIS. *Computer & Geosciences*, 26 (2000), 51-56.

United Nations Environment Programme (UNEP). (1999). *Report on the development and harmonization of environmental standards in East Africa*, August 22, 2011, <http://www.unep.org/padelia/publications/VOLUME2K32.htm>.

Weather Phenomena. (2009). *Monsoon*, Jan 2, 2010, <http://www.kjc.gov.my/english/education /weather/monsoon01.html>

Weber, E. (1982). *Air pollution: Assessment Methodology and Modelling*, Plenum Press, New York.

World Health Organization (WHO). (1999). *Guidelines for Air Quality*, WHO, Geneva.

Zhang, C.& McGrath, D. (2003). Geostatistical and GIS analyses on soil organic carbon concentrations in grassland of southeastern Ireland from two different periods. *Geoderma*. 119 (2004), 261-275.

9

Bio-Monitoring of Air Quality Using Leaves of Tree and Lichens in Urban Environments

M. Maatoug, K. Taïbi, A. Akermi, M. Achir and M. Mestrari

Faculty of Natural Sciences,
Ibn Khaldoun University, Tiaret
Algeria

1. Introduction

Air pollution is the presence of pollutants in the atmosphere from anthropogenic or natural substances in quantities likely to harm human, plant, or animal life; to damage human-made materials and structures; to bring about changes in weather or climate; or to interfere with the enjoyment of life or property. In developing countries, industrial growth and population increase, together with rising standards of living will probably lead to patterns of motorization that resemble those of industrialized countries. Many urban areas have high concentrations of air pollution sources resulting from human activities; sources such as motor vehicle traffic, power generation, residential heating and industry. Urban air pollution is a major environmental problem in the developing countries of the world. It not only represents a threat to human health and the urban environment, but it can also contribute to serious regional and global atmospheric pollution problems (Maatoug 2010). Motor vehicles are now recognized as the major contributor due to their emissions of suspended particulate matter including heavy metals. Meteorological and topographical conditions affect dispersion and transport of these pollutants, which can result in ambient concentrations that may harm people, structures, and the environment. Air pollution is worse in locations with unfavorable topographical or meteorological characteristics. Meteorological factors such as thermal inversions restrict dispersion of pollutants and result in high ambient pollutant concentrations. Unfavorable topography and wind direction have similar effects.

Current scientific evidence indicates that urban air pollution, which is derived largely from combustion sources, causes a spectrum of health effects ranging from eye irritation to death. Recent assessments suggest that the impacts on public health may be considerable. The health effects of pollutants depend on many factors, including the number and age group of exposed people and their health status, ambient concentrations and types of pollutants, and dose-response functions. Quantifying the magnitude of these pollutants and health impacts in cities worldwide, however, presents considerable challenges owing to the limited availability of information on both effects on health and on exposures to air pollution in many parts of the world.

Heavy metals in ambient air also originate from emissions from coal combustion and various trace metals-based industries. They are the subject of special attention because of the risks they may pose to human health and dangers associated with their persistence in ecosystems. Chronic sources of trace metals in urban highway have two origins: vehicles and road infrastructures. The pollutants emissions from vehicles are due in part to abrasion and corrosion of materials of vehicle and in other part to the use of different fluids (Delmas-Gadras, 2000). The two main metallic pollutants emitted, lead and zinc, are mainly found in the exhaust gas and brake linings (75% of lead content in gasoline is emitted in engine exhaust), but zinc is also present in tires, lubricants and especially in the guardrail (Deletraz, 2002). However, the brakes are an important source of copper. The trace metals are naturally present in soil in small quantities. They are partially released during the degradation of bedrock and form the endogenous pool called pedogeochemical bottom (Baize, 1997). A second pool, more or less important depending on geographic situation and exogenous supplies comes mainly from the industries, transport and agricultural activities (Fernandez-Cornudet, 2006).

	Copper	Zinc	Lead
Agricultural wastes	55%	61%	20%
Municipal wastes	28%	20%	38%
Atmospheric deposition	16%	40%	40%

Table 1. Percentages of different sources in the annual average land enrichment by heavy metals (Fernandez-Cornudet, 2006).

Biomonitoring is the technique base on use of organisms that have the ability to store and accumulate contaminants in their tissues, "bioaccumulation", under the control of several mechanisms of setting and transfer. This technique has become attractive complement of traditional methods for measurements of air quality. The identification of pollution within sensitive organisms can also allow detection of air quality degradation before it severely affects the biota or humans. Sensitive plants can be real bioindicators of pollution. Urban vegetation can interact, directly and indirectly, with local and regional air quality by altering the urban atmospheric environment. Trees can change local meteorology, alter pollution concentrations in urban areas, and remove gaseous air pollution or intercepting airborne particles. Phytoremediation is a method of environmental treatment that makes use of the ability of some plant species to accumulate certain elements, including heavy metals, in amounts exceeding the nutrition requirements of plants. Some plant species have fairly high specificity towards individual pollutants and can be used to show variations in pollutant concentration, either with time at one place, or from place to place.

Our study constitutes an attempt to evaluate the air quality in semiarid region, Tiaret city, from Algeria. The first aim of this work (case I) was to assess the most frequent atmospheric heavy metal accumulated in leaves of different urban plant species in the context of their usefulness in urban environment as bioindicator or in phytoremediation. In another version of this idea (case II), in order to determine the main source of pollutants from a part and their spatial repartition in the city in the other part (cause of the immobility of urban trees and their heterogeneous repartition), lichens have been used. Lichens have no root system to take up minerals from the substrate on which they grow, and accumulate materials such as heavy metals from wet and dry deposition and also they are easy to transplant and transport.

2. Case I Use of two bioaccumulative plant species (plane on maple leaves *Platanus acerofolia* and Cypress evergreen *Cupressus sempervirens*) to assess the impact of heavy metals (Pb, Zn, Cu) originated from urban traffic in the city of Tiaret (Algeria)

Optimal use of plants as a means of investigation of air pollution concerns cases where the path of soil contamination is either nil, negligible or known. This is why most plants used for these studies are mosses and lichens with, among others, to their lack of root system.

Indeed, the major contaminations of plant leaves by heavy metals are of two kinds:

- Diffuse air intakes, by deposition on the aerial parts of plants, according to the bioclimatic conditions (wind, rain ...), the volatility and transport of pollutants. They are deposited on the leaf area of the plant (cuticle and epidermal hairs). This is an external assimilation of metal pollutants. In general, the accumulation of metals is different depending on the plant considered and plant parts analyzed. The levels of pollutants measured recently in leaves are usually higher than those detected in the stems and roots (Kupper et al., 2001).
- Inputs via the ground. The trace metals can be released into the atmosphere and fall again by dry or moist way into soils along road. They are also likely to be deposited on the floor and then be driven by the storm water runoff. The trace metals present in surface water can migrate into the soil during the infiltration of water and contribute to the degradation of soil quality, reaching groundwater and impair water resources (Delmas-Gadras, 2000). In the city of Tiaret, the main contribution to trace metals is done through the atmosphere, including emissions from road traffic.

Edwards (1986) showed that the metal enters the plant through the roots to the leaves, and it is chelated by organic molecules that support it so it does not disturb the functioning of the plant cell. This is an internal assimilation of metal pollutants. The transfer of a trace element from the soil to harvest depends on both parameters related to soil and plant-specific factors. It follows as a kind of meeting between ground supply and plant demand, which are not totally independent. In particular, plant can modify the soil offer by changing, including root exudation of various compounds (protons, complex organic molecules for example), the physical and chemical conditions that govern the solubility of the element on the solid phase and its speciation in solution. This feedback effect of the plant availability of trace elements is currently difficult to quantify (Agency for Environment and Energy Management ADEME, 2005).

As part of this work, the study is focused on the importance of using plants as bio-indicators of airborne contamination by heavy metals issue from road traffic, the important source of contamination in the city. It aims to establish that the leaves of trees growing in urban areas have the potential to serve as indicators for relative quantification of metal air pollution resulting from road traffic. The emissions of Pb, Zn and Cu original traffic are mainly in the form of fine particles which are then collected by the leaf surfaces.

2.1 Materials and methods

2.1.1 Site selection and plant material

Five sites were selected in this investigation, four urban sites designated by S1, S2, S3 and S4 are located in the center of Tiaret[1] where the road sector is important and a control site. Each site consists of six healthy trees, consistent and having substantially the same age. There are three planes on maple leaves (*Platanus acerofolia*.Willd)[2] and three evergreen cypresses (*Cupressus sempervirens*.L)[3]. The control site is away from sources of contamination of air pollution and serves as a reference when comparing with the contaminated sites.

Note that the plane leaves attach and accumulate heavy metals during periods of summer and spring before falling in autumn. However, the cypress leaves accumulate heavy metals throughout the year. These two species are widely grown in the study sites and the concentrations of lead, zinc and copper issues from road traffic were determined in leaves. It was designed to quantify the levels of pollution air by these three metals from the process of bioaccumulation in these species.

It is important to note that the two sites S1 and S2 are located in a steep area assigned a very high traffic where the frequency of braking and idling is extremely important. However, sites S3 and S4 are located in a road sector relatively less important than sites S1 and S2 (Figure 1).

2.1.2 Samples collection

For each tree, about fifty shoots were taken from ground level to avoid contamination due to projections from the ground.

The material was taken to limit contamination: without losses or pollution, avoiding the use of tools or containers that may contaminate the sample (tool steel or stainless steel containers whose walls contain pigments of trace elements, such as PVC).

---

[1] The study was conducted in the city of Tiaret, which is located in the northwest of Algeria between the mountainous Tell chain in the north and the mountainous Atlas chain in the south, at an altitude of 980 m on average. The climate is Mediterranean semi-arid with average annual rainfall of 400 mm/year. The prevailing winds are from the west and northwest, their average speeds range from 3 to 4m/second. The city of Tiaret has over 200 km of urban roads. Its automobile park contains 6284 vehicles of all types, of which 11% of them are new cars (National Agency for the Development of territories ANAT, 2005: personnel communication). In 2006, the automobile park of the Wilaya (department) of Tiaret contains 8015 registered vehicles. This park is highly heterogeneous due to the variety of vehicles (individual or utility vehicle, gasoline or diesel, old or recent, etc.).. Of these, new cars (from 0 to 5 years) represent only 11% of the park, however, the cars of more than 11 years account for 74%. Yet it is precisely these older vehicles are more polluting [10].

[2] Hybrid tree between *Platanus orientalis* and *Platanus occidentalis*. The plane tree on leaf of maple is a tree commonly reaching 30 m high. Often planted as an ornamental and alignment tree. Palmately lobed leaves are alternated, with deciduous large stipules and hairy. Leaf hairs are particularly abundant on the underside of young leaves.

[3] Coniferous tree, up to 30 m high. It was firstly used as an ornamental plant and in public parks, and then it was also used as windbreaks and reforestation. It consists of two distinct varieties: the variety *horizontalis* with branches almost perpendicular to the trunk and crown basically conic, the variety *pyramidalis* has a fastigiate crown with branches leaning against the trunk, short or long. The leaves are opposite, evergreen, with small scales of 0.5 to 1 mm, slender, finely serrated and closely applied to the branches.

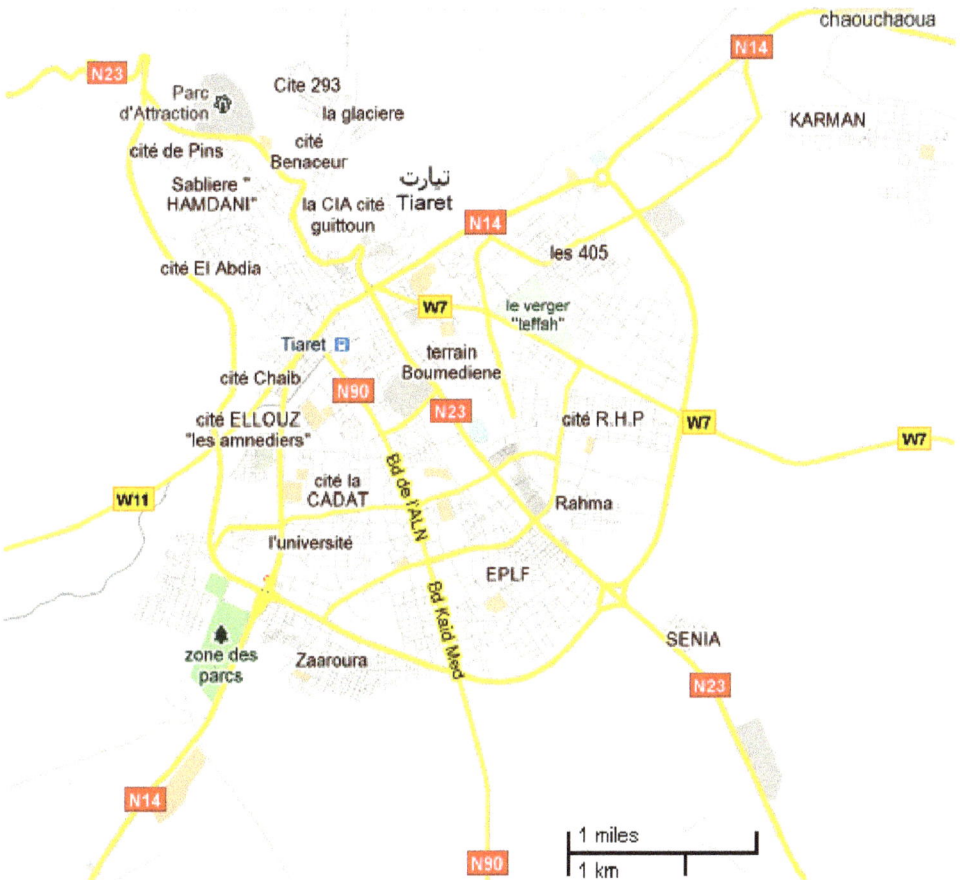

Fig. 1. Study site.

The samples were taken after a period without rain, one month after the release of young leaves and needles (Hébrard-Labit et al, 2004). Harvested leaves previously measured have been conserved in plastic bags with a rubber band, to ensure the conservation of water evaporation.

2.1.3 Samples treatment

In the laboratory, shoots were harvested without previous washing, were the subject of a series of transactions that are:

- Weighing: for the weight of the fresh material FM, about 200 mg of fresh shoots were harvested in each tree,
- Drying: the usual method is drying in an oven at 40±2 °C then dried leaves were weighed for the dry matter weight DM determination,

- Grinding: this step is highly critical as it can be a source of contamination or losses. For the leaves, in this case, the coffee grinder that allows incorporating the wad to the powder plant is used. The materials of the grinder are made of titanium, steel guaranteed without heavy metal.
- Mineralization and dissolved: the fine powder obtained is placed in an acid solution and oxidant (0.5 ml of nitric acid) and then heated in a water bath for 24 h, until the destruction of organic matter. Tubes that have been supplemented by boiling 10 ml of distilled water. This method allows the determination of all trace minerals.

The quantification of metals in solution is carried out by atomic absorption spectrometry with electrothermal atomization mode (Spectrometer Perkin Elmer 100).

2.2 Results and discussion

2.2.1 Comparison Fresh Matter/Dry Matter (FM/DM) ratio between control and polluted sites

Many authors have demonstrated that metals induce, from a certain threshold, biochemical and physiological dysfunctions. The macroscopic effects of aerial parts appear for high concentrations. It is essentially a reduction in leaf length and biomass produced (Vazquez et al, 1992; Brown et al, 1994; Ouzounidou et al, 1995; Ouzounidou et al, 1997).

In contact with a plant, the pollutant can be deposited on the surface then causing an alteration in the activities of photosynthesis and respiration (dust deposits), or into the tissues of the plant, causing tissues lesion, metabolic dysfunctions and disorders of the regulatory mechanisms.

The damages observed, mainly external, accompanied by a decrease in crop production and consequently a certain decline in agricultural and forestry yields (Belouahem, 1993). The excess of lead, for example, in plants induced physiological and biochemical injuries decreasing photosynthesis and transpiration thus leading to growth retardation (Bazzaz et al., 1974).

The report on fresh/dry matter is one of the indicators of air quality in a given urban areas. When the air is healthy, the plant development is normal, however, if the air is contaminated, the plant development is disturbed, resulting in manifestations of chlorosis, necrosis, etc ... at the expense of the fresh matter. The ratio FM/DM of a polluted area is lower than that recorded in an unpolluted area.

It should also be noted that very often this pollution is not the major cause. Indeed, in the city, trees are subjected to many other stressors that affect their health: mechanical stress, climate stress: high temperature, wind, light stress and poor soil quality (dry, mineral poverty, uneven texture, canalization, etc). It is recognized that the poor health of trees in town usually comes from poor soil quality (Domergue-Abak, 1981; Peulon, 1988).

Data analysis (Table 2) allows, at first, to see that the ratio FM/DM is higher in the control site compared to contaminated sites by pollutants issues from road traffic (7.52±0.004 and 3.85±0.004 respectively in the plane tree and the cypress). The lower FM/DM ratio is found in the leaves of plane tree, these leaves seem to assess the air quality in the city better than Cypress (see Table 2).

Site	Species	Trees	Measures/Tree	N	Mean	Min	Max	Stand.deviation
S1	Plane	3	2	6	3.11	3,10	3,12	0.007
	Cypress	3	2	6	4,42	4,40	4,44	0,110
S2	Plane	3	2	6	2,68	2,68	2,69	0,002
	Cypress	3	2	6	2,58	2,58	2,59	0,004
S3	Plane	3	2	6	2,94	2,94	2,96	0,007
	Cypress	3	2	6	3,91	3,91	3,92	0,003
S4	Plane	3	2	6	2,72	2,72	2,729	0,002
	Cypress	3	2	6	2,65	2,65	2,659	0,001
Control	Plane	3	2	6	7,52	7,52	7,532	0,004
	Cypress	3	2	6	3,85	3,85	3,865	0,004

Table 2. Comparison FM/DM ratio of polluted sites to the control site.

In towns and neighborhoods of trunk road traffic, tree leaves constitute a very good collectors of dust and heavy metals. This deposition causes a high decrease in photosynthetic activity of the tree, growth reduction, leaf necrosis and leaf discoloration. The result is a subsequent reduction of the weight of the fresh matter of harvested leaves, and consequently, a low ratio FM/DM.

Bhatti et al. (1988) and Joumard et al. (1995) showed that the growth of new shoots in the Norway spruce was reduced by 25% along the highway compared to control trees, which results in a decrease in the surface and the length of the sheet.

These results are almost seen in Figure 2, showing a significant negative correlation between Pb ($r = -0.29^*$) and the ratio FM/DM and a highly significant negative correlation for Zn ($r = -0.36^{**}$), however, it has no meaning for Cu metal. Thus, when the ratio is low, it means that the weight of the fresh matter is low and levels of trace elements are high. The effects of pollutants related to road traffic in urban areas on plants growth and development should be to induce alterations in their physiology.

2.2.2 Trace metals contents (Pb, Zn, Cu) in different sampling sites

Different concentrations accumulated by the plane and the cypress tree are shown in Figure 3, in the form of box plots with histogram, from which we find that:

In the leaves of plane, the levels of lead, zinc and copper (average values for all sites, respectively: 0.28±0.05, 3.97±0.58 and 0.11±0092 µg/g) are higher about 50%, compared to trees growing in the control site (recorded values, respectively: 0.13±0.001, 1.31±0.005 and 0.01±0.00 µg/g). Atmospheric deposition of road traffic are, in this case, the main sources of Pb and Zn, knowing that the concentration of metal in the soil solution is extremely small compared to the content of the fixed phase.

Soil lead is not very mobile and not available to plants. It accumulates mainly in the roots which, in some cases, it precipitates as phosphate (ADEME, 2005). Zinc also has a predominantly anthropogenic origin. As an illustration, 96% of zinc air emissions are anthropogenic (Colandini, 1997). It comes mainly from the leaching of roofing, corrosion of pipes and galvanized materials as well as tire wear (Chocat, 1997). Its presence is permanent and generalized to all urban sites, and does not present seasonality.

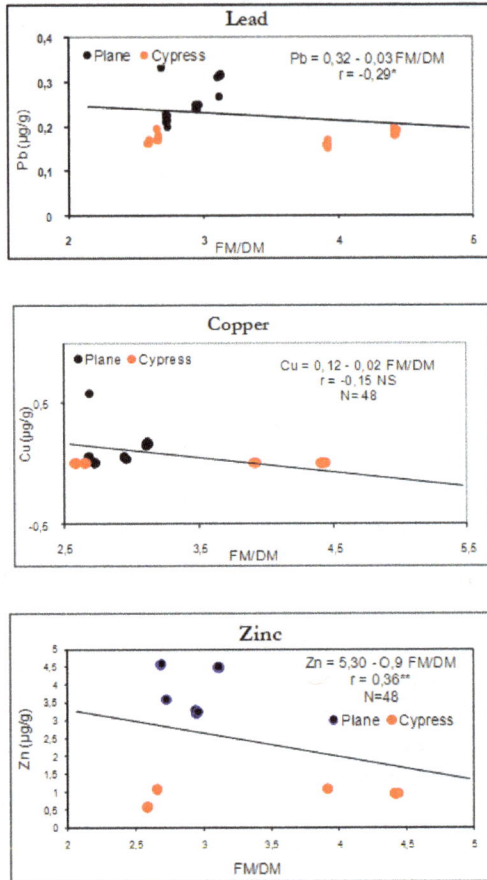

*: *Significant correlation at 5%*
**: *Significant correlation at 1%*
NS: *No significant correlation*

Fig. 2. Correlation: ratio FM/DM – trace metals (Pb, Zn, Cu) contents.

In the case of cypress leaves, concentrations of lead and zinc recorded in the city were moderately low compared with plane tree leaves (average values for all sites, respectively: 0.17±0013 and 0.93±0.20 µg/g compared to the control site in the order of 0.10±0.005 and 0.4±0.00 µg/g). This significant difference in concentrations of metals accumulated in plane and cypress trees is partly due to the leaf surfaces collector of particles (mainly in the epicuticular waxes) on a hand and to the difference between particles of lead and zinc on the other hand. Garrec et al. (2002) showed that gaseous and particular pollutants accumulated mainly on leaf surfaces following the highly lipophilic properties of cuticular waxes, with the key a bit of disturbing effects on plants.

Note that high concentrations for the three metals measured were noted in sites S1 and S2, the characteristics of mountain areas affected by a relatively steep slope and very important road traffic, the slope causes the engine to develop more power and reject more pollutants. The slope therefore causes a significant increase of emissions (Madany et al, 1990).

Moreover, for all sites measured, equal concentrations of copper were observed (0.01 ± 0.00 µg/g) between the urban and the control cypress, where it can be assumed that copper accumulation from road traffic is absent in the leaves of this species.

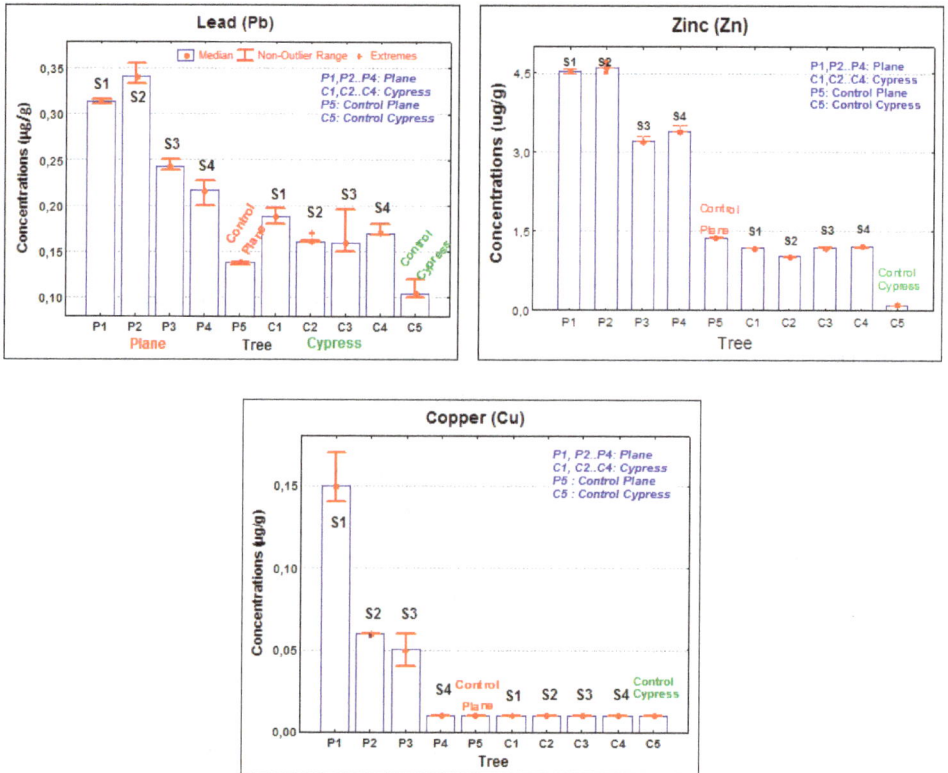

Fig. 3. Comparison of trace metals contents (Pb, Zn, Cu) between control and polluted sites.

2.2.3 Relationship between trace metals contents (Pb, Zn, Cu) and urban sites

In this section, we conducted a correspondence analysis by addressing the relationships between variables: concentrations of trace metals (Pb, Zn, Cu) and contaminated sites by these elements. First of all, we have a table with one entry consisting of data from four sites (two measurements per tree, six measurements per site), then the variables said lines (four sites) were linked to the values concentrations of trace metals called variable columns with a cross table.

For this analysis, the percentages of inertia explained by each axis, were considered, including the relative differences have to know the number of axes that can be interpreted. The results of this Factorial Analysis of Correspondence are shown in Figure 4.

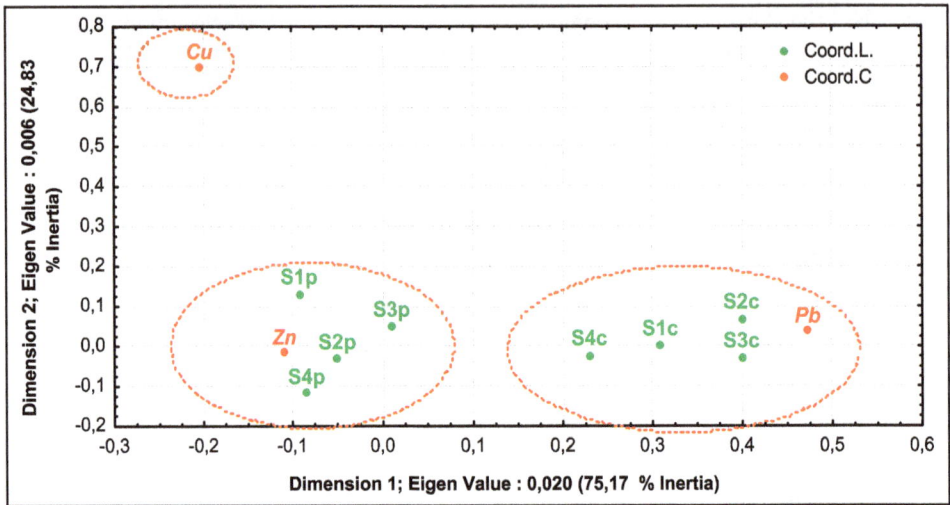

S1p...S4p : Downtown sites (bioaccumulative species: plane-tree)
S1c...S4c : Downtown sites (bioaccumulative species Cypress-tree)

Fig. 4. Results of Factorial Analysis of Correspondence between traces metals content and polluted sites.

Two axes may be interpreted to show that about 75% and 25% of the dispersion of the cloud variables made respectively in the plane of the two axes:

Axis No. 1 (75.17% of the total inertia of the cloud):

Variables that contribute substantially to the inertia of this axis are:

The negative side

- *Variable lines:* the four urban sites S1p, S2p, S3p and S4p where the plane tree was found to be bio-accumulative essence,
- *Variables-columns:* one modality was found: zinc (Zn). This association between the four urban sites and accumulation of zinc explains that the plane tree leaves that are smooth and wide accumulate more zinc than lead, despite the high levels of lead stored in these sites.

In this context and in hyper-accumulator plants, accumulation is observed in epidermal cells for zinc in *Arabidopsis halleri* (Dahmani-Muller, 2000) and for nickel in *Alyssum lesbiacum* and *Thlaspi goesingsense* (Kupper et al., 2001). This distribution has also been demonstrated for zinc in barley when the exposure level of the plant increases (Brown et al, 1994).

The positive side

Examination of the axis on a positive side to define the association of the four urban sites S1c, S2c, and S3c S4c using cypress with lead (Pb). So the cypress has a tendency to accumulate more lead than zinc.

Several factors may be involved in the susceptibility of leaves to pollutants such us the particles emitted by road traffic which are better captured by rough surfaces, and the presence of a hair growth promoting their retention by the smooth epidermis. It should be noted that for the same site and same exposure to traffic pollution, leaves of *Terminalia catapa* that are rough and large accumulated about two fold more lead than *Nerium oleander* leaves that are smooth and wide (Vazquez et al, 1992).

Axis No. 2 (24.83% of the total inertia of the cloud)

In this axis, a single modality was found: copper (Cu). It was confirmed from Figure 4 of subsection 3.2 that copper is weakly accumulated by the plane, for against; it does not accumulate at all in the leaves of cypress. These results confirm the observations of the AFC in this axis. Note also that copper is a trace element essential to the physiological functioning of the plant, in this case, it is likely that low levels of copper were used by the trees.

Results show that the ratio fresh weight/dry weight that is an indicator of air quality in the area of investigation is negatively correlated with the concentration of trace elements. It is extremely low in the contaminated sites, reflecting the effect of pollutants originated from the traffic on the vigor and the physiological activity of trees nearby. These trace metals accumulated in leaf surfaces following the micro-relief created by cuticular waxes, or enters the leaf through stomata.

Furthermore, we found that the plane tree leaves accumulate more zinc than lead. Nevertheless, among the cypress, the degree of accumulation of lead is greater than that of zinc, furthermore, very low concentrations or absent of copper were observed in both species. In addition, these results show the interest of the use of leaves of woody plants as bioindicators of air contamination. Thus, urban trees can form networks of plant bioindicator of air pollution.

3. Case II Mapping of atmospheric pollution by heavy metals originated from road traffic using a transplanted bioaccumulative lichen *Xanthoria parietina* in the city of Tiaret (Algeria)

The monitoring of trace metals pollution obeys at specific constraints that require the deployment of sophisticated and expensive techniques. These constraints have led many countries to promote the use of living organisms in which monitored contaminants are assayed. Thus, since the early 1990s, thirty of European countries have established a network of biomonitoring of metal atmospheric deposition using plants, lichens, mosses, etc... (Leblond, 2004).

Lichens are organisms that are particularly well suited for the study of gaseous and particulate pollutants. They owe this effectiveness due to their anatomical particularities (vegetative structure in the form of thallus leading to a high ratio surface/volume, absence of waxy cuticle, stomata and conducting vessels, presence of a cortex often rich of mucilage and pore) and their physiological characteristics. They will therefore be subject to the effects of pollutants in both dry and wet deposition (Garrec and Van Haluwyn, 2002).

Deruelle (1984) transplanted lichens at distances ranged between 5 and 10 m from the road. After exposure more or less prolonged, transplants are returned to their native communities under unpolluted atmosphere and assays are conducted at regular intervals in order to test decreases of lead concentration.

Semadi and Deruelle (1993) indicated two methods of lichen transplantation. The first is to graft a disk supporting bark lichen on a tree of the same species, if possible, or on a board. The second is to expose branches covered with lichen thalli. The authors affirm that the latest method was used for *Hypogymnia physodes* and *Ramalina duriarei*. The latter method was chosen because it is easier. Indeed, it is not necessary to conduct sampling to the board. Lichens were transplanted six months to a distance of 5 m and 10 m of roads characterized by heavy traffic along predefined transects. Lead assays are performed each month on all graft at every site. Assays were performed using a method similar to that employed by (Deruelle, 1981).

3.1 Materials and methods

3.1.1 Transplantation of the lichen *Xanthoria pariteina*

The experimental protocol used in this study is to transplant the lichen *Xanthoria pariteina* in the city, this lichen species is abundant in the region of Tiaret and has been the subject of numerous scientific studies (excellent accumulator of heavy metals). This technique developed by (Brodo, 1961) based on transplantation into a site to evaluate, lichens originating from a reference site, exempt of contamination. Transplanted samples were collected from uncontaminated areas (the cedar forest of the national park of Theniet al Had at 120 km Northern of Tiaret) which ecological conditions are similar as possible. Lichens controls, far from any source of contamination of air pollution, have been used to serve as references when comparing with contaminated lichens.

Transplantation was performed on selected stations in the city, whose location is determined from a mesh of the sampled area (Figure 3). The study sites are located either at the intersection of the mesh, in the middle of each mesh (mesh size of the territory of 0.5 km x 0.5 km) (Garrec and Van Haluwyn, 2002). On total, 48 lichen samples were transplanted in different sites in the city, and each lichen sample is set in a tree trunk at 1.30 m high (Maatoug, 2010). The duration of transplantation was determined by one month (from April 21 to May 22, 2008).

3.1.2 Samples collect

The material is taken to limit contamination: neither losses nor pollution, avoiding the use of tools or containers that may contaminate the sample (tool steel or stainless steel containers whose walls contain pigments based on trace elements, such as PVC). The geographical coordinates (x, y) of each transplant were recorded using a GPS.

The lichen samples are collected, registered and transported in dry paper on the day of harvest.

3.1.3 Sample processing

Most often, methods of preparation and determination of metals in soil are the same for plant leaves, lichens and fungi (Alfani and Baldantoni, 2000). The most commonly used is the dosage in the thalli of lichen collected from the study site. Lichen samples were taken on the same day for each study area, facing the highway and at 1.3 m above the ground. The assay was performed after dehydration at least 72 hours at 105 °C, weighing and cleaning with boiling hydrogen peroxide to mineralize the lichen. The assay is performed in a solution of HCl decinormal (Deruelle, 1981).

In the laboratory, harvested thalli without prior washing have been subject to a series of transactions that are:

- Dehydration of thalli: the usual method is dehydration in an oven at 105±2 °C during 72 hours. Dried thalli were weighed for dry matter DM determination, about 0.2 to 0.3 g.
- Grinding: This step is highly critical because it can be a source of contamination or losses. For lichens, the grinder used is an agate mortar. Grinder materials are made of titanium, steel guaranteed free of heavy metals. The resulting powder is calcinated in an oven whose temperature is gradually increased up to 500 °C, using quartz capsules.
- Mineralization and dissolved: the fine powder obtained after calcination, is placed in an acidic and oxidizing solution (0.5 ml mixture of nitric acid HNO_3, hydrofluoric acid HF and perchloric acid $ClHO_4$) and then heated in a water bath for 24 h until the complete destruction of organic matter. Tubes that have been boiling are supplemented by 10 ml of distilled water. This method allows the determination of all trace minerals.

The determination of lead and Zinc is performed by atomic absorption spectrometry with electrothermal atomization mode (Perkin Elmer spectrometer 100).

3.2 Results and discussion

The determination of lead and Zinc was performed on 46 samples. Initially, we sought to consolidate the concentrations that are most like them, to achieve this objective; we submitted data to an automatic classification (hierarchical cluster where the distance measurements are those of the Euclidean distances).

The automatic classification allows identifying three distinguishable classes of pollution by lead and four polluted classes by zinc. The results of this classification are shown in Fig. 5 and 6:

3.2.1 Lead

The concentrations of lead accumulated by the lichen *Xanthoria parietina* are presented in Table 1, as classes of pollution defined by automatic classification method.

	N	Mean	Median	Min	Max	1st Quartile	3rd Quartile	Stand. Dev
Class 1	5	237,60	238,00	229,00	248,00	234,00	239,00	7,02
Class 2	4	194,00	192,00	184,00	208,00	185,00	203,00	11,19
Class 3	37	76,31	76,90	43,80	119,00	61,50	89,60	19,02
Control	4	28,35	28,20	27,80	29,20	28,00	28,70	0,59

Table 3. Descriptive statistics of different classes of lead pollution ($\mu g/g$) in the city of Tiaret (Maatoug, 2010).

The reading of Table 3 shows that:

- Lead levels (mean values for all classes, ranges from 76.31 ± 19.02 µg/g, with extreme values 43.80 and 119.00 µg/g, at 237.60 ± 7.02 µg/g with extreme values 229.00 and 248 µg/g) are higher than the lichens control (28.35 ± 0.59 µg/g with extreme values 27.8 and 29.2 µg/g). Semad and Deruelle (1993) show that the average lead content in thalli of *Ramalina farinacea*, transplanted and harvested in the region of Annaba (Algeria) at 5 m from the floor, is about 60 µg/g (µg/g) (transplantation for 1 month). Our results indiquate that *Xanthoria pareitina* accumulates much more lead than *Ramalina farinacea* (Maatoug, 2010),

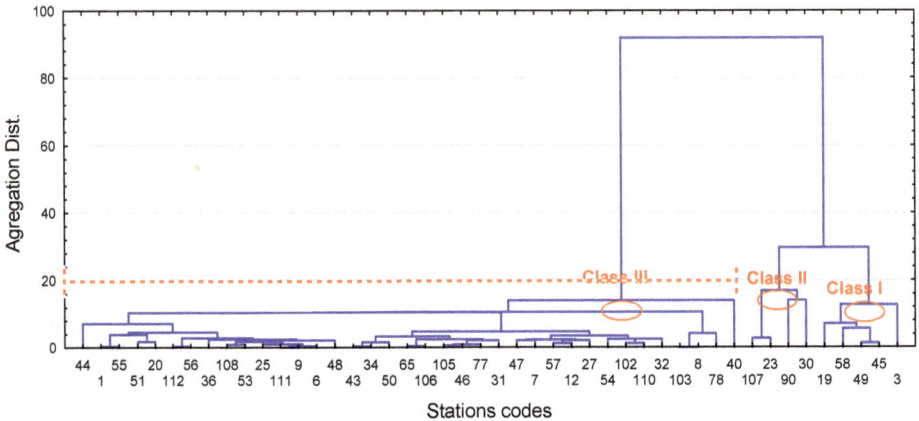

Fig. 5. Three lead polluted classes according to cluster analysis method.

- High concentrations of lead are registered in the class I, the maximum value can reach 248 µg/g. This class, called hot class, brings together the most polluted sites assigned a very high traffic and steep slope where the frequency of braking and idling is extremely important (especially the north and center of the city). Maatoug (2007) found high concentrations of lead in central and north side of this city, by the use of cypress with leaves evergreen and plane with leaves of maple.
- Classes 2, regroups, in a regressive way, sites or emissions of lead relatively less important as sites of Class 1,
- Low concentrations were noted in class 3 around the value 76.31 ± 2.19 µg/g, but still elevated compared with control sites. We can also divide this class into 3 subclasses whose lead levels are varied from 49.80 ± 4.78 at 60 ± 5.30 µg/g for the 1st subclass, from 60 ± 5.30 to 90 ± 2.80 µg/g in the 2nd subclass and from 90 ± 2.80 at 119 ± 5.20 for the 3rd subclass. It is assumed that this class, including subclasses, includes the remaining stations (37 observations) which are located in a relatively low road area (the south side of town) as the sites of classes 1 and 2. It is important to note that these sites are fairly open in urban areas and have relative protection, movement of air masses influencing the dispersal of pollutants (Maatoug, 2010)

3.2.2 Zinc

Automatic classification defines four classes of pollution by zinc indicated by the lichen *Xanthoria parietina*. Results are shown in the following table 4.

	N	Mean	Median	Min	Max	1st Quartile	3rd Quartile	Stand. Dev
Class 1	6	1160,72	1013	882	1665	963,85	1311	272,28
Class 2	6	693,362	695	621,5	740	665,2	732,5	42,24
Class 3	11	338,42	306,5	262,2	515	293	360	77,83
Class 4	21	101,4	97,7	10,3	190	63,9	144,6	55,29
Control	4	40.50	41.00	30.00	50.00	33.50	47.50	8.81

Table 4. Descriptive statistics of different classes of zinc pollution ($\mu g/g$) in the city of Tiaret.

By the same way as lead, Zinc content (average values for all classes, range from 101.4 ± 55.29, with extreme values of 10.3 and 190 $\mu g/g$, to 1160.72 ± 272.28 $\mu g/g$ with extreme values of 882 and 1665 $\mu g/g$) is higher in sampling sites than those of the control (40.50 ± 8.81 $\mu g/g$ with extreme values 30.00 and 50.00 $\mu g/g$). Recorded values are also higher when comparing with those found in lichens by Van Haluwyn and Cuny (1997) between 30 and 50 $\mu g/g$.

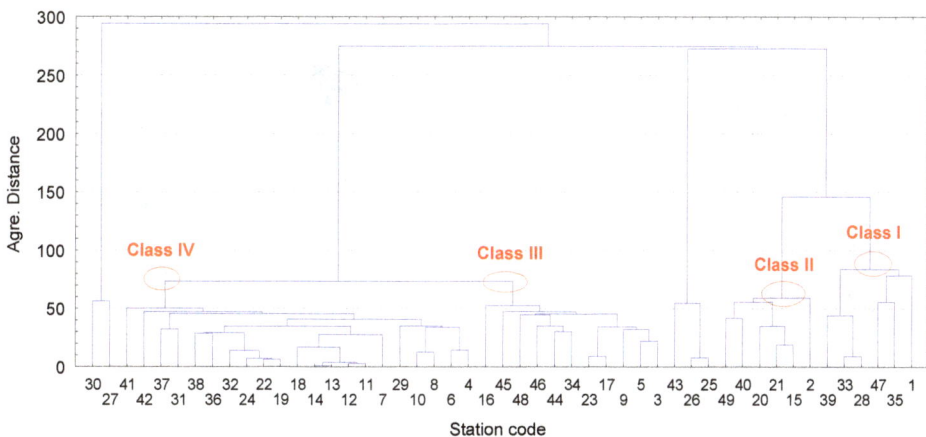

Fig. 6. Four Zinc polluted classes according to cluster analysis method.

High concentrations of zinc are recorded in Class I with a maximum value of 1665 $\mu g/g$. This class hot regroups together the most polluted sites by high road traffic and situated in steep slope where the frequency of braking and tire wear is important.

Class 2 includes, in regressive way, sites supporting zinc emissions less than the first class. The concentrations of zinc found in Class 3 range around a value of 338.42 ±77.83 µg/g but still high compared to the sites of class 4 which includes the sites with low traffic levels. It is important to note that these sites are quite open in urban areas, is what promoted the dispersion of pollutants.

3.2.3 Establishment of map pollution

Mapping of spatial lead and zinc pollution in the city of Tiaret was produced by the automatic method, interpolation/extrapolation, of data collected *in situ* of lead pollution knowing the extent of pollutants concentration in some points using the software MapInfo© and Vertical mapper (™) (Garrec and Van Haluwyn, 2002). The geographical coordinates of each observation, were obtained by a GPS. We finally got a file of 46 readings with the coordinates (x, y, z), z is the concentration of lead in the site. The details of this mapping are illustrated in Figure 5, taking into account the results of automatic classification.

3.2.4 Map reading

Looking at the map of Figure 7a, therefore we can define three classes of lead pollution, from an average of 76.31 µg/g to 237.60 µg/g. The highest concentrations of lead are found in Class 1, indicated by a red color. In this class, the most sites are located in the main axe North-Center-South of the city. These observations have led to distinguish the deteriorating of air

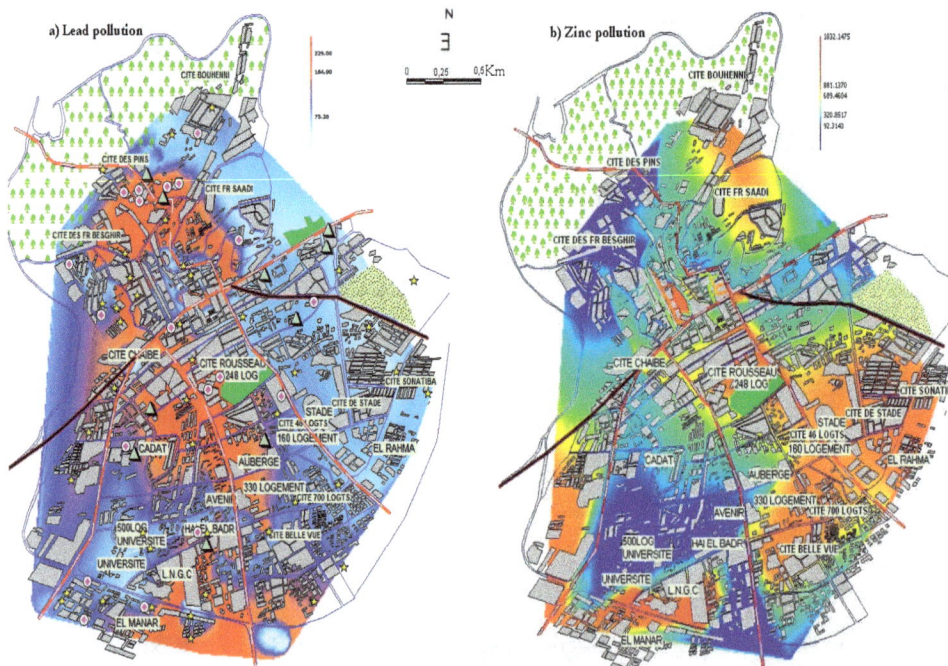

Fig. 7. Map of atmospheric pollution by lead and zinc from road traffic in Tiaret city.

quality in these sites and that atmospheric deposition of road traffic are, in this case, the main sources of lead pollution, knowing that the ground contribution of this metal are nil. However, sites of this class, called hotspots, are mountainous areas affected by a relatively steep slope and a very important traffic, slope forces the engine to develop more power and emit more pollutants, so it causes a significant increase in emissions (Madany et al., 1990).

High levels have also been observed in class 2, these results suggest, in a first approach, the environmental conditions, including pollution of lead, have an important responsibility on the poor air quality in these cities. It is important to note that a significant number of schools are located in these hotspots that constitute a danger on the students health.

In Class 3, low values were observed, these sites are fairly open, promoting the dispersion of atmospheric deposition that are transported by wind.

Zinc is a very ubiquitous. It is used in a very large number of materials in the form of oxide, sulfate, chloride or organic. It is usually emitted into the atmosphere in the form of small particles from road traffic and various industries. Our results concerning polluted zinc repartition indicate the presence of four classes according to the automatic classification method;

As shown in Figure 7b, we can define four classes of zinc pollution, from an average of 101.4 µg/g to 1160.72 µg/g. The highest concentrations are grouped in class 01, represented also by the red color, were observed at the southern part corresponds to a high movement especially in peak hours because it constitutes a principal way in the city.

There are also hot spots in the north where we can explain the accumulation of pollutants by the relief nature more than the rate of traffics because of the steep slope which reduces the dispersion of pollutants and thus promotes their accumulation in the medium, and therefore, by lichens.

Sites of the Class 2 support the same road traffics than the first except that these sites are open environments that lead the wind to disperse these pollutants.

For Class 3, low values are recorded corresponding to a temporal and intermittent circulation in these sites during the day.

Finally the Class 4 includes sites of the lowest values of zinc concentrations, these stations are open and wide characterized by very low traffic movements.

4. Case III Comparison between levels of lead accumulated by trees to those accumulated by the lichen *Xanthoria parietina* transplanted into the city of Tiaret

In towns and neighborhoods of trunk road traffic, plants leaves are also collectors of dust and heavy metals.

In this context, a quantification of lead and zinc deposition originated from road traffic has been made in the same city, from the leaves of two types of trees growing in the center of the city. There are six trees *Platanus acerofolia*.Willd and six trees *Cupressus sempervirens*.L divided into two sites S1 and S2 (Fig. 5). For each tree, fifty leaves are collected, they were the subject of lead and zinc determination (Maatoug, 2007).

On the same trees (the same sites S1 and S2), lichens *Xanthoria pareitina* were transplanted following the steps described in previous paragraph. Results of this comparison are illustrated in Figure 8.

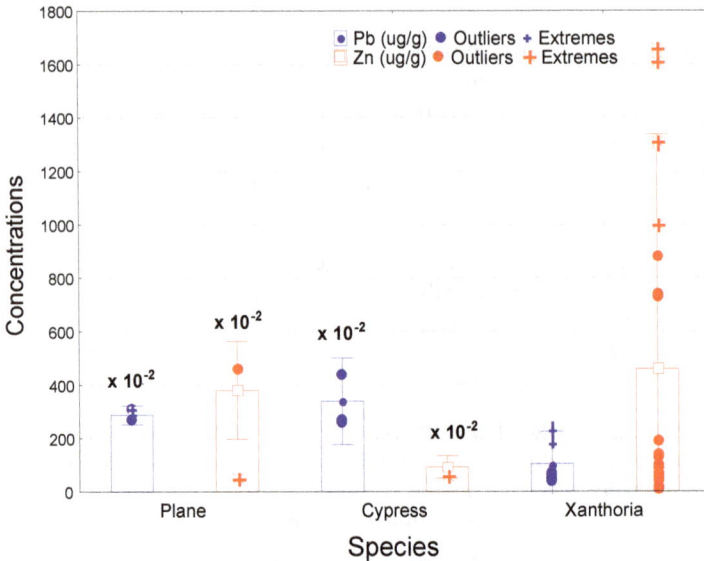

Fig. 8. Comparison between contents of lead and zinc accumulated by trees (Plane and Cypress) to those accumulated by the lichen Xanthoria in Tiaret City.

We notice very clearly that lead storage capacity of Xanthoria is very high compared to that of the trees, the values are 1.04 ± 0.60 µg/g in Xanthoria against $2.86 \pm 0.17 \times 10^{-2}$ µg/g and $3.39 \pm 0.01 \times 10^{-2}$ µg/g respectively in the plane and cypress.

By the same way, the Xanthoria accumulates 460.38 ± 43.85 µg/g against $380.45 \pm 91.53 \times 10^{-2}$ µg/g and $93.20 \pm 20.96 \times 10^{-2}$ µg/g respectively in plane and cypress leaves (Maatoug, 2010).

Lichens absorb indiscriminately by their pseudocyphelles all the nutrients as toxic substances. Absorption capacity of metals by lichens is directly related to their morphological and anatomical structures. Lichens accumulate more metals when their surfaces increase (Garty et al., 1996).

Sharing some biological traits (Asta et al., 2003), lichens have a relatively quick response to the deteriorating of air quality and are extremely sensitive to other types of environmental changes such as climate change and eutrophication (Galun, 1988), (Garrec and Van Haluwyn, 2002).

In trees, the emissions of lead and zinc originated from road traffic are mainly in the form of fine particles that are then collected by leaves surfaces. It is an assimilation of external metal pollutants. The significant difference in concentrations of lead and zinc accumulated between the plane and cypress is certainly due to the nature of leaves surfaces (mainly on epicuticular waxes) of each tree, the sycamore leaves, which are smooth and wide, accumulate substantially lead better than the cypress leaves.

Generally, optimal use of plants as means of investigation of air pollution concerns cases where the route of contamination by the soil is either nil, negligible or known. The transfer of a trace element from the soil to a harvest organ depends on both parameters related to soil and plant-specific factors. It results in some sort of meeting between ground supply and plants demand that are not completely independent. Therefore, the plants most used for these studies are mosses and lichens through, inter alia, their lack of root system.

5. Conclusion

Our study constitutes an attempt to evaluate the air quality in semiarid region, Tiaret city, from Algeria. The first aim of this work (case I) was to assess the most frequent atmospheric heavy metal, especially (Pb, Zn, Cu), accumulated in leaves of two urban plant species, plane and cypress, in the context of their usefulness in urban environment as bioindicator or in phytoremediation. Results show that the ratio fresh weight/dry weight (FM/DM) of trees leaves, which is an indicator of air quality in the area of investigation, is negatively correlated with the concentration of trace elements. It is extremely low in the contaminated sites, reflecting the effect of urban pollutants on trees vigor and their physiological activity. Furthermore, we found that the plane tree leaves accumulate more zinc than lead against very low concentrations of copper. Nevertheless, among the cypress, the degree of accumulation of lead is greater than that of zinc, furthermore, very low concentrations or absent of copper were observed in both species. In addition, these results show the interest of the use of leaves of woody plants as bioindicators of air contamination and the impact of pollution on urban plants.

In order to identify the main source of lead and zinc accumulation in trees leaves then, in the intend of estimate concentrations of these pollutants and mapping their spatial repartition in the city (case II), lead and zinc was measured in 48-lichen *Xanthoria parietina* samples, transplanted for while into different sites located near major highways and in urban areas of Tiaret city. Automatic classification of data has shown three classes of lead-polluted sites and four classes of zinc-polluted sites. These results established that the degree of lead pollution including traffic and roads infrastructures is causing a major source of heavy metals in this city. All the observations of pollution have been mapped to reveal spatial pollution originated from road traffic and spot on the vulnerable sites. Hot points sites were locating on land with relatively steep slope and high traffic, so, this forces the vehicle to release a lot of smoke. Sites with low traffic are also open enough to promote the dispersion of atmospheric deposition. These maps could answer questions about the air pollution problem in Tiaret city. These biological methods contribute to a health risk assessment because they assist to identify areas potentially exposed to urban air pollution.

These maps could answer questions about the air pollution problem in Tiaret city. These biological methods contribute to a health risk assessment because they assist to identify areas potentially exposed to air pollution. Thus, urban trees can form networks of plant bioindicator of air pollution; the use appears to be particularly simple, economical and efficient method. The second part of the study has demonstrated the perfect suitability of lichens bioaccumulators for mapping different elementary deposits and location of sources of metals. Lichens and urban trees can form networks of plant bioindicators of air pollution, the use appears to be also very simple, flexible, economical and efficient to build good repartition in space and time maps of pollution. The identification of pollution within sensitive organisms can also detect the degradation of air quality before it severely affects the biota or humans.

Today, the use of motor vehicles has increased substantially in Algeria. However, in recent decades, urban and industrial developments have accelerated the use of motor vehicles. Unfortunately, at present, air pollution generated by traffic is taken into account only through its effects on human health. On the other hand, the reduction of pollutant emissions from road traffic demands renewal of Automobile Park and improvement of combustion engines using less pollutant fuels.

In the end, it appears that pollution related to road traffic has an impact on urban environment. Plant contamination comes mainly from the aerial parts, which are often the beginning of food chains with all the problems that may cause, particularly on human health. This study confirms the need to control the automobile park and to reduce emissions from road traffic by the renewal of the automobile park, improvements in the accuracy of regulating combustion engines and the use less pollutant fuels.

6. References

[1] ADEME,. Agence de l'Environnement et de la Maîtrise de l'Energie. Dérogations relatives à la réglementation sur l'épandage des boues de stations d'épuration. Comment formuler une demande pour les sols à teneurs naturelles élevées en éléments traces métalliques ?. Guide technique. J. Béraud et A. Bispo (Coordinateurs). D. Baize, T. Sterckeman, A. Piquet, H. Ciesielski, J. Béraud et A. Bispo (2005). 120p

[2] Alfani, A. et Baldantoni. D. Temporal and spatial variation of C, N, S and trace element contents in the leaves of Quercus ilex within the urban area of Naples. Environ. Pollut. 2000 (109): 119-129.

[3] Asta, J., Erhardt, W., Ferreti, M., Fornassier, F., Kirschbaum, U., Nimis, P.L., Purvis, O.W., Pirintsos, S., Scheidegger, C., van Haluwyn, C., Wirth, V. (2003). European guideline for mapping lichen diversity as an indicator of environmental stress. 20p.

[4] Baize D., 1997. Teneurs totales en éléments traces métalliques dans les sols (France). INRA de France. Editions, Paris, 408p.

[5] Bazzaz, F.A., Rolfe, G.L., et Carlson, R.W., Effect of cadmium on photosynthesis and transpiration of excised leaves of corn and sunflower. Physiologia Plantarum, 1974 (32), 373-377.

[6] Belouahem D., Détection de la pollution atmosphérique fluorée d'origine industrielle à l'aide de certaines espèces végétales bioaccumulatrices dans les régions de Annaba et Taraf. Thèse de Magister. Institut National Agronomique INA (Algérie) 1993. 165p.

[7] Bhatti G.H., Iqbal M.Z, 1988. Investigations into the effect of automobile exhausts on the phenology, periodicity and productivity of some roadside trees, Acta Societatis Botanicorum Poloniae, 57 (3). 395-399.

[8] Brodo .I M. Transplant experiments with corticolous lichens using a new technique. Ecology 1961 (42): 838-841.

[9] Chocat, B.,. Eurydice. Encyclopédie de l'hydrologie urbaine et de l'assainissement. Paris : Tec et Doc Lavoisier. 1997, 1136 p.

[10] Colandini, V., Effets des structures réservoirs à revêtement poreux sur les eaux de ruissellement pluviales : qualité des eaux et devenir des métaux lourds. Thèse de doctorat. Pau. Université de Paut des pays de l'Adour (France) 1997, 161 p. + annexes

[11] Dahmani-Muller H., Phytoréhabilitation des sols pollués par des éléments métalliques : facteurs et mécanismes de prélèvement dans les sols et d'accumulation par les espèces métalliques. Thèse de l'Ecole Nationale de Génie rural, des eaux et des Forêts ENGREF (France) 2000, 151 p.

[12] Deletraz G., Géographie des risques environnementaux lies aux transports routiers en montagne. Incidences des émissions d'oxydes d'azote en vallées d'Aspe et de Biriatou (Pyrénées). Thèse de Doctorat en Géographie - Aménagement. Université de Pau et des pays de L'Adour. Institut de Recherche sur les Sociétés et l'Aménagement (France) 2002, 564 p.

[13] Delmas – Gadras C., Influence des conditions physico-chimiques sur la mobilité du plomb et du zinc dans un sol et un sédiment en domaine routier. Thèse de docteur de l'université de Pau et des Pays de l'Adour (France) 2000, 191 p.

[14] Deruelle .S. Effets de la pollution atmosphérique sur la végétation lichénique dans le bassin Parisien. Convention de recherche n° 79-15, Ministère de l'Environnement et du Cadre de Vie 1981 : 91-112.

[15] Deruelle .S. L'utilisation des lichens pour la détection de la pollution par le plomb. Bull.Eco 1984 :1-6.

[16] Domergue–Abak. M.F., Etude de synthèse sur les causes de dépérissement de la végétation en milieu urbain, et notamment les arbres d'alignement. Rapport du Ministère de l'Urbanisme et du Logement (France) 1981, N° 81/47142/00, 52p.

[17] Edwards N.T., 1986. Uptake, translocation and metabolism of anthracene in bush bean (Phaseolus vulgaris L.). Environ. Toxicol. Chem.. (5),659-65.

[18] Fernandez-cornudet,C., Devenir du Zn, Pb et Cd issus de retombées atmosphériques dans les sols, à différentes échelles d'étude. Influence de l'usage des sols sur la distribution et la mobilité des métaux. Thèse de doctorat de l'INA-PG (France) 2006, 232p.

[19] Galun, M. Handbook of lichenology. Springer, 1988; 181p.

[20] Garrec j.p. et Van Haluwyn c., Biosurveillance végétale de la qualité de l'air. Concepts, méthodes et applications. Editions Tec et Doc Lavoisier, Paris (France) 2002,118 p.

[21] Garty, J., Kauppi, M., Kauppi, A., Accumulation of airborne elements from vehicles in transplanted lichens in urban sites. Journal of Environmental Quality 1996 (25) 265-272.

[22] Hébrard-Labit. C et Meffray.L, Comparaison de méthodes d'analyse des éléments traces métalliques (ETM) et des hydrocarbures aromatiques polycycliques (HAP) sur les sols et les végétaux. Guide technique. CETE Nord Picardie (France) 2004, 121 p.

[23] Joumard R., Lamure C., Lambert J., Politiques de transport et qualité de l'air dans les agglomérations. LEN n° 9515. Bron : INRETS (France) 1995, 125 p.

[24] Kupper H, Lombi E, Zhao F J, Wieshammer G, McGrath S P., Cellular compartimentation of nickel in the hyperaccumulators Alyssum lesbiacum, Alyssum bertolonii and Thlaspi goesingense. Journal of Experimental Botany; 2001 (52), 2291-2300.

[25] Leblond. S. Etude pluridisciplinaire du transfert des métaux de l'atmosphère vers les mousses (Scleropodium purum (Hedw.) Limpr.) : Suivi sur un site rural (Vouzon, France). Thèse de doctorat en Chimie de la Pollution Atmosphérique et Physique de l'Environnement, Université Paris 7 - Denis Diderot 2004. 212p.

[26] Maatoug M. Détection de la pollution de l'air d'origine routière par certaines espèces végétales bioaccumulatrices de quelques métaux lourds (Pb, Zn, Cu). Revue pollution atmosphérique 2007 ; 196 : 385-394.

[27] Maatoug M. Cartographie de la pollution atmosphérique par le plomb d'origine routière à l'aide de transplantation d'un lichen bioaccumulateur Xanthoria parietina dans la ville de Tiaret (Algérie). Revue pollution atmosphérique 2010 ; 205 : 93-101.

[28] Madany IM, Ali SM, Akhter MS. Assessment of lead in roadside vegetation in Bahrain. Environment International 1990 (16): 123-126.

[29] Ouzounidou G, Ciamporova M, Moustakas M, Karataglis S.,. Responses of Maize (Zea-Mays L) Plants to Copper Stress .1. Growth, Mineral-Content and Ultrastructure of Roots. Environmental and Experimental Botany; 1995 (35), 167-176.

[30] Ouzounidou G, Moustakas M, Eleftheriou E P.,. Physiological and ultrastructural effects of cadmium on wheat (Triticum aestivum L) leaves. Archives of Environmental Contamination and Toxicology; 1997 (32),154-160.

[31] Peulon V., Le dépérissement des arbres en ville. Edition du STU, Ministère de l'Equipement (France) 1988, 61p.

[32] Semadi .A et Deruelle .S Détection la pollution plombique à l'aide de transplants lichéniques dans la région de Annaba (Algérie). Revue pollution atmosphérique Octobre-Décembre 1993 : 86-101.

[33] Vazquez M D, Poschenrieder C, Barcelo J., Ultrastructural Effects and Localization of LowCadmium Concentrations in Bean Roots. New Phytol; 1992 (120), 215-226.

10

Fugitive Dust Emissions from a Coal-, Iron Ore- and Hydrated Alumina Stockpile

Nebojša Topić[1] and Matjaž Žitnik[2]
[1]Luka Koper, d.d., Koper
[2]Jožef Stefan Institute, Ljubljana and
Faculty of Mathematics and Physics,
University of Ljubljana
Slovenia

1. Introduction

Dust control measures must be implemented in order to reduce the detrimental and harmful effects on material and human resources. According to Mohamed & Bassouni (2006), besides causing additional cleaning of homes and vehicles, fugitive dust can affect visibility and, in severe cases, it can interfere with plant growth by clogging pores and reducing light reception. In addition, dust particles are abrasive to mechanical equipment and damaging to electronic equipment such as computers.

Of the total suspended material, the PM10 fraction (the inhaled dust) is particularly problematic. The US Environmental Protection Agency (US EPA, 1996) reports that a 50 $\mu g/m^3$ increase in the 24 hr average PM10 concentration was statistically significant in increasing mortality rates by 2.5%–8.5% and hospitalization rates due to chronic obstructive pulmonary disease by 6%–25%. It also describes how children can be affected by short-term PM10 exposure by increasing the number of cases of chronic cough, chest illness, and bronchitis (US EPA, 1996). The chronic effects from PM10 depend on the composition of the dust and on the amount of exposure to PM10 over a person's lifetime. A wealth of epidemiological data support the hypothesis that on average, for every 10 $\mu g/m^3$ rise of the total mass concentration of PM10 in the air there is an 1 % increase in cardiovascular mortality on a day-to-day basis (Routledge & Ayers, 2006). In addition to health effects there are other adverse impacts from PM10 exposure. For instance, it is thought that the amount of PM10 contributes to climate change, because the small particles in the atmosphere absorb and reflect the sun's radiation, affecting the cloud physics in the atmosphere (Andrae, 2001).

One of the strong dust emitter sources are large stockpiles of material such as coal and iron ore that are stored and manipulated in the open and exposed to the varying weather conditions. Coal is a mixture of compounds and contains mutagenic and carcinogenic polycyclic aromatic hydrocarbons. Exposure to coal is considered as an important non-cellular and cellular source of reactive oxygen species that can induce DNA damage as shown for wild rodents in an open coal mining area (Leon et al., 2007). In addition, coal is a well known respiratory toxicant and employees who work around coal stockpiles are susceptible to black lung (pneumoconiosis) (Peralba, 1990). With respect to the workers handling the material at above ground facilities, the risk of suffering a death of a lung cancer or pneumoconiosis for an underground worker is expected to be much higher due to additional presence of radon and more limited means of ventilation (Archer, 1988). Iron ore is often found in a form of hematite, the mineral form of iron (III) oxide (Fe_2O_3). Chronic inhalation of excessive concentrations of iron oxide fumes or dusts may result in development of a benign pneumoconiosis, called siderosis (Nemery, 1990). Inhalation of excessive concentrations of iron oxide may enhance the risk of lung cancer development in workers exposed to pulmonary carcinogens. According to a study by Boyd et al. (1970) who investigated the mortality of 5811 Cumberland iron-ore miners who died between 1948 and 1967, the miners who work underground suffer a lung cancer mortality about 70% higher than "normal". However, besides the carcinogenic effects of iron oxide the risk may be also due to radioactivity in the air of the mines (average radon concentration of 100 pCi/l), since the same study found no evidence of any excess mortality from lung cancer among surface workers. The study of Lawler et al. (1985) of 10403 Minnesota iron ore (hematite) miners found no excesses of lung cancer mortality among either underground or above ground miners. This "no risk" result is different from other studies and, according to some authors can be explained by the apparent absence of significant radon exposure, a strict smoking prohibition underground, an aggressive silicosis control programme, and the absence of underground diesel fuel at the location. Hydrated alumina (Aluminium hydroxide: $Al_2O_3.3H_2O$) is known to cause mild irritation to eyes, skin and the upper respiratory tract. It can lead to aggravation of medical conditions such as asthma, chronic lung disease and skin rashes. King et al. (1955) studied the exposure to guinea pigs to alumina and found no harmful effects in lungs after one year.

In a typical open air coal terminal dust arises through activities such as loading, unloading, storage and transport of the millions of tonnes of coal that the terminal handles on average each year. To estimate reliably the quantity and composition of dust emitted by stored material upon different wind conditions, a characterisation of the basic dusting processes is required. This consists in measuring quantity of the material lost and concentration of dust emitted by suitably prepared surface at different local wind velocities as a function of material type, granule size and moisture. Below we report on dust emission measurements of a test unit surface, performed on a wind stream test track set up at the open air terminal (EET) of the Port of Koper. Time dependence of mass loss from the test surface was measured together with concentrations of inhalable particulate matter (PM10) and black carbon (BC) in the vicinity of the test surface for a range of different wind velocities. Threshold wind velocities were measured for four different size fractions of coal and for three different fractions of iron ore stored at the terminal, and for hydrated alumina - a strong dust emitter that is also handled by the port. The same set up was employed to study the suppression of dust emitted by the finest coal fraction and iron ore upon watering of the dry test surface. Although CFD calculations of a large real scale physical configuration are necessary to obtain the final result concerning emissions from the realistic stockpile, the precise measurements of the "dusting" potential generated by the "unit" test surface under different conditions are of significant value.

2. Scientific background

Dusting continues to be an important topic in environmental research. Globally, scientists are studying the induction of dust in order to devise accurate models of dusting as part of efficient dust mitigation and control measures. A significant corpus of literature now exists concerning fugitive dust emissions, dust erosion, wind erosion, and importantly studies of dusting from stockpiles. The most important of the early studies is considered that of Bagnold (1941), who initiated oriented aeolian research as a result of investigating dune formation. Since Bagnold's initial studies, a number of efforts have been devoted to understanding the erosion mechanisms of wind-blown particles in regards desert expansion (White, 1998; Parsons et al., 2004), farmland erosion (Saxton et al., 2000) and the assessment of coal and mineral dust pollution (Lee & Park, 2002; Gillette, 1977; Cowherd et al., 1988; Chane Kon et al., 2007). Other research has focused on dusting from open coal mining (Chakraborty et al., 2002), dusting from unpaved and paved roads (US EPA 1995; Gillette, 1977; Ferrari, 1986), and dusting from open stockpiles located at ports and power stations (Lee & Park 2002; Nicol & Smitham, 1990; Smitham & Nicol, 1990). Within the port, most loose bulky materials have the potential to generate dust during handling. It can be generated either while the material is being transported to storage, or while the material is in a static storage state like a stockpile. During stockpiling, the coal in the surface layers experiences a range of climatic conditions. Intuitively, one would expect that the factors affecting dusting are the surface area of the coal exposed, the particle size of the coal in the surface layer, the velocity of the prevailing wind, and the effect of moisture in the surface layer.

2.1 Erosion modeling

Today, most research is oriented towards developing accurate wind erosion models (Gillette, 1977). As a result, it is known that fine particle erosion from material stored in open yards is a consequence of the wind acting on the region next to the pile surface. The forces acting on the particles include gravity, pressure and viscosity. The gravity force depends on the diameter of the material and its mass; the forces of pressure and viscosity depend upon the flow field generated around the pile. The sum of these forces, if resolved in the direction of the flow and in a perpendicular direction to it, results in the so-called aerodynamic forces: lift force and drag force, respectively. Bagnold (1941) was one of the first to identify the three modes of particulate transport:

- **Creep:** Large particles that are too heavy to be lifted from the surface by the wind can roll along the surface. This is known as creep or surface creep motion. Creep constitutes between 5% and 25% of the total particle transport during a wind erosion event. As particles roll, they abrade the surface and produce or liberate dust particles. Nonetheless, the amount of dust emitted from this mode of transport is relatively small.
- **Saltation:** Particles that are lifted from the ground by aerodynamic forces, but are too heavy to be dispersed into the atmosphere by the wind turbulence, hop across the surface. This type of motion is referred to as saltation. Saltation is the principal process of wind erosion and accounts for roughly 50% to 80% of the bulk of the total particle transport.
- **Suspension:** this is the transport of fine dust particles that are dispersed into the atmosphere and carried away over large distances. Wind alone usually cannot entrain the fine dust particles directly into suspension. This is due to the strong cohesive forces that bind fine particles together (Shao et al., 1993).

These different transportation mechanisms are important as they act to remove the material from the stockpile test area. The measurements that are reported are designed to investigate the removal of different size fractions of material from a test area. The measurements included: i) the suspension loss i.e., detection of PM10 fraction and, ii) the total removal of particles from the test area (mass loss) with all three transportation mechanisms at different wind speeds. The wind field situation is characterized by both, geometry and wind source strength. To interpret the data and compare the findings from this study with previous ones, which apply to a flat terrain situation without obstacles, it is important to know the flow characteristics around the test area.

2.2 The emission factor according to the US EPA

The study of dust is complicated by the many variables that need to be considered. Therefore, any theoretical/numerical simulation would probably result in the forced assumption of several simplifications (Witt et al., 2002). Alternatively, an empirical valuation covers only a limited range of working conditions; even more, for a given location if the conditions are in flux (Kinsey et al., 2004). In trying to establish a level of airborne dust generated from an open pile, one of the most extended methodologies is the US EPA method (US EPA, 1998), which provides the emission factors and procedures required to estimate total emissions. For stockpiles, the US EPA gives several parameters and inputs for cone and flat top oval configurations (Stunder & Arya, 1988). Emissions from stockpile activities may be estimated using the information from sections 13.2.4 and 13.2.5 of the AP-42 Compilation of Air Pollutant Emission Factors document (US EPA, 1995). The methodology applied in section 13.2.5, Industrial Wind Erosion, requires that the initial wind parameters be established; this being the fastest mile used to convert the values obtained from reference anemometers to friction velocity u^*,

$$u^* = 0.053 \cdot u_{10}^+ \tag{1}$$

The fastest mile of wind at a height of 10 m is denoted by u_{10}^+. This equation only applies to flat piles or those with little penetration into the surface wind layer. The wind speed profile in the surface boundary layer follows the following logarithmic distribution:

$$u(z) = \frac{u^*}{K} \ln \frac{(z - Z)}{z_0}, \qquad z \geq Z \tag{2}$$

where u is the wind speed, z is height above the test surface, z_0 is the roughness height, Z is the zero plane displacement and K=0.4 is von Karman's constant. The friction velocity u^* is a measure of wind shear stress on the erodible surface, as determined from the slope of the logarithmic velocity profile. The roughness height z_0 is a measure of the roughness of the exposed surface. According to EPA methodology the emission factor E_f for surface airborne dust subject to disturbances may be obtained in units of grams per square metre per year (g/m²/year) as follows:

$$E_f = k \sum_{i=1}^{N} P_i S_i \tag{3}$$

where k is the particle size multiplier, N is the number of disturbances per year, P_i is the erosion potential corresponding to the observed fastest mile of wind for the i-th period between disturbances in g/m^4, and S_i is the pile surface area in m^2. The particle size multiplier in Eq. (3) varies with aerodynamic particle size and is reported in Table 1.

Particle Size	30 μm	< 15 μm	< 10 μm	< 2.5 μm
Multiplier k	1.0	0.6	0.5	0.2

Table 1. Aerodynamic particle size multipliers in Eq. (3) (After US EPA, 1995).

The erosion potential P for a dry, exposed surface is calculated from

$$P = 58\,(u^* - u_t^*)^2 + 25\,(u^* - u_t^*), \quad u^* \geq u_t^*,$$
$$P = 0, \quad u^* < u_t^* \tag{4}$$

where u_t^* is the threshold friction velocity in m/s. Therefore, erosion is affected by particle size distribution and by the frequency of a disturbance; because of the nonlinear form of the erosion, each disturbance must be treated separately. Eq. (1) assumes a typical roughness height of 0.5 cm and height to base ratio not exceeding 0.2. If the pile exceeds this value, it would be necessary to divide the pile area into sub-areas of different degrees of exposure to wind. When working with higher height ratios the following formulas must be employed:

$$u^* = 0.10\,u_s^+ \tag{5}$$

with

$$u_s^+ = \frac{u_s}{u_r} u_{10}^+ \tag{6}$$

US EPA (1995) suggests, for representative cone and oval top flat pile shapes, the ratios of surface wind speed (u_s) versus the approaching wind (u_r) are derived from wind tunnel studies. Hence the total surface of the pile is subdivided into areas of constant u^* where Eqs. (4) – (6) can be used to determine u^*. A convenient methodology therefore exists to generate rough estimates for the gross dust output of the large stock area if the material stored and the asymptotic wind conditions are known. However, the aim of this work is to obtain data, which can yield accurate and detailed estimate of the dust output of a real stockpile, in particular PM10. This study aims to measure the dependence of dust emissions on the wind velocity for coal, iron-ore and hydrated alumina to provide supplemental generic parameters of the corresponding erosion potentials.

2.3 Evaluation of dusting from transport operations

An important information is also how much dust the port produces during the loading of trucks or wagons with truck loaders as well as during the massive unloading of ships. Chakraborty et al., (2002) has developed twelve empirical formulae to calculate the suspended particulate matter emission rate from various opencast coal/mineral mining activities. The measured and calculated values of the emission rate have been compared for each activity and, as reported, they agree to 77.2 – 80.4%. Many of the activities quoted in

Table 10 of their paper are similar to those at the open air terminal but there are additional site-specific assumptions that must be taken into account. For example, the dust emitted during loading of crusted material is different from the case of completely dry and crushed material. It is obvious, when assessing the final dusting situation, such activities can not be neglected. For example, the measurements at the open air terminal in Port of Koper revealed a base level weekly concentration of Fe in the PM10, determined by the sole presence of the stockpile, over which several times larger concentrations are superimposed during short times of intensive unloading activities of the iron-ore (Žitnik et al., 2005). However, the estimation of emission levels generated by different port activities is not in the focus of the present work.

2.4 Wind tunnel studies

In wind tunnel studies particulate emissions from cargo unloading and handling activities are usually not considered, although they can make an important contribution to the total amount of dust emitted. Borrego et al. (2007) made a maquette (1/333) of the port of Leixoes in North Portugal – a scrap metal pile, in order to reduce the particulate emission from the site. They found that while large particles emitted during material handling mostly settled before reaching residential areas, higher daily mean concentrations of PM10 were detected inside nearby houses (200 m away) than at the site (30 m). The proposed technical solutions, the upwind and downwind barrier were tested in a wind tunnel. The use of a smoke and laser sheet allowed the authors to visualize the turbulent structures of the local wind field for wind velocity magnitudes of 2 - 11 m/s. From the scaled down data and using the EPA's compilation (US EPA, 1995), they estimate that about 65% less PM10 is emitted from the top of the pile, mostly due to the reduced wind velocity and redirection of the wind flow for both the upwind porous barrier and downwind container barriers. Unfortunately, the authors do not report the effects in a 1:1 situation, which was realized on account of their findings (with solid windbreaks). In a similar wind tunnel study, Ferreira & Oliveira (2008) make use of an electronic balance with precision 0.05 g, and a model box filled with dusting test material. The authors filled a transparent parallelepiped box (99 x 420 x 145 mm^3) with dry sand with grain diameter of d ≈ 0.5 mm. In each experiment the test box was filled to form, initially, a flat "free-surface" levelled at a depth of 10 mm below the box surface. An undisturbed wind velocity of 11.4 ± 0.4 m/s was used in all the experiments. The maximum emission rate occurred in the initial stage, when the box was still full but became practically zero after one hour. The maximum mass loss rate per unit surface q can be compared with a flat bed situation for which Bagnold (1941) suggested the following formula

$$q = C\sqrt{d/D}\frac{\rho}{g}(u_t^*)^3 \qquad (7)$$

where C=1.5 m^{-1} for a uniform sand, D is a grain diameter of the standard 0.25 mm sand, ρ is the air density, g is acceleration of gravity, and u_t^* is the threshold friction velocity. There is a good agreement between the calculated q with the maximum observed values of Ferreira & Oliveira (2008). The authors followed the modification of the free surface due to wind erosion with time and found that the release of particles from inside the box occurs mainly at the leading edge. In the present research, this was avoided by having no sharp edges near the test surface in order to approach a flat bed situation. In addition, in place of a well defined particulate sample in terms of grain sizes, the mass loss rate for real materials at different wind velocities was measured. The same group had previously studied the problem of coal dust release from rail wagons during transportation using a 1/25 scale model with a locomotive and four wagons placed into the wind tunnel using an undisturbed wind velocity of 13.4 m/s (Ferreira & Vaz, 2004). The experiments show that the use of covers reduces the amount of dust released by more than 80% compared to uncovered wagons. Most of the coal grains in the experiment have a diameter in the range 0.2 - 1.1 mm with the cut at 2 mm and the total mass loss was determined using an electronic balance. Roney & White (2006) measured the PM10 emission rates for different types of soil. The PM10 concentrations c and wind velocities were measured before and after the test track at different heights z and the total emission from the control volume was determined. Horizontal PM10 emission rates as high as 20 mg/m²/s were measured when the wind velocity extrapolated to 10 m height was 24.3 m/s. Roney & White (2006) also measured the total mass of fugitive dust using a sand trap. Typical values of 0.001- 0.0001 for the ratio of PM10 versus total removed mass were obtained. The authors also considered the vertical PM10 flux, which was derived from the gradient of the vertical concentration c according to the diffusion formula

$$F_a = - K \, u_t^* z \frac{dc}{dz} \tag{8}$$

This vertical PM10 flux was directly related to the horizontal flux of dust measured in the wind tunnel (Roney & White, 2006). Due to the locally restricted wind field it is expected that in present work most of the dust from the test surface is transported away in the horizontal direction. In setting-up wind tunnel studies of a scaled-down physical situation, it is important to consider the similarity principles. By obeying such principles, one can expect to grasp the information that is relevant to a real 1:1 situation. For example, Xuan & Robins (1994) studied the influence of turbulence and the complex terrain on dust emissions from a model coal pile including the subsequent dispersion and deposition. Although this is one possible approach for determining emissions, the present study does not deal with the wind field experiments on a scaled-down EET area. Emissions from a given test area at different wind velocities were measured instead. In the future, these data may be used in combination with computational fluid dynamics (CFD) wind simulation of the whole stockpile area to predict erosion potential of the whole stockpile area.

2.5 Watering

One of the simplest and effective means to reduce dusting is by adding water to the stock material. Due to the surface tension of the water, the inter-particle cohesive forces are larger and additional work is required to eject dust particles into the atmosphere. The influence of moisture on coal losses is known, Cowherd et al., (1988) have shown that a watering intensity of about 1 l/m²/h for an unpaved road surface results in an approximate control efficiency of 50%, presumably for total suspended PM and under normal wind conditions. Another study focuses on the pickup of soil by a belly scraper at a landfill site. The authors demonstrate that watering is effective in controlling PM10 emission at wind velocities of 18 m/s (Fitz & Bumiller, 2000). An increase of 4% in the moisture content in the upper layer of the soil was responsible for a reduction in PM10. The authors assume that the water penetrated to a depth of 0.02 m into a soil having a density of about 2800 kg/m³. This means that a 90% reduction in the PM10 concentration at the sampling point was achieved with a watering rate of $6.5 \, l/m^2/h$.

3. Experimental design

The benchmark twenty-four hour measurements of dust made using filter-based gravimetric methods are costly and time consuming. The filter handling involves a large number of steps including pre-conditioning, weighing of blanks, filter installation and filter removal on the sampling site, post-conditioning and weighing of dustloaded filters. However, to follow up our dust emission studies time resolution of the order of at least one minute is needed.

3.1 Tapered element oscillating microbalance

The measurements of the PM10 concentrations were made using a Patashnick & Rupprecht's Series 1400a tapered element oscillating microbalance (TEOM). The most important part of the TEOM is its mass detector or microbalance, which utilizes an inertial mass weighing principle. The detector consists of a filter cartridge placed on the end of a hollow tapered tube while the other end is fixed rigidly to a base. The tube with the filter on the free end then oscillates in a clamped-free mode at its resonant frequency. This frequency depends on the physical characteristics of the tube and the mass on its free end. A particle laden air steam is drawn through the filter where the particles deposit themselves. As the amount of particulates builds up, the mass of the filter cartridge increases which decreases the frequency of the system. By accurately measuring this frequency change, the accumulated mass can be determined. In essence, the system can be considered a simple harmonic oscillator, through which the following equation can be derived,

$$\Delta m = K_0 \left(\frac{1}{f_f^2} - \frac{1}{f_i^2} \right) \tag{9}$$

where f_i and f_f are the initial and final frequency, respectively, of the system, Δm is the mass change of the system from initial value, and K_0 is the calibration (spring) constant of the tapered element. Combining the accumulated mass with the volume of air drawn through the system during sampling then yields the particle mass concentration. For ambient PM measurements the mass sensor provides a minimum mass detection limit of 0.01 μg. To avoid a problem of air humidity the air is heated typically to 50°C prior to passing through the filter. A more complete description of the instrument and conditions under which it can be used reliably is given by Patashnick & Rupprecht (1991).

3.2 Black carbon concentration by Aethalometer

As an independent device – a portable Aethalometer (Magee Scientific, AE 42-2ER-P3) was used to measure coal dust. The instrument comprised two-channels with two diode types: IR and UV. The Aethalometer provides a real-time readout of the concentration of 'Black' or 'Elemental' carbon aerosol particles ('BC' or 'EC') in an air stream and was first described by Hansen et al. (1984). Black carbon is defined by the "blackness" of an optical measurement. Physically, the Aethalometer measures attenuation of a specific wavelength of light through a quartz fibre filter as it loads over time. The 'EC' definition is based on a thermal chemical measurement. There is no accepted definition of 'elementarily' and the different thermal parameters used by different EC analysis protocols yield different EC numerical results, even on portions of the same filter samples whereas the optical analysis for BC is consistent and reproducible (Weingartner et al., 2003). In present work the Aethalometer draws the air through the inlet port without an impactor stage. The flow rate was set to 3 l/min by a small internal pump. The dust sample is collected on a quartz fibre filter tape and photomultiplier measures the signals of diode light passing the clean exposed part of quartz filter tape, and a light beam attenuation and BC concentrations is calculated.

3.3 Auxiliary equipment

The fan is of centrifugal type 6Cv6 and produced by Klima Celje. It delivers a maximum flux of 9400 m³/h at 3 kW. A frequency inverter Watt drive L2000 - 030HFE was additionally installed to control the power in a continuous manner. Wind velocity was measured using a wind sensor VMT 107 connected to the read-out station AMES. A hand held sensor served as a check for the wind velocity around the test area. An electronic balance My-weigh HD-150 with a capacity of 60 kg and mass resolution of 0.02 kg was employed to measure the total rate of dust emission from the test pile. The total force exerted by the effective weight of the test vessel was recorded to obtain the desired time dependence of the mass loss with good time resolution.

3.4 The wind test track

The test track was setup in the open air near to the western terminal wind barrier in order to shelter the apparatus from local wind field perturbations. Additional smaller side fences were installed to reduce the effects of any side wind. All measurements were performed during periods of stable and calm weather during June and July 2008.

The test track starts with an air stream source. The centrifugal fan produces an air stream with an adjustable air velocity through a 0.38 x 0.35 m output opening. During testing the air stream velocity at 1.30 m from the opening in the middle of the test track reached a maximum of 18 m/s and was a linear function of fan frequency, $u_{.20} = 0.352 v - 0.398$ (v is inserted in Hz and $u_{.20}$ in m/s). The experimental wind velocities ($u_{.20}$) are referred to as the velocity measured at 0.20 m above the centre of the test vessel. On the flat plate (track), was mounted the electronic balance with the test vessel secured on the top. The vessel's dimensions are 0.60 x 0.60 m, and to reduce wind turbulence the edges were angled at 45° so that the surface of material exposed to wind erosion was 0.46 x 0.46 m. An additional wind screen was fixed on the track to reduce the effect of buoyancy on the mass measurement. The test vessel was then filled with the test material and the material levelled to the upper edge of the vessel; the layer of material in the vessel was approximately 7 cm thick. At the end of the test track was positioned the TEOM PM10 and the Aethalometer with the inlets placed 1.30 m after the vessel. The measurement consisted of submitting the vessel and test material, to airflows of different wind velocities. Each test at each velocity lasted for 10 minutes and the mass of the test vessel was recorded every 2 seconds. Every 10 seconds the PM10 concentration was recorded (10 s average composed of 5 readings) and every minute the black carbon (IR and UV) concentration. The wind speed sensor was placed behind the inlets to monitor wind speed stability.

3.5 Sample preparation

The coal and iron ore samples were separated into four and three size fractions, respectively, while hydrated alumina, which has a narrow grain size distribution, was taken directly from the cargo. 40.3 kg of the test coal, named ABK, which originates from Indonesia, was taken from the pile and fractionated as follows:

- Fraction 1 (FR 1): grain size larger than 15 mm, 27.1% of the total mass,
- Fraction 2 (FR 2): grain size between 5 and 15 mm, 25.8 % of total mass,
- Fraction 3 (FR 3): grain size between 3 and 5 mm, 9.1 % of the total mass,
- Fraction 4 (FR 4): grain size less than 3 mm, 38.0% of total mass of the coal sample.

The majority of the coal with density of 1100 kg/m^3 in the pile was present in the finest, the FR4 fraction.

The particle size distribution (relative number N of particles as a function of coal particle diameter d) of FR4 coal fraction was estimated by the following simple method. The particles were allowed to disperse according to size by dropping them through a horizontally directed stable air stream with a restricted cross section on to a horizontal metal plate. An average diameter of particles, d_i, deposited along the metal plate in each bin was determined by optical microscopy and the mass m_i of the deposit in each bin was measured accurately using a microbalance. In the log-log plot the N(d) is approximately a linear function of d suggesting that N is proportional to $d^{-1.6}$. The median diameter of the particle distribution, defined as the diameter that "splits" the number of particles into two halves, was d_{50}=40 µm. Prior to making tests on the wind test track, a maximum cone angle for three of the finest coal fractions was measured. This was achieved by building up a cone on a flat surface until the angle at the top assumed the largest value. As the angle increases, the component of the weight force acting on a coal particle along the cone surface is increased. When this force becomes larger than the friction force, exerted by the surrounding particles, the particle will start to move downhill. This is important for estimating the threshold friction velocity on inclined surfaces. Obviously, the component of the weight force "helps" the wind to lift the particles if this is blowing downhill and vice-versa.

	FR4	FR3	FR2
Base diameter (cm)	20	20	20
Height (cm)	13.0	15.5	18.0
Angle	33.0°	37.8°	42.0°

Table 2. Top cone half-angle α_t for different coal fractions.

The iron ore with 5000 kg/m^3 density was separated into three fractions:

- FR 1: d > 7 mm, 25.5 % of the total mass,
- FR 2: 3 mm < d < 7 mm, 29.2 % of the total mass,
- FR 3: d < 3 mm, 45.2 % of the total mass.

4. Results and discussion

4.1 Coal

For FR1 coal fraction no mass loss larger than 0.02 kg and no increase in PM10 or the BC signal for test wind velocities up to 18 m/s were detected during a time interval of 10 minutes and for an initial coal mass of 9 kg.

4.1.1 FR2 coal

The average background concentration of PM10 during the FR2 measurement was 73 µg/m³ but this did not correlate with the detected mass loss. The latter was detected only at the highest wind speeds i.e. > 45 Hz (16 m/s). The Aethalometer signal was also stable during the measurements, displaying average values of 1470 ng/m³ and 1320 ng/m³ for the IR and UV diode, respectively. The effect of buoyancy is to make the test vessel effectively lighter when the air stream is on and is larger for higher air velocities. This results in an offset of the mass signal and can be easily accounted for in the data analysis.

4.1.2 FR3 coal

For FR3, a mass loss on a time-scale of a few minutes at a wind speed of $u_{.20}$=13.5 m/s (40 Hz) is observed (Figure 1). The effect of 18 m/s wind velocity was such as to remove approx. 20% of the material from FR3 from the vessel in 10 minutes.

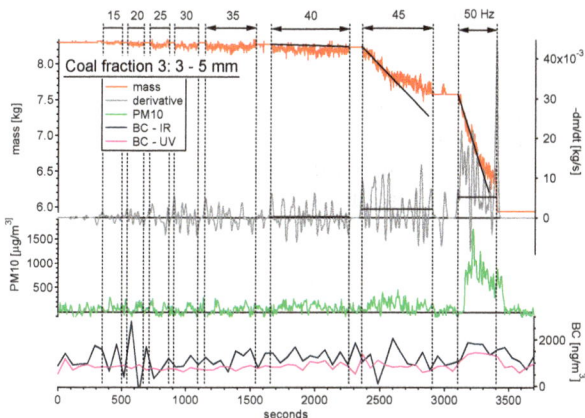

Fig. 1. Fraction 3 (FR3) measurement with a dry coal reporting the total mass (red, kg), the negative of its time derivative (gray, kg/s), PM10 concentration (green, µg/m³), BC-IR (black) and BC-UV (pink) concentrations (ng/m³) as a function of time. Narrow intervals between vertical dashed lines correspond to wind velocity zero. For the intervals between, the wind velocity is denoted by a frequency of the converter v depending on $u_{.20}$ in a linear fashion as presented in Section 3.4.

The results also show a correlation between the increasing PM10 concentration and the negative time derivative of the mass (-dm/dt), which is not so prominent in the BC detection channel. In fact, there should be no PM10 signal since, in principle, FR3 should not contain particles with a diameter smaller than 3 mm. There are two possible reasons for the nonzero PM10 signal. The first is that the sieving efficiency was not 100% and some residual FR4 remained in FR3 and second, which is more probable, is that small particles are produced by creep and saltation processes. As mentioned above, this is when larger particles are eroded while rolling or jumping along the surface under the force of the wind field. If the second option prevails, then an estimate of this effect can be obtained by comparing the relative ratio of PM10 to the mass time derivative signal for FR4 (see below) and FR3 measurement: it appears that these indirect processes produce about 10% of the concentration, which would be measured in a direct PM10 removal (as dominant for FR4) for the same total mass loss.

4.1.3 FR4 coal

As expected, FR4 coal fraction is most affected by wind erosion. Figure 2 shows that while at 8 Hz the surface is stable i.e., no directly detectable change of mass in 5 min, at 15 Hz (corresponding to $u_{,20} = 5$ m/s) there is a detectable 20 g mass loss in 360 s. At 20 Hz the loss is larger: 240 g in 1050 s. Looking at the graph one clearly sees how mass loss per unit time increases with increasing wind velocity. Typically, after switching on the wind source the mass outflow is at maximum but later drops to almost zero. In these measurements, saturation is reached at 25 Hz only. This means that after 1800 s and after a 26% mass loss (2.54 kg/9.60 kg) the surface of the coal dust in the test vessel had altered in such a way that the friction velocity of the wind was smaller than the threshold velocity for all particle sizes on the exposed surface. The oscillations in the mass loss are a probable result of oscillations in the wind velocity and are affected by the different stages through which the surface morphology alters as it reaches equilibrium at saturation. At 30 Hz the measurement had not lasted long enough to reach saturation, although 44% (4.30 kg/9.70 kg) of the initial mass m_0 had been lost. At 22 Hz the much smaller relative mass loss of 3% (0.28 kg/9.54 kg) indicates that the results are not significantly affected by changes in surface morphology. It is interesting to note that the mass time derivative, the PM10 concentration and both BC concentrations correlate with time – so that if the proportionality factors would be known

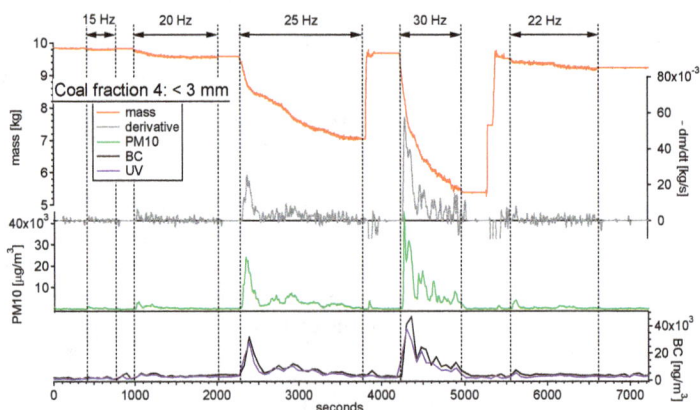

Fig. 2. The same as Figure 1 but for FR4 coal fraction.

only one of these three would suffice, since the same dynamics is observed with the three detection channels. In this case, the PM10 should be chosen for detection as it demonstrates the largest sensitivity.

4.1.4 None size-fractioning

Finally, measurements were performed on an 11 kg of dry sample of unfractionated ABK coal, taken directly from the surface of an actual pile (Figure 3). The dynamics of the mass outflow at different wind speeds reflects the dynamics of the mass outflow of the different fractions previously studied. In Figure 4, which summarizes the results for coal, the two main fraction thresholds are clearly observable, but the mass of the mixture is removed more slowly above the FR4 threshold than for the pure FR4 fraction. Instead of a rapid increase in the signal, stabilization is observed soon after the FR4 threshold; here the finer FR4 particles are being screened by the larger particles in the mixture. Alternatively, the mixture's mass is removed at a rate 3-times faster than the mass of the pure FR2 fraction but about 15 times slower than the mass of the pure FR3 fraction at equivalent wind speeds above the FR3 threshold. Below the FR3 threshold, the PM10 signal is proportional to the mass loss of the mixture, but there is relatively more PM10 component in the total mass removed than for the pure FR4 fraction: the larger particles (> 3 mm) are as efficiently removed from the mixture due to the rougher surface. However, considering the absolute concentration values and a mass weight of 38% for FR4, the PM10 signal is much smaller than in the pure FR4 case. Above the FR3 threshold, the mass outflow of the mixture is somewhere between the pure FR3 and FR2 (PM10 follows the FR3 case), while below the FR3 and above the FR4 threshold the outflow is different from zero and stable with a relatively large PM10 component.

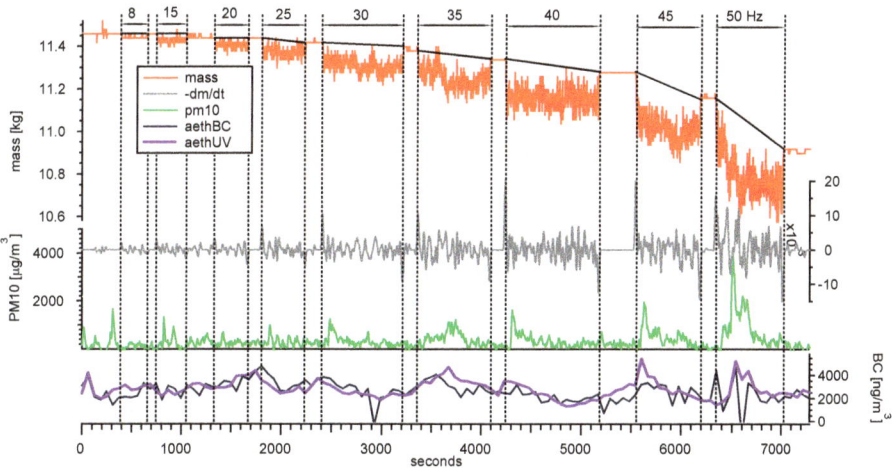

Fig. 3. The same as Figure 1 but for coal with no size-fractioning.

The threshold $u_{t.20}$ velocities for different fractions i are found by fitting the mass outflow FR_i (kg/s) with the functional form of Eq. (4) for the erosion potential:

$$FR_i = B_i \left(u_{.20} - u_{t.20} \right)^2 + A_i \left(u_{.20} - u_{t.20} \right), \quad u_{.20} > u_{t.20},$$
$$FR_i = 0, \quad u_{.20} \leq u_{t.20}$$

(4)

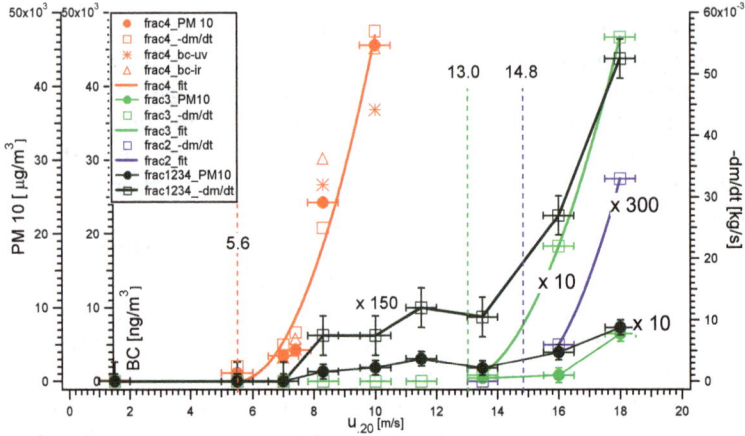

Fig. 4. The mass outflow, PM10 and BC concentrations from a 0.21 m^2 "unit" surface for FR2 (blue), FR3 (green), FR4 (red) and FR1234 (black) as a function of the measured wind velocity $u_{.20}$. Dashed vertical lines denote threshold velocities $u_{t.20}$ for different coal fractions.

	$A_i[10^{-5}$ kg/m]	$B_i[10^{-5}$ kg s/m^2]	$u_{t.20}[$ m/s]
FR$_2$	0.6	0.9	14.8
FR$_3$	14	19	13.0
FR$_4$	84	270	5.6

Table 3. Parameter values to fit the measured mass outflow.

A comparison between the observed mass losses as a function of time for different fractions of coal reveals a similar development in surface profiles with time as that reported for sand by Ferreira & Oliveira (2008), whereas Eq. (7), which was successful when dealing with sand, greatly underestimates the measured values for coal.

4.2 Iron ore

The setting-up of the test track for iron ore and alumina measurements was similar to that for coal. The test surface area was 0.35 x 0.51 m^2 and only mass loss was recorded. As expected from the density ratio, the iron ore is a less intensive source of dust than coal. For FR1 there was no mass loss detected up to and including $u_{.20}$ = 18 m/s.

4.2.1 Iron ore fraction 2 and 3

For FR2, the detected loss is not statistically significant. For FR3 a none zero mass loss was observed only for the largest values of $u_{.20}$, at 16 m/s and 20 m/s (Figure 5-left).

Measurements therefore show that threshold velocity $u_{t.20}$ for FR2 must be larger than 18 m/s. According to the Greeley-Iversen scheme (described below) the threshold velocity $u_{t.20}$ equal to 26.7 m/s is predicted if d_{50} = 5 mm and u^*_t = 1.5 m/s. For FR3 the measured threshold wind velocity $u_{t.20}$ is 14.0 m/s (Figure 5–right). Since this is less than 18 m/s, according to the G-I scheme the FR3 aerodynamic roughness length should be less than 300 µm and consequently, d_{50} < 9 mm. This is not in contradiction with our measurements since the maximum diameter of iron ore particles in FR3 is 3 mm.

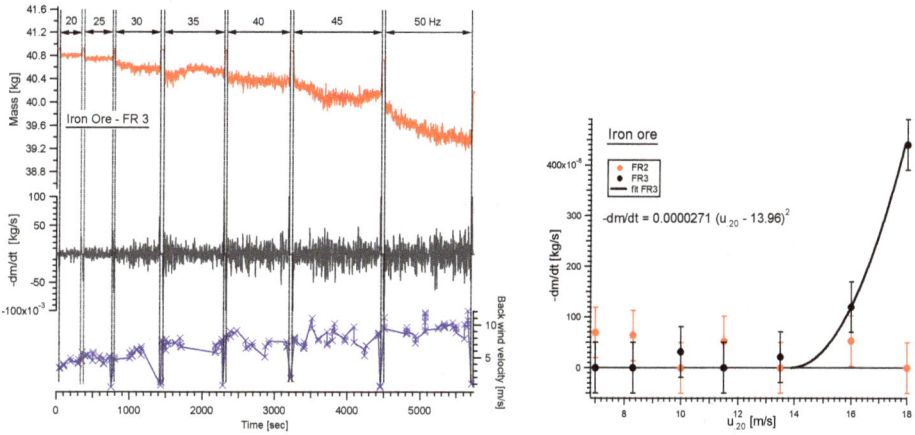

Fig. 5. -Left, FR3 iron ore mass loss measurement (red), its negative time derivative (gray) and back wind velocity (blue). -Right, the mass loss from the test surface as a function of $u_{.20}$ air velocity for FR2 (red) and FR3 (black).

4.3 Hydrated alumina

Hydrated alumina is a fine material with large erosion potential. It is usually extracted from bauxite, which is mined for the production of aluminium. In the port, the alumina is unloaded from ships using a continuous unloader and transported to silos where it is stored. Experience tells that only a small wind velocity is sufficient to lift particles of alumina into the air. Indeed, the alumina particles are small with diameters from 20-100 µm and 3200 kg/m^3 density.

The fraction of alumina tested was taken directly from the storage facility. Already at about 2.5 m/s (8 Hz) there is an observable mass loss (Figure 6 -left). It is notable that this mass loss is constant in time at any given air velocity, suggesting that no major alteration in surface morphology occurred during the test. Figure 6-right shows the results and reveals that the threshold velocity for alumina at 0.20 m above the centre of the test vessel is 2.45 m/s.

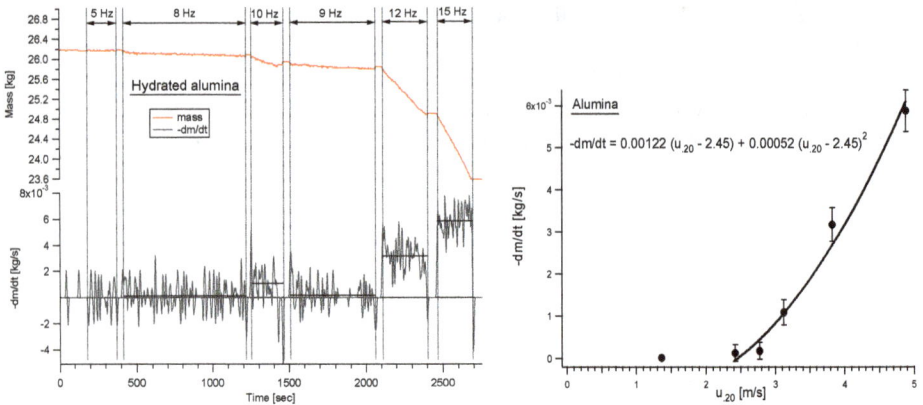

Fig. 6. -Left, the mass loss and its negative time derivative from alumina test surface with respect to time at different air velocities. -Right, the threshold wind velocity at 0.20 m above the pile is 2.45 m/s. Assuming z_0=30 μm, this gives u^*_t=0.11 m/s.

4.4 Comparison with results in the literature

Before quantifying and comparing the results for different fractions, it is important to note a few basics about the dynamics of the mass outflow. It can safely be assumed that the surface S exposed to erosion stays unchanged for a short while after the measurement begins and that the mass flux is governed by the equation dm/dt = - p S, where p is some erosion potential of the surface S. If the wind field, the shape of the surface and the composition of the surface material layer are independent of time, then the mass outflow is a constant and the mass loss will be obviously a linear function of time m(t) = m_0 - p S t. Such a linear drop is indeed observed in the starting time interval of each measurement. The "active" surface (affected by wind erosion) in the vessel inevitably diminishes with time and finally results in saturation (a zero mass outflow). This is reflected in the measured time dependence of the mass loss. The mass outflow diminishes with time because the wind friction velocity u^* gradually approaches (from above) the local threshold velocity u^*_t. Consistent with measurements and according to the US EPA Eq. (4), p depends quadric-linearly on the friction velocity difference u^*- u^*_t but the time evolution of u^* is an unknown a priori since this is a non-local component that depends on the wind field that in turn depends on surface morphology. To predict the time dependence of the mass loss m(t) the coupled wind field - surface erosion problem must be solved. On the one hand, the erosion potential at a given point depends on the wind field, but the wind field itself depends on the erosion effect, which modifies the surface. If one is able to calculate the wind field for a particular geometry (given by function g of the coal surface z(x,y,t) in the vessel at certain time t), it is possible to estimate the alteration of the surface due to erosion for the small time interval. The wind field is then calculated again and with this knowledge a new shape of the surface is calculated, until u^* drops below u^*_t everywhere at the surface. Formally, the equations, which should be propagated in time, starting from t = 0, when the wind field is turned on, are:

$$z(x,y,t+\Delta t) = z(x,y,t) - p(u^*(x,y,t) - u^*_t)\Delta t,$$
$$u^*(x,y,t+\Delta t) = g(z(x,y,t+\Delta t))$$

(10)

If friction velocity u^*_t at any given time is the same everywhere along the surface (as assumed above), a reasonable assumption for a homogeneous flat terrain composed of uniform particle sizes submitted to the uniform wind field, erosion will proceed infinitely long with the same speed resulting in a constant mass outflow signal. Any reduction in the mass outflow and finally the saturation is an effect of screening, which occurs for example, when a finite quantity of material is placed in the vessel, or even for the flat terrain situation if this is composed of different particle sizes. The aim of this work was not to follow up the mass outflow with substantial modifications of surface morphology. However, for the "finite" test surface, the erosion potential p can still be measured by changing the wind velocity $u_{.20}$ above an initially flat test surface and considering only the first, linear part of the mass loss. This is true, if in this initial and relatively short period of time (so that surface morphological changes can be neglected), u^* can be thought as being frozen to a value which is the same along the whole surface area (negligible edge effects and large scale variation of u^* over the surface due to the localized wind source) and if the way that u^* is related to the measurable quantity, in our case $u_{.20}$, is known.

According to US EPA (1995), for a flat terrain with no obstacles, the wind friction velocity is related to u_{10}, the wind velocity measured at 10 m above the terrain (Eq. 1). The wind speed profile in the surface boundary layer follows a logarithmic distribution and becomes zero at the aerodynamic surface roughness height z_0 (Eq. 2). Accordingly, u^*_t is the wind speed just above the surface at $z-Z = z_0 \exp(0.4) \sim 1.5 \, z_0$. Even in the case of a finite flat terrain (as in the vessel), the surface boundary behaviour is similar to that of the flat terrain if the vertical profile is limited to the region close to the surface and away from the edges. The idea is to determine the actual wind friction velocity from the simulations of the wind field for the described experimental set-up. First, the input parameters of the wind field simulation are chosen in such a way to obtain the wind velocity $u_{.20}$ value at 0.20 m above the test surface as actually measured. Then the vertical wind velocity profiles close to the surface is extracted in order to identify the region with logarithmic z/z_0 dependence and their possible differences at different parts of the surface. By comparing this local behaviour to the behaviour of the flat terrain close to the surface, the relation between the local and flat terrain situation can be established. Let it be assumed for the moment that 0.20 m is sufficiently close to the 0.46 x 0.46 m² surface so that the wind profile is defined solely by the surface characteristics, i.e. the central point 0.20 m above the surface is still in the boundary layer. As for the aerodynamic surface roughness, this is usually smaller than the particle diameter. The experimental parametrisation (Byrne, 1968) formula is

$$z_0 = Q \exp(d / \lambda) \tag{11}$$

It applies for a uniform bed of quartz particles with diameter d and parameters $Q = 2.0 \times 10^{-6}$ m and $\lambda = 6.791 \times 10^{-4}$ m. According to Eq. (11), the average d=4 mm of FR3 coal corresponds to a $z_0 = 0.72$ mm. If the vertical profile of the wind velocity obeys the large flat terrain law (Eq. 1 and 2) and $z - Z \approx z$, then the relation between the measured $u_{.20}$ and wind friction velocity for FR3 is given by

$$u^* = \frac{K u_{.20}}{\ln(286)} \approx 0.070 \, u_{.20} \tag{12}$$

Since at threshold the wind friction velocity u* equals the threshold velocity, the threshold velocity for FR3 coal would be, according to the measured $u_{t.20}$, approximately $u*_t \approx 0.91$ m/s. Alternatively, for a bed composed of several grain sizes the overall roughness length is given as $z_0 = d_{50}/30$ (Chane Kon et al., 2007 and references therein). For FR4 coal this means that approximately 50% of the FR4 volume is made up of particles with a diameter smaller than 1 mm so that the majority of particles have much a smaller diameter than this. Taking $d_{50} \sim 40$ μm, it comes out that $z_0 \sim 1$ μm for FR4 so that

$$u^* = \frac{K u_{.20}}{\ln(200\,000)} \approx 0.034\, u_{.20} \tag{13}$$

and consequently $u*_t = 0.18$ m/s for FR4. This result is not greatly affected by the choice of z_0. If a 50 times larger z_0 is selected, $u*_t$ would increase to 0.27 m/s. Prior to any corrections due to the local wind field situation this simple scheme produces results similar to those reported by Barrett & Upton (1988). They investigated the erodibility of industrial materials in a wind tunnel. For a coal with a density of 1300 kg/m³ and a median particle diameter of 1.67 mm the threshold value of wind friction velocity was 0.35 m/s. After removing those particles with a diameter larger than 1 mm, the friction velocity reduced to 0.22 m/s. This second measurement is a good match to FR4 estimates found in this research, while FR3 deals with larger coal particles. A large difference in the threshold velocity for FR3, i.e., 0.91 m/s versus 0.34 m/s can be to some extent explained by the difference in the aerodynamic roughness lengths. Indeed, in this case Eq. (11) needs to be extrapolated outside of actually measured range of diameters. At the upper edge of the studied interval is a particle diameter of 2.5 mm which results, according to Eq. (11), in an aerodynamic roughness length of $z_0 = 0.08$ mm. The error introduced by extrapolation to diameters greater than 2.5 mm is not known. Alternatively, although FR3 is a relatively well-defined fraction regarding particle diameter distribution, it is still composed of different grain sizes. Employing median diameter $d_{50} = 3$ mm one arives at $z_0 = 0.1$ mm. Taking this as a more realistic value, the threshold velocity for FR3 calculated by Eq. (12) is reduced to $u*_t = 0.68$ m/s. Finally, assuming that $d_{50} = 10$ mm for FR2 (5-15 mm), the aerodynamic roughness length becomes $z_0 = 0.33$ mm so that corresponding $u*_t = 0.94$ m/s for a measured threshold velocity $u_{t.20}$ equal to 15 m/s.

4.5 Greeley-Iversen scheme

The measured threshold velocities published by Barrett & Upton (1988) fit well into the so-called semi empirical Greeley-Iversen scheme (Greeley & Iversen 1985) that considers all physical forces acting upon a particle exposed to the wind. Chane Kon et al. (2007) gave a brief description of the scheme. Here the scheme is used to estimate the threshold friction velocity $u*_t$ (d) for particles with diameters up to 10 mm and for materials investigated. In short, at each d the minimum of the expression

$$u_t^*(d) = A(R(u^*,d))F(R(u^*,d)) \cdot \sqrt{1 + \frac{1.9 \cdot 10^{-10}}{\rho_a g d^{2.5}}} \sqrt{\frac{g d(\rho_p - \rho_a)}{\rho_a}} \tag{14}$$

is sought by changing the friction velocity u*. This enters the Reynolds number $R = u*d/v$, where the kinematic viscosity of air $v = 1.57 \times 10^{-5}$ m²/s. The empirical functions A and F in Eq. (14) are given by

$$0.03 \leq R < 0.3 : A=0.20, F=(1-2.5\ R)^{-0.5},$$

$$0.3 \leq R < 10 : A=0.13, F=(1.9828\ R^{0.092} - 1)^{-0.5}, \qquad (15)$$

$$10 \leq R : A=0.12, F=1-0.0858\ \exp(-0.0617(R-10))$$

The air density ρ_a =1.24 kg/m^3 and ρ_p is the particle density. Under the first square root in Eq. (14) the diameter d must be inserted in meters. The results for coal show a good match with the scheme predictions for FR3. Obviously, the extrapolation of Eq. (11) for aerosol roughness length with d=4 mm does not lead to the right result for FR3 coal. A reasonable agreement of the Greeley-Iversen (G-I) estimate of u^*_t with the measured value is obtained also for FR2. Alternatively, FR4 has a broad particle diameter size distribution, while the G-I scheme applies to pure size fractions. Nevertheless, the measured value of 0.21 m/s is inside the velocity range 0.09 - 0.50 m/s suggested by G-I scheme. While for pure fractions the G-I scheme reliably predicts the threshold friction velocities, in the case of mixtures such as FR4 fraction of coal and FR3 fraction of iron ore it gives only a rough estimate.

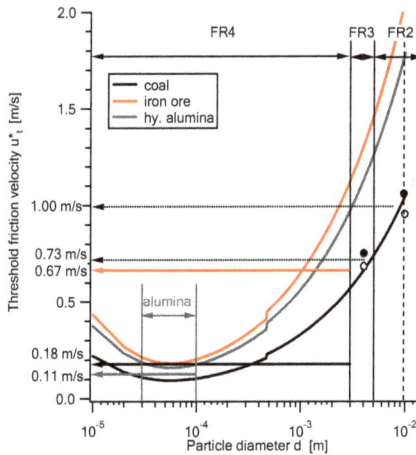

Fig. 7. Curves denote threshold friction velocity with respect to particle diameter for coal (black), iron ore (black) and hydrated alumina (gray) according to the Greeley-Iversen scheme. Hollow circles denote measured threshold friction velocities, deduced from the measured wind speed at 0.20 m above the pile centre using estimated aerodynamic roughness lengths for different fractions and materials. Full circles represent the corresponding wind field corrected data. Vertical lines mark particle diameter ranges for different materials and horizontal arrows the corresponding threshold friction velocities.

4.6 Geometric wind correction

The equations quoted above apply to flat terrain with no obstacles and a vertical wind profile independent of the base position. Because measurements were made in a restricted geometry and with a localised wind source it is important to estimate the effect that this particular set-up may have on the threshold friction velocities extracted from measured u_{-20} dependence using the "flat terrain" equations.

Two-dimensional wind simulation of the set-up does not take into account surface roughness effects. This does not seriously affect the wind field except close to the surface, in the so-called surface layer. The CFD simulation by Fluent 6.2 shows, that because of the pile, the wind velocity at 0.20 m above the pile centre has increased to 13 m/s compared to the value of 10 m/s at the same height above the flat terrain before the pile. While both profiles in the boundary layer: the "pile" and the "flat", fit well the functional form of Eq. (2), the flat profile surpasses the logarithmic extrapolation outside the boundary layer and the "pile" profile undershoots the logarithmic extrapolation. The extrapolation of fit from the boundary layer suggests the value of the wind velocity at 0.20 m above the base in the absence of geometrical and wind source constraints. According to this procedure, a measured 13 m/s wind velocity at 0.20 m above the pile is equivalent to a velocity of 14.6 m/s in a flat terrain situation. This translation is important since formulas quoted previously apply for the flat terrain situation. This simulation suggests that the measured $u_{.20}$ should be increased by a factor of 1.12 and this corrected value then used to determine the wind friction velocity. In conclusion, a particular geometry of the set-up does not affect more than 10% the threshold velocity values in the sense that these can be related to a wind velocity $u_{.20}$, measured at 0.20 m above the centre of the test surface using just the flat terrain formulae. So, when $u_{.20}$ velocity distributions for any pile, modelled smoothly at a 0.5 m length are given, the experimental $u_{.20}$ data presented in Figure 4 can be used to estimate the total mass outflow and the corresponding PM10 and BC concentrations. Table 3 summarises our results for dry materials.

	d_{50} (mm)	z_0 (mm)	$u_{t.20}$ (m/s)	u^*_t (m/s)	u^+_{t10} (m/s)
FR2 coal	10	0.33	14.8	0.93	23.8
FR3 coal	3	0.10	13.0	0.68	19.7
FR4 coal	0.04	0.0014	5.6	0.18	7.4
FR1234 coal		0.1	5.6, 13.0		7.6, 19.7
FR2 iron		>0.3	>18		
FR3 iron		0.05	14.0	0.67	20.6
Alumina		0.03	2.5	0.11	3.6

Table 3. Threshold wind velocities $u_{t.20}$ for different fractions of coal, iron ore and alumina as determined by our measurements. Also reported are u^*_t and u^+_{t10} threshold velocities which are calculated from measured $u_{t.20}$ by using the estimates for d_{50} (z_0).

4.7 FR4 coal watering

Dusting from the coal stockpile can be significantly reduced by spraying the surface with water. The water effectively increases the mass of the particles by binding the particles to the surface layer of the coal. However, the protection afforded by a single spraying is not permanent because the water evaporates.

We measured that watering of FR4 coal by 0.5 l/m² every two hours is sufficient to suppress dusting at $u_{.20}$ = 8.3 m/s and at 95% air humidity. Compared to a dry FR4, after 364 seconds more than 100 times less mass was removed from the test vessel (1.16 kg versus 0.01 kg). To achieve the same effect at $u_{.20}$ = 11.3 m/s, 2 l/m² would be required. The estimates of a water quantity needed to suppress the dusting from FR4 surface should be considered as upper estimates when applied to realistic surface FR1234. The results are consistent with the belly scraper case at a landfill described by Fitz & Bumiller (2000) and with the unpaved road surface watering efficiency determined by Cowherd et al. (1988). On the other hand, it is difficult to compare this study's results with those of Nicol & Smitham (1990), because neither exposure time nor wind velocity is reported. While they made wind tunnel tests with uniformly humid and conditioned coal, in our case the water was sprayed directly onto a dry coal surface just prior to switching on the air stream. Measurements show that mass loss is not a steady function of time, except when the watering is sufficient to establish a lasting linear dependence of mass loss with time. For the two linear cases with u_{-20} equal to 8.3 m/s and 11.3 m/s the mass losses for the coal (+ water) system were 480 g/m^2/h and 1400 g/m^2/h, respectively. In both cases water evaporation contributed mostly to the mass loss, as shown by a separate wind track experiment with water sample only.

4.8 Iron ore watering

Concerning iron ore, the following phenomenon was observed: after watering, the surface of the iron ore (FR4, FR3) developed a firm crust upon drying. The formation of the crust strongly reduces the dusting potential and a much lower mass loss per unit time is observed for crusted material than for freshly sieved iron ore under the same experimental conditions. Only 0.06 kg loss in 1548 seconds was observed for the crusted FR3 iron ore at u_{-20}=18 m/s, which is 9-times less than for sieved FR3. In view of this finding, it is highly desirable to crust the iron ore, which is stored for a significant period. Different to the coal, watering the iron ore only once is sufficient to immobilize it, provided the crust remains intact and is not disrupted by subsequent movement of heavy machinery.

The mechanism behind the hardening of iron ore is attributed to agglomeration of iron ore particles. PIXE analysis, as well as x-ray diffraction of an iron ore sample shows that the ore consists of 98% pure hematite (Fe_2O_3) with small impurities of K, Ca, Cl, Ti, Mn and Sr. According to specifications, other impurities may be mixed with the ore in small quantities ($SiO_2 < 3\%$, $Al_2O_3 \approx 1\%$), but in this authors opinion it is the content of water which is responsible for binding. When hematite is watered and left to dry a uniform crust develops over the sample. The crust can be destroyed by mechanical action or by re-watering the iron ore. As found out previously, the first two water layers near the hematite surface are highly ordered, and effectively are adsorbed to the surface through hydrogen bonding and further from the surface, water still displays some ordering in the form of layering (Catalano et al., 2009). Different hematite particles are bound between themselves by the interaction of these water layers, which are not removed in the process of drying. Upon heating an iron ore sample to 600 °C the water is completely removed and the mass percentage of this "residual" water is about 1%. No such effect is observed for coal. Upon drying, the chemical structure of iron ore is not changed, as shown by x-ray diffraction measurements.

5. Conclusions

Time dependence of mass loss from the test surface was measured together with concentrations of inhalable particulate matter (PM10) and black carbon (BC) in the vicinity of the test surface for wind velocities ranging from 1 - 18 m/s at 0.20 m above the centre of the test surface. Threshold wind velocities $u_{t.20}$ were measured for four different size fractions of coal and for three different fractions of iron ore stored at the terminal, and for hydrated alumina - a strong dust emitter. The same set up was employed to study the suppression of dust emitted by the finest coal fraction (d < 3 mm) upon watering of the dry test surface. By CFD modelling the similarity of our size limited test track was examined with respect to a flat terrain situation. The analysis shows that measured wind velocity thresholds for different fractions are consistent with the predictions of the Greley-Iversen scheme. Although the dusting potential of the non-fractioned (realistic) coal sample reflects dusting potentials of its fractions, measurements show that the former cannot be estimated simply by a weighted average of different fractions. Watering (1.0 l/m²/h) of the coal surface reduced the mass outflow by one hundred times with respect to the dry coal surface when $u_{.20} = 11.3$ m/s. Alternatively, stabilisation of the iron ore surface can be achieved by a single act of watering; upon drying agglomeration of hematite particles occurs and a firm crust forms on the surface strongly reducing further dusting.

Although CFD calculations of a large real scale physical configuration are necessary to obtain the final result concerning emissions (Loredo-Souza et al., 2004; Torano et al., 2007; Diego et al., 2008), the precise measurements of the "dusting" potential generated by the "unit" test surface under different conditions are important as they provide relevant input parameters for the simulation. To demonstrate further the use of the results, the measured erosion potential is employed to estimate the PM10 emission from a real stockpile in a single disturbance.

In Port of Koper the open air (EET) coal terminal stores approximately 660000 m³ of coal. If the height of the pile is 8 m and the top cone angle is about 40°, the surface area covered by the pile is about 100000 m². Based on the data obtained, and prior to any CFD simulations it is possible to roughly estimate the intensity of dusting from the EET if several simplifying assumptions are invoked. First, we neglect the effect of the fence which actually surrounds the pile, as well as any deviations from the flat terrain situation, i.e the pile edges. Then, according to the experimental data for FR1234, the dry coal pile does not emit PM10 particles until $u_{.20} < 5$ m/s. For wind velocities 5 m/s < $u_{.20}$ < 18 m/s the ten-minute average PM10 concentration, denoted by ‹c› is found to increase from 100 µg/m³ at the lower limit to 700 µg/m³ at the higher velocity limit. To maintain the average concentration for t = 10 minutes in the rectangular box with a cross section of s = 0.4 x 0.4 m² and length of l=$u_{.20}$ t the test surface released ‹c› s t $u_{.20}$ mass of the PM10 coal particles. This means that the test unit surface of 0.21 m² in 10 minutes releases from 0.048 g (at $u_{.20}$ = 5m/s) to 1.2 g (at $u_{.20}$ = 18 m/s) of PM10. The whole (and dry) coal stockpile at ETT would then release 22 kg to 570 kg of PM10 matter in 10 minutes. The source strength of PM10 emission is not constant but diminishes with time (see for example Figure 2); after 10 minutes it becomes small compared to the source strength at the beginning, i.e., when the air stream is just switched on. The US EPA Eq. (4) employs a concept of a number of disturbances per year: the duration of disturbance is unimportant; it is just the highest wind velocity that determines the quantity of dust emitted in a single disturbance. In fact, what is described above is just a 10 minutes long disturbance and it is interesting to see, how the US EPA estimate correlates with our.

To generate an EPA estimate it is important to express the difference between the friction velocity and the threshold friction velocity. Assuming that the roughness height z_0 is known, both threshold velocities can be expressed by the corresponding velocities at 0.20 m above the surface:

$$u^* - u_t^* = K / \ln[0.20 / z_0](u_{.20} - u_{.20}^*) \tag{16}$$

In an extreme case when $u_{.20} = 18$ m/s ($u^+_{10} = 27.2$ m/s, assuming $z_0 = 0.0001$ m) and the threshold velocity value (FR3) of $u_{t.20} = 13.0$ m/s, EPA Eq. 3 gives 5.3 g/m^2 of PM10, which translates to 530 kg of material removed from the real stockpile by a single disturbance. This result matches very closely our estimate of 570 kg.

The lifted material is carried away by the wind and dispersed into the atmosphere. The PM10 concentration around the pile depends on the exact location. A detailed compilation of the PM10 concentration 3D distribution can be derived from knowing the actual wind field around the real stockpile area, which CFD simulation can provide. There is an important caveat, within the measured PM10 emission strengths it is not possible to distinguish between PM10 particles that are lifted directly by the force of the wind (suspension) and those emitted due to the saltation of heavier and larger particles along the surface. It is believed that the majority of the PM10 material is emitted due to the latter processes since the cohesion forces between the finest particles are too strong to allow direct wind entrainment (Shao et al., 1993). At sufficiently high wind speeds these large particles enter the neighbouring unit surfaces on the pile, effectively increasing PM10 emission strength of the neighbouring unit's surfaces with respect to the "parent" or "self" test surface contribution. In fact, as seen for FR1234, the mass lost into PM10 is more than one hundred times lower than the total mass, which "vanished" from the test surface by moving to the neighboring area. Most of this mass may contribute to PM10 emissions from the neighbouring surfaces. This saltation process, in fact, seems to be the origin of the nonzero PM10 emission in the case of FR3 coal (Figure 1), while in the case of FR4 and FR1234 the creeping and saltating particles which form the total "missing" mass, generate a major part of observed PM10 concentration. In this sense, the single-unit-surface PM10 emission strengths are the lowest estimates. However, for FR1234 the saltating particles are expected not to propagate far from the parent unit since the surface is relatively rough – this is required by another assumption that the stockpile keeps its shape under the force of wind.

A comment is necessary about particle removal efficiency on slanted surfaces. We saw that on an inclined surface particles are effectively lighter, if the wind blows downhill and vice versa, they are more difficult to remove when the wind blows uphill. When the top cone half-angle is at the threshold value α_t (Table 2) the particles move downhill by themselves even when wind friction velocity is zero. The crudest approximation is obtained by linearization: if the threshold velocity for downhill wind direction is zero, for an uphill wind direction it is twice the value which applies to a flat surface. The threshold velocity for particles on a "critically" slanted surface can be therefore approximated by

$$u_{.20}(\varphi) = u_{.20} + u_{.20}(2\varphi / \pi - 1) \qquad (17)$$

where φ is an angle between the wind direction and local downhill direction in the plane of the inclined surface. If the inclination is less than critical, i.e. the top cone half-angle α is larger than threshold angle α_t, $u_{.20}(\varphi)$ deviates less from the $u_{.20}$ of the flat horizontal terrain because in this case an additional factor $(\cos\alpha/\cos\alpha_t)$ multiplies second term on the left of Eq. (17). A more accurate description of slanted situation could be obtained by rederiving the G-I scheme. However, due to other deficiencies of the scheme (it applies for pure size fractions), the above correction may be accurate enough for samples with mixed size fractions.

Finally, it is interesting to estimate how much of the mass in total is removed by the action of the wind. From Figure 5 it can be seen that the FR1234 mass loss increases from 0.5 g/s/m² to 2.5 g/s/m² when the wind speed $u_{.20}$ increases from 5 m/s to 18 m/s. In 10 minutes, the total mass displaced over the flat surface of the pile would be 30 to 150 tonnes respectively. At the same time this action would generate at least 22 kg and 570 kg of PM10 dust, respectively.

6. Acknowledgment

We gratefully acknowledge the support of Jožef Stefan International Postgraduate School and of Environmental Agency of the Republic of Slovenia (ARSO).

7. References

Andrae, M. O. (2001). The dark side of aerosols, *Nature,* Vol.409, pp. 671-672

Archer, V. E. (1988). Lung cancer risks of underground miners: cohort and case-control studies, *Yale Journal of Biology and Medicine,* Vol.61, pp. 183–193

Bagnold, R. A. (1941). The Physics of Blown Sands and Desert Dunes, ISBN 0-486-43931-3 (Mineola: Dover Publications 2005).

Barrett, C. F. & Upton S. L. (1988). Erodibility of stockpiled materials – a wind tunnel study, *Report no.* LR 656 (PA) M, Warren Spring Laboratory: Stevenage, UK

Borrego, C; Costa, A. M.; Amorim, J. H.; Santos, P.; Sardo, J.; Lopes, M. & Miranda, A.I. (2007). Air quality impact due to scrap-metal handling on a sea port: A wind tunnel experiment, *Atmospheric Environment,* Vol.41, pp. 6396-6405

Boyd, J. T.; Doll, R.; Faulds, J. S. & Leiper, J. (1970). Cancer of the lung in iron ore (haematite) miners, *British Journal of Industrial Medicine,* Vol.27, pp. 97-105

Byrne, R. J. (1968). Aerodynamic roughness criteria in aeolian sand transport, *Journal of Geophysical Research,* Vol.73, pp. 541-547

Catalano, J. G.; Fenter, P. & Park, C. (2009). Water ordering and surface relaxations at the hematite (110) – water interface, *Geochimica and Cosmochimica Acta,* Vol.73, pp. 2242-2251

Chakraborty, M. K.; Ahmad, M.; Singh, R. S.; Pal, D.; Bandopadhyay, C. & Chaulya, S. K. (2002). Determination of the emission rate from various opencast mining operations, *Environmental Modelling & Software,* Vol.17, pp. 467–480

Chane Kon, L.; Durucan, S. & Korre A. (2007). The development and application of a wind erosion model for the assessment of fugitive dust emissions from mine tailings dumps, *International Journal of Mining, Reclamation & Environment,* Vol.21(3), Taylor & Francis, pp. 198–218

Cowherd, C.; Muleski, G. E. & Kinsey, J.S. (1988). Control of Open Fugitive Dust Sources; EPA-450/3-88/008; *Prepared by Midwest Research Institute,* Kansas City, MO, for Office of Air Quality Planning and Standards, U.S. Environmental Protection Agency: Research Triangle Park, NC

Diego, I.; Pelegry, A.; Torno, S.; Torano, J. & Menendez, M. (2008). Simultaneous CFD evaluation of wind flow and dust emission in open storage piles, *Applied Mathematical Modelling,* Vol.33, pp. 3197-3207

Ferrari, L. M.; Pender, E. & Lundy, R. (1986). Dust control on coal haul roads, *Institution of Engineers Australia,* Workshop on coal fugitive dust control of coal industry works, pp. 19-26.

Ferreira, A. D. & Olivieira, R. A., (2008). Wind erosion of sand placed inside a rectangular box, *Journal of Wind Engineering and Industrial Aerodynamics*, Vol.97, pp. 1-10

Ferreira, A. D. & Vaz, P. A. (2004). Wind tunnel study of coal dust release from train wagons, *Journal of Wind Engineering and Industrial Aerodynamics*, Vol.92, pp. 565-577

Fitz, D. R. & Bumiller, K. (2000). Evaluation of Watering to Control Dust in High Winds, *Journal of the Air & Waste Management Association*, Vol.50, pp. 570-577

Gillette, D.A. (1977). Fine particulate emissions due to wind erosion. *Transactions of the American Society of agricultural Engineers*, Vol.20, pp. 890–987

Greeley, R. & Iversen, J.D. (1985). Wind as a Geological Process on Earth, Mars, Venus and Titan, ISBN 0-521-24385-8 (*Cambridge University Press: New York*)

Hansen, A. D. A.; Rosen, H. & Novakov T. (1984). The aethalometer – an instrument for real time measurement of optical absorption by aerosol particles, *The Science of Total Environment*, Vol.38, pp. 191-196

King, E. J.; Harison, C. V.; Mohanty, G. P. & Nagelschmidt, G. (1955). The effect of various forms of alumina on the lungs of rats, *Journal of Pathology and Bacteriology*, Vol.69, pp. 81-93

Kinsey, J. S.; Linna, K. J.; Squier, W. C.; Muleski, G. E. & Cowherd, C. (2004). Characterization of the fugitive particulate emissions from construction mud/dirt carryout, *Journal of the Air & Waste Management Association*, Vol.54, pp. 1394-1404

Lawler, A. B.; Mandel, J. S.; Schuman, L. M. & Lubin J. H. (1985). A retrospective cohort mortality study of iron ore (hematite) miners in Minnesota, *Journal of Occupational Medicine*, Vol.27, pp. 507-517

Lee, S. J.; Park, K. C. & Park, C. W. (2002). Wind tunnel observations about the shelter effect of porous fence on the sand particle movements, *Atmospheric Environment*, Vol.36, pp. 1453-1463

Leon, G.; Perez, L. E.; Linares, J. C.; Hatmann, A. & Quintana M. (2007). Genotoxic effect in wild rodents (Rattus rattus and Mus musculus) in an open coal mining area, *Mutation Research*, Vol.630, pp. 42-49

Loredo-Souza, A. M. & Schettini E. B. C. (2004). Wind Tunnel Studies on the Shelter Effect of Porous Fences on Coal Piles Models of the CVRD – Vitoria, Brazil, In: *Proceedings of the A&WMA's 98th Annual Conference & Exhibition – Exploring Innovative Solutions*, Minneapolis, Minnesota, USA, June 21-24

Mohamed, A. M. O. & Bassouni K. M. E. (2006). Externalities of Fugitive Dust, *Environmental Monitoring and Assessment*, Vol.130, pp. 83-98

Nemery, B. (1990). Metal toxicity and the respiratory tract, *European Respiratory Journal*, Vol.3, pp. 202-219

Nicol, S. K. & Smitham J. B. (1990). Coal stockpile dust control, *Institution of Engineers Australia*, International coal engineering conference Sydney, pp. 154-158

Patashnick, H. & Rupprecht E. G. (1991). Continuous PM10 measurements using a tapered element oscillating microbalance, *Journal of the Air & Waste Management Association*, Vol.51, pp. 1079-1083

Parsons, D. R.; Wiggs, G.; Walker, I.; Ferguson, R. & Garvey B. (2004). Numerical modelling of airflow over an idealised traverse dune, *Environmental Modelling & Software*, Vol.19, pp. 153–162

Peralba, M. C. R. (1990). Caracterizacao quimicia dos hidrocarbonatos de betumens de carvoes sul-brasileiros, In: *Dissertacao de Doutorado, Instituto di Fisica e Quimica de Sao Carlos*, Brasil, http://hdl.handle.net/10183/28215

Roney, J. A. & White, B. R. (2006). Estimating fugitive dust emission rates using an environmental boundary layer wind tunnel, *Atmospheric Environment*, Vol.40, pp. 7668-7685

Routledge, H. C. & Ayers, J. G. (2006). Cardiovascular Effects of Particles, *Air Pollution Reviews*, Vol.3, pp. 19-42

Saxton, K.; Chandler, D.; Stetler, L.; Lamb, B.; Claiborn, C. & Lee, B. (2000). Wind Erosion and Fugitive Dust Fluxes on Agricultural Lands in the Pacific Northwest, *Transactions of the ASAE 2000*, Vol. 43(3), pp. 623-630

Shao, Y.; Raupach, M. R. & Findlater, P. A. (1993). The effect of saltation bombardement on the entrainment of dust by wind, *Journal of Geophysical Research*, Vol.98, pp. 12719–12726

Smitham, J. B. & Nicol, S. K. (1990). Physico-chemical principles controlling the emission of dust from coal stockpiles, *Powder Technology*, Vol.64, pp. 259-270

Stunder, B. J. B. & Arya, S. P. S. (1988). Windbreak effectiveness for storage pile fugitive dust control: a wind tunnel study, *Journal of Air Pollution Control Association*, Vol.38, pp. 135–143

Torano, J. A.; Rodriguez, R.; Diego, I.; Rivas, J. M. & Pelegry, A. (2007). Influence of the pile shape on wind erosion CFD emission simulation, *Applied Mathematical Modelling*, Vol.31, pp. 2487-2502

US EPA (1995). AP 42, Chapter 13: Miscellaneous Sources, (Section 13.2.1: Dusting From Unpaved And Paved Roads, Section 13.2.4: Aggregate Handling And Storage Pile, Section 13.2.5: Industrial Wind Erossion), Fifth Edition, vol.I, http://www.epa.gov/ttn/chief/ap42/ch13/index.html

US EPA (1996). Executive summary. In: Air quality criteria for particulate matter. Vol.1, Research Triangle Park, NC, U.S. Environmental Protection Agency, Natinal Center for Environmental Assessment, EPA Publication No. EPA/600/P-95/001aF, pp. 1-21

US EPA (1998). Revised Draft – User's Guide for the AMS/EPA Regulatory Model – AERMOD, Office of Air Quality Planning and Standards, Research Triangle Park, NC

Weingartner, E.; Saathoff, H.; Schnaiter, M.; Streit, N.; Bitnar B. & Baltensperger U. (2003). Absorption of light by soot particles: determination of the absorption coefficient by means of aethalometers, *Aerosol Science*, Vol.34, pp. 1445-1463

White, B. & Tsoar, H. (1998). Slope effect on saltation over a climbing sand dune, *Geomorphology*, Vol.22, pp. 159–180

Witt, P. J.; Carey, K. & Nguyen, T. (2002). Prediction of dust loss from conveyors using CFD modelling, *Applied Mathematical Modelling*, Vol.26 (2), pp. 297– 309

Xuan, J. & Robins, A. (1994). The effects of turbulence and complex terrain on dust emissions and depositions from coal stockpiles, *Atmospheric Environment*, Vol.28, pp. 1951-1960

Žitnik, M.; Jakomin, M.; Pelicon, P.; Rupnik, Z.; Simčič, J.; Budnar, M.; Grlj, N. & Marzi, B. (2005). Port of Koper – Elemental concentrations in aerosols by PIXE, *X-Ray Spectrometry*, Vol.34, pp. 330-334.

11

Strategies for Estimation of Gas Mass Flux Rate Between Surface and the Atmosphere

Haroldo F. de Campos Velho[1], Débora R. Roberti[2],
Eduardo F.P. da Luz[1] and Fabiana F. Paes[1]
*[1]Laboratory for Computing and Applied Mathematics (LAC),
National Institute for Space Research
(INPE) São José dos Campos (SP)
[2]Department of Physics, Federal University of
Santa Maria (UFSM) Santa Maria (RS)
Brazil*

1. Introduction

A relevant issue nowadays is the monitoring and identification of the concentration and rate flux of the gases from the greenhouse effect. Most of these minority gases belong to important bio-geochemical cycles between the planet surface and the atmosphere. Therefore, there is an intense research agenda on this topic. Here, we are going to describe the effort for addressing this challenging. This identification problem can be formulated as an inverse problem. The problem for identifying the minority gas emission rate for the system ground-atmosphere is an important issue for the bio-geochemical cycle, and it has being intensively investigated. This inverse problem has been solved using regularized solutions (Kasibhatla, 2000), Bayes estimation (Enting, 2002; Gimson & Uliasz, 2003), and variational methods (Elbern et al., 2007) – the latter approach coming from the data assimilation studies.

The inverse solution could be computed by calculating the regularized solutions. Regularization is a general mathematical procedure for dealing with inverse problems, looking for the smoothest (*regular*) inverse solution. For this approach, the inverse problem can be formulated as an optimization problem with constrains (*a priori* information). These constrains can be added to the objective function with the help of a regularization parameter. For adding a regularization operator, the ill-posed inverse problem becomes in a well-posed one. The maximum second order entropy principle was used as a regularization operator (Ramos et al., 1999), and the regularization parameter is found by the L-curve technique. A recent approach for solving inverse problems is provided by the use of artificial neural networks (ANN). Neural networks are non-linear mappings, and it is possible to design an ANN to be one inverse operator, with robustness to deal on noisy data.

We have applied two different strategies for addressing the inverse problem: formulated as an optimization problem, and neural network. The optimization problem has been solved by using a deterministic method (quasi-Newton), implementation used in the E04UCF routine (NAG, 1995) – see Roberti et al. (2007), and applying a stochastic scheme (Particle Swarm Optimization: PSO) (Luz et al., 2007). The PSO is a meta-heuristic based on the collaborative behaviour of biological populations. The algorithm is based on the swarm theory, proposed by Kennedy & Eberhart (1995) and is inspired in the flying pattern of birds which can be achieve by manipulating inter-individual distances to synchronize the swarm behaviour. The swarm theory is then combined with a social-cognitive theory. The multilayer perceptron neural network (MLP-NN) with 2 hidden layers (with 15 and 30 neurons) was applied to estimate the rate of surface emission of a greenhouse gas (Paes et al., 2010). The input for the ANN is the gas concentration measured on a set of points.

2. Techniques for solving inverse problems

There is an aphorism among mathematicians that the theory of partial differential equation could be divided into two branches: an easy part, and a difficult one. The former is called the *forward* (or *direct*) *problem*, and the latter is called *inverse problem*. Of course, the anecdotal statement tries to translate the pathologies found to deal with inverse problem.

For a given mathematical problem to be characterized as a *well-posed problem*, Jacques Hadamard in his studies on differential equations (Hadamard 1902, 1952) has established the following definition:

A problem is considered well-posed if it satisfies all of three conditions:

a. *The solution exists (existence);*
b. *The solution is unique (uniqueness);*
c. *The solution has continuous dependence on input data (stability).*

However, inverse problems belong to the class of ill-posed problems, where we cannot guaranty existence, uniqueness, nor stability related to the input data. A mathematical theory for dealing with inverse problems is due to the Russian mathematician Andrei Nikolaevich Tikhonov during the early sixties, introducing the regularization method as a general procedure to solve such mathematical problems. The regularization method looks for the smoothest (*regular*) inverse solution, where the data model would have the best fitting related to the observation data, subjected to the constrains. The searching by the smoothest solution is an additional information (or *a priori* information), which becomes the ill-posed inverse problem in a well-posed one (Tikhonov & Arsenin, 1977).

Statistical strategies to compute inverse solution are those based on Bayesian methods. Instead to compute a set of parameter or function as before, the goal of the statistical methods is to calculate the probability density function (PDF) (Tarantola, 1987; Kaipio & Somersalo, 2005). Bayesian methodology will not be applied here.

With some initial controversy, artificial neural networks are today a well established research field in artificial intelligence (AI). The idea was to develop an artificial device able to emulate the ability of human brain. Sometimes, this view is part of program of so called strong AI. Some authors have used the expression *computational intelligence* (CI) to express a set of methodologies with inspiration on the Nature for addressing real world problems. Initially, the CI includes fuzzy logic systems, artificial neural networks, and evolutionary computation. Nowadays, other techniques are also considered to belong to the CI, such as swarm intelligence, artificial immune systems and others. Different from other methodologies for computing inverse solutions, neural networks do not need the knowledge on forward problem. In other words, neural networks can be used as *inversion operator* without a mathematical model to describe the direct problem. Several architectures can be adopted to implement a neural network, here will be adopted the multi-layer perceptron with back-propagation algorithm for learning process. Other approaches to implement artificial neural network for inverse problem are described by Campos Velho (2011).

2.1 Regularization theory

Regularization is recognized as the first general method for solving inverse problems, where the regularization operator is a constrain in an optimization problem. Such operator is added to the objective function with the help of a Lagrangian multiplier (also named the *regularization parameter*). The regularization represents an *a priori* knowledge from the physical problem on the unknown function to be estimated. The regularization operator moves the original ill-posed problem to a well posed one. Figure 1 gives an outline of the idea behind the regularization scheme, one of most powerful techniques for computing inverse solutions.

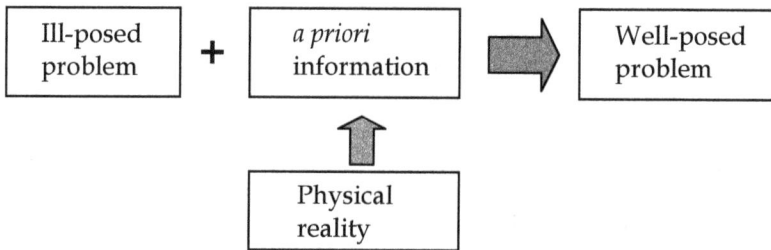

Fig. 1. Outline for regularization methods, where additional information is applied to get a well-posed problem.

The regularization procedure searches for solutions that display *global* regularity. In the mathematical formulation of the method, the inverse problem is expressed as optimization problem with constraint:

$$\min_{x \in X} \left\| A(x) - f^{\delta} \right\|_2^2, \text{ suject to } \Omega[x] \le \rho \tag{1}$$

where $A(x) = f^{\delta}$ represents the forward problem, and $\Omega[x]$ is the regularization operator (Tikhonov & Arsenin, 1977). The problem (1) can be written as an optimization problem without constrains with the help of a Lagrange multiplier (penalty or regularization parameter):

$$\min_{x \in X} \left\{ \left\| A(x) - f^{\delta} \right\|_2^2 + \alpha \Omega[x] \right\} \tag{2}$$

being α the regularization parameter. The first term in the objective function (2) is the fidelity of the model with the observation data, while the second term expresses the regularity (or smoothness) required from the unknown quantity. Note that for $\alpha \to 0$, the fidelity term is overestimated; on the other hand, for $\alpha \to \infty$ all information in the mathematical model is lost. A definition of a family of regularization operator is given below.

Definition 1: A family of continuous regularization operators $R_\alpha : F \to U$ is called a regularization scheme for the inverse operation of $A(x) = f^\delta$, when

$$\lim_{\alpha \to 0} R_\alpha \{A(x)\} = x \qquad (3)$$

where $x \in X$, and α is regularization parameter.

The expression (2) is a practical implementation of the Definition 1. Several regularizations operators have been derived from the pioneer works. Here, only one class of these regularization operators will be described.

2.1.1 Entropic regularization

The maximum entropy (MaxEnt) principle was firstly proposed as a general inference procedure by Jaynes (1957), on the basis of Shannon's axiomatic characterization of the amount of information (Shannon & Weaver, 1949). It has emerged at the end of the 60's as a highly successful regularization technique. Similar to others regularization techniques, MaxEnt searches for solutions with *global* regularity. Employing a suitable choice of the penalty or regularization parameter, MaxEnt regularization identifies the smoothest reconstructions which are consistent with the measured data. The MaxEnt principle has successfully been applied to a variety of fields: pattern recognition (Fleisher et al., 1990), computerized tomography (Smith et al., 1991), crystallography (de Boissieu et al., 1991), non-destructive testing (Ramos & Giovannini, 1995), magnetotelluric inversion (Campos Velho & Ramos, 1997).

For the vector of parameters x_i with nonnegative components, the discrete entropy function S of vector u is defined by

$$S(x) = -\sum_{q=1}^{N} s_q \log(s_q), \quad \text{with} \begin{cases} x = \begin{bmatrix} x_1 & \cdots & x_N \end{bmatrix}^T \\ s_q = x_q / \sum_{q=1}^{N} x_q \end{cases} \qquad (4)$$

The (nonnegative) entropy function S attains its global maximum when all s_q are the same, which corresponds to a uniform distribution with a value of $S_{\max} = \log N$, while the lowest entropy level, $S_{\min} = 0$, is attained when all elements s_q but one are set to zero.

Following the Tikhonov regularization, new entropic higher order regularization techniques have been introduced: MinEnt-1 – applied to identify the 2D electric conductivity from electro-magnetic data (Campos Velho & Ramos, 1997), MaxEnt-2 – employed to the retrieval of vertical profiles of temperature in the atmosphere from remote sensing data (Ramos et al., 1999). They represent a generalization of the standard MaxEnt regularization method, allowing a greater flexibility to introduce information about the expected shape of the true physical model, or its derivatives, for the inverse solution.

For defining higher order entropy functions, two approaches can be derived from the maximization of the entropy of the vector of *first-* and *second-differences* x. Under the assumption $x_{min} < x_i < x_{max}$, (i=1, ..., N), the methods of *maximum entropy of order 1* (MaxEnt-1) and *maximum entropy of order 2* (MaxEnt-2) are defined below (Muniz et al., 2000):

$$p_i = \begin{cases} x_i - x_{min} + \zeta & \text{(zeroth order)} \\ x_{i+1} - x_i + (x_{max} - x_{min}) + \zeta & \text{(first order)} \\ x_{i+1} - 2x_i + x_{i-1} + 2(x_{max} - x_{min}) + \zeta & \text{(second order)} \end{cases} \tag{5}$$

where ζ is a small parameter ($\zeta = 10^{-10}$). The argument in the Eq. (4) is to be considered with p_i, instead of x_i.

The non-extensive formulation for the entropy proposed by Tsallis (Tsallis, 1988) can be used to unify the Tikhonov and entropic regularizations (Campos Velho et al., 2006, see also Campos Velho, 2011). The Tsallis' non-extensive entropy is expressed as

$$S_q(x) = \frac{k}{q-1}\left[1 - \sum_{i=1}^{M} x_i^q\right] \tag{6}$$

The non-extensive parameter q plays a central role in the Tsallis' thermostatistics. The Boltzmann-Gibbs-Shannon's entropy is recovered setting q=1. However, for q=2, the maximum non-extensive entropy principle is equivalent to the standard Tikhonov regularization. Of course, for $q \neq 1$ and 2, we have a new regularization operator. In context of inverse problem, the Boltzmann constant (k) is assumed to be equal to 1.

In order to have a complete theory, the regularization operator should be known, and it is also necessary to have a scheme to compute the regularization parameter. Several methods have been developed to identify this parameter: Morosov's discrepancy criterion, Hansen's method (L-curve method), and generalized cross validation are the most used schemes.

The Morosov's criterion is based on the difference between data of the mathematical model and observations. It should have the same magnitude as measurement errors (Morosov, 1984; see also Morosov & Stessin, 1992). Therefore, if δ is the error in the measure process, α is the root of the following equation:

$$\left\{\left\|A(x) - f^\delta\right\|_2^2\right\}_{\alpha^*} \approx \delta \tag{7}$$

For determining N parameter in an inverse problem, and assuming that measurement errors could be modeled by a Gaussian distribution with σ^2 variance, the discrepancy criterion for independent measures could expressed as:

$$\left\|A(x) - f^\delta\right\|_2^2 \approx N\sigma^2 \tag{7.1}$$

The discrepancy principle can also be applied to the higher entropy regularization (Muniz et al., 2000). However, if the probability density function of measurement errors follows a distribution where the second statistical moment is not defined (Lévy or Cauchy distributions, for example) the *generalized Morosov's discrepancy principle* can be used (Shiguemori et al., 2004).

If the statistics on the observational data is not available, the generalized cross-validation method can be applied (Bertero & Bocacci, 1998; Aster et al., 2005). Considering now a linear forward problem: $A(x) \equiv Ax$, being A a matrix, the goal of the cross-validation scheme is to minimize the *generalized cross-validation function* (Aster et al., 2005):

$$V(\alpha) = \frac{N\|A(x_\alpha) - f^\delta\|_2^2}{\left[Tr\{I - B(\alpha)\}\right]^2} \tag{8}$$

being $Tr\{C\}$ the trace of matrix C, and $B(\alpha)$ is the following matrix:

$$B(\alpha) \equiv AA^*(AA^* + \alpha I)^{-1} \tag{9}$$

where A^* is the *adjoint matrix*: $(Af, g) = (f, A^*g)$ and I is the identity matrix.

Another scheme to compute the regularization parameter is the L-curve method. The L-curve criterion is a geometrical approach suggested by Hansen (1992) (see also Bertero & Bocacci, 1998; Aster et al., 2005). The idea is to find the point of maximum curvature, on the corner of the plot: $\Omega[x_\alpha] \times \|A(x) - f^\delta\|_2^2$. In general, the plot smoothness × fidelity shows a L-shape curve type.

2.1.2 Optimization methods

This is a central issue in science and engineering, where several methods have been developed. They can classified as a deterministic and stochastic schemes. Here, only the methods used in the application treated will be described.

2.1.2.1 Quasi-Newton optimization

The minimization of the objective function $J(x)$ given by equation (2), subjected to simple bounds on x, is solved using a first-order optimization algorithm from the E04UCF routine, NAG Fortran Library (1993). This routine is designed to minimize an arbitrary smooth function subject to constraints (simple bounds, linear and nonlinear constraints), using a sequential programming method. For the n-th iteration, the calculation proceeds as follows:

1. Solve the forward problem to x and compute the objective function $J(\alpha, x)$,
2. Compute the gradient $\nabla J(\alpha, x)$ by finite difference,
3. Compute a positive-definite approximation (quasi-Newton) to the Hessian H^n:

$$H^n = H^{n-1} + \frac{b^n(b^n)^T}{(b^n)^T u^n} - \frac{H^{n-1}u^n(u^n)^T H^{n-1}}{(u^n)^T H^{n-1}u^n}$$

where

$$b^n = x^n - x^{n-1} \quad \text{and} \quad u^n = \nabla J(\alpha, x^n) - \nabla J(\alpha, x^{n-1})$$

4. Compute the search direction d^n as a solution of the following quadratic programming subproblem:

$$\text{minimize: } [\nabla J(\alpha, x^n)]^T d^n + \frac{1}{2}(d^n)^T H^{n-1} d^n$$

$$\text{under constrain: } (x_{\min})_q - x_q^n \le d_q \le (x_{\max})_q - x_q^n$$

5. Set: $x^{n+1} = x^n + \beta^t d^n$, where the step length β^t minimizes $\nabla J(x^n + \beta^t d^n)$.
6. Test of convergence: Stop, if x satisfies the first-order Kuhn-Tucker conditions and $\beta^t ||d|| < \varepsilon^{1/2}(1+||x||)$, where ε specifies the accuracy to which one wishes to approximate the solution of the problem. Otherwise, return to step 1.

2.1.2.2 Particle swarm optimization

The particle swarm optimization (PSO) is a heuristic search method the uses the behavior of biological collective behavior, like birds, fishes or bees to drive the artificial particles in a search pattern that equalizes a global search with a local search, in a way that an optimum, or near to optimum, result is found (Kennedy & Eberhart, 1995, 2001). There is a socio-cognitive theory for supporting the particle swarm scheme. The cultural adaptive process encloses two components: high-level (pattern formation trough the individual and the ability of problem solving), and a low-level (individual behavior), with some skills for evaluating, comparing, and emulating (Kennedy & Eberhart, 2001).

The tendency to evaluate is, maybe, the behavioral feature more present in several living beings. This is intrinsic for the learning process. The comparison among individuals from a population is a behavior process used to identify patterns to find a way to improve each one quality. The emulation behavior gives the living beings the ability to learn through the observations of others actions. This behavior is not so much common in the nature. The emulation behavior is closer to the human society, but it can be used to improve the performance of the algorithms. These three concepts can be combined, even in simplified computational rules, providing a scheme for solving optimization problems.

PSO searches for the optimum in an infinite ($\Re^\infty$) or $\Re^N$ (N-dimensional space of real numbers,). Indeed, computational solutions can only obtain a projection of an infinite space into $\Re^N$ space. The algorithm allows the particles to fly in the search space, looking for the optimization of a predetermined function. A special position x^* of a particle in the search space represents the solution of the problem. This position can be represented by an vector of real numbers $x = (x_1, x_2, ..., x_N)$. In the PSO, the solution x^* is searched in a iterative procedure. At each time step, or iteration of the algorithm, the position of the each particle x^k is updated, by the addition of a speed vector v^k, in each dimension (space direction). The iteration could be given by

$$x^k(t) = x^k(t-1) + v^k(t) \tag{10}$$

where $x^k(t-1)$ is the previous position of the particle. The velocity that is applied to the particle is function of two basic behaviors: a cognitive component, and a social component. The velocity is up dated taken into account the behaviors mentioned, as following:

$$v^k(t) = wv^k(t-1) + c_1 rand_1(p^k - x^k) + c_1 rand_2(p^g - x^k) \tag{11}$$

where $c_{1,2}$ are constants (real numbers) that weights the cognitive and social components, respectively, $rand_{1,2}$ are random numbers from a uniform distribution in the interval [0,1], p^k is the best position achieved by the particle and p^g is the best position of the swarm, with x^k actual position of the particle k. The parameter w is used to give more stability to the algorithm (Shi & Eberhart, 1998). The selection of the w value follows a heuristic rule:

- If $w < 0.8$: the algorithm works in a local search mode, in a exploration mode;
- If $w > 1.2$: the algorithm works in a exploitation mode, where the global is more explored;
- If $0.8 < w < 1.2$: a balance point, where exploration and exploitation are well divided.

2.2 Artificial neural networks

The artificial neuron is a simplified representation of a biological neuron, where a weighted combination of the inputs is the value for the non-linear activation function (in general, a sigmoid one). In artificial neural networks (ANN), the identification of the connection weights is called the learning (or training) phase.

Several architectures can be used for the ANN. Related to the learning process, ANNs can be described into two important groups: supervised, and unsupervised NNs. For the supervised scheme, the connection weights are selected to become the output from the ANN as close as possible to the target. For example: the weights could be found by minimizing a functional of the square differences between the ANN output and the target values. Some authors have suggested the use of a Tikhonov's functional (Poggio & Girosi, 1990). The regularization suggested by Poggio & Girosi (1990) is applied on the output values x of the NN:

$$J[w] = \sum_{i=1}^{N} [x_i^{Target} - x_i(w)]^2 + \alpha \|Px\| \tag{12}$$

where x is the NN output, w is the connection weight matrix, and P is a linear operator. Shiguemori et al. (2004) employed three different architectures for the NNs, with focus on determination of the initial profile for the heat conduction problem. The NNs used were: multilayer perceptron (MLP), RBF-NN, and cascade correlation. The backpropagation algorithm was used for the learning process (without regularization). Figure 2 shows a cartoon to represent a MLP-NN. Cybenko (1989) has proved that the standard multilayer NN with a single hidden layer with a sigmoid activation function, for a finite number of hidden neurons, are universal approximators on a compact subset of $\mathfrak{R}^N$ (Cybenko theorem).

The backpropagation training is a supervised learning algorithm that requires both input and output (desired) data. Such pairs permit the calculation of the error of the network as the difference between the calculated output and the desired vector: $\varepsilon^{1/2} \equiv x^{target} - x(w)$. The weight adjustments are conducted by backpropagating such error to the network, governed by a change rule. The weights are changed by an amount proportional to the error at that unit, times the output of the unit feeding into the weight. Equation below shows the general weight correction according to the so-called delta rule (Haykin, 1999)

$$\Delta w_{ji} = \eta \frac{\partial \varepsilon}{\partial w_{ji}} x_i + \gamma [w_{ij}(n) - w_{ij}(n-1)] = \eta \delta_i x_i + \gamma [w_{ij}(n) - w_{ij}(n-1)] \tag{13}$$

where, δ_j is the local gradient, x_i is the input signal of neuron j, η is the learning rate parameter that controls the strength of change, and γ the momentum coefficient, which decides to what degree this previous adjustment is to be considered so as to prevent any sudden changes in the direction in which corrections are made. The learning rate η and momentum γ were set to 0.1 and 0.5, respectively. Nevertheless, the activation test is an important procedure, indicating the performance of a NN. The effective test is defined using an unknown vector x that did not belong to the training set. This action is called the *generalization* of the NN.

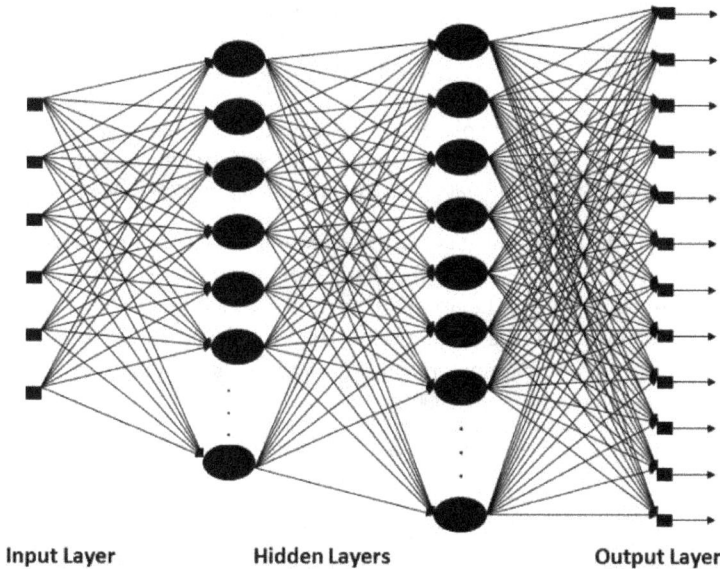

Input Layer   **Hidden Layers**   **Output Layer**

Fig. 2. Outline of fully connected MLP-NN with 2 hidden layers.

The up date for the weight matrix of neural connections is obtained with the innovation described in Equation (13):

$$w_{ij}(n) = w_{ij}(n-1) + \Delta w_{ji}$$

where the iterative process follows up to reach a maximum number of epochs (iterations) or the required precision to be caught. The delta rule can be justified if the diagonal matrix ηI (the learning parameter multiply by identity matrix) could be understood as an approximate inverse of Hessian matrix (a matrix C is defined as an approximate inverse if: $||I - CA|| <$

1, see Conte and de Boor (1980)). Therefore, the free parameter η should determined to be an approximation for the inverse of the Hessian matrix, producing similar results as the Newton method. However, the Newton method, or even the steepest descent scheme, is not good to deal if a shallow local minimum. The momentum term ($\gamma \neq 0$, in Eq. (13)) is added to circumvent some instabilities (Rumelhart et al., 1986) and avoid pathological behaviour with shallow error surface (Haykin, 1999). However, any other optimization method can be used instead of delta rule. Indeed, the PSO scheme described in Section 2.1.2.2 can be employed to determine the weights of the neural network (Li and Chen, 2006).

3. The lagrangian stochastic model LAMBDA

The Lagrangian particle model LAMBDA was developed to study the transport process and pollutants diffusion, starting from the Brownian random walk modeling (Anfossi & Ferrero, 1997). In the LAMBDA code, full-uncoupled particle movements are assumed. Therefore, each particle trajectory can be described by the generalized three dimensional form of the Langevin equation for velocity (Thomson, 1987):

$$du_i = a_i(\mathbf{x}, \mathbf{u}, t)dt + b_{ij}(\mathbf{x}, \mathbf{u}, t)dW_j(t) \tag{14a}$$

$$d\mathbf{x} = (\mathbf{U} + \mathbf{u})dt \tag{14b}$$

where $i, j = 1, 2, 3$, and $\mathbf{x}$ is the displacement vector, $\mathbf{U}$ is the mean wind velocity vector, $\mathbf{u}$ is the Lagrangian velocity vector, $a_i(\mathbf{x}, \mathbf{u}, t)$ is a deterministic term, and $b_{ij}(\mathbf{x}, \mathbf{u}, t)dW_j(t)$ is a stochastic term and the quantity $dW_j(t)$ is the incremental Wiener process.

The deterministic (drift) coefficient $a_i(\mathbf{x}, \mathbf{u}, t)$ is computed using a particular solution of the Fokker-Planck equation associated to the Langevin equation. The diffusion coefficient $b_{ij}(\mathbf{x}, \mathbf{u}, t)$ is obtained from the Lagrangian structure function in the inertial subrange, $(\tau_K \ll \Delta t \ll \tau_L)$, where τ_K is the Kolmogorov time scale and τ_L is the Lagrangian de-correlation time scale. These parameters can be obtained employing the Taylor statistical theory on turbulence (Degrazia et al., 2000).

The drift coefficient, $a_i(\tilde{\mathbf{x}}, \tilde{\mathbf{u}}, t)$, for forward and backward integration is given by

$$a_i P_E = \frac{\partial}{\partial u_j}\left(B_{i,j} P_E\right) + \varphi_i\left(\mathbf{x}, \mathbf{u}, t\right) \tag{15a}$$

with,

$$\frac{\partial \varphi_i}{\partial u_i} = -\frac{\partial P_E}{\partial t} - \frac{\partial}{\partial x_i}\left(u_i P_E\right) \text{ and } \varphi_i \to 0 \text{ when } \mathbf{u} \to \infty \tag{15b}$$

where $c_v = -1$ for forward integration and $c_v = 1$ for backward integration, $P_E = P(\mathbf{x}, \mathbf{u}, t)$ is the non-conditional PDF of the Eulerian velocity fluctuations, and $B_{i,j} = (1/2)b_{i,k}b_{j,k}$. Of course, for backward integration, the time considered is $t' = -t$, and velocity $\mathbf{U}' = -\mathbf{U}$, being $\mathbf{U}$ the mean wind speed. The horizontal PDFs are considered Gaussians, and for the vertical coordinate the truncated Gram-Charlier type-C of third order is employed (Anfossi

& Ferrero, 1997). The diffusion coefficients, $b_{ij}(\mathbf{x},\mathbf{u},t)$, for both forward and backward integration is given by

$$b_{ij} = \delta_{ij} \left[2\frac{\sigma_i^2}{\tau_{Li}} \right]^{1/2} \tag{16}$$

where δ_{ij} is the Kronecker delta, σ_i^2 and τ_{Li} are velocity variance at each component and the Lagrangian time scale (Degrazia et al., 2000), respectively. With the coordinates and the mass of each particle, the concentration is computed (see below).

The inverse problem here is to identify the source term $S(t)$. A source-receptor approach can be employed for reducing the computer time, instead of running the direct model, Eq. (14), for each iteration. This approach displays an explicit relation between the pollutant concentration of the i-th receptor related the j-th sources:

$$C_i = \sum_{j=1}^{N_s} M_{ij} S_j \tag{17}$$

where the matrix M_{ij} is the *transition matrix*, and each matrix entry given by

$$M_{ij} = \begin{cases} \left(V_{S,j}/V_{R,i}\right)\left(\Delta t/N_{S,j}\right)N_{R,i,j} & \text{- forward model;} \\ \left(\Delta t/N_{R,i}\right)N_{S,i,j} & \text{- backward model.} \end{cases} \tag{18}$$

where $V_{R,i}$ and $V_{S,j}$ are the volume for the i-th receptor and j-th source, respectively; $N_{S,j}$ and $N_{R,i}$ are the number of particle realised by the j-th source and i-th sensor, respectively; $N_{R,i,j}$ and $N_{S,i,j}$ are the number of particle released by the j-th source and detected by the i-th receptor.

4. Determining gas flux between the ground and the atmosphere

For testing the procedure to estimate the emission rate procedure, it is considered the area pollutant sources placed in a box volume, where the horizontal domain and vertical height are given by: (1500 m × 1000 m) × 1000 m. There are two embedded regions R_1 and R_2 into computational domain, with following horizontal domain (600 m × 600 m) for each region, and 1 m of height, and they are realising contaminants with two different emission rates. Figure 1 shows the computational scenario in a two-dimensional projection (x,y): the six sensors are placed at 10 m height and spread horizontally with the coordinates presented in Table 1. The domain is divided into sub-domains with 200 m × 200 m × 1 m. The emission rates for each sub-domain (S_{A_k}) is as following:

$$R_1 = S_{A_2} = S_{A_3} = S_{A_4} = S_{A_7} = S_{A_8} = S_{A_9} = 10 \text{ gm}^{-3}\text{s}^{-1} \text{ (Region-1)}$$

$$R_2 = S_{A_{12}} = S_{A_{13}} = S_{A_{14}} = S_{A_{17}} = S_{A_{18}} = S_{A_{19}} = 20 \text{ gm}^{-3}\text{s}^{-1} \text{ (Region-2)}$$

$$S_{A_1} = S_{A_5} = S_{A_6} = S_{A_{10}} = S_{A_{11}} = S_{A_{15}} = S_{A_{16}} = S_{A_{20}} = S_{A_{21}} = S_{A_{22}} = S_{A_{23}} = S_{A_{24}} = S_{A_{25}} = 0$$

The dispersion problem is modelled by the LAMBDA backward-time code, where data from the Copenhagen experiment are used, for the period (12:33 h - 12:53 h, 19/October/1978) (Gryning & Lyck, 1984, 1998). It is assumed a logarithmic vertical profile for the wind field

$$U(z) = \frac{u_*}{\kappa} \ln\left(\frac{z}{z_0}\right) \tag{19}$$

being $U(z)$ the main stream, u_* the fiction velocity, κ the von Kármán constant, z the height above the ground, and z_0=0.06 m the roughness. The wind speed was measured at 10 m, 120 m, and 200 m. The numerical value for u_* was obtained from the best fitting with the measured wind speed – see Table 2 – and the equation (19).

Sensor-number	x (m)	y (m)
1	400	500
2	600	300
3	800	700
4	1000	500
5	1200	300
6	1400	700

Table 1. Position of sensors in physical domain.

For this experiment the wind direction had an angle θ=180°, and the boundary layer height is h=1120 m (Gryning & Lyck, 1998). The turbulence parameterization follows Degrazia et al. (2000), for computing the wind variance (σ_i^2) and the Lagrangian decorrelation time scale (T_{L_i}). These two turbulent parameters are considered constant for the whole boundary layer, and their numerical values were calculate at z=10 m level. This characterises a stationary and vertically homogeneous atmosphere.

The sensors dimension, where the fictitious particles have arrived, was de 0,1 m × 0,1 m × 0,1 m, centred in the computational cells presented in Table 1. Next, the reverse trajectories are calculated for 1000 particles per sensor. The parameters for numerical simulations were 1800 time-steps with Δt = 1 s, meaning 30 min for the whole simulation. After 10 min, the concentration was computed at each 2 min, for the remaining 20 min of simulation. The mean concentration is found using the following expression:

$$C^{Mod}(x_j, y_j) = \frac{1}{10} \sum_{n=1}^{10} \left[\sum_{i=1}^{25} S_i \frac{\Delta t}{N_{PES,j}} N_{PVF,i,j,n} \right] \tag{20}$$

where (x_j, y_j) is position the each sensor (the number of sensors is 6), $N_{PES,j}$=1000 is the number of particles arriving at each sensor. $N_{PVS,i,j,n}$ is the quantity of fictitious particles in the i-th volume source, at the n-th instant.

Time	Speed U (ms⁻¹)			Wind direction (°)		
(h:min)	10 m	120 m	200 m	10 m	20 m	200 m
12:05	2.6	5.7	5.7	290	310	310
12:15	2.6	5.1	5.7	300	310	310
12:25	2.1	4.6	5.1	280	310	320
12:35	2.1	4.6	5.1	280	310	320
12:45	2.6	5.1	5.7	290	310	310

Table 2. Measured wind speed for the Copenhagen experiment (Gryning & Lyck, 1998).

The inverse approach is tested using synthetic measured concentration data The synthetic observations are emulated as:

$$C^{Exp}(x_j, y_j) = C^{Mod}(x_j, y_j)[1 + \lambda\mu] \tag{21}$$

where $C^{Mod}(x,y)$ is the gas concentration computed from the forward model (14), μ is a random number associated to the Gaussian distribution with zero mean and unitary standard deviation, and λ is the level of the noise (Gaussian white noise). In our tests, $\lambda=0.05$ (Experiment-1) and $\lambda=0.1$ (experiment-2) were used. In other words, two numerical experiments are performed, using two different maximum noise levels: 5% and 10%.

The optimization method is used to find the minimum for the functional:

$$F(\alpha, \mathbf{S}) = \sum_{j=1}^{6} \left[C_j^{Exp} - C_j^{Mod}(\mathbf{S}) \right]^2 + \alpha\Omega(\mathbf{S}) \tag{22}$$

where the vector $\mathbf{S} = [S_{A_1}, S_{A_2}, ..., S_{A_{25}}]^T$ represent the emission rate from 25 sub-domains, presented in Figure 3. However, some sub-domains are considered releasing nothing. Therefore, the number of the unknown parameters to be estimated by inverse method decreases to 12 cells: $\mathbf{S} = [S_{A_2}, S_{A_3}, S_{A_4}, S_{A_7}, S_{A_8}, S_{A_9}, S_{A_{12}}, S_{A_{13}}, S_{A_{14}}, S_{A_{17}}, S_{A_{18}}, S_{A_{19}}]$. Another *a priori* information is smoothness of the inverse solution (regularization operator). As mentioned before, the second order maximum entropy is applied, Eq. (13c). The use of entropic regularization for this type of inverse problem was independently proposed by Roberti (2005) and Bocquet (2005a, 2005b) – see also Roberti et al. (2007) and Davoine & Bocquet (2007).

There are several methods to compute the regularization parameter. Here, a numerical experimentation was employed to determine this parameter. Roberti (2005) had found the values $\alpha = 10^{-6}$ for Experiment-1, and $\alpha = 10^{-5}$ for Experiment-2.

The first inverse solution was obtained by applying the quasi-Newton optimization method (described in Section 2.1.1). In the Experiment-1, a global error around 2% is found for the inverse solution. For the region-1 (lower emission rate), the error in the estimation is 10%, and for region-2 the error is 3%. The biggest errors for estimation are verified for sub-domains A_9 and A_{14}, with error of 55% and 42%, respectively.

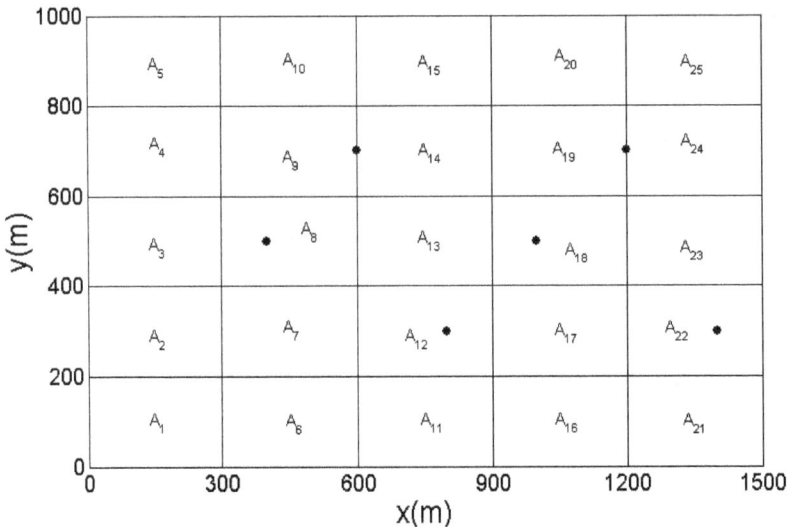

Fig. 3. Sub-domains representation, in a two-dimensional projection (x,y), with different emission rates, and black points (•) represent the sensor position at 10 m height.

In order to try some improvement for the inverse solution, the minimum value for the functional (22) is computed using the PSO scheme. This method is a stochastic technique and it deals with a population (candidate inverse solutions). Therefore, the numerical results represent the average from 25 seeds used by the PSO algorithm, working with 12 particles, with the following parameters: $w = 0.2$, $c_1 = 0.1$, and $c_2 = 0.2$. The total error obtained with the PSO was less than inverse solution computed from the quasi-Newton method. It is also important to point out that for the PSO approach no regularization operator was used ($\alpha = 0$ in Eq. (22)).

The inverse problem was also addressed by neural network. The MLP-NN was designed and trained to find the gas flux between the soil and the atmosphere. However, there are two previous steps before the application of MLP-NN. Firstly, it is necessary to determine the appropriated activation function, number of hidden layers for the NN, and the number of neurons for each layer. Secondly, the values of the connection weights (learning process) must be identified.

The hyperbolic tangent was used as activation function. Different topologies for MLP-NN were tested with 6 inputs (measured points for 6 different sensors) and 12 outputs (emission for each subdomain). Good results were achieved for two neural networks with 2 hidden layers: NN-1 with 6 and 12 (6:6:12:12), and NN-2 with 15 and 30 (6:15:30:12) neurons for the hidden layers, respectively.

Table 3 shows the numerical values computed for the emission rate for each marked cell. Good results were obtained for all methodologies. The cells 12 and 14 had a worst estimation using optimization solution. Inversions with MLP-NN were better, in particular for the emission of the cells 12 and 14. Figures 4 and 5 show the emission estimation for the experiment-1 (5% of noise level) and experiment-2 (10% of noise level), only results obtained with NN-2 (6:15:30:12) are shown (the best results with MLP-NN).

Sub-domain	Exact	Q-N	PSO	MLP-NN-1	MLP-NN-2
A_2	10	8.97	9.83	10.33	10.11
A_3	10	9.97	10.40	10.11	10.11
A_4	10	12.53	10.79	10.12	10.20
A_7	10	7.99	10.50	10.27	10.20
A_8	10	10.15	12.06	10.24	10.06
A_9	10	11.56	11.28	10.21	10.17
A_{12}	**20**	**13.85**	**14.56**	**21.28**	**20.97**
A_{13}	20	22.65	22.67	21.32	21.00
A_{14}	**20**	**14.14**	**15.85**	**21.47**	**20.95**
A_{17}	20	19.99	21.56	21.47	20.99
A_{18}	20	21.17	20.05	21.38	20.99
A_{19}	20	24.90	21.74	20.96	20.96

Table 3. Estimation of gas flux emission by the methodologies: quasi-Newton (Q-N), PSO, MLP-NN-1 (6:6:12:12) MLP-NN-2 (6:15:30:12).

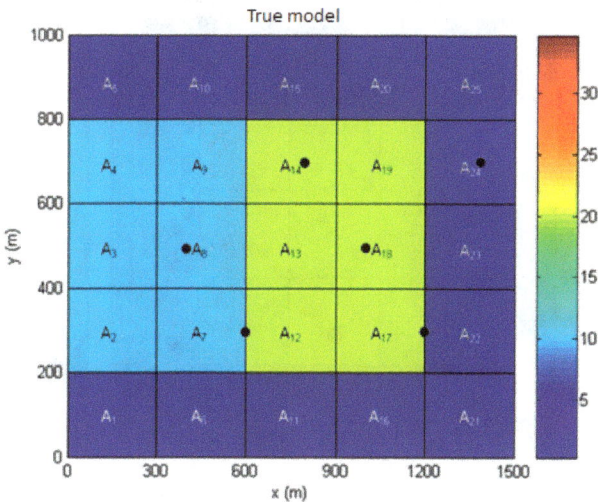

(a)

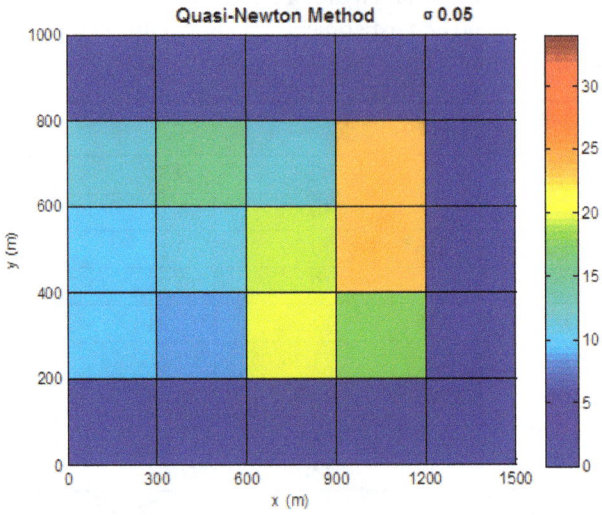

(b)

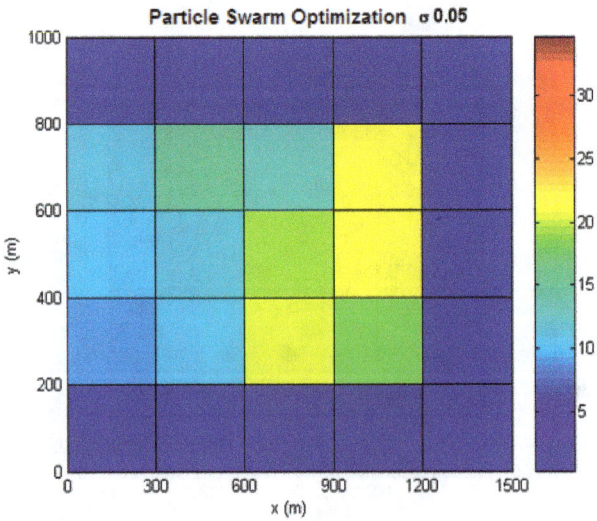

(c)

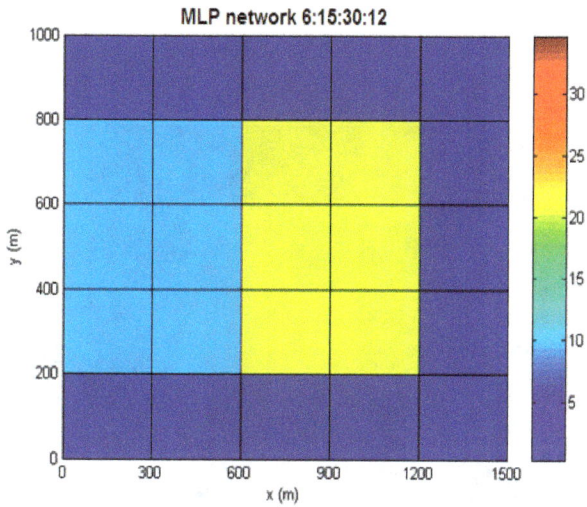

(d)

Fig. 4. Gas flux rate for Experiment-1: (a) true values, (b) quasi-Newton (Q-N), (c) PSO, and (d) MLP-NN.

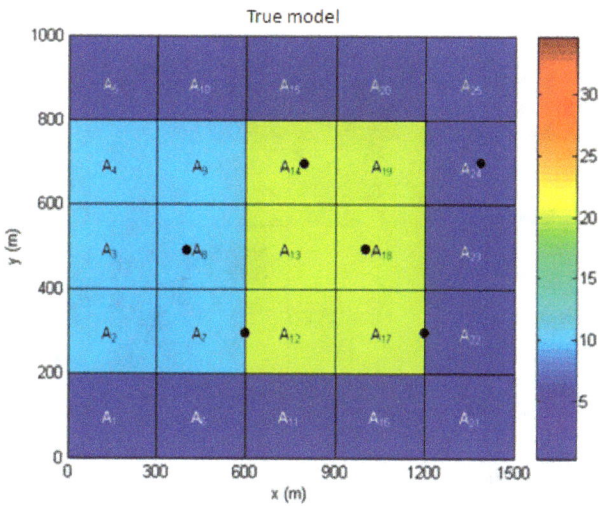

(a)

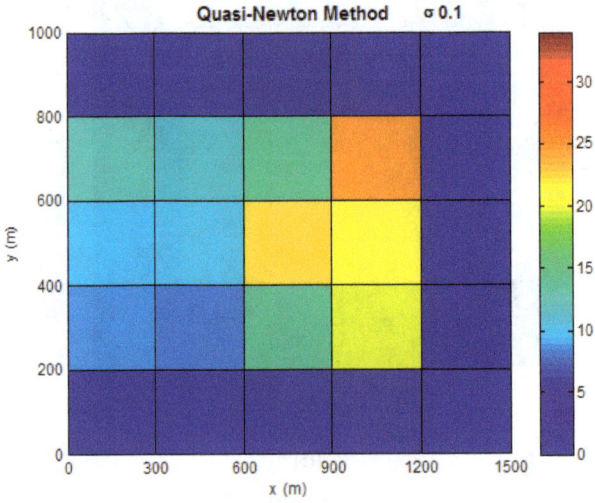

(b)

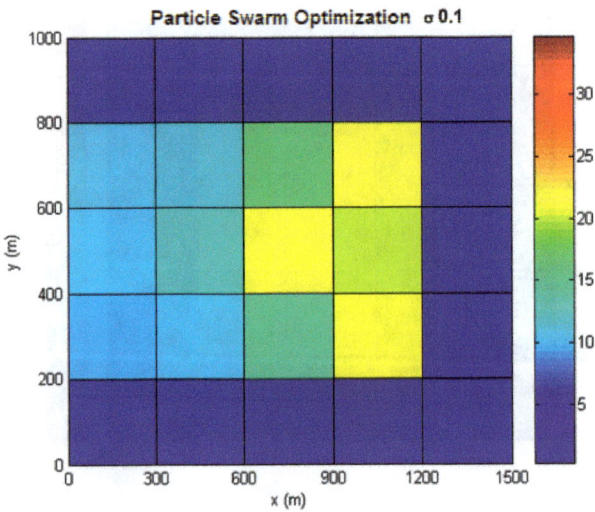

(c)

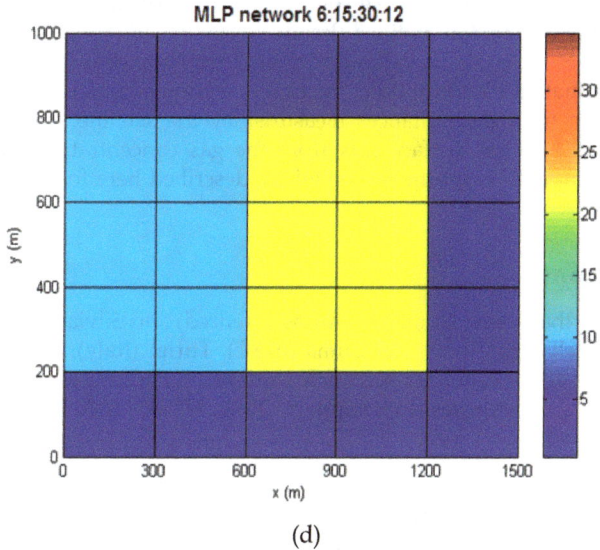

(d)

Fig. 5. Gas flux rate for Experiment-1: (a) true values, (b) quasi-Newton (Q-N), (c) PSO, and (d) MLP-NN.

5. Conclusions

The estimation of gas flux between the surface and the atmosphere is a key issue for several objectives: air pollution quality, climate change monitoring, short and medium range weather forecasting, and ecology. This lecture presents two formulations to address this important inverse problem: optimization problem approach, and artificial neural networks.

When the inverse problem is formulated as an optimization problem, a regularization operator is, in general, required. The optimal solution could be computed using a determinisc or a stochastic searching methods. Deterministic methods has a faster convergence, but stochastic schemes can overcome the local minima. The regularization strategy requires the estimation of the regularization parameter, responsible by a fine balance between adhesion term (square difference between measurements and model data) and of the regularization function. For the present estimation, the PSO inverse strategy does not use the regularization. This is an advantage, but there are much more calculations of the objective function than the determinisc method.

The artificial neural networks produced the better results for the worked example. More investigations are needed to confirm this good performance. We are generating different weather conditions, for all seasons of the year and for different latitudes, to provide more robust conclusions. Any way, the results are very encouraging. As a final remark, two topologies for MLP neural network were identified. Of course, the best result is an obvious criterion to select the topology, but an additional criterion could be based on the complexity: less complex NN (less neurons) is better. Using these two criteria, the NN-1 (6:6:12:12) could be better than NN-2 (6:15:30:12).

The methodology for computing inverse solution was described for *in situ* measurements. The new wave of environmental satellites already started (Clerbaux et al., 1999; Buchwitz et al., 2004; Crevoisier et al., 2004; Carvalho & Ramos, 2010). The next step is to use a cascade of inversions to estimate the surface flux. The first inversion is to determine the gas vertical profile concentration from the radiances measured by the satellite. The second inversion would be to identify the gas surface flux from the gas concentration found by the first inversion. We are going to adapt the methodology described here for this new observation data.

6. Acknowledgments

The authors want to thank Drs. Domenico Anfossi (retired) and Silvia Trini Castelli from the Istituto di Scienze dell'Atmosfera e del Clima (ISAC), Turim (Italy), to allow us to use the LAMBDA code, and for scientific cooperation. Author H. F. Campos Velho acknowledges the CNPq, Brazilian agency for research support – Proc. 311147/2010-0.

7. References

Anfossi, D. & Ferrero, E. (1997). Comparison among Empirical Probability Density Functions of the Vertical Velocity in the Surface Layer Based on Higher Order Correlations, *Boundary-Layer Meteorology*, Vol. 82, pp. 193-218. ISSN 0006-8314.

Aster, R. C., Borchers, B. & Thurber, C. H. (2005). *Parameter Estimation and Inverse Problems*, Academic Press, ISBN 0-12-065604-3, USA.

Bertero, M. & Boccacci, P. (1998). Introduction to Inverse Problems in Imaging, Taylor & Francis, ISBN 0-7503-0435-9.

Bocquet M. (2005a). Reconstruction of an atmospheric tracer source using the principle of maximum entropy, I: Theory. *Quarterly Journal of the Royal Meteorological Society*, Vol. 131, No. 610, pp. 2191-2208. ISSN 0035-9009

Boissieu, J. De, Papoular, R. J. & Janot C. (1991). Maximum entropy method as applied in quasi-crystallography. *Europhysics Letters*, Vol. 16, pp. 343-347.

Buchwitz, M., Noel, S., Bramstedt, K., Rozanov, V. V., Eisinger, M., Bovensmann, H., Tsvetkova, S. & Burrows, J. P. (2004). Retrieval of trace gas vertical columns from SCIAMACHY/ENVISAT near-infrared nadir spectra: first preliminary results. *Advances in Space Research*, Vol. 34, No. 809–814.

Carvalho, A. R. & Ramos, F. M. (2010). Retrieval of carbon dioxide vertical concentration profiles from satellite data using artificial neural networks. *TEMA: Tendências em Matemática Aplicada e Computacional*, Vol. 11, pp. 205-216.

Clerbaux, C., Hadji-Lazaro, J., Payan, S., Camy-Peyret, C. & Megie G. (1999). Re-trieval of CO columns from IMG/ADEOS spectra. *IEEE Transactions on Geo-science and Remote Sensing*, Vol. 37, pp. 1657–1661. ISSN: 0196-2892

Crevoisier, C. ,Chedin, A. & Scott, N. A. (2004). AIRS channel selection for CO2 and other trace-gas retrievals, *Quarterly Journal of the Royal Meteorogical Society*, Vol. 129, pp. 2719-2740.

Campos Velho, H. F. & Ramos, F. M. (1997). Numerical Inversion of Two-Dimensional Geoelectric Conductivity Distributions from Eletromagnetic Ground Data. *Brazilian Journal of Geophysics*, Vol. 15, pp. 133-143. ISSN: 0102-261X.

Campos Velho, H. F., Ramos, F. M., Shiguemori, E. H., & Carvalho J. C. (2006). A Unified Regularization Theory: The Maximum Non-extensive Entropy Principle, *Computational and Applied Mathematics*, Vol. 25, pp. 307-330. ISSN: 1807-0302.

Campos Velho H. F. (2011). Inverse Problems and Regularization (Chapter 8, Part II). In *Thermal Measurements and Inverse Techniques* (H. R. B. Orland, O. Fudym, D. Maillet, R. M. Cotta. (Org.)), pp. 283-313, CRC Press, New York. ISBN: 10 1436845557

Conte, S. D., C. de Boor (1980). *Elementary Numerical Analysis, An Algorithmic Approach.* McGraw Hill, New York, ISBN 07001224477.

Cybenko G. (1989). Approximations by superpositions of sigmoidal functions. *Mathematics of Control, Signals, and Systems*, Vol. 2, No. 4, pp. 303-314.

Degrazia, G. A., Anfossi, D., Carvalho, J. C., Mangia, C., Tirabassi, T. & Campos Velho H. F. (2000). Turbulence parameterization for PBL dispersion models in all stability conditions. *Atmospheric Environment*, Vol. 34, pp. 3575-3583.

E04UCF routine (1995). NAG Fortran Library. Oxford: Mark 17.

Enting I. G. (2002). *Inverse Problems in Atmospheric Constituent Transport*, University Press, Cambridge, ISBN 0521812100.

Elbern, H., Strunk, A., Schmidt, H., & Talagrand O. (2007). Emission rate and chemical state estimation by 4-dimensional variational inversion. *Atmospheric Chemistry and Physics*, Vol. 7, pp. 1725–1783. ISSN: 16807367

Ferrero, E., Anfossi, D. & Brusasca G. (1995). Lagrangian particle model Lambda: evaluation against tracer data, *International Journal of Environment and Pollution*, Vol. 5, pp. 360-374.

Fleisher, M., Mahlab, U. & Shamir J. (1990). Entropy optimized filter for pattern recognition. *Applied Optics*, Vol. 29, pp.2091-2098.

Gimson, N. R. & Uliasz, M. (2003). The determination of agricultural methane emissions in New Zealand using inverse modeling techniques. *Atmospheric Environment*, Vol. 37, pp. 3903- 3912.

Gryning, S. E. & Lyck, E. (1984). Atmospheric dispersion from elevated source in an urban area: comparison between tracer experiments and model calculations. *Journal of Climate Applied Meteorology*, Vol. 23, pp. 651-654.

Hadarmard J. (1902). Sur les problèm aux dérivées partielles et leur signification physicque, *Bull. Princeton Univ.*, Vol. 13, pp. 49-52.

Hadarmard, J. (1952). Lectures on Cauchy's Problem in Linear Partial Differential Equations. Dover Publications, N. Y.

Hansen P.C. (1992). Analysis of discrete ill-posed problems by means of the L-curve. *SIAM Review*, Vol. 34, pp. 561-580.

Haykin, S. (1999). *Neural Networks, a Comprehensive Foundation.* 2nd Edition, Prentice Hall, New Jersey, ISBN 0132733501.

Jaynes E. T. (1957). Information theory and statistical mechanics. *Physical Review*, Vol. 106, 620-630.

Kaipio, J. & Somersalo, E. (2005). Statistical and Computational Inverse Problems, Springer, ISBN 0387220739, USA.

Kennedy, J., & Eberhart, R. C. (1995). Particle swarm optimization. *Proc. IEEE Int'l. Conf. on Neural Networks*, IV, 1942–1948. Piscataway, NJ: IEEE Service Center.

Luz, E. F. P. (2007). *Estimation of atmospheric pollutant source by particle swarm.* M.Sc. Thesis, National Institute for Space Research (INPE), Applied Computing, (in Portuguese).

Luz, E. F. P., Campos Velho, H.F. de, Becceneri, J.C. & Roberti, D. R., (2007), Estimating Atmospheric Area Source Strength Through Particle Swarm Optimization, *Inverse Problems, Design and Optimization Symposium*, vol. 1, pp. 354-359. ISBN: 978-1-59916-279-9.

Li, Y. & Chen, X. (2006). A New Stochastic PSO Technique for Neural Network Training. *International Symposium on Neural Networks* (ISNN), Chengdu (China), pp. 564-569.

Morozov, V.A. (1984). Methods for Solving Incorrectly Posed Problems. Springer Verlag. ISBN 0387960597.

Morozov, V. A., & Stessin M. (1992). Regularization Methods for Ill-Posed Problems. CRC Press.

Muniz, W. B., Campos Velho, H. F. & Ramos, F. M. (1999). A comparison of some inverse methods for estimating the initial condition of the heat equation. *Journal of Computational and Applied Mathematics*, Vol. 103, pp.145-163. ISSN: 0377-0427

Muniz, W. B., Ramos, F. M. & Campos Velho H. F., (2000). Entropy- and Tikhonov-based Regularization Techniques Applied to the Backwards Heat Equation, *Computres & Mathematics with Applications*, Vol. 40, pp. 1071-1084. ISSN: 08981221.

Paes, F. F., Campos Velho, H.F. de, & Ramos, F. M. (2010). Land fluxes for the bio-geochemical cycles estimated by artificial neural networks, *Inverse Problems, Design and Optimization Symposium*. Proceedings in CD-Rom.

Ramos, F. M. & Giovannini A. (1995). Résolution d'un problème inverse multidimensionnel de diffusion de la chaleur par la méthode des eléments analytiques et par le principe de l'entropie maximale. *International Journal of Heat and Mass Transfer*, Vol. 38, pp. 101-111. ISSN: 0017-9310.

Ramos, F. M., Campos Velho, H.F. de, Carvalho, J.C. & Ferreira, N.J. (1999). Novel approaches on entropic regularization. *Inverse Problems*, Vol 15, No. 5, pp. 1139-1148. ISSN: 0266-5611.

Roberti, D.R., Anfossi, D., Campos Velho, H.F. de, & Degrazia, G.A. (2005). Estimation of Location and Strength of the Pollutant Sources, *Ciencia e Natura*, pp. 131-134. ISSN: 0100-8307.

Roberti D. R. (2005). *Inverse Problems in Atmospheric Physics*. D.Sc. Thesis, Federal University of Santa Maria.

Roberti, D.R., Anfossi, D., Campos Velho, H.F. de, & Degrazia G.A. (2007). Estimation of Pollutant Source Emission Rate from Experimental Data, *Il Nuovo Cimento*, Vol. 30 C, No. 02, pp. 177-186. ISSN: 0369-3554.

Rumelhart, D. E., G. E. Hinton, R. J. Willians (1986). Learning representations of back-propagation erros. *Nature*, Vol. 323, pp. 533-536.

Shannon, C.E. & Weaver W. (1949). The Matemathical Theory of Communication. University of Illinois Press.

Shiguemori, E. H., Campos Velho, H. F., & Silva J. D. S. Da, (2004). Generalized Morozov's principle. *Inverse Problems, Design and Optimization Symposium*, Vol. 2, pp. 290-298.

Shi, Y. & Eberhart R. C. (1998). A modified particle swarm optimizer. *Proceedings of the IEEE International Conference on Evolutionary Computation*. pp. 69-73, IEEE Press, Piscataway, NJ (USA).

Smith, R. T., Zoltani, C. K., Klem, G. J. & Coleman, M. W. (1991). Reconstruction of tomographic images from sparse data sets by a new finite element maximum entropy approach. *Applied Optics*, Vol. 30, pp. 573-582. ISSN: 0003-6935.

Tarantola A. (1987). *Inverse Problem Theory*. Elsevier Science Publishers, Amsterdam.

Tikhonov, A.N. & Arsenin V.I. (1977). *Solutions of Ill-posed Problems*, John Wiley & Sons.

Thomson, D. J. (1987). Criteria for the selection of stochastic models of particle trajectories in turbulent flows, *Journal of Fluid Mechanics*, Vol. 180, pp. 529-556.

Tsallis C. (1988). Possible generalization of Boltzmann-Gibbs statistics, *Journal of Statistical Physics*, Vol. 52, p. 479-487. ISSN: 0022-4715.

12

Methodology to Assess Air Pollution Impact on Human Health Using the Generalized Linear Model with Poisson Regression

Yara de Souza Tadano[1], Cássia Maria Lie Ugaya[2]
and Admilson Teixeira Franco[2]
[1]State University of Campinas – Sao Paulo,
Department of Mechanical Engineering
[2]Federal University of Technology – Paraná,
Department of Mechanical Engineering
Brazil

1. Introduction

The growth of urban areas increased the access to many facilities such as transportation, energy, education, water supply, etc. As a consequence, there was a vehicular and industrial growth that combined with unfavorable meteorological conditions caused several worldwide episodes of excessive air pollution with life losses and health damage. Some examples were the well known Donora disaster in October, 1948 and fog episodes occurred after December, 1952 in London (Lipfert, 1993).

Since then, the researchers started to worry about air pollutants impact. Many epidemiological studies of air pollution have been conducted showing that air pollution affects human health, especially in respiratory and cardiovascular diseases, even where concentration levels of pollutants are below the air quality standard levels (Braga et al., 1999; Braga et al., 2001; Burnett et al., 1998; Ibald-Mulli et al., 2004; Peng et al., 2006; Peters et al., 2001; Pope III et al., 2002; Samet et al., 2000a, 2000b).

The evaluation of the impact of air pollution on human health is complex due to the fact that several personal characteristics (age, genetics, social conditions, etc.) influence on the response to a given air pollutant concentration. For instance, several studies have shown that a higher air pollution concentration increases the number of respiratory diseases in elderly and children (Braga et al., 1999; Braga et al., 2001). These studies show that children are more susceptible because they need twice the amount of air inhaled by adults and the elderly are more affected due to their weak immune and respiratory systems in addition to the fact they have been exposed to a great amount of air pollution throughout their lives. Another characteristic is genetics. The studies showed that people with chronic diseases or allergies, such as bronchitis and asthma are more sensitive to air pollution.

In this chapter, it will be presented a summary of four kinds of studies usually used to assess air pollution impact on human health and emphasizing the most used one: the time series studies. In time series studies, a model very useful is the Generalized Linear Model (GLM). Then, the steps to apply the GLM to air pollution impact on human health studies will be presented in details, including a case study as an example. The results have shown that the GLM with Poisson regression fitted well to the database of the case study considered.

It is relevant to emphasize that the concepts included in this chapter are available in the literature, but the methodology presented to assess air pollution impact on human health employing the GLM with Poisson regression has no precedents.

2. Statistical methods

To assess air pollution impact on human health, epidemiological studies often use statistical methods that are extremely useful tools to summarize and interpret data.

The health effects (acute or chronic), type of exposure (short or long term), the nature of the response (binary or continuous) and data structure lead to model selection and the effects to be estimated. Regression models are generally the method of choice.

The exposure to ambient air pollution varies according to temporal and/or spatial distribution of pollutants. Most air pollution studies have used measures of ambient air pollution instead of personal exposure because estimating relevant exposures for each person can be daunting. According to this approximation; "misclassification of exposure is a well-recognized limitation of these studies" (Dominici et al., 2003).

According to Dominici et al. (2003), epidemiological studies of air pollution fall into four: time series; case-crossover; panel and cohort. The time series, case-crossover and panel studies are more appropriate for acute effects estimation while the cohort studies are used for acute and chronic effects combined.

2.1 Case-crossover studies

Case-crossover studies are conducted to estimate the risk of a rare event associated with a short-term exposure. It was first proposed by Maclure (1991) cited in Dominici et al. (2003) to "study acute transient effects of intermittent exposures". In practice, this design is a modification of the matched case-control design. The difference between a case-crossover and a case-control design is that in case-control designs, each case acts as his/her own control and the exposure distribution is then compared between cases and controls and in case-crossover design "exposures are sampled from an individual's time-varying distribution of exposure". In particular, "the exposure at the time just prior to the event (the *case* or *index time*) is compared to a set of *control* or *referent times* that represent the expected distribution of exposure for non-event follow-up times". In such a way, the unique characteristics of each individual such as gender, age and smoking status; are matched, reducing possible confounding factors (Dominici et al., 2003).

According to Maclure & Mittleman (2000) cited in Dominici et al. (2003), "in the last decade of application, it has been shown that the case-crossover design is best suited to study intermittent exposures inducing immediate and transient risk, and abrupt rare outcomes".

2.2 Panel studies

Panel studies collect individual time and space varying exposures, outcomes counts and confounding factors. Consequently they include all other epidemiological designs which are based on temporally and/or spatially aggregated data. Actually, panel studies also rely on group-level data.

In panel designs, the goal is to follow a cohort or panel of individuals to investigate possible changes in repeated outcome measures. This design shows to be more effective in short-term health effects of air pollution studies, mainly for a susceptible subgroup of the population. Usually, panel studies involve the collection of repeated health outcomes measures for all considered subjects of a susceptible subpopulation over the entire time of study. The measure of pollution exposure could be from a fixed-site ambient monitor or from personal monitors (Dominici et al., 2003).

Some care should be taken when designing a panel study, because the main goal of estimating the health effect of air pollution exposure sometimes can be less clear. It happens whenever the panel members do not share the same observation period, so parameterization and estimation of exposure effects need to be considered with much care (Dominici et al., 2003).

2.3 Cohort studies

The cohort studies are frequently used to associate long-term exposure to air pollution with health outcomes. Prospective or retrospective designs are possible. The first one consists of participants' interview at the beginning of the research containing particular information such as age, sex, education, smoking history, weigh, and so on. After that, the participants are followed-up over time for mortality or morbidity events. The retrospective design consists of using already available database information. Cohort designs are frequently used to multicity studies, as it ensures "sufficient variation in cumulative exposure, particularly when ambient air pollution measurements are used" (Dominici et al., 2003).

2.4 Time series studies

The time series impact studies are often used as they demand simply data such as the amount of hospital admission or mortality in a given day, being easy to obtain on health government departments. So it is unnecessary to follow-up the group of people involved in the study, which demands much time (Schwartz et al., 1996).

Another key advantage of the time series approach is the use of daily data and while the underlying risk on epidemiological studies of air pollution varies with some factors such as age distribution and smoking history, these factors will not have influence on the expected number of deaths or morbidity on any day, since they do not vary from day to day (Schwartz et al., 1996).

Regression models are usually chosen in time series studies, as they are useful tools to assess the relationship between one or more explanatory variables (independent, predictor variables or covariates) (x_1, x_2,...,x_n) and a single response variable (dependent or predicted variable) (y) (Dominici et al., 2003). The simplest regression analysis consisting of more than one explanatory variable is the multiple linear regression and is given by:

$$y = \beta_0 + \beta_1 x_1 + \beta_2 x_2 + \cdots + \beta_n x_n + \varepsilon , \qquad (1)$$

where y is the response variable and x_i (i = 1, 2, …, n) are the explanatory variables. β_0 represents the value of y when all the explanatory variables are null, β_i terms are called regression coefficients and the residual (ε) is the prediction error (the difference between measured and adjusted values of the response variable).

The regression models goal is to find an expression that better predicts the response variable as a combination of the explanatory variables. It means to find the $\beta's$ that better fits to the database.

Due to the non-linearity of the response variable in time series studies of air pollution impacts on human health, the Generalized Linear Models (GLM) with parametric splines (e.g. natural cubic splines) (McCullagh & Nelder, 1989) and the Generalized Additive Models (GAM) with non-parametric splines (such as smoothing splines or lowess smoothers) (Hastie & Tibishirani, 1990) are usually applied.

According to the studies conducted in the last decade, GAM was the most widely applied method as it allows for non-parametric adjustment of the non-linear confounding factors such as seasonality, short-term trends and weather variables. It is also a more flexible approach than fully-parametric models like GLM with parametric splines. Nevertheless, recently the GAM implementation in statistical softwares, like S-Plus has been called into question (Dominici et al., 2003).

To evaluate the impact of default implementation of the GAM software on published analyses, Dominici et al. (2002) reanalyzed the National Morbidity, Mortality, and Air Pollution Study (NMMAPS) data (Samet et al., 2000a, 2000b) using three different methods: The GLM (Poisson regression) with natural cubic splines to achieve nonlinear adjustments for confounding factors; the GAM with smoothing splines and default convergence parameters; and the GAM with smoothing splines and more stringent convergence parameters than the default settings. The authors found that "estimates obtained under GLMs with natural cubic splines better detect true relative rates than GAMs with smoothing splines and default convergence parameter". The authors also added that: "although GAM with nonparametric smoothers provides a more flexible approach for adjusting for nonlinear confounders compared with fully parametric alternatives in time series studies of air pollution and health, the use and implementation of GAMs requires extreme caution".

In such a way, in this chapter it will be presented all the steps that should be followed to conduct a time series study using GLM with Poisson regression, from data collection to measure the goodness of fit. More details about design comparisons between time series; case-crossover; panel and cohort studies are in Dominici et al. (2003).

3. Generalized Linear Models (GLM)

The GLMs are a union of linear and non-linear models with a distribution of the exponential family, which is formed by the normal, Poisson, binomial, gamma, inverse normal distributions including the traditional linear models, as well as logistic models (Nelder & Wedderburn, 1972).

Since 1972, many researches on GLMs where conducted and as a consequence several computational skills were created such as, GLIM (Generalized Linear Interactive Models), S-Plus, R, SAS, STATA and SUDAAN (Dobson & Barnett, 2008; Paula, 2004).

The GLMs are defined by a probability distribution of the exponential distribution family, and are formed by the following components (McCullagh & Nelder, 1989):

-   Random component: n explanatory variables $(y_1,...,y_n)$ of a response variable which follows a distribution of the exponential family with expected value $E(y_i) = \mu$;
-   Systematic component: concerns a linear structure for the regression model $(\eta = \beta x^T)$, called linear predictor, where $x^T = (x_{i1}, x_{i2}, \cdots, x_{ip})^T$, $i = 1, \cdots, n$ are the so-called explanatory variables and;
-   Link function: a monotone and differentiable function g, called link function, capable of connecting the random and systematic components, relating the response variable mean (μ) to the linear structure, defined in GLMs as $g(\mu) = \eta$, where:

$$\eta = \beta_0 + \beta_1 x_1 + \beta_2 x_2 + \cdots + \beta_n x_n ,\tag{2}$$

or in matrix form:

$$\eta = \beta x^T ,\tag{3}$$

where the regression coefficients $\beta = (\beta_1, \beta_2, \cdots, \beta_n)$ represents the vector of parameters to be estimated (McCullagh & Nelder, 1989).

Each distribution has a special link function, called canonical link function which occurs when $\eta = \theta_i$, where θ is called the local or canonical parameter. Table 1 shows the canonical function for some distributions of the exponential family (McCullagh & Nelder, 1989).

Distribution	Canonical link function (η)
Normal	μ
Poisson	$\ln(\mu)$
Binomial	$\ln\{\mu/(1-\mu)\}$
Gamma	μ^{-1}
Inverse Gaussian	μ^{-2}

Table 1. Canonical link functions of some distributions of the exponential family (McCullagh & Nelder, 1989).

According to Myer & Montgomery (2002), using the canonical link function implies some interesting properties, although it does not mean it should be always used. This choice is convenient because, besides the simplification of the estimative of the model parameter, it also becomes easier to obtain the confidence interval of the response variable mean. However, the convenience do not necessarily implies in goodness of fit.

In studies of air pollution impact on human health with non-negative count data as response variable, the GLM with Poisson regression is broadly applied (Dockery & Pope III, 1994; Dominici et al., 2002; Lipfert, 1993; Metzger et al., 2004).

The GLM with Poisson regression consists in relating the response variable (y) (mortality or morbidity), which can take on only non-negative integers, with the explanatory variables ($x_1, x_2, \ldots, x_n$) (pollutants concentration, weather variables, etc.) according to:

$$\ln(y) = \beta_0 + \beta_1 x_1 + \beta_2 x_2 + \cdots + \beta_n x_n,$$
(4)

Usually, the regression coefficients ($\beta's$) are estimated using the Fisher score method of maximizing the likelihood function (maximum likelihood method), which is the same as the Newton-Raphson method when the canonical link function is considered. For the Poisson regression, the likelihood density function is given by (Dobson & Barnett, 2008; McCullagh & Nelder, 1989):

$$f_y(y;\theta;\phi) = \exp\{y\ln(\mu) - \mu - \ln(y!)\},$$
(5)

where y is the response variable, θ is the canonical parameter and ϕ is the dispersion parameter. When the link function is the canonical link ($\ln \mu = \eta$).

One feature of the GLM with Poisson regression is that even if all the explanatory variables were known and measured without error, there would still be considerable unexplained variability in the response variable. This is a result of the fact that even if the response variable is more precise, the Poisson process ensures stochastic variability around that expected count. In a classic stationary Poisson regression, the variance is equal to the mean ($\phi = 1$), but in many actual count processes there is overdispersion, when the variance is greater than the mean or, underdispersion the other way round. In these cases, it is still possible to apply the GLM with Poisson regression (Everitt & Hothorn, 2010; Schwartz et al., 1996). One way to adjust the over or underdispersion is to assume that the variance is a multiple of the mean and estimate the dispersion parameter using the quasi-likelihood method. Details of this method are in McCullagh & Nelder (1989).

4. Steps to fit GLM with poisson regression

To apply the GLM with Poisson regression, four main steps should be followed: development of the database; adjustment of the temporal trends; goodness of fit analysis and results analysis. The details of each step are at the following topics.

4.1 Database

Usually, in time series studies of air pollution impact on human health using the GLM with Poisson regression, the data used are air pollutants concentration, weather measures, outcome counts and some confounding terms. The data must be collected daily and for at least two years, to capture the seasonal trends.

Pollutants concentrations are usually obtained by fixed-site ambient monitors. The weather measures frequently used are temperature (or dewpoint temperature) and air relative humidity.

The outcome depends on the purpose of the study. For example, in some studies mortality is used, in others, morbidity. The outcome can be stratified by type of disease (such as respiratory or cardiovascular diseases), age (children, young and elderly people) and any other factor of interest. The confounders can be of long-term (such as seasonality) or short-term (day of the week, holiday indicator, etc.) and will be shown in Section 4.2.

4.2 Temporal trends adjustment

A common feature of epidemiological studies is biases due to confounding factors and correlations among covariates that can never be completely ruled out in observational data. Confounding factors are present when a covariate is associated with both the outcome and the exposure of interest but is not a result of the exposure. So, in all epidemiological studies, a basic issue in modeling is to control properly for all the potential confounders. Time series studies have some unique features in this regard (Dominici et al., 2003; Peng et al., 2006).

The intercorrelation of different pollutants in the atmosphere is one source of biases. One way to address this intercorrelation "has been to conduct studies in locations where one or more pollutants are absent or nearly so" (Dominici et al., 2003).

The sources of potential confounding factors in time series studies of air pollution impact on human health can be broadly classified as measured or unmeasured. Measured confounding factors such as weather variables (temperature; dewpoint temperature; humidity and others) are of unique importance in this kind of studies. Some studies have demonstrated a relationship between temperature and mortality being positive for warm summer days and negative for cold winter days, like in Curriero et al. (2002) cited in Peng et al. (2006). One approach to adjust confounding factors by temperature or humidity is to include non-linear functions or a mean of current and previous days temperature (or dewpoint temperature) in the model (Peng et al., 2006). Unmeasured confounding factors are those factors that have influence in outcome counts and have a similar variation in time as air pollutants concentration. These confounding factors produce seasonal and long-term trends in outcome counts that can confound its relationship with air pollution. Some important examples are influenza and respiratory infections (Peng et al., 2006).

4.2.1 Seasonality

In time series studies, the primarily concern is about potential confounding by factors that vary on timescales in a similar manner as pollution and health outcomes. This attribute is usually called seasonality. "A common approach to adjust this trend is to use semi-parametric models which incorporate a smooth function of time". The smooth function serves as a linear filter for the mortality (morbidity) and pollution series and "removes any seasonal or long-term trends in the data" (Peng et al., 2006). Several methods to deal with this trend are being used such as smoothing splines, penalized splines, parametric (natural cubic) splines and less common LOESS smoothers or harmonic functions (Dominici et al., 2002; Peng et al., 2006; Samet et al., 2000a, 2000b; Schwartz et al., 1996).

The spline function provides an approximation for the behavior of functions which has local and abrupt changes. The most used spline to smoothing curves in GLMs is the natural cubic spline (Chapra & Canale, 1987; Samoli et al., 2011; Schwartz et al., 1996), the other ones are usually applied in GAM.

Using splines, polynomial functions will be provided for each defined interval instead of a single polynomial for the whole database. The natural cubic spline is based on third order polynomials derived for each interval between two knots at fixed locations throughout the range of the data (Chapra & Canale, 1987; Peng et al., 2006). The choice of knots locations can result in substantial effect on the resulting smooth. So, in Peng et al. (2006) study the authors "provided a comprehensive characterization of model choice and model uncertainty in time series studies of air pollution and mortality, focusing on confounding factors adjustment for seasonal and long-term trends". According to their results, for natural splines, the bias drops suddenly between one and four degrees of freedom (df) per year and is stable afterwards, suggesting that at least 4 degrees of freedom per year of data should be used. In such way, in time series studies of air pollution and mortality (or morbidity) usually is used four to six knots per year, as the seasonality trend is due to the different behavior of variables during the seasons of the year (Tadano, 2007). Their results show that "both fully parametric and nonparametric methods perform well, with neither preferred. A sensitivity analysis from the simulation study indicates that neither the natural spline nor the penalized spline approach produces any systematic bias in the estimates of the log-relative-rate β " (Peng et al., 2006).

The smooth functions of time accounts only for potential confounding factors which vary smoothly with time, such as seasonality. Some potential confounders which vary on shorter timescales are also important, as they confound the relationship between air pollution and health outcomes, such as day of the week and holiday indicator (Peng et al., 2006).

4.2.2 Day of the week and holiday indicator

Important potential confounding factors that may bias time series studies of air pollution and mortality (or morbidity) are factors which vary on shorter timescales like calendar specific days, such as day of the week and holiday indicator (Lipfert, 1993). These trends are not necessarily present, but they occur often enough that they should be checked (Samoli et al., 2011; Schwartz et al., 1996). For example, on weekends the number of hospital admissions can be lower than on weekdays and can also be lower during holidays.

One way to adjust according the week day trend is to add qualitative explanatory variable for each day of the week (varying from one to seven) starting at Sundays. To adjust the holiday indicator, it can be considered an additional binomial explanatory variable in which one means holidays and zero means workdays (Tadano et al., 2009).

Adding all the time trends mentioned and explanatory variables in the GLM with Poisson regression, the expression used in some studies of air pollution impact on population's health is as follows (Tadano, 2007):

$$\ln(y) = \beta_0 + \beta_1 T + \beta_2 RH + \beta_3 PC + \beta_4 H + \beta_5 dow + \beta_6 ns, \tag{6}$$

where y = health outcome of interest; T = air temperature or dewpoint temperature (°C); RH = air relative humidity (%); PC = pollutant concentration ($\mu g/m^3$); H = time trend variable for holidays; *dow* = time trend variable for days of the week; *ns* = natural cubic spline to adjust for seasonality.

Some of these short-term trends can lead to autocorrelation between data from one day to previous days, even after its adjustment. In this regard, partial autocorrelation functions are used.

4.2.3 Partial autocorrelation functions

The short-term trends such as days of the week and holiday indicator can lead to an autocorrelation between data from one day and previous days, even using the adjustment. One way to analyze this time trend is plotting the partial autocorrelation function (Partial ACF) against lag days.

The autocorrelation function of the model's residuals is as follows:

$$ACF = \frac{c_k}{c_0}, \tag{7}$$

where $c_k = \frac{1}{n} \sum_{i=1}^{n-k} (y_i - \mu)(y_{i+k} - \mu)$, with n = number of observations and k = lag days (Box et al., 1994). In the partial autocorrelation function plot, the residuals should be as smaller as possible, ranging from $-2n^{-1/2}$ to $2n^{-1/2}$ (dashed lines) as shown in Fig. 1.

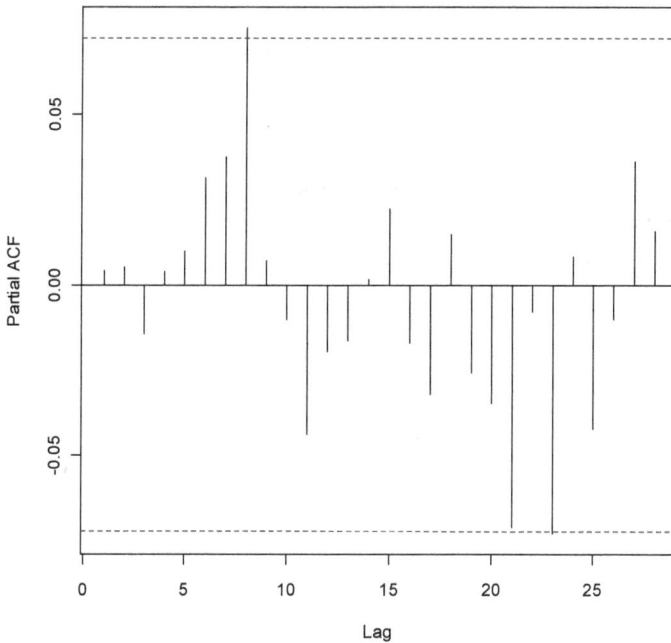

Fig. 1. Example of the partial autocorrelation function (Partial ACF) plot against lag days where there are no autocorrelations between data for less than five lag days.

In epidemiological studies of air pollution, the important autocorrelations are those occurring in the first five days, which are usually caused by the decrease of health outcomes in weekends and holidays (Tadano, 2007). If the database has autocorrelations, then the model should consider them by including the residuals in the model.

In R or S-Plus language, the residuals to be included are the working residuals. These residuals are returned when extracting the residuals component directly from the *glm* comand. They are defined as:

$$r_i^W = \left(y_i - \hat{\mu}_i\right)\frac{\partial\eta}{\partial\hat{\mu}_i}, \tag{8}$$

where η = the link function which in canonical form of the Poisson regression $\eta = \ln\mu$; y_i = measured values of the response variable; $\hat{\mu}_i$ = adjusted value by modeling and $i = 1,2,...,n$ with n = number of observations.

After adjusting the GLM with Poisson regression including all time trends and explanatory variables, the fitting model need to be tested to assure that this is the best model to be applied to the database.

4.3 Goodness of fit

The GLM with Poisson regression has been widely applied in epidemiological studies of air pollution (Dockery & Pope III, 1994; Dominici et al., 2002; Lipfert, 1993; Metzger et al., 2004; Tadano et al., 2009) but it needs caution, as sometimes this model may not fit well to the database. There are two statistical methods that can be used to evaluate goodness of fit in GLMs, as follows.

4.3.1 Pseudo R²

One interesting and easy to apply goodness of fit test for GLM with Poisson regression is the statistic called pseudo R² which is similar to the determination coefficient of classic linear models. It is defined as:

$$Pseudo\,R^2 = \frac{l\left(b_{\min}\right)-l\left(b\right)}{l\left(b_{\min}\right)}, \tag{9}$$

where l =log-likelihood function; $l(b_{\min})$ = maximum value of the log-likelihood function for a minimal model with the same rate parameter for all y's and no explanatory variables (null model) and $l(b)$ = maximum value of the log-likelihood function for the model with p parameters (complete model) (Dobson & Barnett, 2008).

This statistic measures the deviance reduction due to the inclusion of explanatory variables and can be applied in R (R Development Core Team, 2010) throughout the Anova Table with chi-squared test where the residual deviance values indicates the maximum value of the log-likelihood function for the complete model and the null one.

According to Faraway (1999), a good value of R² depends on the area of application. The author suggests that in biological and social sciences it is expected lower values for R².

Values of 0.6 might be considered good, because in these studies the variables tend to be more weakly correlated and there is a lot of noise. The author also advises that it is a generalization and "some experience with the particular area is necessary for you to judge your R^2's well".

4.3.2 Chi-squared statistic

Another statistical test used as goodness of fit in GLM with Poisson regression is the chi-squared (χ^2) or Pearson statistic, which is used to evaluate the model fit comparing the measured distribution to that obtained by modeling. The expression that represents the chi-squared statistic is:

$$\chi^2 = \sum_{i=1}^{n} (y_i - \hat{\mu}_i)^2 / \hat{\mu}_i , \qquad (10)$$

where y_i = measured values of the response variable; $\hat{\mu}_i$ = adjusted value by modeling and $i = 1, 2, ..., n$ with n = number of observations.

The chi-squared statistic is the sum of the Pearson residual of each observation. According to this statistic a model that fits well to the data has a chi-squared statistic close to the degrees of freedom (df) ($\chi^2/df{\sim}1$), where df = n – p (n = number of observations and p = number of parameters) (Wang et al., 1996). There is no evidence of which goodness of fit (Pseudo R^2 or χ^2) is preferred.

After the confirmation of GLM with Poisson regression fitting, the results must be analyzed to find the nature of the correlation between air pollutants concentration and health outcomes.

4.4 Results analysis

In epidemiological studies of air pollution it is common to find a relation between the air pollutants concentration of one day to the health outcomes of the next day, two days later or even after one week. Then, researchers usually fit the model to different arrangements of the same database with lags. In time series studies, lags of one day to seven days are frequently applied and then the one that best fits is chosen. One criteria to select the best option is the Akaike Information Criterion.

4.4.1 Akaike information criterion

The Akaike Information Criterion (AIC) is very useful when choosing between models from the same database. The smallest is the AIC, the better is the model. The AIC is automatically calculated in R software when applying the GLM algorithm and is calculated by:

$$AIC = -2l(b) + 2(df)\hat{\phi} , \qquad (11)$$

where $l(b)$ = maximum log-likelihood value for the complete model; df = degrees of freedom of the model and $\hat{\phi}$ = estimated dispersion parameter (Peng et al., 2006).

After choosing for the model that better fits the database and which has the best relationship between air pollution and health outcomes, a method to verify the strength of this relationship is applied. One method frequently applied (used in S-Plus and R software) is the Student t Test.

4.4.2 Student t test for statistical significance

In time series studies, the confirmation of any relation between air pollution and health outcomes is obtained throughout a hypothesis test that can show if the regression coefficients are statistically significant or not.

The statistical hypothesis to be tested is the null one (H_0) expressed by an equality. The alternative hypothesis is given by an inequality (Mood et al., 1974).

The hypothesis test has several goals; one of them is to verify if the estimated regression coefficient can be discredited. In this case, the following hypotheses are considered H_0: $\beta = 0$ and H_1: $\beta \neq 0$. The statistical test used to verify these hypotheses is given by:

$$t_0 = \beta/\varepsilon, \tag{12}$$

where ε is the standard error of the estimated regression coefficient (β). The rejection of the null hypothesis occurs when $|t_0| > t_{\alpha/2,n-k-1}$ (n = number of observations, k = number of explanatory variables, α is the considered significance level), indicating that the estimated regression coefficient is statistically significant. In other words, the explanatory variable influences in the response variable (Bhattacharyya & Johnson, 1977; Mood et al., 1974).

The values $t_{\alpha/2,n-k-1}$ are presented in Student t distribution table, where n-k-1 is the degrees of freedom (df) and α is the considered significance level (Bickel & Doksum, 2000).

If the study results in a statistically significant relation between air pollutant concentration and the health outcome of interest, then some analysis and projections are made using the relative risk (RR).

4.4.3 Relative risk

The relative risk (called rate ratio by statisticians) (Dobson & Barnett, 2008) is used to estimate the impact of air pollution on human health, making some projection according to pollutants concentration.

The relative risk is a measure of the association between an explanatory variable (e.g. air pollutant concentration) and the risk of a given result (e.g. the number of people with respiratory injury) (Everitt, 2003).

In a specific way, the relative risk function at level x of a pollutant concentration, denoted as RR(x), is defined as (Baxter et al., 1997):

$$RR(x) = \frac{E(y|x)}{E(y|x=0)}. \tag{13}$$

It is the ratio of the expected number of end points at level x of the explanatory variable to the expected number of end points if the explanatory variable was 0 (Baxter et al., 1997). For the Poisson regression, the relative risk is given by:

$$RR(x) = e^{\beta x}, \tag{14}$$

indicating, for example, that the risk of a person exposed to some pollutant concentration (x) having a specific injury is $RR(x)$ times greater than someone who has not been exposed to this concentration. A $RR(x) = 2$ for a pollutant concentration of 100 $\mu g/m^3$, indicates that a person exposed to this concentration has two times more chance to get a health problem than someone who has not been exposed to any concentration.

5. Case-study

To exemplify the appliance of GLM with Poisson regression, a case study will be presented. It will be evaluated the impact of air pollution on population's health of Sao Paulo city, Brazil, from 2007 to 2008. Sao Paulo is the largest and most populated city of Brazil, and one of the most populated in the world.

In this study, it was evaluated the impact of PM_{10} (particles with an aerodynamic diameter less or equal to 10 μm) on the number of hospital admissions for respiratory diseases, according to the International Classification of Diseases (ICD-10).

PM_{10} was chosen because according to WHO (2005) as cited in Schwarze *et al.* (2010), particulate air pollution is regarded as a serious health problem and some studies reported that reductions in particulate matter levels decrease health impact of air pollution. According to Braga *et al.* (2001) the health outcomes had high correlation with PM_{10} concentration in Sao Paulo (Brazil) population.

5.1 Case-study database

The data collected in this study consisted of daily values from January 1st, 2007 to December 31st, 2008 to Sao Paulo city, Brazil.

The hospital admissions for respiratory diseases, according to the ICD-10, were considered as response variable. The data was obtained from the Health System (SUS) website (2011). The explanatory variables consisted of PM_{10} concentration and weather variables (air temperature and air relative humidity), also including parametric splines for long-term trend (seasonality), qualitative variable for days of the week and binomial variable for holiday indicator.

The PM_{10} concentration, air temperature and humidity where obtained from QUALAR system in Cetesb (Environmental Company of Sao Paulo State) website (2011).

The fixed-site monitoring network of Sao Paulo city, held by Cetesb, has twelve automatic stations, and PM_{10} concentration is collected in all but one of the stations and temperature and humidity data are acquired in eight of them. This network also contains nine manual stations, but none of them monitors PM_{10} concentration, just TSP (Total Suspended Particles

- particles with an aerodynamic diameter less or equal to 50 μm) and $PM_{2.5}$ (particles with an aerodynamic diameter less or equal to 2.5 μm).

The PM_{10} concentration was monitored in eight fixed-site automatic monitoring stations during the study period (from 2007 to 2008); the air temperature and the air relative humidity were monitored only at two stations. The daily data of these variables used in this study comprise the mean of the available data.

The descriptive analysis of the variables considered in this study is presented in Table 2. The values in Table 2 show that the maximum daily PM_{10} concentration (103 μg/m³) did not overcome the national air quality standard (150 μg/m³).

Variable	Mean	Standard Deviance	Minimum	Maximum
RD	165.53	42.43	57.00	267.00
PM_{10}	39.25	18.27	11.00	103.00
Temperature	20.49	2.94	10.10	27.20
Humidity	73.36	9.13	40.40	99.50

Table 2. Descriptive statistics for hospital admissions for respiratory diseases (RD), concentration of PM_{10} and weather variables.

To have an initial idea of the relation between the response variable and the explanatory ones, the Pearson correlation matrix was constructed (Table 3). The Pearson correlation between hospital admissions for respiratory diseases (RD) and PM_{10} was positive and statistically significant. It means the number of RD increases as PM_{10} concentration increases. This table also shows that the number of RD increases as temperature and humidity decreases, but the Pearson correlation was not statistically significant in this case. Consequently, as shows Table 3, the PM_{10} concentration increases in days with low temperature and humidity indexes.

	RD	PM_{10}	Temperature	Humidity
RD	1.00			
PM_{10}	0.41*	1.00		
Temperature	-0.14	0.01	1.00	
Humidity	-0.15	-0.57*	-0.27*	1.00

* statistically significant ($p<0.05$)

Table 3. Pearson correlation matrix between hospital admissions for respiratory diseases (RD), concentration of PM_{10} and weather variables.

5.2 Long and short-term trend adjustment

The long-term trend usually included in time series studies of air pollution impact on human health is seasonality and in this case study it was considered a natural cubic spline, the most used parametric smooth in GLMs.

To apply the natural cubic spline in GLM with Poisson regression, an explanatory variable for the days is added to the model, consisting of values from 1 to 731, comprising all two years of data.

The short-term trends usually considered in epidemiological studies of air pollution are the day of the week and holiday indicator. The day of the week variable was considered as a qualitative variable which varies from one to seven, starting at Sundays. The holiday indicator was adjusted adding a binomial variable in which one means holidays and zero means workdays.

According to the considerations above, Table 4 brings an example of the first lines of the database used.

Data	RD	PM10	T	RH	day	dow	H
01/01/2007	104	15	21.7	85.7	1	2	1
02/01/2007	171	14	22.1	82.7	2	3	0
03/01/2007	140	17	21.9	87.2	3	4	0
04/01/2007	155	20	22	92	4	5	0
05/01/2007	130	13	21.5	89.9	5	6	0
06/01/2007	93	11	21.9	84.2	6	7	0
07/01/2007	101	18	23.2	80.4	7	1	0

Table 4. First values of the database considered in this study (RD = daily number of hospital admissions for respiratory diseases, PM10 = concentration of PM_{10} in $\mu g/m^3$; T = air temperature in °C; RH = air relative humidity; day = variable to consider seasonality; dow = days of the week; H = holidays indicator).

With the database considered, the expression used to apply the GLM with Poisson regression in R software (R Development Core Team, 2010) is:

$$m.name < -glm\left(\begin{array}{l} RD \sim ns(day, df) + as.factor(dow) + as.factor(H) + T + RH + PM10, \\ data = database.name, family = poisson, na.action = na.omit \end{array} \right), (15)$$

where $m.name$ = is the name given to the analysis; ns = natural cubic spline; df = degrees of freedom; $database.name$ = name given to the database file.

To apply this model, one important decision is about the number of degrees of freedom (df) to be considered in the natural cubic spline of days of study. In epidemiological studies of air pollution, the common values are four, five or six degrees of freedom per year of data. To decide which one to use, three analyses where made considering four, five and six degrees of freedom (df) in Equation (15) and the results were compared using the AIC, as shown in Table 5.

Number of df per year	AIC
4	6,911.4
5	6,909.2
6	6,798.7

Table 5. Comparison of models with different numbers of degrees of freedom for seasonality adjustment.

According to the results indicated in Table 5, the model with 6 degrees of freedom per year of data is the one that better fits the data. Then, in the following analyses, it was considered $df = 6$ in Equation (15).

The short-term trends considered in this study (days of the week and holiday indicator) can lead to autocorrelation between data from one day and the previous days, so the Partial ACF plot against lag days must be analyzed. The lines of each lag day until five lags must be between $-2n^{-1/2}$ and $2n^{-1/2}$. In this case study the number of observations (n) is equal to 731, so the lines in Partial ACF plot out of the range (-0.07;0.07) indicates a strong autocorrelation between data from one day and previous days.

The Partial ACF plots against lag days for the model with six degrees of freedom and considering the effects of PM_{10} concentration on the same day for the model with no residual inclusion is shown Fig. 2 and Fig. 3 shows for the model after including residuals.

For the model with 6 degrees of freedom, the Partial ACF plot (Fig. 2) shows autocorrelations between one day and the previous 1, 2, 3 and 4 days, as the lines for these lag days are out of the range. To adjust for this time trend, it is necessary to include the residuals for these lag days in the model.

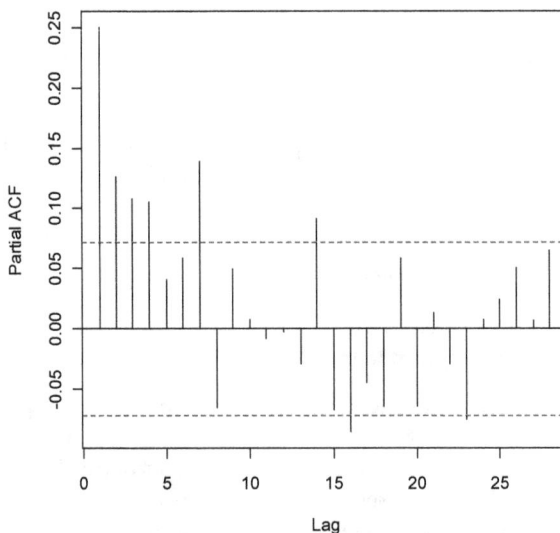

Fig. 2. Partial ACF plot against lag days with no residuals included.

After do so, the Partial ACF plot (Fig. 3) shows no more autocorrelations between data for the first five days, indicating that it is the best fitted model.

After adjusting the GLM with Poisson regression including all the time trends and explanatory variables and choosing the degrees of freedom that better fits the data; the fitted model was tested using the pseudo R^2 and the chi-squared statistic to assure that it is the right one to be applied to the case-study.

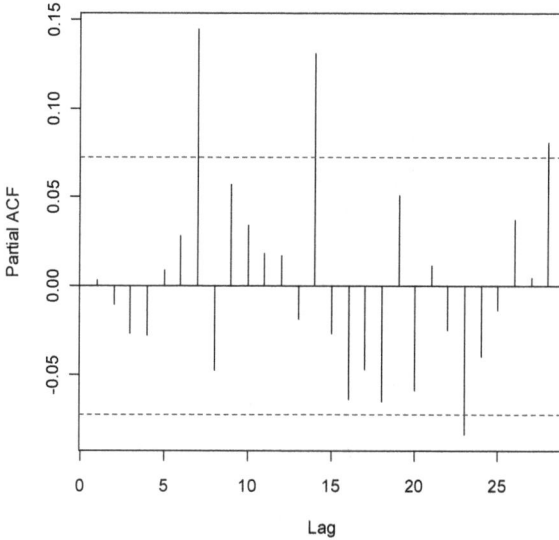

Fig. 3. Partial ACF plot against lag days including residuals.

5.3 Model adjustment results

In epidemiological studies of air pollution it is common to find a relation between the air pollutants concentration of one day to the health outcomes of some lag days. In this case-study, analyses of the relation between PM_{10} concentration of one day and the number of hospital admission for respiratory diseases for the same day until one week later was performed.

All models were fitted with no residual inclusion and also with inclusion of residuals due to autocorrelation.

The goodness of fit to the analyses from no lag to seven lag days is shown in Table 6 (A) without residual and, (B) with residuals. The ACF plots for all models adjusted will not be shown, but they are similar to that of Fig. 2. All of them (from no lag to seven lag days) indicated the need of residuals inclusion for 1, 2, 3 and 4 lag days. The models with residuals inclusion did not show anymore autocorrelations.

A	Lag days	R^2	χ^2	χ^2/df
	0	0.82	1,509	2.16
	1	0.82	1,530	2.19
	2	0.81	1,528	2.19
	3	0.81	1,526	2.19
	4	0.81	1,529	2.20
	5	0.81	1,531	2.20
	6	0.81	1,531	2.21
	7	0.81	1,525	2.20

B	Lag days	R^2	χ^2	χ^2/df
	0	0.80	1,701	2.40
	1	0.79	1,718	2.43
	2	0.79	1,709	2.42
	3	0.79	1,708	2.42
	4	0.79	1,715	2.44
	5	0.79	1,721	2.45
	6	0.79	1,721	2.45
	7	0.79	1,724	2.46

Table 6. Goodness of fit results for the analyses from no lag to seven lag days for models with no residual inclusion (A) and including residuals in the model (B).

According to the analysis of the goodness of fit shown in Table 6, all the models presented a pseudo R^2 greater than 0.6, showing that the models fitted well to the data, but the chi-squared statistic analysis showed values much greater than the degrees of freedom. As there is no evidence of which statistic is suitable for each situation, we can conclude the model fitted well to the data, according to Pseudo R^2 statistic results.

After verifying the models fitted well to the data, the analysis of the regression coefficients was held. The results are shown in Table 7 without residuals and Table 8 with residuals. Analyzing the AIC, the model that include the residuals seems to fit better than the one which was not included and the model considering the effect of seven days lag shows better results, but the regression coefficient did not show statistical significance.

Furthermore, the AIC value with three days lag is lower than for two, one or no lags; showed no autocorrelation and with a regression coefficient statistically significant.

In conclusion, the chosen model was the one with the effect of three days lag in which residuals was included. The relative risk results are therefore only presented for this model (with # symbol in Table 8).

Lag days	AIC	β	ε	t-value
0	6,798.7	1.60×10^{-03}	2.54×10^{-04}	6.30***
1	6,809.6	9.88×10^{-04}	2.17×10^{-04}	4.56***
2	6,793.6	9.07×10^{-04}	1.98×10^{-04}	4.59***
3	6,786.0	8.91×10^{-04}	1.96×10^{-04}	4.55***
4	6,785.8	6.20×10^{-04}	1.94×10^{-04}	3.19**
5	6,785.0	3.78×10^{-04}	1.91×10^{-04}	1.97*
6	6,778.2	3.46×10^{-04}	1.91×10^{-04}	1.82
7	6,774.9	7.36×10^{-05}	1.92×10^{-04}	0.38

***= 0; **= 0.001; *= 0.01 (Statistical significance level - α).

Table 7. Results analysis for no lag to seven lag days for models with no residual inclusion.

Lag days	AIC	β	ε	t-value
0	6,588.0	1.45×10^{-03}	2.54×10^{-04}	5.71***
1	6,602.4	6.98×10^{-04}	2.18×10^{-04}	3.20**
2	6,593.7	6.22×10^{-04}	2.00×10^{-04}	3.12**
3#	6,584.8	5.82×10^{-04}	1.98×10^{-04}	2.94**
4	6,581.2	2.71×10^{-04}	1.96×10^{-04}	1.38
5	6,576.1	5.50×10^{-05}	1.94×10^{-04}	0.29
6	6,569.6	6.54×10^{-05}	1.93×10^{-04}	0.34
7	6,556.6	-1.94×10^{-04}	1.94×10^{-04}	-1.00

***= 0; **= 0.001; *= 0.01 (Statistical significance level - α); # = the better model

Table 8. Results analysis for no lag to seven lag days for models including residuals.

5.4 Relative risk analysis

To analyze and estimate the PM_{10} impact on Sao Paulo's population health, the relative risk for the model considering the effects of three lag days including residuals was calculated. The expression that represents it is given by:

$$RR(x) = e^{0.000582x} .$$

(16)

The relative risks where calculated according to Equation (16). The plot of it against PM_{10} concentration is shown in Fig. 4.

In Fig. 4 it can be seen that the RR has a linear relation with PM_{10} concentration, then the greater the PM_{10} concentration, the higher the RR. Thus, when the PM_{10} concentration increases from 10 to 100 $\mu g/m^3$, the RR increases 5%. It may mean someone exposed to a concentration ten times greater has 5% more chance of getting a respiratory disease.

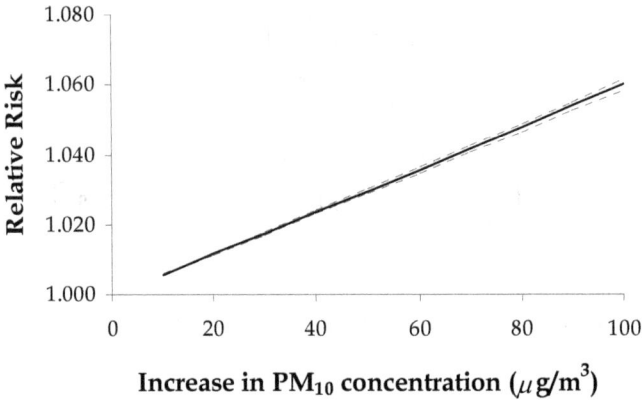

Fig. 4. Estimates of relative risk for the model considering the effect of three days lag and including residuals according to the increase in PM_{10} concentration (the dashed lines are the confidence interval).

6. Conclusion

Concluding, previous studies have not found yet a single model that can explain the impact of all kinds of air pollution on human health. Lipfert (1993) made a comparison among approximately 100 studies involving air pollution and demands for hospital services and concluded that this comparison is hampered due to the diversity encountered. The studies vary in design, diagnoses studied, air pollutant investigated; lag periods considered and the ways in which potentially confounding variables are controlled.

Lipfert (1993) also concluded that study designs have evolved considerably over the 40 years of published findings on this topic. The early studies tended to emphasize the need to limit the populations studied to those living near air pollution monitors, but more recent studies employed the concept of regional pooling, in which both hospitalization and air monitoring data are pooled over a large geographic area.

In this chapter it was emphasized times series studies appliance using Generalized Linear Models (GLM) with Poisson regression. This model is often used when response variables are countable, which demands less time then studies of follow-up kind.

The four steps to be followed to fit GLM with Poisson regression (database construction, temporal trends adjustment, goodness of fit analysis and results analysis) were applied to a case study comprising the PM_{10} concentration impact on the number of hospital admissions for respiratory diseases in Sao Paulo city from 2007 to 2008. The results showed that GLM with Poisson regression is useful as a tool for epidemiological studies of air pollution.

According to the case study, the model fitted well to the data as the pseudo R^2 statistic has shown good results (around $0.8 > 0.6$) for all adjustments. The models without residual inclusion for effects of PM_{10} concentration of the same day (no lag) to five days later (five lag days) showed regression coefficients statistically significant, but autocorrelations for one, two, three and four lag days was identified, suggesting a correlation between data from one day until four days later even after adding variables for day of the week and holiday confounders. So, the models that fitted well to the data were those with residuals inclusions for one, two, three and four lag days for effects of PM_{10} concentration for the same day (no lag) to three days later (three lag days), but the better fit was for the effect after three days of exposure according to Akaike Information Criterion (AIC) analysis that has shown the lowest AIC (6584.8).

In this way, an analysis of the relative risk (RR) for the model with residual inclusion and considering the effects of exposures after three days (three lag days) showed that the risk of someone get sick with a respiratory disease increases 5% as the concentration goes from 10 to 100 $\mu g / m^3$.

The results of the study showed that the risk of getting sick due to PM_{10} concentration can occur up to three days after the exposure and the more concentration, the higher the risk.

Finally, the steps to apply the GLM with Poisson regression to studies of air pollution impact on human health presented in this chapter had not been found in the literature and can be extended to all air pollutants and health outcomes.

7. Acknowledgment

This chapter was developed with financial support of CNPQ (Conselho Nacional de Desenvolvimento Científico e Tecnológico) and ANP (Agência Nacional do Petróleo).

8. References

Baxter, L.A.; Finch, S.J.; Lipfert, F.W. & Yu, Q. (1997). Comparing estimates of the effects of air pollution on human mortality obtained using different regression methodologies. *Risk Analysis*, Vol. 17, No. 3, pp. 273-278.

Bhattacharyya, G.K. & Johnson, R.A. (1977). *Statistical Concepts and Methods*, John Wiley & Sons, Inc., ISBN: 0471072044, USA.

Bickel, P.J. & Doksum, K.A. (2000). *Mathematical Statistics: Basic Ideas and Selected Topics. (Volume I)* (second edition), Pearson Prentice Hall, ISBN: 013850363X, USA.

Box, G.E.P.; Jenkins, G.M. & Reinsel, G.C. (1994). *Time Series Analysis: Forecasting and Control* (third edition), Prentice-Hall, ISBN: 0470272848, USA.

Braga, A. L. F.; Conceição, G. M. S.; Pereira, L. A. A.; Kishi, H. S.; Pereira, J. C. R.; Andrade, M. F.; Gonçalves, F. L. T.; Saldiva, P. H. N. & Latorre, M. R. D. O. (1999). Air pollution and pediatric respiratory hospital admissions in Sao Paulo, Brazil. *Journal of Occupational and Environmental Medicine*, Vol. 1, pp. 95-102.

Braga, A.L.F.; Saldiva, P.H.N.; Pereira, L.A.A.; Menezes, J.J.C.; Conceição, G.M.S.; Lin, C.A.; Zanobetti, A.; Schwartz, J. & Dockery, D.W. (2001). Health effects of air pollution exposure on children and adolescents in Sao Paulo, Brazil. *Pediatric Pulmonology*, Vol. 31, pp. 106-113.

Burnett, R.T.; Cakmak, S. & Brook, J.R. (1998). The effect of the urban ambient air pollution mix on daily mortality rates in 11 Canadian cities. *Canadian Journal of Public Health*, Vol. 89, pp. 152-156.

Cetesb – Environmental Company of Sao Paulo. June 5th, 2011. Available from: <http://www.cetesb.sp.gov.br/ar/qualidade-do-ar/32-qualar>: In Portuguese.

Chapra, S. C. & Canale, R. P. (1987). *Numerical Methods for Engineers with Personal Computer Applications* (second edition), McGraw-Hill International Editions, USA.

Dobson, A.J & Barnett, A.G. (2008). *An Introduction to Generalized Linear Models* (third edition), Chapman & Hall. ISBN: 1584889500, USA.

Dockery, D.W. & Pope III, C.A. (1994). Acute respiratory effects of particulate air pollution. *Annual Review of Public Health*, Vol. 15, pp. 107-132.

Dominici, F.; McDermott, A.; Zeger, S.L. & Samet, J.M. (2002). On the use of Generalized Additive Models in time-series studies of air pollution and health. *American Journal of Epidemiology*, Vol. 156, No. 3, pp. 193-203.

Dominici, F.; Sheppard, L. & Clyde, M. (2003). Health effects of air pollution: A statistical review. *International Statistical Review*, Vol. 71, No. 2, pp. 243-276.

Everitt B. S. (2003). *Modern Medical Statistics*, Oxford University Press Inc., 0340808691, USA.

Everitt, B.S. & Hothorn, T. (2010). *A Handbook of Statistical Analyses Using R* (second edition), Chapman & Hall, ISBN: 9781420079333, USA.

Faraway, J.J. *Practical Regression and Anova using R*. (1999), June 15th, 2006. Available from: <http://cran.r-project.org/doc/contrib/Faraway-PRA.pdf>.

Hastie, T.J. & Tibishirani, R.J. (1990). *Generalized Additive Models* (first edition), Chapman & Hall, ISBN: 0412343908, USA.

Ibald-Mulli, A.; Timonen, K. L.; Peters, A.; Heinrich, J.; Wölke, G.; Lanki, T.; Buzorius, G.; Kreyling, W. G.; Hartog, J.; Hoek, G.; Brink, H. M & Pekkanen, J. (2004). Effects of particulate air pollution on blood pressure and heart rate in subjects with cardiovascular disease: A multicenter approach. *Environmental Health Perspectives*, Vol. 112, No. 3, pp. 369-377.

Lipfert, F.W. (1993). A critical review of studies of the association between demands for hospital services and air pollution. *Environmental Health Perspectives Supplements*, Vol. 101, Suppl. 2, pp. 229-268.

McCullagh, P. & Nelder, J.A. (1989). *Generalized Linear Models* (second edition). Chapman & Hall. ISBN: 0412317605, USA.

Metzer, K.B.; Tolbert, P.E.; Klein, M.; Peel, J.L.; Flanders, W.D.; Todd, K.; Mulholland, J.A.; Ryan, P.B. & Frumkin, H. (2004). Ambient air pollution and cardiovascular emergency department visits. *Epidemiology*, Vol. 15, Iss. 1, pp. 46-56.

Mood, A.M.; Graybill, F.A. & Boes, D.C. (1974). *Introduction to the Theory of Statistics* (third edition), McGraw-Hill, ISBN: 0070428646, USA.

Myers, R.H. & Montgomery, D.C. (2002). *Response Surface Methodology: Process and Product Optimization Using Designed Experiments*, John Wiley & Sons, USA.

Nelder, J. A. & Wedderburn, R. W. M. (1972). Generalized linear models. *Journal of the Royal Statistical Society A*, Vol. 135, No. 2, pp. 370-384.

Paula, G. A. *Regression Models with Computational Support*. Sao Paulo: Math and Statistic Institute, Universidade de Sao Paulo (2004). January 25th, 2006. Available from: <http://www.ime.usp.br/~giapaula/livro.pdf>: In Portuguese.

Peng, R.D.; Dominici, F. & Louis, T.A. (2006). Model choice in time series studies of air pollution and mortality. *Journal of the Royal Statistical Society Series A*, Vol. 169, pp. 179-203.

Peters, A.; Dockery, D. W.; Muller, J. E. & Mittleman, M.A. (2001). Increased particulate air pollution and the triggering of myocardial infarction. *Circulation*, Vol. 103, pp. 2810-2815.

Pope III, C. A.; Burnett, R. T.; Thun, M. J.; Calle, E. E.; Krewski, D.; Ito, K. & Thurston, G. D. (2002). Lung cancer, cardiopulmonary mortality, and long-term exposure to fine particulate air pollution. *Journal of the American Medical Association*, Vol. 287, No. i9, pp. 1132-1142.

R Development Core Team. (2010). *R: a language and environment for statistical computing/* R version 2.12.0. The R Foundation for Statistical Computing.

Samet, J.M.; Zeger, S.L.; Domini, F.; Curriero, F.; Coursac, I.; Dockery, D.; Schwartz, J. & Zanobetti, A. (2000a). *The National Morbidity, Mortality, and Air Pollution Study, Part I, Methods and Methodological Issues (Report No. 94, I)*, Cambridge: Health Effects Institute.

Samet, J.M.; Zeger, S.L.; Dominici, F.; Curriero, F.; Coursac, I.; Dockery, D.; Schwartz, J. & Zanobetti, A. (2000b). *The National Morbidity, Mortality, and Air Pollution Study, Part II, Morbidity and Mortality from Air Pollution in the United States (Report No. 94, II)*, Cambridge: Health effects Institute.

Samoli, E.; Nastos, P.T.; Paliatsos, A.G.; Katsouyanni, K. & Priftis, K.N. (2011). Acute effects of air pollution on pediatric asthma exacerbation: evidence of association and effect modification. *Environmental Research*, Vol. 111, pp. 418-424.

Schwartz, J.; Spix, C.; Touloumi, G.; Bachárová, L.; Barumamdzadeh, T.; Tetre, A. Le; Piekarksi, T.; Ponce de Leon, A.; Pönkä, A.; Rossi, G.; Saez, M. & Schouten, J.P. (1996). Methodological issues in studies of air pollution and daily counts of deaths or hospital admissions. *Journal of Epidemiology and Community Health*, Vol. 50, Suppl. 1, pp. S3-S11.

Schwarze P.E.; Totlandsdal A.I.; Herseth J.I.; Holme J.A.; Lag M.; Refsnes M.; Øvrevik J.; Sandberg W.J. & Bølling A.K. (2010). Importance of sources and components of particulate air pollution for cardio-pulmonary inflammatory responses, In: *Villanyi, V. Intech*, p. 47-74.

SUS – Health System. June 6th, 2011, Available from: <http://www2.datasus.gov.br/DATASUS/index.php?area=0701&item=1&acao=11>: In Portuguese.

Tadano, Y.S. (2007) *Analysis of PM$_{10}$ Impact on Population's Health: Case Study in Araucaria, PR*, Parana, Brazil, 120p, Thesis (Master in Mechanical and Material Engineering), Federal University of Technology - Parana: In Portuguese.

Tadano, Y.S.; Ugaya, C.M.L. & Franco, A.T. (2009). Methodology to assess air pollution impact on the population's health using the Poisson regression method. *Ambiente & Sociedade*, Vol. XII, No. 2, pp. 241-255: In Portuguese.

Wang, P.; Puterman, M.L.; Cockburn, I. & Le, N. (1996). Mixed Poisson regression models with covariate dependent rates. *Biometrics*, Vol. 52, pp. 381-400.

13

Developing Neural Networks to Investigate Relationships Between Air Quality and Quality of Life Indicators

Kyriaki Kitikidou and Lazaros Iliadis
Democritus University of Thrace,
Department of Forestry and
Management of the Environment
and Natural Resources, Orestiada
Greece

1. Introduction

Quality of life (QOL) is an integral outcome measure in the management of diseases. It can be used to assess the results of different management methods, in relation to disease complications and in fine-tuning management methods (Koller & Lorenz, 2003). Quantitative analysis of quality of life across countries, and the construction of summary indices for such analyses have been of interest for some time (Slottje et al., 1991). Most early work focused on largely single dimensional analysis based on such indicators as per capita GDP, the literacy rate, and mortality rates. Maasoumi (1998) and others called for a multidimensional quantitative study of welfare and quality of life. The argument is that welfare is made up of several distinct dimensions, which cannot all be monetized, and heterogeneity complications are best accommodated in multidimensional analysis. Hirschberg et al. (1991) and Hirschberg et al. (1998) identified similar indicators, and collected them into distinct clusters which could represent the dimensions worthy of distinct treatment in multidimensional frameworks.

In this research effort we have considered the role of air quality indicators in the context of economic and welfare life quality indicators, using artificial neural networks (ANN). Therefore in this presentation we have obtained the key variables (life expectancy, healthy life years, infant mortality, Gross Domestic Product (GDP) and GDP growth rate) and developed a neural network model to predict the air quality outcomes (emissions of sulphur and nitrogen oxides). Sustainability and quality of life indicators have been proposed recently by Flynn et al. (2002) and life quality indices have been used to estimate willingness to pay (Pandey & Nathwani, 2004). The innovative part of this research effort lies in the use of a soft computing machine learning approach like the ANN to predict air quality. In this way, we introduce the reader to a technique that allows the comparison of various attributes that impact the quality of life in a meaningful way.

2. Materials and methods

It is well known that the quality of the air in a locale influences the health of the population and ultimately affects other dimensions of that population's welfare and its economy. As a simple example, in cities where pollution levels rise significantly in the summer, worker absenteeism rates rise commensurately and productivity is adversely impacted. Other dimensions of the economy are influenced on "high pollution days" as well. For example, when outdoor leisure activity is restricted this may have serious consequences for the service sector of the economy (Bresnahan et al., 1997). In this chapter, we have introduced two measures of environmental quality or air quality as quality of life factors. A feature of these indices is the fact that these types of pollution are created by some of the very activities that define economic development. The two factors under investigation here are sulfur oxides (SOx) and nitrogen oxides (NOx) (million tones of SO_2 and NO_2 equivalent, respectively). Sulphur oxides, including sulphur dioxide and sulphur trioxide, are reported as sulphur dioxide equivalent, while nitrogen oxides, including nitric oxide and nitrogen dioxide, are reported as nitrogen dioxide equivalent. They are both produced as byproducts of fuel consumption as in case of the generation of electricity. Vehicle engines also produce a large proportion of NOx. SOx is primarily produced when high sulphur content coal is burned which is usually in large-scale industrial processes and power generation. Thus, the ratio of these emissions to the population is an indication of pollution control.

The following attributes of QOL have been used:

- Life expectancy at birth: The mean number of years that a newborn child can expect to live if subjected throughout his life to the current mortality conditions (age specific probabilities of dying).
- Healthy life years: The indicator Healthy Life Years (HLY) at birth measures the number of years that a person at birth is still expected to live in a healthy condition. HLY is a health expectancy indicator which combines information on mortality and morbidity. The data required are the age-specific prevalence (proportions) of the population in healthy and unhealthy conditions and age-specific mortality information. A healthy condition is defined by the absence of limitations in functioning/disability. The indicator is also called disability-free life expectancy (DFLE). Life expectancy at birth is defined as the mean number of years still to be lived by a person at birth, if subjected throughout the rest of his or her life to the current mortality conditions (WHO, 2010).
- Infant mortality: The ratio of the number of deaths of children under one year of age during the year to the number of live births in that year. The value is expressed per 1000 live births.
  - Gross Domestic Product (GDP) per capita: GDP is a measure of the economic activity, defined as the value of all goods and services produced less the value of any goods or services used in their creation. These amounts are expressed in PPS (Purchasing Power Standards), i.e. a common currency that eliminates the differences in price levels between countries allowing meaningful volume comparisons of GDP between countries.

- GDP growth rate: The calculation of the annual growth rate of GDP volume is intended to allow comparisons of the dynamics of economic development both over time and between economies of different sizes. For measuring the growth rate of GDP in terms of volumes, the GDP at current prices are valued in the prices of the previous year and the thus computed volume changes are imposed on the level of a reference year; this is called a chain-linked series. Accordingly, price movements will not inflate the growth rate.

Data were extracted for 34 European countries, for the year 2005, from the Eurostat database (Eurostat, 2010). Descriptive statistics for all variables are given in **Table 1**.

Statistics	Emissions of sulphur oxides (million tones of SO_2 equivalent)	Emissions of nitrogen oxides (million tones of NO_2 equivalent)	Infant Mortality	GDP (Purchasing Power Standards, PPS)	GDP Growth Rate	Life Expectancy At Birth (years)	Healthy Life Years
Valid N*	34	34	34	33	33	33	27
Missing**	0	0	0	1	1	1	7
Mean	0.503	0.372	5.721	95.921	4.206	77.535	60.448
Std. Deviation	0.648	0.482	4.227	46.620	2.521	3.244	5.443
Min	0.00	0.00	2.30	28.50	0.70	70.94	50.10
Max	2.37	1.63	23.60	254.50	10.60	81.54	69.30

*Number of observations (countries) for each variable.
**Number of countries that didn't had available data.

Table 1. Descriptive statistics for all variables used in the analysis.

For the performance of the analyses, multi-layer perceptron (MLP) and radial-basis function (RBF) network models were developed under the SPSS v.19 statistical package (IBM, 2010). We specified that the relative number of cases assigned to the training:testing:holdout samples should be 6:2:1. This assigned 2/3 of the cases to training, 2/9 to testing, and 1/9 to holdout. For the MLP network we employed the back propagation (BP) optimization algorithm. As it is well known in BP the weighted sum of inputs and bias term are passed to the activation level through the transfer function to produce the output (Bishop, 1995; Fine, 1999; Haykin, 1998; Ripley, 1996). The sigmoid transfer function was employed (Callan, 1999; Kecman, 2001), due to the fact that the algorithm requires a response function with a continuous, single valued with first derivative existence (Picton, 2000).

Before using the input data records to the ANN a normalization process took place so that the values with wide range do not prevail over the rest. The autoscaling approach was applied. This method outputs a zero mean and unit variance of any descriptor variable (Dogra, Shaillay, 2010). Thus, each feature's values were normalized based on the following equation:

$$Z_i=(X_i-\mu_i)/\sigma_i$$

where X_i was the ith parameter, Z_i was the scaled variable following a normal distribution and σ_i, μ_i were the standard deviation and the mean value of the ith parameter.

These networks were trained in an iterative process. A single hidden sub layer architecture was followed in order to reduce the complexity of the network, and increase the computational efficiency (Haykin, 1998). Two units were chosen in the hidden layer. The schematic representation of the neural network is given in **Fig. 1**.

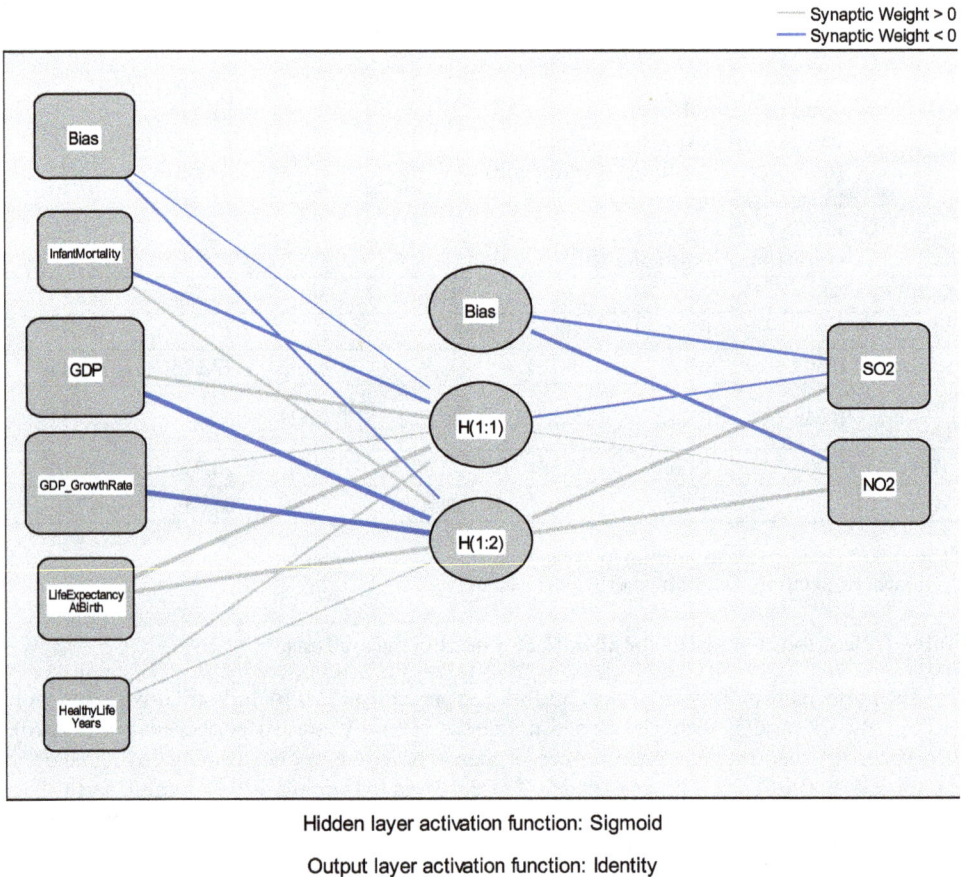

Hidden layer activation function: Sigmoid

Output layer activation function: Identity

Fig. 1. Multi-layer perceptron network structure.

As regards the RBF network (Bishop, 1995; Haykin, 1998; Ripley, 1996; Tao, 1993; Uykan et al., 2000), the architecture that was developed included nine neurons in the hidden layer. The transfer functions (hidden layer activation functions and output function) determine the output by depicting the result of the distance function (Bors & Pitas, 2001; Iliadis, 2007). The schematic representation of the neural network with transfer functions is given in **Fig. 2**.

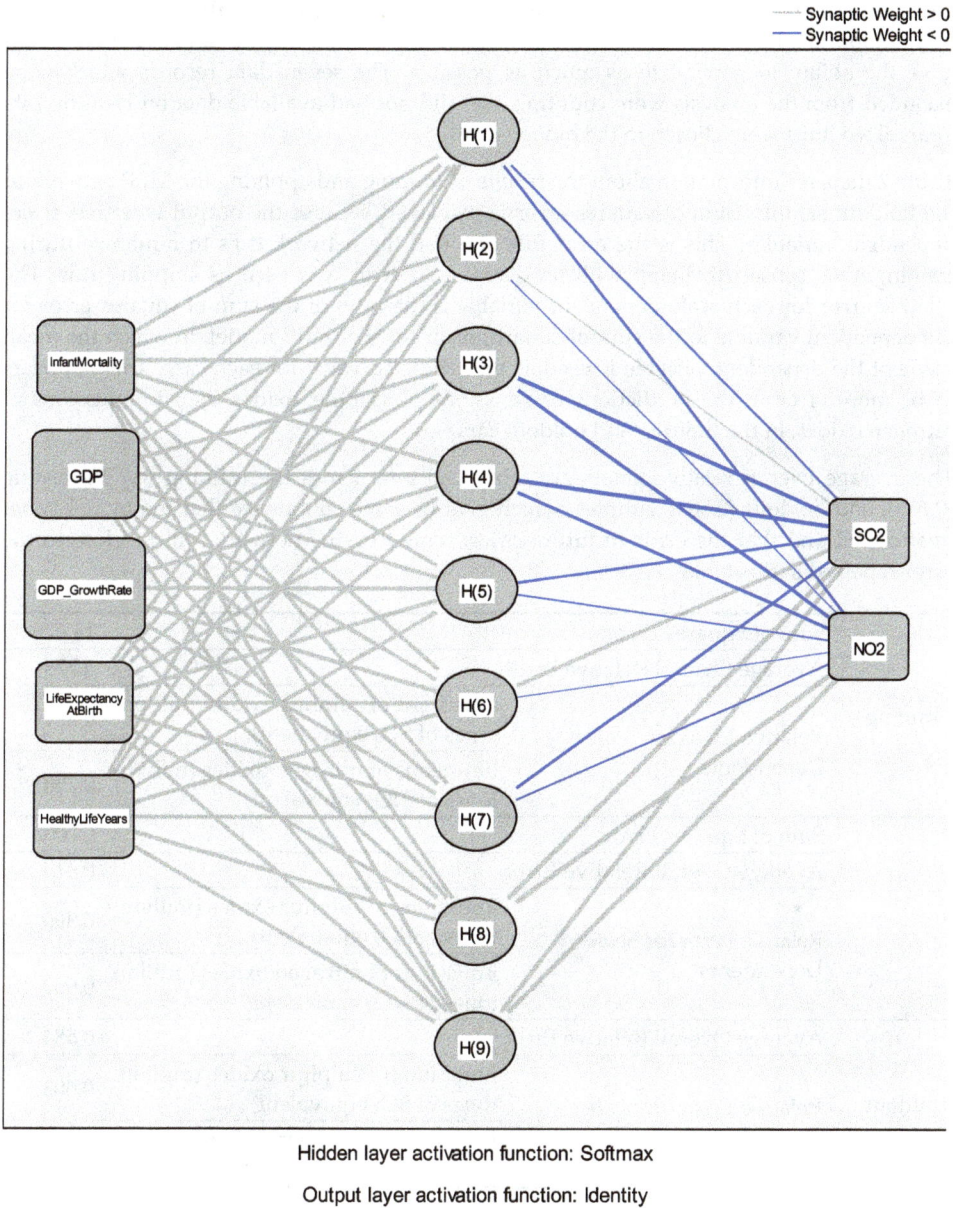

Fig. 2. Radial-basis function network structure.

3. Results – Discussion

From the MLP analysis, 19 cases (70.4%) were assigned to the training sample, 2 (7.4%) to the testing sample, and 6 (22.2%) to the holdout sample. The choice of the records was done in a random manner. The whole effort targeted in the development of an ANN that would have the ability to generalize as much as possible. The seven data records which were excluded from the analysis were countries that did not had available data on Healthy Life Years. Two units were chosen in the hidden layer.

Table 2 displays information about the results of training and applying the MLP network to the holdout sample. Sum-of-squares error is displayed because the output layer has scale-dependent variables. This is the error function that the network tries to minimize during training. One consecutive step with no decrease in error was used as stopping rule. The relative error for each scale-dependent variable is the ratio of the sum-of-squares error for the dependent variable to the sum-of-squares error for the "null" model, in which the mean value of the dependent variable is used as the predicted value for each case. There appears to be more error in the predictions of emissions of sulphur oxides than in emissions of nitrogen oxides, in the training and holdout samples.

The average overall relative errors are fairly constant across the training (0.779), testing (0.615), and holdout (0.584) samples, which give us some confidence that the model is not overtrained and that the error in future cases, scored by the network will be close to the error reported in this table

Training	Sum of Squares Error		14.029
	Average Overall Relative Error		0.779
	Relative Error for Scale Dependents	Emissions of sulphur oxides (million tones of SO_2 equivalent)	0.821
		Emissions of nitrogen oxides (million tones of NO_2 equivalent)	0.738
Testing	Sum of Squares Error		0.009
	Average Overall Relative Error		0.615
	Relative Error for Scale Dependents	Emissions of sulphur oxides (million tones of SO_2 equivalent)	0.390
		Emissions of nitrogen oxides (million tones of NO_2 equivalent)	0.902
Holdout	Average Overall Relative Error		0.584
	Relative Error for Scale Dependents	Emissions of sulphur oxides (million tones of SO_2 equivalent)	0.603
		Emissions of nitrogen oxides (million tones of NO_2 equivalent)	0.568

Table 2. MLP Model Summary.

In the following **Table 3** parameter estimates for input and output layer, with their corresponding biases, are given.

Predictor		Predicted			
		Hidden Layer 1		Output Layer	
		H(1:1)	H(1:2)	SO$_2$	NO$_2$
Input Layer	(Bias)	-0.119	-0.537		
	Infant Mortality	-0.805	0.752		
	GDP	1.033	-3.377		
	GDP Growth Rate	0.318	-3.767		
	Life Expectancy At Birth	1.646	1.226		
	Healthy Life Years	0.567	0.358		
Hidden Layer 1	(Bias)			-0.635	-0.877
	H(1:1)			-0.518	0.116
	H(1:2)			1.396	1.395

Table 3. MLP Parameter Estimates.

Linear regression between observed and predicted values ($SO_2 = a + b\hat{SO_2} + error$, $NO_2 = a + b\hat{NO_2} + error$) showed that the MLP network does a reasonably good job of predicting emissions of sulphur and nitrogen oxides. Ideally, linear regression parameters a and b should have values 0 and 1, respectively, while values of the observed-by-predicted chart should lie roughly along a straight line. Linear regression gave results for the two output variables $SO_2 = 0.114 + 0.918\hat{SO_2} + error$ (**Fig. 3**) and $NO_2 = 0.005 + 1.049\hat{NO_2} + error$ (**Fig. 4**), respectively. There appears to be more error in the predictions of emissions of sulphur oxides than in emissions of nitrogen oxides, something that we also pointed out in Table 2. **Figs 3 and 4** actually seem to suggest that the largest errors of the ANN are overestimations of the target values.

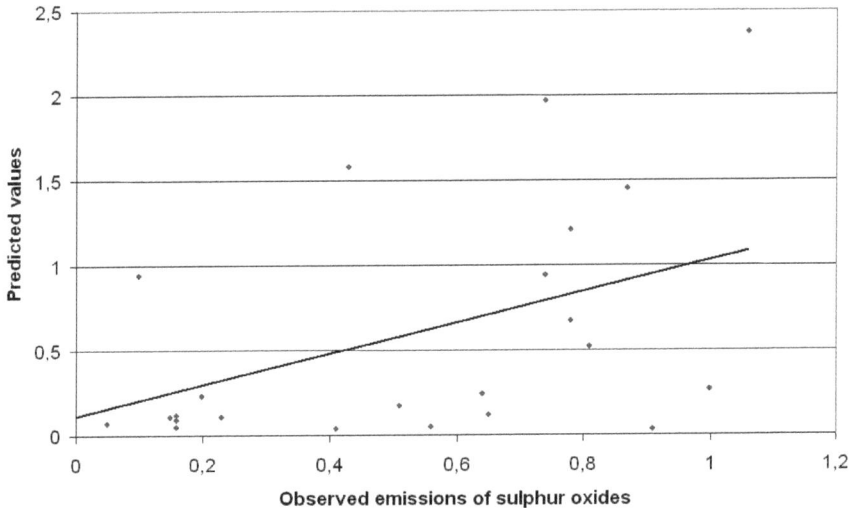

Fig. 3. Linear regression of observed values for emissions of sulphur oxides by predicted values of MLP.

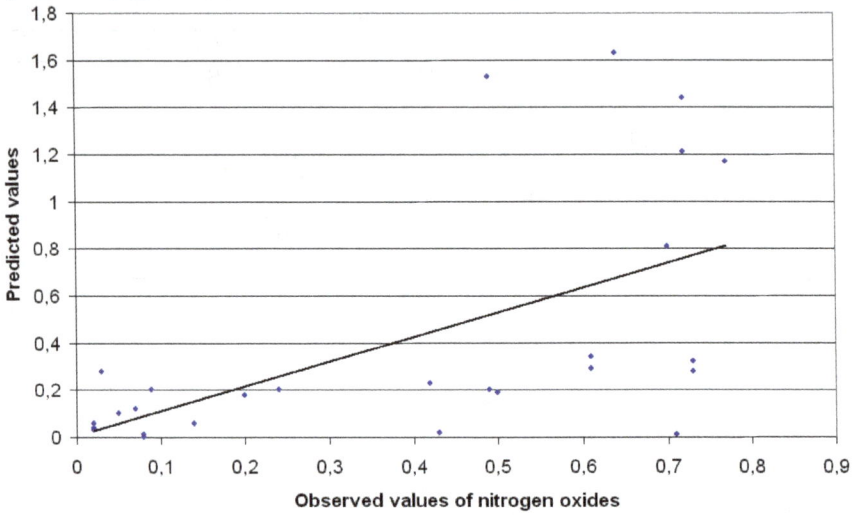

Fig. 4. Linear regression of observed values for emissions of nitrogen oxides by predicted values of MLP.

The importance of an independent variable is a measure of how much the network's model-predicted value changes for different values of the independent variable. A sensitivity analysis to compute the importance of each predictor is applied. The importance chart (**Fig. 5**) shows

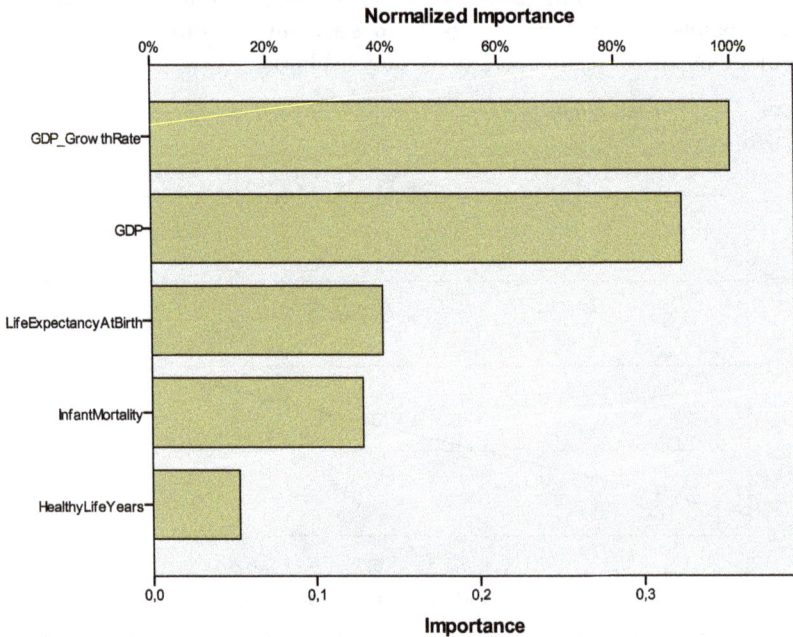

Fig. 5. MLP independent variable importance chart.

that the results are dominated by GDP growth rate and GDP (strictly economical QOL indicators), followed distantly by other predictors.

From the RBF analysis, 19 cases (70.4%) were assigned to the training sample, 1 (3.7%) to the testing sample, and 7 (25.9%) to the holdout sample. The seven data records which were excluded from the MLP analysis were excluded from the RBF analysis also, for the same reason.

Table 4 displays the corresponding information from the RBF network. There appears to be more error in the predictions of emissions of sulphur oxides than in emissions of nitrogen oxides, in the training and holdout samples.

The difference between the average overall relative errors of the training (0.132), and holdout (1.325) samples, must be due to the small data set available, which naturally limits the possible degree of complexity of the model (Dendek & Mańdziuk, 2008).

Training	Sum of Squares Error		2.372
	Average Overall Relative Error		0.132
	Relative Error for Scale Dependents	Emissions of sulphur oxides (million tones of SO_2 equivalent)	0.161
		Emissions of nitrogen oxides (million tones of NO_2 equivalent)	0.103
Testing	Sum of Squares Error		0.081
	Average Overall Relative Error		a
	Relative Error for Scale Dependents	Emissions of sulphur oxides (million tones of SO_2 equivalent)	a
		Emissions of nitrogen oxides (million tones of NO_2 equivalent)	a
Holdout	Average Overall Relative Error		1.325
	Relative Error for Scale Dependents	Emissions of sulphur oxides (million tones of SO_2 equivalent)	1.347
		Emissions of nitrogen oxides (million tones of NO_2 equivalent)	1.267

[a]Cannot be computed. The dependent variable may be constant in the training sample.

Table 4. RBF Model Summary.

In Table 5 parameter estimates for input and output layer are given for the RBF network.

Predictor		Predicted										
		Hidden layer									Output layer	
		H(1)	H(2)	H(3)	H(4)	H(5)	H(6)	H(7)	H(8)	H(9)	SO_2	NO_2
Input Layer	Infant Mortality	1.708	1.517	-1.064	-1.279	-0.562	-0.491	-0.276	-0.204	-0.132		
	GDP	-1.092	-0.986	3.098	0.451	0.667	-0.714	-0.101	-0.076	0.161		
	GDP Growth Rate	1.572	-0.164	0.575	1.390	-0.448	0.924	-0.544	-0.720	-1.212		
	Life Expectancy At Birth	-1.710	-1.640	0.500	1.169	0.578	-0.611	0.211	0.820	0.461		
	Healthy Life Years	-1.245	-1.223	0.497	1.161	1.123	-0.111	-0.346	0.868	-0.831		
Hidden Unit Width		0,606	0.363	0.363	0.363	0.645	0.363	0.576	0.363	0.363		
Hidden Layer	H(1)										-0.552	-0.668
	H(2)										0.463	-0.395
	H(3)										-0.813	-0.773
	H(4)										-0.833	-0.795
	H(5)										-0.617	-0.401
	H(6)										0.970	-0.253
	H(7)										-0.718	-0.547
	H(8)										3.116	3.429
	H(9)										2.698	2.790

Table 5. RBF Parameter Estimates.

Linear regression between observed and predicted values ($SO_2 = a + b\hat{SO_2} + error$, $NO_2 = a + b\hat{NO_2} + error$) showed that the RBF network does also a reasonably good job of predicting emissions of sulphur and nitrogen oxides. Linear regression gave results for the two output variables $SO_2 = -0.0114 + 0.8583\hat{SO_2} + error$ (**Fig. 6**) and $NO_2 = -0.026 + 0.7932\hat{NO_2} + error$ (**Fig. 7**), respectively. In this case, it is difficult to see if there is more error in the predictions of emissions of sulphur or nitrogen oxides.

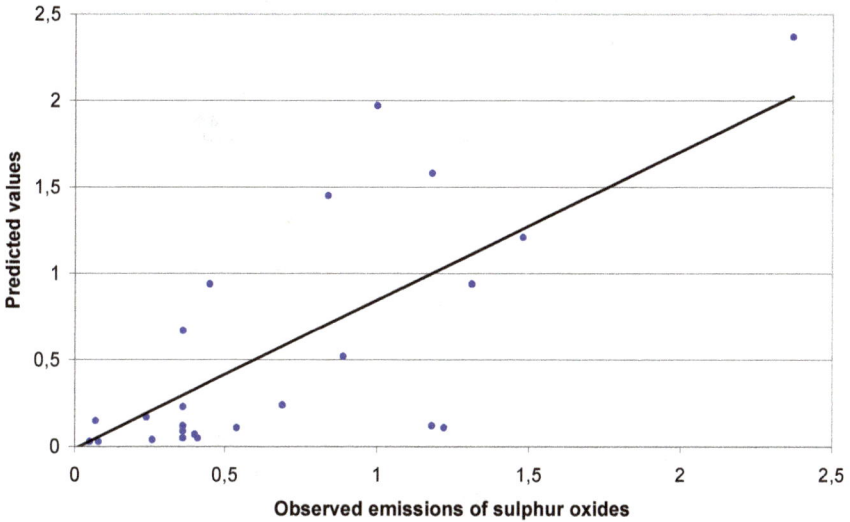

Fig. 6. Linear regression of observed values for emissions of sulphur oxides by predicted values of RBF.

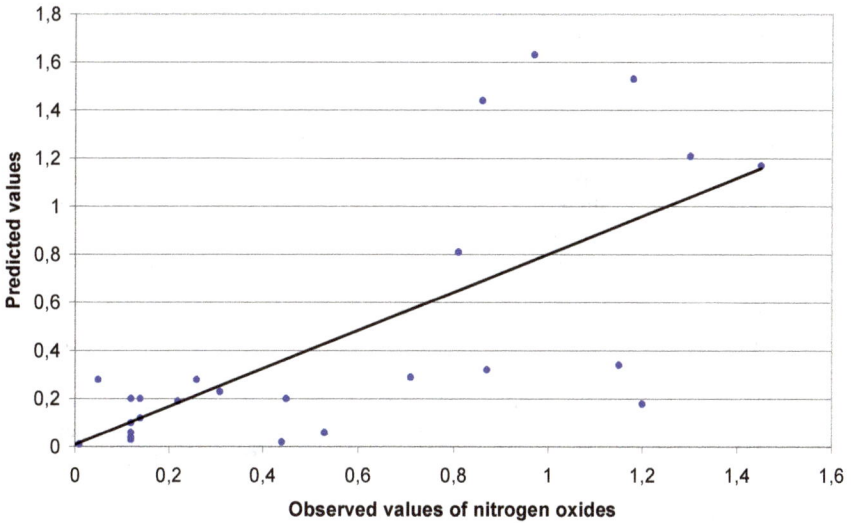

Fig. 7. Linear regression of observed values for emissions of nitrogen oxides by predicted values of RBF.

Finally, the importance chart for the RBF network (**Fig. 8**) shows that, once again, GDP growth rate and GDP are the most important predictors of sulphur and nitrogen oxides emissions.

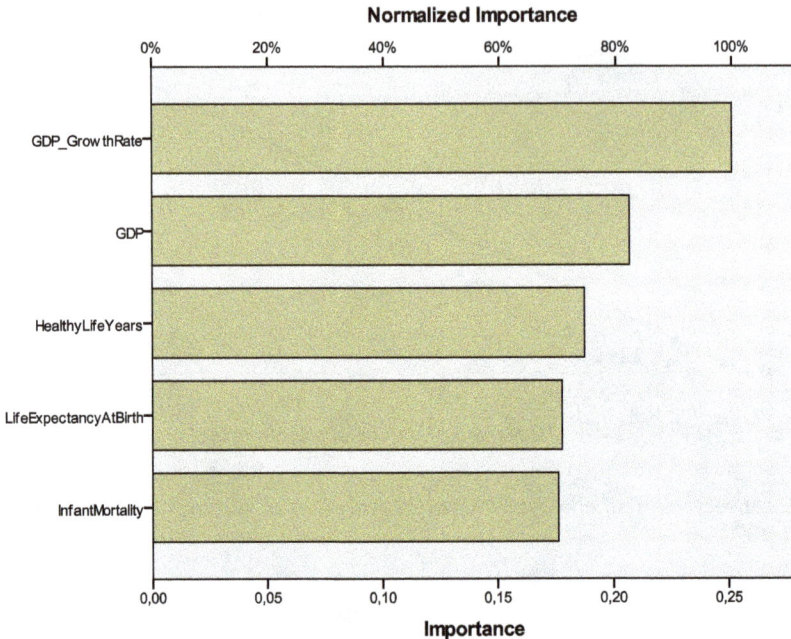

Fig. 8. RBF Independent variable importance chart.

4. Conclusions

The multi-layer perceptron and radial-basis function neural network models, that were trained to predict air quality indicators, using life quality and welfare indicators, appear to perform reasonably well. Unlike traditional statistical methods, the neural network models provide dynamic output as further data is fed to it, while they do not require performing and analyzing sophisticated statistical methods (Narasinga Rao et al., 2010).

Results showed that GDP growth rate and GDP influenced mainly air quality predictions, while life expectancy, infant mortality and healthy life years followed distantly. One possible way to ameliorate performance of the network would be to create multiple networks. One network would predict the country result, perhaps simply whether the country increased emissions or not, and then separate networks would predict emissions conditional on whether the country increased emissions. We could then combine the network results to likely obtain better predictions. Note also that neural network is open ended; as more data is given to the model, the prediction would become more reliable. Overall, we find that predictors that include economic indices may be employed by investigators to represent dimensions of air quality that include, as well as go beyond, these simple indices.

5. References

Bishop, C. (1995). *Neural Networks for Pattern Recognition, 3rd ed.* Oxford University Press, Oxford.

Bors, A., Pitas, I. (2001). Radial Basis function networks In: Howlett, R., Jain, L (eds.). *Recent Developments in Theory and Applications in Robust RBF Networks*, 125-153 Heidelberg, NY, Physica-Verlag.

Bresnahan, B., Mark, D., Shelby, G. (1997). Averting behavior and urban air pollution. *Land Economics* 73, 340–357.

Callan, R. (1999). *The Essence of Neural Networks.* Prentice Hall, UK .

Dendek, C., Mańdziuk, J. (2008). *Improving Performance of a Binary Classifier by Training Set Selection.* Warsaw University of Technology, Faculty of Mathematics and Information Science, Warsaw, Poland.

Dogra, Shaillay, K. (2010). *Autoscaling.* QSARWorld - A Strand Life Sciences Web Resource. http://www.qsarworld.com/qsar-statistics-autoscaling.php

Eurostat. (2010). http://epp.eurostat.ec.europa.eu.

Fine, T. (1999). *Feedforward Neural Network Methodology, 3rd ed.* Springer-Verlag, New York.

Flynn P., Berry D., Heintz T. (2002). Sustainability & Quality of life indicators: Towards the Integration of Economic, Social and Environmental Measures. *The Journal of Social Health* 1(4), 19-39.

Haykin, S. (1998). *Neural Networks: A Comprehensive Foundation, 2nd ed.* Prentice Hall, UK.

Hirschberg, J., Esfandiar, M., Slottje, D. (1991). Cluster analysis for measuring welfare and quality of life across countries. *Journal of Econometrics* 50, 131–150.

Hirschberg, J., Maasoumi, E., Slottje, D. (1998). *A cluster analysis the quality of life in the United States over time.* Department of Economics research paper #596, University of Melbourne, Parkville, Australia.

IBM. (2010). SPSS Neural Networks 19. SPSS Inc, USA.

Iliadis, L. (2007). *Intelligent Information Systems and Applications in Risk Management.* Stamoulis editions, Thessaloniki, Greece.

Kecman, V. (2001). *Learning and Soft Computing.* MIT Press, London.

Koller, M., Lorenz, W. (2003). Survival of the quality of life concept. *British Journal of Surgery* 90(10), 1175-1177.

Maasoumi, E. (1998). Multidimensional approaches to welfare. In: Silber, L. (ed.). *Income Inequality Measurement: From Theory to Practice.* Kluwer, New York.

Pandey, M., Nathwani, J. (2004). Life quality index for the estimation of social willingness to pay for safety. *Structural Safety* 26(2), 181-199.

Picton, P. (2000). *Neural Networks, 2nd ed.* Palgrave, New York.

Narasinga Rao, M., Sridhar, G., Madhu, K., Appa Rao, A. (2010). A clinical decision support system using multi-layer perceptron neural network to predict quality of life in diabetes. *Diabetes & Metabolic Syndrome: Clinical Research & Reviews* 4, 57–59.

Ripley, B. (1996). *Pattern Recognition and Neural Networks.* Cambridge University Press, Cambridge.

Slottje, D., Scully, G., Hirschberg, J., Hayes, K. (1991). *Measuring the Quality of Life Across Countries: A Multidimensional Analysis.* Westview Press, Boulder, CO.

Tao, K. (1993). A closer look at the radial basis function (RBF) networks. In: Singh, A. (ed.). *Conference Record of the Twenty-Seventh Asilomar Conference on Signals, Systems, and Computers.* IEEE Computational Society Press, Los Alamitos, California .

Uykan, Z., Guzelis, C., Celebi, M., Koivo, H. (2000). *Analysis of input-output clustering for determining centers of RBFN.* IEEE Transactions on Neural Networks 11, 851-858.

WHO: World Health Organization. (2010). http://www.who.int.

14

Air Pollution and Health Effects in Children

Yungling Leo Lee[1] and Guang-Hui Dong[2]
[1]Institute of Epidemiology and Preventive Medicine,
College of Public Health, National Taiwan University, Taipei
[2]Department of Biostatistics and Epidemiology and
Department of Environmental and Occupational,
School of Public Health, China Medical University, Shenyang
[1]Taiwan
[2]PR China

1. Introduction

Asthma, which is characterized by reversible airway obstruction and inflammation, is the most frequent chronic disease in children. Recent global studies showed, although the rising trends in occurrence of allergic diseases in children appear to have leveled off or even reversed in many developed countries, while in many developing countries, the increase in prevalence has only recently started (Asher et al. 2006; Bjorksten et al. 2008; Pearce et al. 2007). These findings indicate that the global burden of asthma is continuing to be rising with the global prevalence differences lessening (Pearce et al. 2007). It has been suggested that this increase can not be explained by genetic factors and improvements in diagnostic methods alone. Comparative studies of the population of the same ethnic background living in different environments revealed important environmental risk factors for asthma, especially for ambient air pollution, which may play an important role in the development of asthma symptoms (Asher et al. 2010; Berhane et al. 2011; Samoli et al. 2011; McConnell et al. 2010). A systematic review concluded that, although weather, pollen, and environmental tobacco smoke (ETS) are important risk factors for asthma, each has been found to act independently of air pollution, and thus they do not explain the association between air pollution and asthma (Leikauf 2002).

Comparing with the adults, children appear to be most vulnerable to the harmful effects of air pollutants exposure due to their stage of physical growth, immature immune system, and development of lung function with increased permeability of the respiratory epithelium (Dietert et al. 2000). Children also inhale a higher volume of air per body weight compared with adults (Oyana and Rivers 2005), delivering higher doses of different compositions that may remain in the lung for a longer duration (Bateson and Schwartz 2008). Another source of increased sensitivity of children to air pollution may be the qualitative and quantitative differences in the respiratory, immune, endocrine, and nervous systems during stages of rapid growth and development (Selgrade et al. 2006). Although epidemiologic studies have shown associations between asthma outcomes and routinely monitored air pollutants including particulate matter (PM), sulfur dioxide (SO_2), nitrogen oxides (NO_2) and ozone (O_3), the contribution of ambient air pollution exposure to children's asthma seems to vary in different parts of the world, perhaps because of the difference in spatial and temporal variability of air pollutants sources and composition between different regions (Kim et al. 2005; Kim et al. 2008; Levy et al. 2010), and also, it still remains unclear whether the association of the pollutants with increased asthma prevalence in children is independent of each other or is attributable to other toxic air pollutants, such as combustion-related organic compounds that are not routinely monitored (Delfino et al. 2006).

The purpose of this chapter is to review/summarize the evidence regarding the link between air pollution and children's asthma and asthma related symptoms. We would like to discuss gender difference for air pollution effects, as well as some other health effect susceptibilities, such as age, genetic predisposition, and disease susceptibility. Finally, the potential health effects from preventive strategies for air pollution would be also proposed.

2. Air pollution and asthma in children

2.1 Particulate Matter (PM)

Numerous studies worldwide have reported the associations between particulate matter (PM) and asthma outcomes (Berhane et al. 2011; McConnell et al. 2010; Özkaynak et al. 1987; Pope 1991; Samoli et al. 2011; Spira-Cohen et al. 2011), and most of those studies focused on fine particulate matter < 2.5 μm in average aerodynamic diameter ($PM_{2.5}$) and thoracic particulate matter < 10 μm in average aerodynamic diameter (PM_{10}) that dominate concentrations of fine particles in most urban areas. The health effects of the different particle sizes may differ between various areas as a result of variation in particle origin, chemical composition and biogenic content. The distinction of health effects between the fine and coarse fractions is important to define because these particles may arise from different sources (Kim et al. 2005; Kim et al. 2008; Levy et al. 2010). The coarse fraction of relatively large inhalable particles, ranging from 2.5 to 10 μm in average aerodynamic diameter ($PM_{10-2.5}$), generally consists of particles which are mainly derived from re-entrained road dust (containing soil particles, engine oil, metals, tire particles, sulfates and nitrates), construction and wind-blown dusts (mostly soil particles), and mechanical wear processes. The chemical signatures of $PM_{2.5}$ contains a mixture of particles including carbonaceous material like soot (with possibly adsorbed reactive metals and organic compounds), and secondary aerosols like acid condensates, and sulphate and nitrate particles, and is derived direct or indirect primarily from combustion of fossil fuels used in power generation, and industry, and automobile, and ultrafine particles produced by traffic, coal combustion, and metal, oil, and chemical manufacturing. This fine-mode faction has the largest surface area and contributes also to the total particle mass concentration (Bree and Cassee 2000; Özkaynak et al. 1987; Pope 1991). Based on the previous epidemiologic studies, it was believed that $PM_{2.5}$ fraction of PM_{10} had stronger effects than the $PM_{10-2.5}$ fraction (Dreher 1996; Schwartz and Neas 2000; Tzivian 2011). However, recent analyses are somewhat more ambiguous, and inconsistent results have been reported on the association between the coarse particle fraction and asthma outcomes. Several studies compared the effects of $PM_{10-2.5}$, $PM_{2.5}$, and PM_{10} on hospitalizations of children for asthma (Dreher et al. 1996; Lin et al. 2002; Schwartz and Neas 2000; Strickland et al. 2010).

A study conducted in 41 Metropolitan Atlanta hospitals indicated that the emergency department visits for asthma or wheeze among children was significantly associated only with $PM_{2.5}$ in cold season, however, only with PM_{10} and $PM_{10-2.5}$ in warm season (Strickland et al. 2010). Lin et al. (2002) found an effect for $PM_{10-2.5}$ but not for $PM_{2.5}$ and PM_{10} after adjustment for daily weather conditions. The study in Toronto also found the $PM_{10-2.5}$ fraction to be a better predictor of asthma admissions than the $PM_{2.5}$ and PM_{10} fractions in subjects of all ages. The estimated relative risks for $PM_{10-2.5}$, $PM_{2.5}$, and PM_{10} corresponding to an increase of 10 $\mu g/m^3$ with a 3-day averaging were 1.04, 1.01, and 1.01, respectively (Burnett et al. 1999). Whereas, Tecer et al. (2008) found that $PM_{2.5}$ had larger effects than did $PM_{10-2.5}$, despite that both were significantly associated with hospitalizations for asthma after adjustment for daily weather conditions. Also, a recent study found strong adverse effects of exposure to PM_{10} on emergency hospital admissions for asthma-related diagnosis among children, although analyses for $PM_{10-2.5}$ and $PM_{2.5}$ were not evaluated (Samoli et al. 2011). Besides the emergency hospital admissions for asthma mentioned above, the health effects of the different particle sizes on other asthma outcomes are also inconsistent. In the FACES cohort study of asthmatic children conducted in California of USA, wheeze was associated significantly with $PM_{10-2.5}$ but not with $PM_{2.5}$ (Mann et al. 2010). However, the results from France study conflict in this regard, no association between wheeze episodes or prevalence and PM <13 μm in aerodynamic diameter or black smoke in either the winter time or warm seasons among asthmatic children residing in Paris (Just et al. 2002). Results from a cohort study of school children showed that both $PM_{2.5}$ and PM_{10} were significantly associated with airway inflammation independent of asthma and allergy status, with PM_{10} effects significantly higher in the warm season, however, for the exhaled nitric oxide fraction (FeNO), the marker of important aspects of airway inflammation, the effects of $PM_{2.5}$ was higher than that of PM_{10} (Berhane et al. 2011). However, the pathophysiologic mechanism by which particles exert their health effects has not been well established. Reasons may include: (a) Variability in population characteristics and natural systems; Because populations differ markedly with respect to age, underlying health, and concomitant exposure to other insults, it is also reasonable to anticipate that different populations exposed to air pollution in different places may show a range of distinct responses. (b) Variability in the complex mixture of fine particles with a different level and composition over time and space.

Air pollution is a complex mixture whose level and composition varies over time and space according to a large number of factors ranging in scale from personal microenvironments to regional climate. Such differences in the nature of air pollution are among the possible explanations for the divergent findings about differ size of particles. For example, studies on mice and human airway cells have found that, compared with particles from wood smoke or car exhaust, particles from diesel exhaust have higher capacity to induce proallergenic Th2 cytokine production, increased major histocompatibility complex class II expression, and increased inflammatory cell proliferation (Porter et al. 2007; Samuelsen et al. 2008). Studies have also examined the relative effects of organic and inorganic fractions of particulate matter but have yielded conflicting results. In one, exposure of mice to the carbon core fraction of diesel exhaust particles (DEP) stimulated greater airway hyper-reactivity compared with the organic fraction (Inoue et al. 2007). In another, both organic and elemental carbon fractions of fine and ultrafine ambient particles were capable of stimulating proinflammatory allergic immune responses, as measured by increased secretion of Th2 cytokines and increased infiltration of eosinophils and polymorphonuclear leukocytes (Kleinman et al. 2007). The U.S. Environmental Protection Agency (EPA) has noted that $PM_{10-2.5}$ deposited in the upper airways may be more relevant for asthmatic responses and irritation (U.S. EPA 1995). Whereas, lower respiratory symptoms including respiratory infection may be the results of effects in the deep lung related to deposition of

PM$_{2.5}$ (Loomis 2000). While investigation of the health effects of particulate matter should continue, it would be unwise to focus too narrowly to the exclusion of other components of air pollution. The preoccupation with the smallest particles may be due in part to the tendency to look where the light is: although most researchers live in North America and western Europe and most previous studies on the health effects of air pollution have been conducted in those areas, many of the world's largest cities and much of the population currently exposed to high levels of air pollution are located in developing countries. For public health purposes, it is clearly important to examine the health effects of air pollution in these areas. Research in these settings may also yield far-reaching scientific benefits by allowing the effects of air pollution to be understood in a context more representative of the world's climate and health.

2.2 Sulfur dioxide (SO$_2$)

Sulfur dioxide (SO$_2$), which is commonly from combustion of fuel containing sulfur-mostly coal and oil, metal selting, and other industrial process, is a highly water soluble irritant gas and is rapidly taken up in the nasal passages during normal, quiet breathing. Studies in human volunteers found that, after inhalation at rest of an average of 16 ppm SO$_2$, less than 1% of the gas could be detected at the oropharynx (Speizer and Frank 1966). Penetration to the lungs is greater during mouth breathing than nose breathing, and also is greater with increased ventilation such as during exercise (Costa and Amdur 1996). Since individuals with allergic rhinitis and asthma often experience nasal congestion, mouth breathing is practiced at a greater frequency in these individuals perhaps making them more vulnerable to the effects of water soluble gasses such as SO$_2$ (Ung et al. 1990). Despite the potential for high-level SO$_2$ exposures to cause adverse health effects is well recognized, epidemiologic studies of children have not demonstrated convincing evidence that ambient SO$_2$ exposure at typical current levels are associated with asthma outcomes, even the results from the same nationality's children were also inconsistent. Four Chinese Cities study performed in 1997 did not present any associations between outdoor SO$_2$ exposure and asthma or asthma related symptoms (Zhang et al. 2002); also, Zhao et al. authors reported that indoor SO$_2$, but not outdoor SO$_2$, was a risk factor for wheeze, daytime breathlessness in Chinese children (Zhao et al. 2008). However, the results from the Northeast Chinese Children Health study present some evidence that ambient levels of SO$_2$ were positively associated with children's asthma outcomes (Dong et al. 2011). Multi-pollutant regression analyses indicated that SO$_2$ risk estimates for asthma outcomes were not sensitive to the inclusion of co-pollutants, including PM, NO$_2$, O$_3$ and CO, and concluded that the observed health associations for SO$_2$ might be attributed partly to co-pollutants, with a focus on PM and NO$_2$ as these pollutants tend to be moderately to highly correlated with SO$_2$ and have known respiratory health effects (U.S. EPA 2008). Although the studies show that co-pollutant adjustment had varying degrees of influence on the SO$_2$ effect estimates, among the studies with tighter confidence intervals (an indicator of study power), the effect of SO$_2$ on respiratory health outcomes appears to be generally robust and independent of the effects of ambient particles or other gaseous co-pollutants.

In spite of all the research investigating the relationship between SO$_2$ exposure and responses in individuals with asthma, the mechanism of the SO$_2$ response is still not very clear. SO$_2$ may act as an aspecific bronchoconstrictor agent, similar to inhaled histamine or methacholine: the rapid onset of bronchoconstriction after exposure suggests a neural mechanism of action for SO_2 in asthma, but airway inflammation may result from short-term SO_2 exposure (U.S. 1998). Results from the animal studies indicated that SO_2-induced bronchoconstriction was mediated by the activation of a muscarinic (cholinergic) reflex via the vagus nerves (part of the parasympathelic branch of the autonomic nervous system (Nadel et al. 1965); However, there is some evidence from studies of human asthmatics for the existence of both muscarinic and nonmuscarinic components (Sheppard 1988), with the nonmuscarinic component possibly involving an effect of sulfur dioxide on airway mast cells. Also, injection of atropine, which counteracts the effects of the parasympathetic nervous system, prevented the increase in airway resistance and increased airway conductance in healthy subjects exposed to SO_2 but had no effect on asthmatics (Nadel et al. 1965; Snashall and Baldwin 1982). In conclusion, sulfur dioxide-induced bronchoconstriction and respiratory inhibition appear to be mediated through vagal reflexes by both cholinergic and non-cholinergic mechanisms. Non-cholinergic components include but are not limited to tachykinins, leukotrienes, and prostaglandins. The extent to which cholinergic or non-cholinergic mechanisms contribute to sulfur dioxide-induced effects is not known and may vary between asthmatic and healthy individuals and between animal species (U.S. 1998). So, the mechanism of why SO_2 elicits such a dramatic effect on the bronchial airways of subjects with asthma is still needed further study.

2.3 Nitrogen oxides (NOx)

Nitrogen oxides (NOx), a mixture of nitric oxide (NO) and nitrogen dioxide (NO_2), are produced from natural sources, motor vehicles and other fuel combustion processes. NO is colourless and odourless and is oxidised in the atmosphere to form NO_2 which is an odourous, brown, acidic, highly-corrosive gas. NO_2 are critical components of photochemical smog and produces the yellowish-brown colour of the smog. In urban areas, most of NO_2 in the ambient air arises from oxidization of emitted NOx from combustion mainly from motor engines, and it is considered to be a good marker of traffic-related air pollution.

Many epidemiological studies have reported associations between NOx, NO, NO_2 and asthma outcomes. Nordling et al. (2008) prospectively followed children in Stockholm, Sweden, from birth until 4 years of age and found that exposure to traffic-related air pollution during the first year of life was associated with an excess risk of persistent wheezing of 1.60 (95% confidence interval, 1.09–2.36) for a 44-µg/m3 increase in traffic NOx. Brauer et al. used LUR models to estimate the effect of traffic-derived pollutants on asthma incidence among 4-year-old children in the Netherlands and found odds ratios (ORs) of 1.20 per 10-µg/m^3 increase in NO_2 (Brauer et al. 2007). Also, a population-based nested case-control study in southwestern British Columbia indicated, comparing with the industrial-related pollutants (PM_{10}, SO_2, black carbon), traffic-related pollutants were associated with the highest risks for asthma diagnosis: adjusted OR=1.08 (95%CI, 1.04-1.12) for a 10-µg/m3 increase of NO, and 1.12 (95%CI, 1.07-1.17) for a 10-µg/m^3 increase in NO_2 (Clark et al. 2010). However, epidemiological evidence about these associations is still inconsistent. Four Chinese Cities study did not find any associations between NOx and asthma outcomes (Zhang et al. 2002); No statistically significant NO_2 effect on asthma exacerbation was provided in the study from Greece (Samoli et al. 2011). This inconsistence may be partly due to methodological problems, confounding or effect modification by other pollutants, and a lack of prospective data. To some extent, this inconsistency in epidemiological studies also relates to the differences between the groups of people who have been studied. Despite methodological differences, a systematic review of the health effects caused by environmental NO_2 reported that there was moderate evidence that short-term exposure (24 hours) even for mean values below 50µgm³ NO_2 increased both hospital admissions and mortality. The review also reported that there was moderate evidence that long-term exposure to a NO_2 level below the World Health Organization recommended air quality annual mean guideline of 40µgm³ was associated with adverse health effects including asthma outcomes (Latza et al. 2009).

In addition to the role of exogenous NOx on the development of asthma, the endogenous production of NOx is also associated with the respiratory viral infection in childhood (Everard 2006). Therefore, understanding the mechanisms by which NOx exposure contributes to allergic asthma has important global public health implications. Animal studies have demonstrated that exposure to NO_2 caused epithelial damage, reduced mucin expression, and increased tone of respiratory smooth muscle (Hussain et al. 2004). A mouse model study showed that NO_2 exposure can promote allergic sensitization and cellular damage which induces the elaboration of immunostimulatory molecules by activating TLR2 that can induce Th2 immune responses and promote experimental asthma (Bevelander et al. 2007). Morrow PE concluded that NO_2 exposure can lead to an increased inflammatory cell influx and may affect lung defense mechanisms through reduced mucociliary clearance and changes in alveolar macrophages and other immune cells (Morrow 1984). Clinical experimental studies showed that prolonged NOx exposure induces the generation of mediators such as LTC_4 and influx of pro-inflammatory ILs (GM-CSF, TNF-α and IL-8) in the respiratory tract, with significant differences between cells produced by atopic and nonatopic subjects; In vivo exposure to NO_2 provokes BHR in exposed subjects, notably the asthmatic, with differences between exposure and symptoms of 24h for NO and of 7 days for NO_2 and increase in lymphocytes, macrophages and mast cells of BALF (Devalia et al. 1994). Whereas, contrary to the above results, a study in asthmatic subjects showed that multi-hour exposure to a high ambient concentration of NO_2 does not enhance the inflammatory response to subsequent inhaled allergen as assessed by cell distribution in induced sputum (Witten et al. 2005). So, the mechanism about NO_2-induced allergic sensitization will require future study.

2.4 Ozone (O_3)

Ozone (O_3) is a highly reactive gas that results primarily from the action of sunlight on hydrocarbons and NOx emitted in fuel combustion. Since ozone can provoke both airway hyperreactivity and "prime" epithelial inflammatory responses, it is a likely contributor to the overall burden of asthma (Auten and Foster 2011). Despite the role of O_3 exposure in the initiation of asthma pathophysiology is controversial, it certainly acts as an exacerbating factor and chronic exposure may have durable effects on susceptibility to airway hyperreactivity. Epidemiologic and clinical studies have shown that O_3 exposure is associated with worsening of athletic performance, reductions in lung function, shortness of breath, chest pain with deep inhalation, wheezing and coughing, and asthma exacerbations among those with asthma. A cohort study performed in Mexico City, Mexico, and with a follow-up period of 22 weeks demonstrated that children living in urban high-traffic areas had a greater daily incidence of respiratory symptoms and bronchodilator use, however, in the case of co-pollutant models, only the O_3 effect continued to be significant in regard to symptoms and medication use (Escamilla-Nunes et al. 2008). Silverman and Ito (2010) had investigated the relationship between severe asthma mortality and ozone in 74 New York City hospitals from 1999 to 2006, and found each 22-ppb increase in O_3, there was a 19% (95% CI, 1% to 40%) increased risk for intensive care unit (ICU) admissions and a 20% (95% CI, 11% to 29%) increased risk for general hospitalizations, especially for children age 6 to 18 years who consistently had the highest risk. In one recent study of the short-term associations between ambient air pollutants and pediatric asthma emergency department visits, Strickand et al. (2010) analyzed the daily counts of emergency department visits for asthma or wheeze among children aged 5 to 17 years that were collected from 41 Metropolitan Atlanta hospitals during 1993-2004 (n = 91,386 visits), and the results showed, even at relatively low ambient concentrations, ozone and primary pollutants from traffic sources independently contributed to the burden of emergency department visits for pediatric asthma. One important cohort study suggests that long-term O_3 exposure can increase the chances that children will have asthma. Although acute exposure to O_3 and other outdoor air pollutants clearly exacerbates asthma acutely, the chronic effects of air pollution have been less studied, and air pollution is not generally thought to induce new cases of asthma. However, children exercising outside receive greater doses of outdoor pollutants to the lung than those who do not and thus would be more susceptible to any chronic effects of air pollution. The California Children's Health Study carefully tested the hypothesis that air pollution can cause asthma by investigating the relation between newly diagnosed asthma and team sports in a cohort of children exposed to different concentrations of air pollutants. The relative risk of asthma development in children playing 3 or more sports in the 6 more polluted communities was 3.3 (95% CI, 1.9-5.8) compared with children in these areas playing no sports. Sports had no effect in areas of low O_3 concentration areas (relative risk, 0.8; 95% CI, 0.4-1.6), but time spent outside was associated with a higher incidence of asthma in areas of high O_3 (relative risk, 1.4; 95% CI, 1.0-2.1). Exposure to pollutants other than O_3 did not alter the effect of team sports (McConnell et al. 2002). This study needs to be replicated elsewhere, but it does suggest that higher long-term exposure to air pollution might well cause the induction of asthma.

Despite, the effects of O_3 on innate and adaptive immunologic pathways relevant to human asthma is not yet clear, but the relevance has been observed in animal models. Animal models showed that O_3 exposure has the capacity to affect multiple aspects of the "effector arc" of airway hyperresponsiveness, ranging from initial epithelial damage and neural excitation to neural reprogramming during infancy (Auten and Foster 2011). Exposed to O_3 enhanced responsiveness of airway sensory nerves in rat on specific postnatal days (PDs) between PD1 and PD29 (Hunter et al. 2010). The best evidence from non-human primates suggests that eventual development of asthma in children may also be attributable in part to remodeling of afferent airway nerves and subtle effects on airway structure and function (Auten and Foster 2011). These changes may have implications for lifelong vulnerability to other oxidative and allergic challenges that produce clinical asthma.

3. Air pollution and lung function in children

Lung function as a sensitive marker of respiratory health effects of the lower airway has been documented in previous studies (Anonymous. 1996a; Anonymous. 1996b). Most major pollutants can alter lung function in addition to other health effects when the exposure concentrations are high. However, some studies have indicated no association between ambient air pollution and lung function, especially in ambient low-dose exposure (Dockery et al. 1989; Anonymous. 1995).

Recently, our population-based epidemiologic study in Taiwanese communities was published to show that traffic-related pollutants CO, NOx, NO_2, and NO had chronic harmful effects on lung function in children (Lee et al.). Deficits in lung function indices were not significantly related to the ambient levels of O_3, SO_2, $PM_{2.5}$, and PM_{10}. After adjustment for individual-level confounders, lung function differed only slightly between communities with different levels of air pollution. Our findings indicated that chronic exposure to ambient NOx and CO significantly decreases lung function in children. These findings are in concordance with several previous epidemiologic studies concerning chronic effects of ambient air pollution from Italy (Rosenlund et al. 2009), Finland (Timonen et al. 2002), and California (Peters et al. 1999b). In present data, we found consistent effects of NOx and CO on FVC and FEV1 that represent central airways, whereas the effects on MMEF and PEFR that provide information primarily on damage of the peripheral airways was more limited. Outdoor NO_2 is strongly influenced by local traffic density (Jerrett et al. 2005). Another recent cohort study from California revealed an adverse effect of prenatal exposure to NO_2 and CO on lung function in asthmatic children (Mortimer et al. 2008).

Chronic airway inflammation could produce the decreases in lung function indices and the central airways seem to be mainly affected given the stronger signal that we detected for FVC in the present study. Plausible mechanisms of NO_2 pulmonary toxicity have been well described (Persinger et al. 2002) and may contribute to part of our findings. However, in human exposure studies, adverse pulmonary effects of NO_2 have generally been demonstrated at levels of exposure a magnitude higher than reported here (Kraft et al. 2005). The above experimental results contrast recent epidemiologic findings showing associations of asthma outcomes in children with low levels of indoor NO_2 (Belanger et al. 2006), of personal NO_2 (Chauhan et al. 2003), and of ambient NO_2. The low ambient NO_2 levels we found are more likely to have served as a surrogate for traffic-related air pollutants. Although it is difficult to discuss etiological mechanisms in our cross-sectional design, we believe these pollutants may be causally related to lung function through oxidative stress responses induced by pollutants highly correlated with NO_2 (Seaton et al. 2003; Li et al. 2003). A previous study of Californian children has showed that the acute effect of ambient NO_2 was about -0.4 ml/ppb in contrast to the chronic effect of -1.7 ml/ppb on morning FVC measurements (Linn et al. 1996). Our data also indicated that the chronic effect of NO_2 (-4.81 ml/ppb) was slightly larger than subchronic effect (-4.35 ml/ppb) on FVC. However, subchronic effects of NO_2 were significantly greater than chronic effects on MMEF and PEFR. Comparison of pollutants concentrations between studies is difficult because of differences in the atmospheres and in the measurement techniques.

We further conduct meta-analyses for the present and other available studies of ambient NO_2 effects on lung function indices in children. We accepted a priori all studies with individual as the unit of observation. The fixed-effects models were calculated using the Mantel-Haenszel method with inverse variances of individual effect estimates as weights. We also studied heterogeneity of the study-specific effect estimates by plotting the measures of effect and applying Q-statistics. Meta-analyses were undertaken of previously published studies together with the data from the present study of ambient NO_2 exposure. As shown in Figure 1, the pooled estimates for four lung indices provided consistent evidence of significantly adverse effects. We also elaborated the heterogeneity between the specific estimates by study design and year, but none showed significant Q-statistics. Therefore, only fixed-effect models were systemically used to present pooled effect estimates (Figure 1).

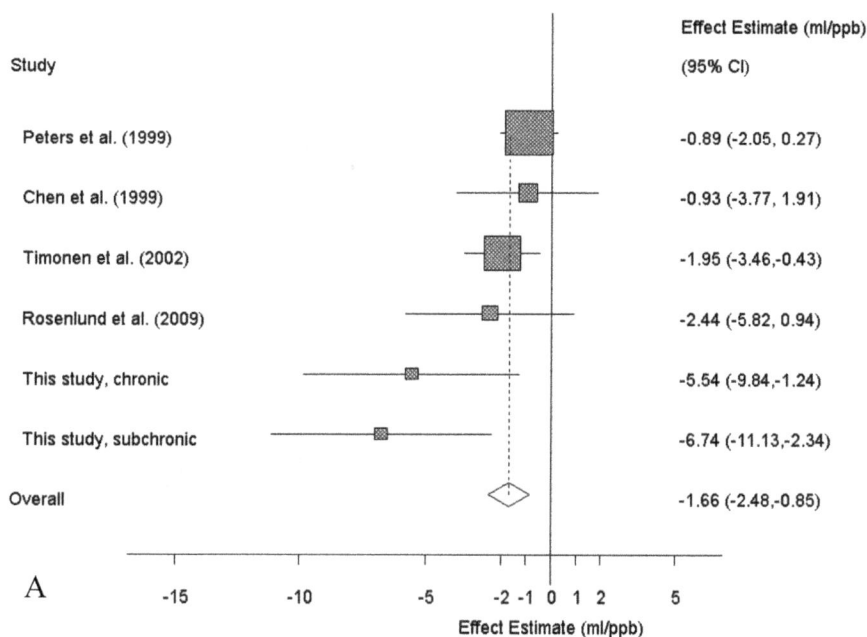

	Effect Estimate (ml/ppb)
Study	(95% CI)
Peters et al. (1999)	-0.89 (-2.05, 0.27)
Chen et al. (1999)	-0.93 (-3.77, 1.91)
Timonen et al. (2002)	-1.95 (-3.46,-0.43)
Rosenlund et al. (2009)	-2.44 (-5.82, 0.94)
This study, chronic	-5.54 (-9.84,-1.24)
This study, subchronic	-6.74 (-11.13,-2.34)
Overall	-1.66 (-2.48,-0.85)

A

Effect Estimate (ml/ppb)

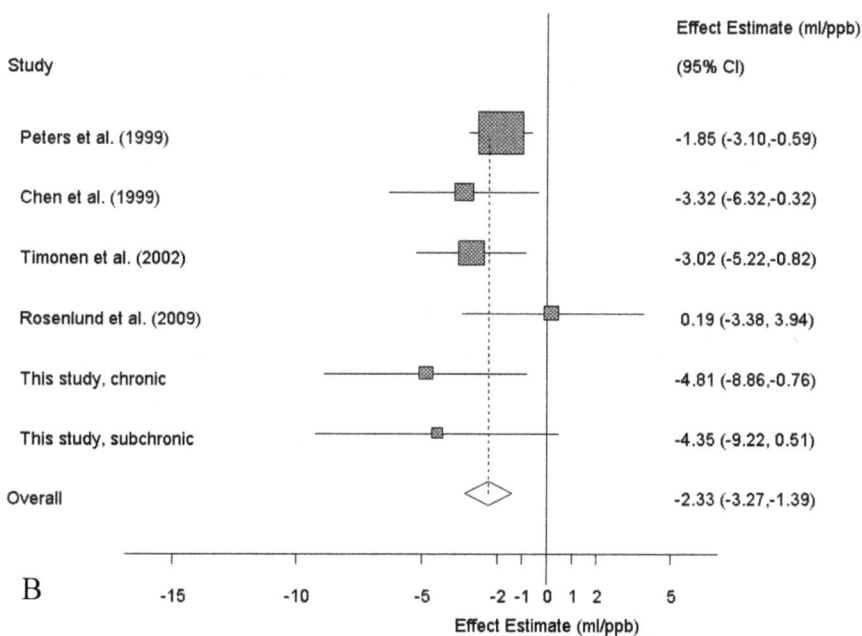

	Effect Estimate (ml/ppb)
Study	(95% CI)
Peters et al. (1999)	-1.85 (-3.10,-0.59)
Chen et al. (1999)	-3.32 (-6.32,-0.32)
Timonen et al. (2002)	-3.02 (-5.22,-0.82)
Rosenlund et al. (2009)	0.19 (-3.38, 3.94)
This study, chronic	-4.81 (-8.86,-0.76)
This study, subchronic	-4.35 (-9.22, 0.51)
Overall	-2.33 (-3.27,-1.39)

B

Effect Estimate (ml/ppb)

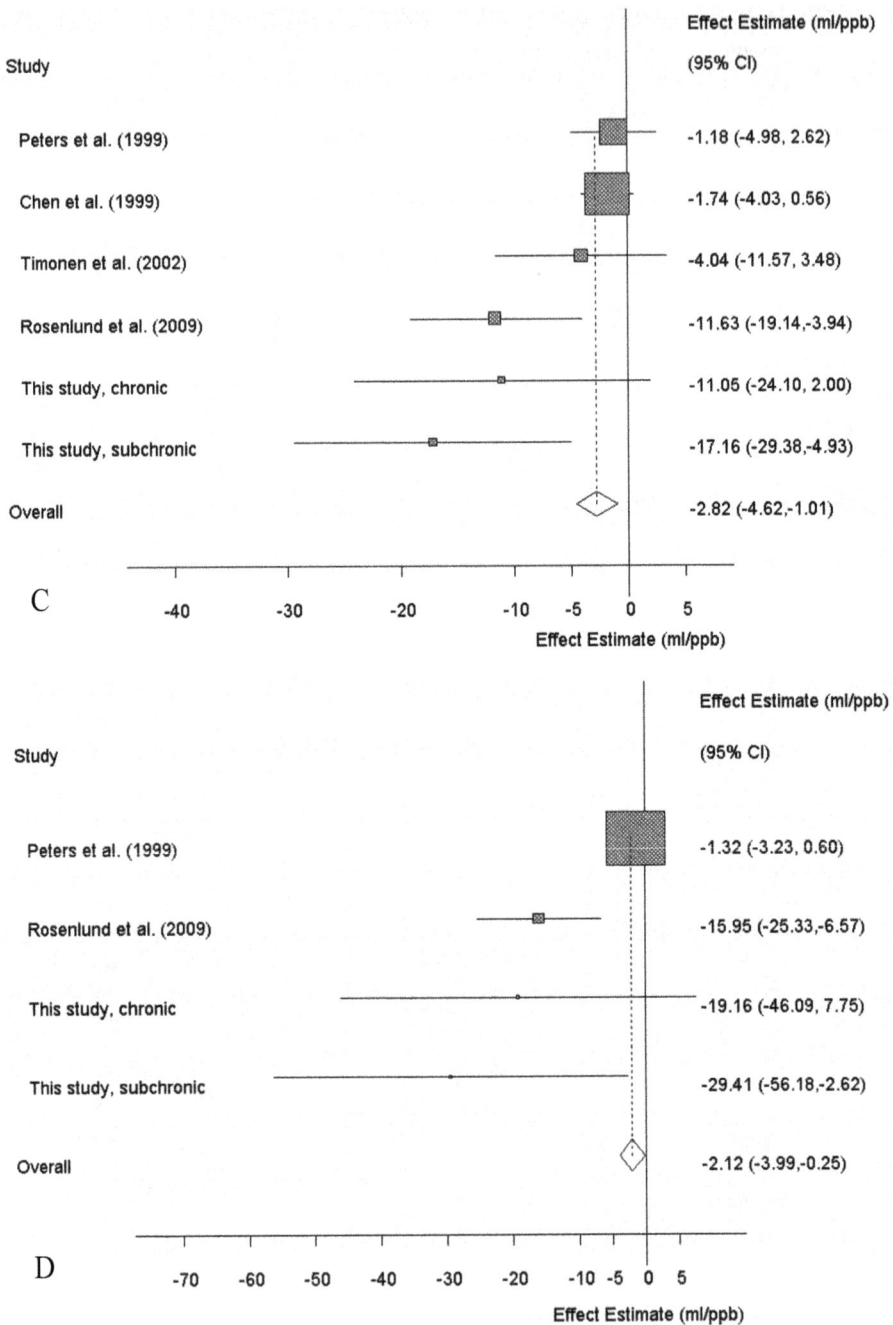

Fig. 1. Meta-analyses of ambient NO_2 effects on lung function in children. (A) FVC (B) FEV_1 (C) MMEF (D) PEFR.

4. Gender difference for the air pollution effects in children

Despite the literature is far from consistent, there is growing epidemiological evidence of the differing associations between air pollution and respiratory health for males and females (Clougherty 2010; Keitt et al. 2004). Generally, many investigations were focused on the differences of gender or sex, however, there is also the distinction between definition of gender and sex (Clougherty 2010). As a social construct, Krieger et al. (2003) had expatiated that gender includes cultural norms, roles, and behaviors shaped by relations among women and men and among girls and boys; however, sex, a biological construct, is based on physiologic differences enabling reproduction, defined by physiologic characteristics (especially reproductive organs) or chromosomal complement. Also, Clougherty et al. (2010) have given recently a summary account that gender, inherently social, varies continuously over multiple dimensions over the life course, whereas sex is normally dichotomous. Gender is shaped at the societal level and varies across nation, culture, class, race, ethnicity, nationality, sexuality, and religion. Gender describes patterns of behavior, place, and role, determining where people spend time and their activities, thereby shaping exposure distributions. Sex-linked traits (e.g., hormonal status, body size) influence biological transport of environmentally derived chemicals (Clougherty 2010).

The gender analyses are more common in occupational epidemiology than in environmental health, because persistent job stratification by gender has produced marked differences in occupational exposures to chemical agents, ergonomic demands, injury, and psychosocial stressors (Clougherty 2010). Compared with the studies among adults, disentangling gender effects in air pollution–health associations among children may be more complicated, because lung function growth rates (critical periods for pollution effects) differ by sex (Berhane et al. 2000). Most air pollution epidemiology studies among children examine chronic exposures, although outcomes considered vary widely, including lung function growth, wheeze, asthma onset and exacerbation, and symptoms.

It has been reported elsewhere that boys were more susceptible than girls to the effects of environmental factors including the effects of ambient air pollution. A 10-yr prospective cohort study of Southern California children in grade 4, 7 and 10 indicated that, based on the 1986–1990 exposure data, prevalence of wheeze was associated with exposure to NO_2 (OR=1.48; 95% CI: 1.08–2.02) and acid (OR=1.55; 95% CI: 1.09–2.21) in males only; also, based on the 1994 exposure measurements, again see a positive association of NO_2 (OR=1.54; 95% CI: 1.08–2.19) and acid (OR=1.45; 95% CI: 1.14–1.83) with wheeze in boys only (Peters et al. 1999a). The results from an international collaborative study on the impact of Traffic-Related Air Pollution on Childhood Asthma (TRAPCA) showed, when stratified analysis by gender, significant associations between residential $PM_{2.5}$, NO_2 and respiratory symptoms (e.g., cough without infection, cough at night) only among boys but not among girls 0–2 years of age (Gehring et al. 2002). In a prospective cohort study of annual mean total suspended particle (TSP) and SO_2 exposures among preadolescent children in Krakow, Poland, Jedrychowski et al. (1999) reported stronger associations with FVC and FEV_1 among boys than among girls. The authors noted sex-differing lung growth rates, producing different critical periods for pollution effects. Also, the cohort study from Southern California suggested that the relation between FEV_1 growth and traffic air pollution was noticeably larger in boys than in girls, although a test of effect modification by sex was non-significant (p=0.10) (Gauderman et al. 2007). Some authors have ever explained that the greater effect in boys may be due to greater time spent outdoors and more physical activity, both of which are factors that would increase exposure to and the respiratory dose of ambient air pollution (Tager et al. 1998). Also, this is plausible as there are differences between male and female airways from early in fetal lung development and throughout life (Becklake and Kauffmann 1999), for example female lungs mature earlier with regard to surfactant production. Throughout life women have smaller lungs than men, but their lung architecture is more advantageous with a greater airway diameter in relation to the volume of the lung parenchyma. Thus, in childhood, airway hyper-responsiveness and asthma are more common among boys than girls.

Despite the existed evidence that boys may be more susceptible to air pollution than girls as mentioned above, however, many studies reported the stronger effects among girls. Using the hospital records for 2768 children aged 0 to 18 years from northern Orange County, California, Delfino et al. authors (2009) estimated the association of local traffic-generated air pollution (CO, NO_2, NOx) with repeated hospital encounters for asthma, and found that associations were stronger for girls but were not significantly from boys. In a dynamic cohort study among Mexican children 8 years of age, Rojas-Martinez et al. (2007) found that negative association of O_3 with lung function growth was stronger on girls than boys in multipollutant models. Also, among Roman children 9-14 years of age, Rosenlund et al. (2009) found associations between chronic residential NO_2 exposure and lung function to be stronger among Roman girls than boys; mean FEV1 and FEF_{25-75} decrements were approximately four times greater in girls than boys. The authors indicated complexities in comparing childhood cohorts differing by age, pubertal status, pollution mixtures, study designs, and susceptibilities and noted that the consistency of results across reporting stronger air pollution effects among girls, meriting further investigation.

In fact, some authors suggested, when focused on the gender difference of the health responses to air pollution, the types of air pollutants should be fully considered. For example, boys may be more susceptible to some kind of pollutants, whereas, girls may be more susceptible to some other pollutants. Results from the Canada showed that respiratory hospitalizations were significantly associated with $PM_{2.5-10}$ among boys and girls, with PM_{10} among boys, and with NO_2 among girls (Lin et al. 2005). Also, Rojas-Martinez et al. (2007) associated elevated PM_{10}, NO_2, and O_3 with reduced lung function among boys and girls. Interquartile range increases in NO_2 predicted FEV1 declines in girls, whereas increases in PM_{10} predicted FEV1 declines among boys. Interesting, results from the Northeast Chinese Children Health Study showed, when stratified by allergic predisposition, amongst children without an allergic predisposition, air pollution effects on asthma were stronger in boys compared to girls; Current asthma prevalence was related to PM_{10} (ORs=1.36 per 31 $\mu g/m^3$; 95% CI, 1.08-1.72), SO_2 (ORs=1.38 per 21$\mu g/m^3$; 95%CI, 1.12-1.69) only among boys. However, among children with allergic predisposition, more positively associations between air pollutants and respiratory symptoms and diseases were detected in girls; An increased prevalence of doctor-diagnosed asthma was significantly associated with SO_2 (ORs=1.48 per 21$\mu g/m^3$; 95%CI, 1.21-1.80), NO_2 (ORs=1.26 per 10$\mu g/m^3$; 95%CI, 1.01-1.56), and current asthma with O_3 (ORs=1.55 per 23$\mu g/m^3$; 95%CI, 1.18-2.04) only among girls (Dong et al. 2011).

The difference of health response to air pollution between boys and girls remains unclear and needs to be explored further. Just as Clougherty (2010) elaborated that careful consideration of gender and sex effects and exploration of nascent methods for quantitative gender analysis may help to elucidate sources of difference. More broadly, exploring the role for gender analysis in environmental epidemiology may provide a model for exploring other social factors that can shape population responses to air pollution.

5. Health effect susceptibility and prevention of air pollution

Air pollution has shown significant effects on asthma development and exacerbation. In a recent Indian study, it was suggested that asthma admissions increased by almost 21% due to the ambient levels of pollutants exceeding the national air quality standards (Pande et al. 2002). In Israel, outdoor air pollution revealed significant correlation with ER visits and 61% of the variance could be explained by air pollutants including NOx, SO_2, and O_3 (Garty et al. 1998). Ambient air pollution was also estimated to attribute 13-15% of the prevalence of childhood asthma in Taiwan, after adjustment for hereditary and indoor factors (Lee et al. 2003). In Central Jakarta and Tangerang, where the average NO_2 concentrations were highest, reduction of NO_2 to a proposed level of 25 ppb could yield savings in mean 15,639-18,165 Indonesian rupiah (6.80-7.90 US dollars) per capita for treatment of the respiratory symptoms, and reduce average work/school days lost per capita by 3.1-5.5 days (Duki et al. 2003). The overall economic cost (both direct and indirect) of air pollution might be higher than what has been estimated in the study.

The worldwide variation in asthma prevalence indicates that environmental factors may be critical. There is now growing evidence to suggest that air pollution is probably responsible, at least in part, for the increasing prevalence of asthma in children. Outdoor ambient air pollutants in communities are relevant to the acute exacerbation and possibly the onset of asthma. From some children's studies, asthma prevalences were different despite of similar prevalences of atopic sensitization. This supported the hypothesis that outdoor air pollution with unchanged levels of allergen could be a risk factor for the increasing prevalence of asthma observed over the past two decades. However, the ecological association with known outdoor air pollutants and asthmatic morbidities remains suggestive. Outdoor air pollutants at levels below the current standards were shown harmful to susceptible individuals. There might be serious public health consequences if small decreases in lung function were associated causally with chronic exposure to air pollution, especially if the size of the exposed population is large. It is possible that some permissible levels of ambient air pollutants in children are not sufficiently low for human health protection. Better technology and public policy are needed to help prevent the enormous suffering and human loss associated with air pollution. What has yet to be documented is whether reduced outdoor pollution will result in decreased morbidities from children's health. Studies to investigate this hypothesis are currently under way.

6. Acknowledgement

This chapter is partially supported by grant #98-2314-B-002-138-MY3 from the Taiwan National Science Council. The authors report no conflicts of interest.

7. Reference

Anonymous. Health effects of outdoor air pollution. Committee of the environmental and occupational health assembly of the american thoracic society. Am J Respir Crit Care Med 1996 (a);153:3-50.

Anonymous. Health effects of outdoor air pollution. Part 2. Committee of the environmental and occupational health assembly of the american thoracic society. Am J Respir Crit Care Med 1996 (b);153:477-498.

Brunekreef B, Dockery DW, Krzyzanowski M. Epidemiologic studies on short-term effects of low levels of major ambient air pollution components. Environ Health Perspect 1995;103:3-13.

Asher MI, Stewart AW, Mallol J, Montefort S, Lai CK, Aït-Khaled N, Odhiambo J; ISAAC Phase One Study Group. Which population level environmental factors are associated with asthma, rhinoconjunctivitis and eczema? Review of the ecological analyses of ISAAC Phase One. Respir Res 2010; 11:8.

Asher MI, Montefort S, Björkstén B, Lai CK, Strachan DP, Weiland SK, Williams H; ISAAC Phase Three Study Group. Worldwide time trends in the prevalence of symptoms of asthma, allergic rhinoconjunctivitis, and eczema in childhood: ISAAC Phases One and Three repeat multicountry cross-sectional surveys. Lancet 2006; 368:733-743.

Auten RL, Foster WM. Biochemical effects of ozone on asthma during postnatal development. Biochim Biophys Acta 2011; 1810(11):1114-9

Bateson TF, Schwartz J. Children's response to air pollutants. J Toxicol Environ Health 2008; 71(3):238-243.

Becklake MR, Kauffmann F. Gender differences in airway behaviour over the human life span. Thorax 1999; 54: 1119-1138.

Belanger K, Gent JF, Triche EW, Bracken MB, Leaderer BP. Association of indoor nitrogen dioxide exposure with respiratory symptoms in children with asthma. Am J Respir Crit Care Med 2006;173:297-303.

Berhane K, Zhang Y, Linn WS, Rappaport EB, Bastain TM, Salam MT, Islam T, Lurmann F, Gilliland FD. The effect of ambient air pollution on exhaled nitric oxide in the Children's Health Study. Eur Respir J 2011; 37:1029-36.

Berhane K, McConnell R, Gilliland F, Islam T, Gauderman WJ, et al. Sex-specific effects of asthma on pulmonary function in children. Am J Respir Crit Care Med 2000; 162:1723-1730.

Bevelander M, Mayette J, Whittaker LA, Paveglio SA, Jones CC, Robbins J, Hemenway D, Akira S, Uematsu S, Poynter ME. Nitrogen dioxide promotes allergic sensitization to inhaled antigen. J Immunol 2007; 179:3680-8.

Björkstén B, Clayton T, Ellwood P, Stewart A, Strachan D; ISAAC Phase III Study Group. Worldwide time trends for symptoms of rhinitis and conjunctivitis: Phase III of the International Study of Asthma and Allergies in Childhood. Pediatr Allergy Immunol 2008; 19:110-24.

Brauer M, Hoek G, Smit HA, de Jongste JC, Gerritsen J, Postma DS, Kerkhof M, Brunekreef B. Air pollution and development of asthma, allergy and infections in a birth cohort. Eur Respir J 2007; 29:879-88.

Bree L, Cassee FR. Toxicity of Ambient Air PM10: A critical review of potentially causative PM properties and mechanisms associated with health effects. May, 2000. RIVM Report number 650010015. www.rivm.nl/bibliotheek/rapporten/650010015.pdf

Burnett RT, Smith-Doiron M, Stieb D, Cakmak S, Brook JR. Effects of particulate and gaseous air pollution on cardiorespiratory hospitalizations. Arch Environ Health 1999; 54:130–139.

Chauhan AJ, Inskip HM, Linaker CH, Smith S, Schreiber J, Johnston SL, Holgate ST. Personal exposure to nitrogen dioxide (NO_2) and the severity of virus-induced asthma in children. Lancet 2003;361:1939-1944.

Clark NA, Demers PA, Karr CJ, Koehoorn M, Lencar C, Tamburic L, Brauer M. Effect of early life exposure to air pollution on development of childhood asthma.Environ Health Perspect. 2010; 118:284-90.

Clougherty JE. A Growing Role for Gender Analysis in Air Pollution Epidemiology. Environ Health Perspect 2010; 118:167-176.

Costa DL, Amdur MO. Air Pollution IN: Klaassen CD (ed) Casarett & Doull's Toxicology: the basic science of poisons, Fifth Edition McGraw Hill, New York, 1996.

Delfino RJ, Chang J, Wu J, Ren C, Tjoa T, Nickerson B, Cooper D, Gillen DL. Repeated hospital encounters for asthma in children and exposure to traffic-related air pollution near the home. Ann Allergy Asthma Immunol 2009; 102:138-144.

Delfino RJ, Staimer N, Gillen D, Tjoa T, Sioutas C, Fung K, George SC, Kleinman MT. Personal and ambient air pollution is associated with increased exhaled nitric oxide in children with asthma. Environ Health Perspect 2006; 114:1736–1743.

Devalia JL, Wang JH, Rusznak C, Calderon M, Davies RJ. Does air pollution enhance the human airway response to allergens? ACI News 1994; 6:80-84.

Dietert RR, Etzel RA, Chen D, Halonen M, Holladay SD, Jarabek AM, Landreth K, Peden DB, Pinkerton K, Smialowicz RJ, Zoetis T. Workshop to identify critical windows of exposure for children's health: immune and respiratory systems work group summary. Environ Health Perspect 2000; 108:483–490.

Dockery DW, Speizer FE, Stram DO, Ware JH, Spengler JD, Ferris BG, Jr. Effects of inhalable particles on respiratory health of children. Am Rev Respir Dis 1989;139:587-594.

Dong GH, Chen T, Liu MM, Wang D, Ma YN, Ren WH, Lee YL, Zhao YD, He QC. Gender Difference for Effects of Compound-Air Pollution on Respiratory Symptoms in Children: Results from 25 Districts of Northeast China. PLoS ONE 2011; 6(8):e20827. DOI: 10.1371/journal.pone.0022470.

Dreher K, Jaskot R, Richards J, Lehmann J, Winsett D, Hoffman A, Costa C. Acute pulmonary toxicity of size fractionated ambient air particulate matter. Am J Respir Crit Care Med 1996; 153:A15.

Duki MI, Sudarmadi S, Suzuki S, Kawada T, Tri-Tugaswati A. Effect of air pollution on respiratory health in Indonesia and its economic cost. Arch Environ Health 2003; 58:135-143.

Escamilla-Nunes M-C, Barraza-Villarreal A, Hernandez-Cadena L, Moreno-Macias H, Ramirez-Aguilar M, Sienra-Monge J-J, Cortez-Lugo M, Texcalas J-L, del Rio-Navarro B, Romieu I. Traffic-related air pollution and respiratory symptoms among asthmatic children, resident in Mexico City: the EVA cohort study. Respir Res 2008; 9:74

Everard ML. The relationship between respiratory syncytial virus infections and the development of wheezing and asthma in children. Curr Opin Allergy Clin Immunol 2006; 6: 56-61.

Garty BZ, Kosman E, Ganor E, Berger V, Garty L, Wietzen T, Waisman Y, Mimouni M, Waisel Y. Emergency room visits of asthmatic children, relation to air pollution, weather, and airborne allergens. Ann Allergy Asthma Immunol 1998; 81:563-570.

Gauderman WJ, Vora H, McConnell R, Berhane K, Gilliland F, Thomas D, Lurmann F, Avol E, Kunzli N, Jerrett M, Peters J. Effect of exposure to traffic on lung development from 10 to 18 years of age: a cohort study. Lancet 2007; 369:571-577.

Gehring U, Cyrys J, Sedlmeir G, Brunekreef B, Bellander T, Fischer P, Bauer CP, Reinhardt D, Wichmann HE, Heinrich J.Traffic-related air pollution and respiratory health during the first 2 yrs of life. Eur Respir J 2002; 19:690-698.

Hunter DD, Wu Z, Dey RD. Sensory neural responses to ozone exposure during early postnatal development in rat airways. Am J Respir Cell Mol Biol 2010; 43:750-7.

Hussain I, Jain VV, O'Shaughnessy P, Businga TR, Kline J. Effect of nitrogen dioxide exposure on allergic asthma in a murine model. Chest 2004; 126:198-204.

Inoue K, Takano H, Yanagisawa R, et al. Effects of components derived from diesel exhaust particles on lung physiology related to antigen. Immunopharmacol Immunotoxicol 2007; 29:403–412.

Jedrychowski W, Flak E, Mroz E. The adverse effect of low levels of ambient air pollutants on lung function growth in preadolescent children. Environ Health Perspect 1999; 107:669–674.

Keitt SK, Fagan TF, Marts SA. Understanding sex differences in environmental health: a thought leaders' roundtable. Environ Health Perspect 2004; 112:604-609.

Kim E, Hopke PK, Pinto JP, Wilson WE. Spatial variability of fine particle mass, components, and source contributions during the regional air pollution study in St. Louis. Environ Sci Technol 2005; 39:4172-4179.

Kim SB, Temiyasathit C, Chen VC, Park SK, Sattler M, Russell AG. Characterization of spatially homogeneous regions based on temporal patterns of fine particulate matter in the continental United States. J Air Waste Manag Assoc 2008; 58:965-975.

Kleinman MT, Sioutas C, Froines JR, et al. Inhalation of concentrated ambient particulate matter near a heavily trafficked road stimulates antigen-induced airway responses in mice. Inhal Toxicol 2007; 19 Suppl 1:117–126.

Kraft M, Eikmann T, Kappos A, Kunzli N, Rapp R, Schneider K, Seitz H, Voss JU, Wichmann HE. The german view: Effects of nitrogen dioxide on human health--derivation of health-related short-term and long-term values. Int J Hyg Environ Health 2005;208:305-318.

Krieger N. Genders, sexes, and health: what are the differences, and why does it matter? Int J Epidemiol 2003; 32:652-657.

Just J, Segala C, Sahraoui F, Priol G, Grimfeld A, Neukirch F. Short-term health effects of particulate and photochemical air pollution in asthmatic children. Eur Respir J 2002; 20:899–906.

Jerrett M, Arain A, Kanaroglou P, Beckerman B, Potoglou D, Sahsuvaroglu T, Morrison J, Giovis C. A review and evaluation of intraurban air pollution exposure models. J Expo Anal Environ Epidemiol 2005;15:185-204.

Latza U, Gerdes S, Baur X. Effects of nitrogen dioxide on human health: Systematic review of experimental and epidemiological studies conducted between 2002 and 2006. Int J Hyg Environ Health 2009; 212:271-87.

Lee YL, Lin YC, Hsiue TR, Hwang BF, Guo YL. Indoor and outdoor environmental exposures, parental atopy, and physician-diagnosed asthma in Taiwanese schoolchildren. Pediatrics 2003; 112:e389-e395 Available from: URL: http://www.pediatrics.org/cgi/content/full/112/5/e389.

Lee YL, Wang WH, Lu CW, Lin YH, Hwang BF. Effects of Ambient Air Pollution on Pulmonary Function among Schoolchildren. Int J Hyg Environ Health (In press)

Leikauf GD. Hazardous air pollutants and asthma. Environ Health Perspect 2002; 110:505-526.

Levy JI, Clougherty JE, Baxter LK, Houseman EA, Paciorek CJ; HEI Health Review Committee. Evaluating heterogeneity in indoor and outdoor air pollution using land-use regression and constrained factor analysis. Res Rep Health Eff Inst 2010; 152:5-80.

Li N, Hao M, Phalen RF, Hinds WC, Nel AE. Particulate air pollutants and asthma. A paradigm for the role of oxidative stress in pm-induced adverse health effects. Clin Immunol 2003;109:250-265.

Lin M, Chen Y, Burnett RT, Villeneuve PJ, Krewski D. The influence of ambient coarse particulate matter on asthma hospitalization in children: case-crossover and time-series analyses. Environ Health Perspect 2002; 110:575–581.

Lin M, Steib DM, Chen Y. Coarse particulate matter and hospitalization for respiratory infections in children younger than 15 years in Toronto: a case-crossover analysis. Pediatrics 2005; 116:235-240.

Linn WS, Shamoo DA, Anderson KR, Peng RC, Avol EL, Hackney JD, Gong H, Jr. Short-term air pollution exposures and responses in Los Angeles area schoolchildren. J Expo Anal Environ Epidemiol 1996;6:449-472.

Loomis D. Sizing up air pollution research. Epidemiology 2000; 11:2–4.

Mann JK, Balmes JR, Bruckner TA, Mortimer KM, Margolis HG, Pratt B, Hammond SK, Lurmann FW, Tager IB. Short-term effects of air pollution on wheeze in asthmatic children in Fresno, California. Environ Health Perspect. 2010; 118:1497-502.

McConnell R, Islam T, Shankardass K, Jerrett M, Lurmann F, Gilliland F, Gauderman J, Avol E, Künzli N, Yao L, Peters J, Berhane K. Childhood incident asthma and traffic-related air pollution at home and school. Environ Health Perspect 2010; 118:1021-1026.

McConnell R, Berhane K, Gilliland F, London SJ, Islam T, Gauderman WJ, Avol E, Margolis HG, Peters JM. Asthma in exercising children exposed to ozone: a cohort study. Lancet 2002; 359:386-91.

Morrow PE. Toxicological data on NOx: an overview. J Toxicol Environ Health 1984; 13:205–227.

Mortimer K, Neugebauer R, Lurmann F, Alcorn S, Balmes J, Tager I. Air pollution and pulmonary function in asthmatic children: Effects of prenatal and lifetime exposures. Epidemiology 2008;19:550-557; Discussion 561-562.

Nadel JA, Salem H, Tamplin B, Tokiway Y. Mechanism of bronchoconstriction during inhalation of sulfur dioxide. J Appl Physiol 1965; 20:164-167.

Nordling E, Berglind N, Melen E, Emenius G, Hallberg J, Nyberg F, Pershagen G, Svartengren M, Wickman M, Bellander T. Traffic-related air pollution and childhood respiratory symptoms, function and allergies. Epidemiology 2008; 19:401–408.

Oyana TJ, Rivers PA. Geographic variations of childhood asthma hospitalization and outpatient visits and proximity to ambient pollution sources at a US-Canada border crossing. Int J Health Geogr 2005; 4:14.

Özkaynak H, Thurston GO. Associations between 1980 US. mortality rates and alternative measures of airborne particle concentration. Risk Anal 1987; 7:449–461.

Pearce N, Aït-Khaled N, Beasley R, Mallol J, Keil U, Mitchell E, Robertson C; and the ISAAC Phase Three Study Group. Worldwide trends in the prevalence of asthma symptoms: phase III of the International Study of Asthma and Allergies in Childhood (ISAAC). Thorax 2007; 62:758-66.

Pande JN, Bhatta N, Biswas D, Pandey RM, Ahluwalia G, Siddaramaiah NH, Khilnani GC. Outdoor air pollution and emergency room visits at a hospital in Delhi. Indian J Chest Dis Allied Sci 2002; 44:13-19.

Persinger RL, Poynter ME, Ckless K, Janssen-Heininger YM. Molecular mechanisms of nitrogen dioxide induced epithelial injury in the lung. Mol Cell Biochem 2002;234-235:71-80.

Peters JM, Avol E, Navidi W, London SJ, Gauderman WJ, et al. A study of twelve Southern California communities with differing levels and types of air pollution. I. Prevalence of respiratory morbidity. Am J Respir Crit Care Med 1999 (a); 159: 760-767.

Peters JM, Avol E, Gauderman WJ, Linn WS, Navidi W, London SJ, Margolis H, Rappaport E, Vora H, Gong H, Jr., et al. A study of twelve southern california communities with differing levels and types of air pollution. II. Effects on pulmonary function. Am J Respir Crit Care Med 1999 (b);159:768-775.

Pope CA III. Respiratory hospital admissions associated with PM10 pollution in Utah, Salt Lake, and Cache valleys. Arch Environ Health 1991; 46:90–97.

Porter M, Karp M, Killedar S, et al. Diesel-enriched particulate matter functionally activates human dendritic cells. Am J Respir Cell Mol Biol 2007;37:706-719.

Rojas-Martinez R, Perez-Padilla R, Olaiz-Fernandez G, Mendoza-Alvarado L, Moreno-Macias H, Fortoul T, McDonnell W, Loomis D, Romieu I. Lung function growth in children with long-term exposure to air pollutants in Mexico City. Am J Respir Crit Care Med 2007; 176:377-384.

Rosenlund M, Forastiere F, Porta D, De Sario M, Badaloni C, Perucci CA. Traffic-related air pollution in relation to respiratory symptoms, allergic sensitisation and lung function in schoolchildren. Thorax 2009;64:573-580.

Samoli E, Nastos PT, Paliatsos AG, Katsouyanni K, Priftis KN. Acute effects of air pollution on pediatric asthma exacerbation: evidence of association and effect modification. Environ Res. 2011; 111:418-24.

Samuelsen M, Nygaard UC, Løvik M. Allergy adjuvant effect of particles from wood smoke and road traffic. Toxicology 2008; 246:124–131.

Seaton A, Dennekamp M. Hypothesis: Ill health associated with low concentrations of nitrogen dioxide--an effect of ultrafine particles? Thorax 2003;58:1012-1015.

Selgrade MK, Lemanske RF, Jr, Gilmour MI, Neas LM, Ward MD, Henneberger PK, Weissman DN, Hoppin JA, Dietert RR, Sly PD, Geller AM, Enright PL, Backus GS, Bromberg PA, Germolec DR, Yeatts KB. Induction of asthma and the environment: what we know and need to know. Environ Health Perspect 2006; 114(4):615–619.

Schwartz J, Neas LM. Fine particles are more strongly associated than coarse particles with acute respiratory health effects in schoolchildren. Epidemiology 2000; 11:6–10.

Sheppard D. Mechanisms of airway responses to inhaled sulfur dioxide. In: Loke J, ed. Lung Biology in Health and Disease. New York, NY: Marcel Dekker, 1988; 34:49-65.

Silverman RA, Ito K. Age-related association of fine particles and ozone with severe acute asthma in New York City. J Allergy Clin Immunol 2010; 125:367-373.

Snashall PD, Baldwin C. Mechanisms of sulphur dioxide induced bronchoconstriction in normal and asthmatic man. Thorax 1982; 37:118-123.

Speizer FE, Frank R. The uptake and release of SO_2 by the human nose. Arch Environ Health 1966; 12:725-758.

Spira-Cohen A, Chen LC, Kendall M, Lall R, Thurston GD. Personal exposures to traffic-related air pollution and acute respiratory health among Bronx schoolchildren with asthma. Environ Health Perspect 2011; 119:559-65.

Strickland MJ, Darrow LA, Klein M, Flanders WD, Sarnat JA, Waller LA, Sarnat SE, Mulholland JA, Tolbert PE. Short-term associations between ambient air pollutants and pediatric asthma emergency department visits. Am J Respir Crit Care Med 2010; 182:307-16.

Tager IB, Kunzili N, Lurmann F, Ngo L, Segal M, Balmes J. Methods development for epidemiologic investigations of the health effects of prolonged ozone exposure: part II. An approach to retrospective estimation of lifetime ozone exposure using a questionnaire and ambient monitoring data (California sites). Cambridge, MA: Health Effects Institute; 1998. Research Report No. 81.

Tecer LH, Alagha O, Karaca F, Tuncel G, Eldes N. Particulate matter ($PM_{2.5}$, $PM_{10-2.5}$, and PM10) and children's hospital admissions for asthma and respiratory diseases: a bidirectional case-crossover study. J Toxicol Environ Health A 2008; 71(8):512–520.

Timonen KL, Pekkanen J, Tiittanen P, Salonen RO. Effects of air pollution on changes in lung function induced by exercise in children with chronic respiratory symptoms. Occup Environ Med 2002;59:129-134.

Tzivian L. Outdoor air pollution and asthma in children. J Asthma 2011; 48:470-81.

Ung N, Koenig JQ, Shaprio GG, Shapiro PA. A quantitative assessment of respiratory pattern and its effect on dentofacial development. Am J Orthod Dentofac Ortho 1990; 98: 523-532.

U.S. Environmental Protection Agency. Air Quality Criteria for Particulate matter. Research Triangle Park. NC: U.S. Environmental Protection Agency, 1995.

U.S. EPA. Integrated Science Assessment for Sulfur Oxides—Health Criteria. EPA/600/R-08/047F. Research Triangle Park, NC:National Center for Environmental Assessment, Office of Research and Development. 2008.

U.S. Department of health and human services . Toxicological profile for sulfur dioxide. Public Health Service, Agency for Toxic Substances and Disease Registry. December 1998.

Witten A, Solomon C, Abbritti E, Arjomandi M, Zhai W, Kleinman M, Balmes J. Effects of nitrogen dioxide on allergic airway responses in subjects with asthma. J Occup Environ Med 2005; 47:1250-1259.

Zhang JJ, Hu W, Wei F, Wu G, Korn LR, Chapman RS. Children's Respiratory Morbidity Prevalence in Relation to Air Pollution in Four Chinese Cities. Environ Health Perspect 2002; 110: 961–967.

Zhao Z, Zhang Z, Wang Z, Ferm M, Liang Y, Norbäck D. Asthmatic symptoms among pupils in relation to winter indoor and outdoor air pollution in schools in Taiyuan, China. Environ Health Perspect 2008; 116:90-97.

15

Exposure to Nano-Sized Particles and the Emergence of Contemporary Diseases with a Focus on Epigenetics

Pierre Madl

*Division of Physics and Biophysics, Department of
Materials Research and Physics University of Salzburg
Austria*

1. Introduction

Mankind has been exposed to airborne nano-sized particles for eons, yet mechanization and industrialization of societies has increased the overall aerosol pollution load to which humans are exposed to. Nano-aerosols with a diameter below 1 μm, can be incorporated via any biological surface structure and in particular when the area available is large enough – as is the case with the skin (approx. 1.5-2 m²), the digestive tract (intestinal villi, approx. 200 m²) or the respiratory tract (alveolar surface area reaching approx. 140 m²).(Raab et al, 2010) Since aerosolised particles are readily inhaled rather than ingested, the lungs represent an ideal gateway with high penetration efficiency rates. From a toxicological rather than from a therapeutic point of view, a deposited xenobiotic particle first interacts with biological tissues on a cellular level. From there it is readily translocated into the cell to interfere with metabolic pathways, eventually inducing inflammatory cellular responses. At an organismic level - along with long-term exposure - these particles become redistributed via the lymphatic or the blood circulatory system to reach sensitive organs or tissues, such as the central nervous system, bone marrow, lymph nodes, spleen, heart, etc.(Oberdörster et al. 2005a) At this level, persistent particle exposure may trigger chronic diseases or when already present, modulate the severity of its course.

1.1 Classification of nano-aerosols

With every breath, with every bite, with every sip, we introduce countless nano-sized particles, viruses and bacteria into our organism. Since the onset of the industrial revolution, some 150 years ago, people are progressively exposed to elevated concentration of arbitrarily shaped nano-aerosols in combination with chemical by-products such as carbon monoxide (CO), nitric oxide (NO), semi-volatile (SVC) and volatile organic compounds (VOCs) of anthropogenic origin – mostly in the form of incompletely combusted by-products that leave stacks or tailpipes as macromolecular clusters (Figure 1). Their minute size grants them easy access even to indoor environments.(Costa & Dreher, 1997)

The terms nano- or ultrafine particles (UFP) comprise particles of less than 100 nm diameter.(Chang, 2010; Alessandrini et al., 2009) Depending on their origin, particles of this size have been categorized as (i) naturally occurring ultrafine particles (UFPs), e.g. volcanic ash, mineral compounds, ocean spray and plant debris, (ii) anthropogenic UFPs, such as diesel exhaust particles (DEPs) as well as environmental tobacco smoke (ETS) and (iii) deliberately tailored nano-particles with designed functions; e.g. DNA- and carbon nano-tubes, fullerenes, nano-wires and nano-coatings, carbon black, synthetic agents contained in whiteners and additives supplemented to food for modification of texture.(Chang, 2010) With recent advances in nanotechnology, these designed and engineered nano-particles are being introduced in ever greater varieties, such as scratch-proof paints, lotus-effect stained glass, suntan lotion, toothpaste, etc.(Raab et al., 2010) Although their release into the environment is not intended in the first place, it will occur sooner rather than later - especially once wear and tear along with product degradation releases contained nano-particles into the environment where they easily can enter the food chain either via aerosolization or washing-out from dumping sites.

Regardless of the particle's origin, the young research field of nano-toxicology deals with effects and potential risks of particle structures <1 µm in size. Indeed, the adverse functions of these aerosol pollutants have been recognized based on their interaction with biological systems on macro-molecular, sub-cellular, cellular, tissue and organism levels. The passage of nano-aerosols from the environment into humans occur via interfacing tissues, e.g. skin, gut and mainly the respiratory system.(Chang, 2010; Traidl-Hoffmann et al., 2009) Due to their small size, these particles possess completely different biological potentials when interacting with living organisms in comparison to particulate matter of increased dimensions. The unique properties are related to their enlarged surface-to-mass ratio in combination with their higher propensity for penetration of biological barriers, for deposition, e.g. in the peripheral lung, and the increased overall retention in biological systems.(Chang, 2010; Alessandrini et al., 2006; Alessandrini et al., 2009) This is particularly true for nano-sized particles as the larger surface area per unit volume compared with their larger-sized siblings renders them biologically more active.(Geiser & Kreyling, 2010) Thus, within these dimensions, the well-known quote of Paracelsus "the dose makes the poison", needs to be revised in that "the dose determines the mechanism".(Oberdörster et al. 2005a)

The minute dimensions of UFPs enable their direct passage by crossing cell barriers to enter blood and lymphatic streams and promote further distribution to various target organs, tissues and interstitial of cells.(Chang, 2010; Alessandrini et al., 2009) At the same time, they also penetrate cells and interact with intracellular structures leading to oxidative stress. Here in particular, insoluble nano-aerosols can extend their multiple adverse effects even further when redistributed from one line of defence to the next as metabolites or cellular debris are released back into the interstitial after cell death.(Casal et al., 2008)

This review aims to provide an overview regarding hazardous effects relevant to humans and at the same time to promote a better understanding of how most contemporary diseases relate to UFPs. Although the review deals with aspects mostly outside the immunological field, and therefore does not take into account virulence of viral agents or immune system conditioning, the absorption of nano-aerosols via various pathways affects essential functions of the immune defense system that include even allergens.(Chang, 2010; Traidl-Hoffmann et al., 2009)

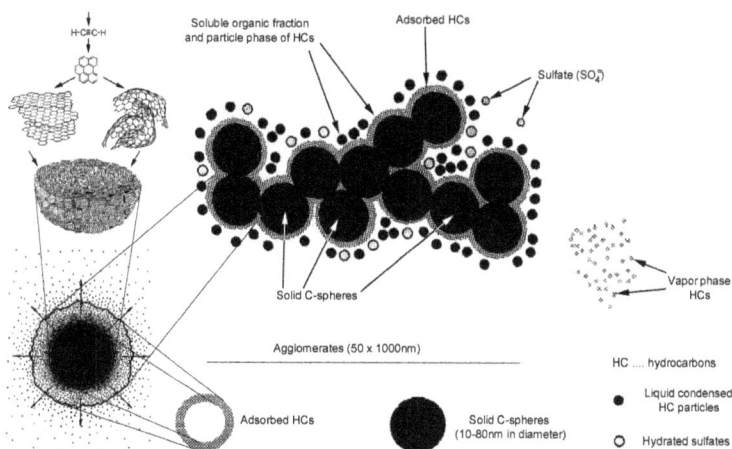

Fig. 1. Schematic representation revealing the formation process and the fractal nature of a nano-sized DEP/MEP cluster and its adsorbed vapour-phase compounds. Model of the carbon core of a DEP that is covered with a layer of inorganic components as well as (bio)organic compounds and pollutants. Solubilisation of the adsorbed fraction is an issue when the semi-/volatile fractions come into contact with the surfactant of the lung.(modified after Madl, 2003)

1.2 Composition of diesel- and ultrafine nano-particles

In the context of health effects of nano-aerosols, diesel exhaust particles (DEPs) as shown in Figure 1, represent a primary source and have extensively been used as a model to investigate the induction of oxidative stress as well as modulation of the immune system.(Chang, 2010) This is also related to the fact, that diesel engines expel about 100 times more particles than gasoline engines.(Riedl & Diaz-Sanchez, 2005) Exhaust particles produced by motorcycles (MEPs) differ from DEPs in their composition, since fuel of two-stroke engines contains a lubricant additive entailing incomplete combustion. Generally, MEPs contain higher amounts of benzene, toluene and xylene than DEPs. Likewise, their polyaromatic hydrocarbon (PAH) spectra differ.(Cheng et al., 2004) DEPs are rather heterogeneous in their composition but share a consensus structure with a carbonaceous core, whereon a multitude of hydrocarbon compounds (10-20,000) can be attached onto,(Riedl & Diaz-Sanchez, 2005; Costa & Dreher, 1997) with the latter encompassing about 20-40% (mass per mass) of DEPs.(Lubitz et al., 2009) Within this attached layer, PAHs (Chang, 2010; Nel, 1998; Takenaka et al., 1995), nitro-PAHs,(Kocbach, 2008) but also bioavailable (transition) metals, such as Fe, Cu, Ni, V, Zn, nitrates, sulfates, (Chang, 2010; Costa & Dreher, 1997) allergens,(Ormstad et al., 1995; Riedl & Diaz-Sanchez, 2005) plant debris, (Solomon, 2002); Taylor et al., 2002) mineral constituents, endotoxins, e.g. lipopolysaccharide (LPS), hopanes, and steranes have been identified.(Kocbach et al., 2008) In the case of wood smoke particles (WSPs), the PAH content is ~150-fold higher (9750 ng/mg) in comparison to DEPs. WEPs slso contained sugars, methoxy-phenols, PAHs, benzene and alkali salts. However, with 31 nm (at the site of formation), the diameter of WSPs is significantly larger than for traffic related UFPs (24-25 nm). For WSPs, the spectrum of core-attached compounds is also influenced by the combustion temperature.(Kocbach et al., 2008) In case of organic dust, as produced by modern farming operations, the nano-aerosol fraction is composed of particulate matter and microbial compounds, including endotoxins, such as LPS, and peptidoglycans (PGNs). The latter compound class represents cell wall constituents particularly of gram-positive bacteria.(Poole et al., 2008)

1.3 Particle deposition and clearance within the human respiratory tract

Deposition of inhaled particles in the human respiratory tract is determined by several biological factors, such as lung morphology and breathing patterns, as well as by physical factors, such as fluid dynamics, particle properties, and deposition mechanisms. Since particle deposition in individual airways cannot be thoroughly analyzed *in-vivo* , particle inhalation and the corresponding deposition patterns are simulated by analytical computer models, which require regular comparison with experiments obtained from human subjects.(Hofmann, 2011) In principle, particle deposition within the respiratory tract is bound to physical principles such as impaction, gravitational settling, diffusion, and electrical attraction (Figure 2).

With the introduction of particle filters and catalytic converters in the exhaust stream, most coarse particles are efficiently removed. Filtering out the coarse fraction usually leaves the smaller without the coarser sibling where the former tend to agglomerate on. Due to their minute dimensions, nano-sized particles largely escape filtering devices and are emitted into the environment where they interact photochemically to form secondary by-products. Upon inhalation, nano-aerosols, along with the adsorbed semi-volatile / volatile chemical cocktail predominantly deposit via Brownian diffusion and electrostatic mechanisms in the nose and the alveolar region where they can unfold their toxicological potential.(Donaldson et al., 1998)

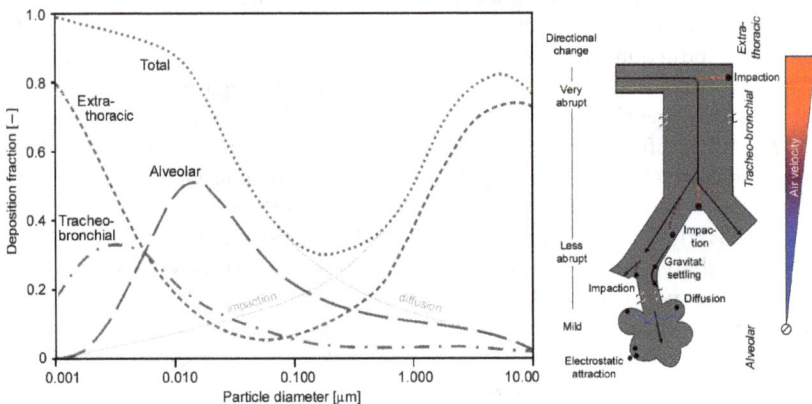

Fig. 2. Total deposition function versus particle diameter for an adult individual with a tidal lung volume of 660 mL at 30 breaths per minute (left). Particle regime <300 nm is dominated by diffusional deposition patterns, those >300 nm by sedimentation and impaction.(modified after Hussain et al., 2011) On the right a schematic view of the pulmonary domains is shown, with the naso-pharyngeal and tracheo-bronchial pathway at the top and the bronchiole and alveolar regime at the bottom, alongside the flow velocities of the in-/exhaled air. (modified after Yip., 2003)

1.4 Potential health effects of nano-aerosol exposure

To a certain degree, macrophages possess the ability to process and detoxify organic constituents adhered to DEP surfaces. Once DEPs are endocytosed (Figure 3a), their clearance can occur by the so-called "*mucociliary escalator*" whereby mucus along with DEP-loaded alveolar macrophages are transferred to the oropharyngeal region by movements of cilia. Their clearing transport might be assisted by an increase in pneumocytes of type-II, which are responsible for the secretion of the alveolar surfactant (see Figure 3b). Finally, the mucus containing macrophages, which are loaded with endocytosed DEPs is either expelled as sputum or swallowed and rerouted over the gastrointestinal tract where DEPs can be reabsorbed over the interstitial lining.(Vostal, 1980) However, cilia-mediated clearance does neither cover terminal bronchioles nor alveoli. Here, nano-particle loaded macrophages readily enter the lymphatic as well as the blood circulatory system. Studies have shown that size and associated surface area do make a difference in terms of penetration efficiency, thereby significantly extending the retention time of inhaled nano-sized particles in comparison with the larger counterparts in alveolar macrophages.(Geiser & Kreyling, 2010; Oberdörster et al. 2005a) Although macrophages represent the most important defense mechanism in the alveolar region against fine and coarse particles, this mechanism is impaired in the case nano-aerosols, which - when inhaled in high abundances – renders phagocytosis inefficient. Subsequent to aerosol exposure, animal studies have shown that only about 20% of the nano-sized fraction (15-20-nm sized particles) can be flushed out by tracheo-bronchial lavage together with the macrophages, whereas lavage efficiency increased to approximately 80% in the case of coarser particles (>0.5 μm in size). (Geiser & Kreyling, 2010; Oberdörster et al., 2005a) As demonstrated in Figure 3a, removal of these particles can only occur via redistribution into the lymphatic or blood stream. With regards to the coarser particle fraction, the larger number of ultrafine siblings leads to particle dispersion into other tissues and organs representing a further burden for the entire organism.

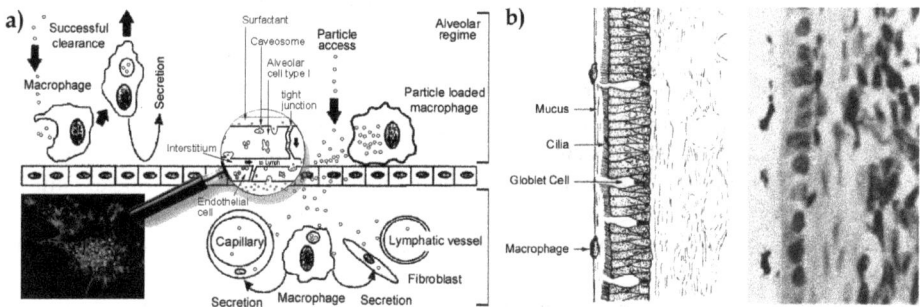

Fig. 3. (a) Schematic drawing of the alveolar tissue showing different clearance mechanism of deposited particles. Pulmonary alveolar macrophage (PAM) mediated removal from the lungs away towards other excretory organs. Colored insert: PAM loaded with 80 nm particles. Approx. 40 nm-sized caveolar openings dis- and re-appear, forming vesicles that constitute pathways through the cells for encapsulated macromolecules.(Madl, 2009, Oberdörster, Donaldson et al., 1998) (b) Schematic and microscopic image of mucociliary escalator transporting macrophages containing DEPs through bronchial tubings.(Vostal, 1980)

Due to their minute size, these particles are also routed through the interstitial compartment between cells or straight through the cells via caveolae; these are openings of around 40 nm in diameter that engulf these nano-sized particles, forming vesicles that are thought to function as transport vehicles for macromolecules across the cells. A similar observation has been described for nano-particle translocation across the olfactoric bulb into the brain, thereby short-cutting the blood-brain-barrier (Figure 4). Studies performed in the 1940s dealt already with this challenge as it was possible to document how 30 nm polio viruses use these nerves as portals of entry into the CNS.[Howe & Bodian, 1940] It is estimated that ~20% of the nano-sized particles deposited onto the olfactoric epithelium translocate to the olfactory bulb within seven days after exposure.[Oberdörster et al., 2004]

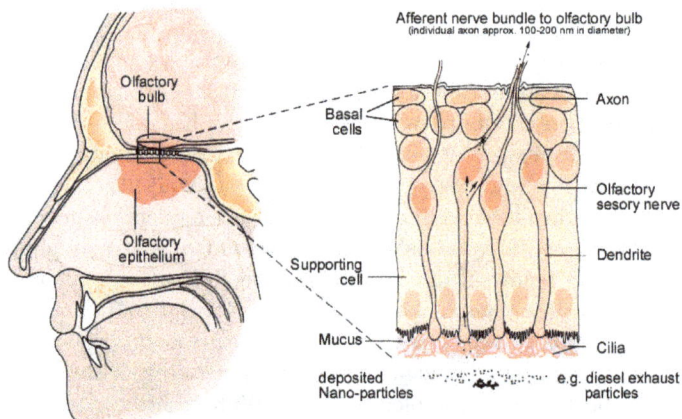

Fig. 4. Nano-particle translocation across the olfactoric bulb into the brain. While approx. 80 % of the nano-particle matter remain attached at the epithelial mucosa, around 20 % of it is capable of transmigration.[modified after Oberdörster et al., 2004, Tortora & Grabowski, 1996]

In the context of nano-sized particle exposure, health consequences become even more challenging when considering that most nano-particles originating from incomplete combustion are of hydrophobic nature. Thus, hygroscopicity and growth via condensation of water vapour is almost inexistent, thereby increasing penetration efficiencies even further - particularly into the alveolar regime of the lungs. In addition, the fractal nature of these incompletely combusted agglomerates are ideal substrates for volatile chemicals or radionuclides to adsorb onto.[Donaldson et al., 1998] In combination, both their low solubility and their role as a "Trojan horse" act threefold: (i) as known with asbestos, insoluble or low soluble matter exerts chronic irritation onto the target cells,[Donaldson et al., 1998] (ii) adsorbed substances on the nano-particle surface unfold their bioreactive properties once in contact with tissues,[Baron & Willeke, 2001] and finally, (iii) UFPs are known to be far more toxic than their coarser siblings [Geiser & Kreyling, 2010; Oberdörster et al. 2005b] - compare with Figure 5. Studies confirmed that nano-particles measuring 20 nm, which were administered directly into the lungs trigger stronger inflammatory reactions than 250 nm particles that are chemically identical. This implies that the toxicological property of the particle is determined by the surface area per unit volume rather than by mass.[Oberdörster et al. 2005a]

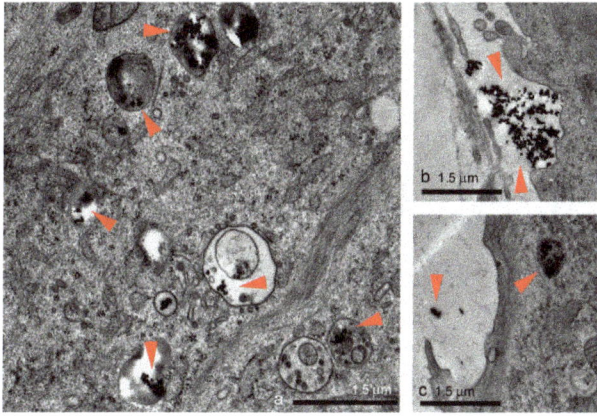

Fig. 5. Exposure of human lung fibroblast to ceria nano-particles (arrows) of 20-50 nm in diameter. (a) Vesicles inside a fibroblast cell with ceria agglomerates. (b) A cluster of nano-particle agglomerates close to the cell membrane. (c) Nano-particles both inside the cell (vesicle) and outside are exclusively found in the form of agglomerates.(adopted from Limbach et al., 2005)

1.5 Toxicity at the cellular level

Adverse effects of nano-aerosols at the cellular level regard (i) inflammation, (ii) oxidative stress, (iii) modulation, enhancement and induction of immune- as well as allergic reactions, e.g. (pulmonary) allergic reactions.(Chan et al., 2008: Alessandrini et al., 2009) Particle interaction with epithelial cells and macrophages are known to trigger inflammatory signalling pathways.(Kocbach et al., 2008)

Figure 6 reveals the cyto-toxicological potential of nano-particles by emphasizing oxidative stress induced by these xenobiotic substances.(Oberdörster et al., 2007) Due to their oxidative properties (step "a" in Figure 6), nano-particles are capable to induce lipid peroxidation. Upon endocytosis (step "b"), these particles exert intracellular oxidative stress and increase cytosolic calcium ion concentration, besides triggering the activation of NADPH oxidase and generation of reactive oxygen species (ROS). Although the latter is essential for normal vital activity,(Voiekov, 2001) improper timing of ROS formation at inappropriate intracellular sites is known to play a crucial role in the initial stage of carcinogenesis.(DeNicola et al., 2011) Depending on ROS concentrations, the effect can be adverse, inducing oxidative stress, damage of DNA, cancer, cardiovascular or neurodegenerative diseases or at appropriate concentrations also protective by beneficial modulation of gene expression.(Chang, 2010)

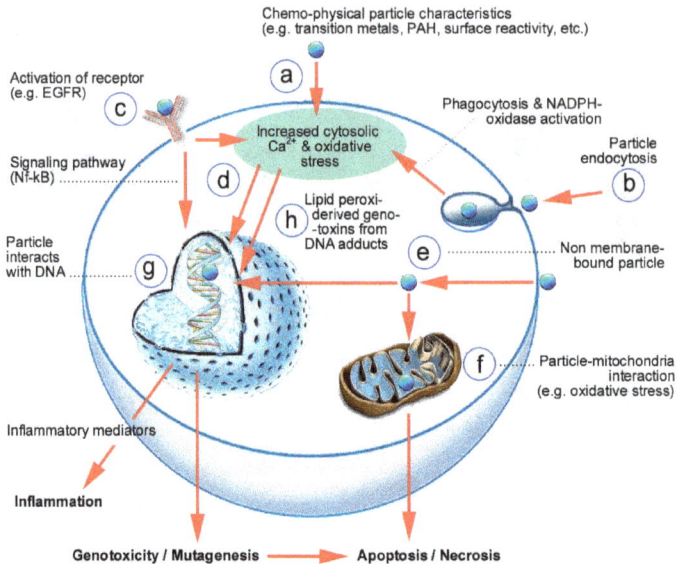

Fig. 6. Hypothetical cyto-toxicological effects of nano-particles range from membrane peroxidation (a), once incorporated formation of reactive oxygen species (ROS) (b), activation of cell-receptors (c), triggering unnecessary cell-metabolic functions, activate expression of inflammatory responses (d), interaction with organelles, such a mitochondria (e), disruption of the electron transport chain (f), interaction with the DNA itself (g) and unfolding genotoxic potential by formation of DNA adducts (h).(modified after Oberdörster et al., 2007)

Both the particles together with the adhered organic fraction and the consecutively induced oxidative stress can activate cell receptors (step "c") - thereby exploiting the energy reservoir of cells for the induction of appropriate compensatory pathways, which trigger several intracellular signalling cascades. These cascades, along with transcription factors activate the expression of pro-inflammatory genes (steps "d"). Apart from interfering with intracellular communication pathways, nano-particles may also enter the cytosol from where they can access mitochondria (steps "e, f") and disrupt normal electron transport, leading to additional oxidative stress (step "f"). Translocation of nano-aerosols into the nucleus may also occur where they interact with the genetic material (step "g"). Eventually, lipid peroxide-derived products can form DNA adducts, which may lead to genotoxicity and mutagenesis (step "h"). In less severe cases, the cell may enter the apoptotic pathway, thereby inducing premature cell death.(Elder et al., 2000) However, apoptosis leaves behind cellular debris together with a toxic particle load that requires clean-up by other cells.

As outlined in Figure 6 (step "g"), genotoxic effects are known to occur also upon MEP exposure as this kind of aerosol induces structural aberrations of chromosomes, formation of micronuclei and DNA adducts. The described mutagenicity seems to be related to PAHs adhered to the surface of DEPs/MEPs.(Cheng et al., 2004) Apart from ROS-mediated activity (step "b"), oxidative DNA damages, such as strand breaks, are associated with the cocktail of UFPs and organic pollutants, e.g. benzene, in ambient air.(Cheng et al., 2004; Avogbe et al., 2005) These

DNA damages have been observed for respiratory as well as for gastrointestinal uptake of UFPs.(Avogbe et al., 2005; Dybdahl et al., 2003) The encountered genotoxity is apparently induced by intracellular ROS, since observed adverse effects were reduced by pre-treatment of cells with anti-oxidants. However, the extent of attenuation depends on the applied anti-oxidant and the protective effects were not complete. Organic fractions of DEPs and MEPs induced ROS are involved in the formation of DNA adducts, e.g. 8-hydroxy-deoxy-guanosine (8-OHdG), at least in the mouse model. This modification is considered a pro/e?-mutagenic lesion.(Cheng et al., 2004; Nagashima et al., 1995) Beside superoxide, also hydroxyl-radicals were formed when challenging mice with DEPs.(Nagashima et al., 1995) Nevertheless, PAHs have to be metabolized into their active forms via the P450-1A1 pathway. Additionally, MEPs contain constituents, which possess direct mutagenic potential and circumvent ROS formation.(Cheng et al., 2004) Furthermore, a genetic pre-disposition has to be considered, since genetic polymorphism in protective protein systems, e.g. glutathione S-transferase, glutathione peroxidase and NAD(P)H:quinone oxireductase-1 seem to be involved as well.(Avogbe et al., 2005)

As mentioned before, organic chemicals adhered to DEPs are apparently directly involved in the formation of reactive nitrogen oxygen species (RNOS) and thus the induction of oxidative stress. Halogenated hydrocarbons and PAHs can induce phase-I drug metabolizing enzymes in alveolar macrophages and epithelial cells, such as cytochrome P450-1A1, which in turn degrade PAHs to redox active metabolites, e.g. quinones and phenols.(Nel et al., 1998; Devouassoux et al., 2002) Production of ROS has been related to the interaction between quinones and DEPs with P450 reductase.(Kumagai et al., 1997) Quinones will then contribute to ROS formation and macrophage activation.(Nel et al., 1998; Devouassoux et al., 2002) Benzo-[a]-pyrene (B[a]P) attached to black carbon enhanced the tumor necrosis factor-α (TNF-α) release of macrophages. Evidently, carbon particles with adhered organic compounds are endocytosed and PAHs get activated by intracellular ROS.(Kocbach et al., 2008) The increase in transcription induced by PAHs has been postulated to be promoted by a cytoplasmatic aryl hydrocarbon receptor (AhR), acting as a nuclear transporter/DNA binding protein PAH complex. Once released from DEPs, PAHs enter adjacent cells due to their hydrophobicity and adhere to the PAH ligand binding part of AhR which is then translocated to the nucleus. There, a heterodimer is created between the AhR and a nuclear translocator. This heterodimer binds to response sequences situated upstream of the target genes promoting e.g. the expression of plasminogen activator inhibitor and interleukin-1β (IL-1β), which would explain the modulating effects of PAHs upon expression. Beside, signalling via AhR also Ca^{++} depending pathways have been discussed as intracellular transmittance channels.(Takenaka et al., 1995; Fahy et al., 1999) Since NO-pathways induce Ca^{++} release to regulate neuronal function, interference via nano-particles adversely affect Ca^{++} release, and ultimately synaptic plasticity.(Kakizawa et al., 2011) As mentioned before, the expression of several enzymatic systems, e.g. the P450-1A1 cytochrome system, are induced. The latter metabolizes PAHs to electrophilic epoxides, which are mutagenic and can interact with DNA. Since 2,3,7,8-tetrachlorodibenzo-p-dioxin (TCDD) is not metabolized by this system, alternative pathways are likely to co-exist.(Takenaka et al., 1995)

1.6 Short term exposure to airborne pollutants

Studies have shown a decrease in pulmonary function associated with short-term exposure to UFPs. These decrements in lung function appear to persist for several weeks after exposure even when the distressing particle load is no longer present. Apart from lung-related pathologies, such as respiratory problems, nocturnal and chronic cough as well as bronchitis and asthma, adverse health effects of UFPs/DEPs at the organismic level that also include cardiovascular disorders.(Chang, 2010; Riedl & Diaz-Sanchez, 2005; Ware et al., 1986) As shown in Table 1, acute exposure to UFPs is associated with increased alveolar inflammation, morbidity, platelet aggregation, accompanied by altered blood coagulation altered heart frequency, myocardial infarction, and including mortality.(Costa & Dreher 1997; Dockery et al. 1982; Brook et al., 2004; Rückerl et al., 2006) However, most noteworthy are thrombogenesis, ischemia (reduction of oxygen supply to target organs due to plaque destabilization and blood clothing) and arrhythmia (disturbance of the electrical activity of the heart muscle).(Bacarelli et al., 2008; Kreyling, 2003)

	Case	Date o. death	Age [yrs]	Sex	Diagnosis 1	Diagnosis 2
	1	07 Dec	76	♀	Heart failure	Bronchitis
	2	23 Jan	61	♂	Bronchitis*	Emphysema
	3	03 Dec	65	♂	Pulmon. embolism	Lung cancer
	4	06 Dec	53	♂	Heart failure	Bronchitis
	5	10 Dec	20 hrs	♂	Prematurity	
	6	12 Dec	54	♀	Emphysema	Hodgkin's disease
	7	17 Dec	51	♀	Sarcoidosis	
	8	19 Dec	53	♂	Heart failure	Bronchitis
	9	25 Dec	51	♂	Pneumonia	Tuberc. meningitis*
	10	04 Jan	60	♂	Heart failure	Syphilitic aortitis
	11	06 Jan	62	♀	Heart failure	Emphysema
	12	12 Jan	0.5	♀	Pneumonia	prob. Cystic fibrosis
	13	14 Jan	55	♂	Bronchitis*	Gastric ulcer
London during the smog	14	17 Jan	64	♀	Esophageal cancer	Aspiration
event in 1952.(Stobbs, 1952)	15	23 Jan	44	♀	Bronchitis*	Pneumonia
	16	28 Jan	62	♀	Lung abscess*	
	17	12 Feb	61	♂	Heart failure	Bronchitis
	18	05 Mar	66	♂	Emphysema*	Myocard. infarction

Table 1. Confirmed deaths of the short-term London smog event (5th - 9th Dec. 1952) with analysis of autopsy, demographics and cause of death. (*) Autopsy note: condition worsened during smog event.(Hunt et al., 2003) London insert: Reduced visibility due to smog-related light scattering and absorption at Nelson's Column in that period.

Following the events of the great London smog, several investigations tried to highlight the adverse health effects of airborne pollutants. The "six cities study" could positively correlate death from cardio-pulmonary diseases and lung cancers with air pollution of $PM_{2.5}$ - that is particle mass with diameters smaller than 2.5 μm.(Dockery et al. 1993) Stimulated by the outcome of this study, and due to the occurrence of several severe air pollution events following thereafter, it was attempted to relate cardiopulmonary mortality as well as lung cancer with long-term exposure to particle-related air pollution. Indeed, exposure to fine particle in combination with sulfur oxide (SO_2) could be associated with the formation of lung cancer and cardiopulmonary mortality. Moreover, each stepwise increment by 10 μg/m^3 of fine particle mass exposure was correlated with an approximate 4% increase in overall mortality, 6% for cardiopulmonary and 8% for lung cancer mortality.(Pope et al., 2002) This relationship was found to be less pronounced for coarse particle fractions than it was for smaller ones.(Perez et al., 2008). This correlation is most obviously related to the cubic relationship between particle mass [μg/cm^3] and number concentration [cm^{-3}] on one side as well as the higher penetrability of the smaller, fine and ultrafine particle fractions for the deeper lung on the other side. While the former (μm-sized class) is efficiently filtered out by the upper respiratory system, the latter (nm-sized class) passes the tracheo-bronchial airway down to the alveolar, gas-exchange regime.(Hussain et al., 2011) Despite their low overall mass concentration, the fractions of nano-particles can reach high concentrations in terms of particle numbers – a tribute that is related to the improved combustion efficiencies and the applied filtering technologies. This is in accordance with the findings of a Dutch research group, which correlated a shortened life expectancy with exposure to nano-sized particles originating from vehicle exhaust.(Hoek et al., 2002). When exposed to long-term elevated doses of nano-sized particles, such as diesel fumes and the corresponding by-products of nitrogen dioxide (NO_2), the authors documented organismic-wide effects, such as cardio-pulmonary mortality, which was significantly increased by at least a factor of two. This disturbing observation is not correlated to intense short-term air pollution events that last for a few days, but more related to chronic and long-term exposure of significantly lower dosages that cover several weeks or even months.

1.7 Long-term exposure to airborne pollutants

As outlined previously, the olfactoric system offers a more straightforward option for nano-particle uptake. Apart from deteriorating effects on the olfactory bulb and alterations of the blood-brain barrier in response to chronic exposure to these aerosols, some neuropathologic effects have been observed and include (i) degeneration of cortical neurons, (ii) apoptotic glial white matter cells, as well as formation of (iii) non-neuritic plaques, and (iv) neurofibrillary tangles.(Calderón-Garcidueñas et al., 2004) There is mounting evidence that neurodegenerative disorders, such as Multiple Sclerosis,(Bizzozero et al., 2005) Alzheimer's-,(Hautot et al., 2003, Sayre et al., 1997) Parkinson's-disease,(Zhang et al., 1999) Amyotrophic Lateral Sclerosis (Shibata et al., 2001), and even Creutzfeldt-Jakob disease (Oberdörster & Utell, 2008) are favoured or at least promoted by chronic exposure to a cocktail of nano-sized particles and their associated chemical by-products.

The combination of UFPs, which are wrapped with a soluble organic fraction (e.g. SVCs/VOCs), with gaseous irritants (usually O_3 and NO_X) is known to increase the susceptibility, sensitization and chronic allergic inflammation that are associated with changes in the epithelial structure.(Traidl-Hoffmann et al., 2009; Galli et al., 2008) Over the past decades, a pronounced increase in allergic disorders has been witnessed particularly in western industrialized countries.(Ring et al., 2001) This prompted allergists to coin the so-called *hygiene hypothesis*, which assumed that frequent contact to pathogenic agents, particularly in early infancy, reduces the likelihood of the immune-system for a polarization towards immunoglobulin (IgE)-based responses (i.e. allergies). Such triggers include bacteria, molds, microbial agents (such as LPS), viruses, and potential antigens. With improved domestic hygiene standards in affected populations, contact with these agents is progressively prevented in favor of the propensity for allergies.(Galli et al., 2008; Ring et al., 2001; Yazdanbakhsh et al., 2001)

To shed some light on this intrinsically interwoven network, synergism between two types of aerosols shall be demonstrated – the combined effect of carbonaceous nano-particles, such as UFPs, and pollen allergens. Allergens, e.g. from grass pollen, attach to the surface of starch granules (amyloplasts) of plants or other plant fragments as so-called pollen-related micro-aerosols.(Solomon, 2002) Although such starchy granules are considerable larger (600 nm - 2.5 μm) than UFPs, they still are inhalable aerosol particles.(Taylor et al., 2002) Remarkably, allergen-carrying micro-aerosols have been shown to interact with DEPs resulting in stable aggregates.(Solomon, 2002) The release of allergens, amyloplasts and cytoplasmic debris occur by a rupture of pollen grains triggered by a cycle of wetting - e.g. dew, fog, gentle rain - and drying events.(Taylor et al., 2002) Additionally, airborne house dust is present as suspended particulate matter (5-10 μm), and can also act as a carrier of allergens; e.g. of the cat allergen Fel-d1, which becomes airborne from saliva and sebaceous glands of cats.(Chang, 2010; Ormstad et al., 1995) Similar carrier functions for Fel-d1 might also be assigned for DEPs. Fel-d1 was visualized on dust particles by scanning electron microscope (SEM) inspection of particles which have been labelled by monoclonal antibodies.(Ormstad et al., 1995) Furthermore, mite allergens, e.g. Der-p1, have been encountered up to 22.8 μg/g in carpet dust.(Warner, 2000) Although these data are not referring to UFPs,(Warner, 2000) release as airborne particulate matter cannot be ruled out. This is of relevance as DEPs readily sneak through the ventilation system thereby entering indoor environments, whereby they readily interact with suspended dust particles.(Guo, 2010) Since about 30% of inhaled 80-90 nm DEP-cluster can deposit in the alveolar region (Ormstad et al., 1995; Vostal, 1980) - see Figure 2 – UFPs that adsorb onto micro-aerosols are deposited in the upper respiratory system (typically in the naso-pharynx and tracheo-bronchial region with deposition efficiencies topping at least 80%),(Hofmann, 2011) where they unfold adverse synergistic or even novel effects.(Donaldson, 2009; Oberdörster & Utell, 2008)

In such cases, an exposure to carbonaceous nano-aerosols prior to challenge with allergens exerts strong adjuvant effects on the manifestation of allergic airway inflammation.(Alessandrini et al., 2006) Hence, allergen-sensitized individuals are more susceptible to the detrimental health effects of nano-particle exposure. Similar studies investigating the conditioning effects of DEPs with adsorbed volatile organics on the immune system confirmed the pro-allergenic potential. Thereby, a crucial role for these pollutants mediating the allergic breakthrough in atopic individuals, who have not yet developed an allergic disease, was suggested.(Lubitz et al., 2009) In this context it has to be stressed, that the conditioning effects of the immune system can occur within relatively short periods of time.(Alessandrini et al., 2006)

Furthermore, there is evidence that UFPs induce autoimmune diseases. Mice exposed to ambient PM on a weekly basis showed an accelerated onset of diabetes type-1. Adequately, similar risks were shown for lupus and collagen induced arthritis in murine models. Crohn's disease was found to be related to micro-particle contamination of food. However, one has to be aware that the apparent UFP-related autoimmune effects are multi-factorial. In case of diabetes type 1, correlated influences of O_3 co-exposure with cigarette smoke or particulate pollutants are known.(Chang, 2010)

Beside PAHs, bioavailable, ionizable metals might induce inflammation. Fly ash from domestic oil-burning furnaces was found to contain up to 166 μg total metal content per mg particulate matter. Metals increase recruitment of macrophages and neutrophils but also eosinophils - the latter are also involved in parasitic and allergic inflammation. An induced cell influx of UFPs was found to be correlated with an elevated intracellular metal content. O_2^- and H_2O_2 release during inflammation can occur via the Fenton reaction with transition metals boosting the oxidative burden by the creation of ROS.(Chang, 2010; Costa & Dreher, 1997)

As will be demonstrated in the following section, long-term exposure to elevated levels of PM mixed with inhalable nickel and arsenic vapors induces also epigenetic changes in human subjects.(Cantone et al., 2011) Thus, the information due to conditioning of the immune system leads to memorization of long-term exposure events at cellular level and is somatically passed on to progeny cells via epigenetic means. As will be exemplified, this also enables cells to leave marks on reproductive cells so that these events can be passed on the filial generation. This rather new field in genomic research is about to unravel - apart from the more rigid lower genome level (nucleotide sequence) - a second, very plastic level of information processing, which employs the epigenome and follows a kind of Lamarckian rule of inheritance.(Bird, 2007)

Passing from milder to more severe, chronic disease pattern, it can easily be deduced that (epi-)genotoxic properties of long-term nano-aerosol exposure replaces acute symptoms by unfolding the full spectrum that obviously includes even cancer cases. Table 2 summarizes some of the major findings of the epidemiological investigation made in Mexico City, known for its notorious pollutant laden air.(Calderón-Garcidueñas et al., 2004) Although the list is not extensive, the high occurrence of various types of cancers is striking. Just by considering the fact that only about 5-10% of cancer and cardiovascular cases can be attributed to heredity, it becomes obvious that 90-95% might be controlled by our lifestyle.(Willet, 2002) Hence, malignancies are derived from environmentally induced epigenetic alterations and not defective genes.(Jones, 2001; Seppa, 2000; Baylin, 1997) While (epi-)genetic factors determine the tendency towards malignancy, environmental factors largely contribute to heart-rate-variabilities, cardiovascular diseases, cancers, and other major causes of mortality.(Bacarelli et al., 2008, Baccarelli et al., 2010b, Willet, 2002)

Pollution level	Age [yrs]	Sex	Occupation	Schooling [yrs]	Clinical diagnosis
Low	34	♀	Housewife	14	Undiff. carcinoma
Low	46	♀	Housewife	10	Lung embolism
Low	49	♀	Housewife	10	Cervic. carcinoma
Low	53	♂	Carpenter	12	Myocard. infarction
Low	58	♂	Farmer	6	Renal carcinoma
Low	66	♂	Farmer	7	Gastric carcinoma
Low	68	♂	Laborer	6	Myocard. infarction
Low	73	♀	Housewife	6	Myocard. infarction
Low	76	♀	Fruit seller	9	Cervical carcinoma
High	32	♂	Policeman	13	DOA accident
High	38	♀	Secretary	15	DO A accident
High	39	♂	Office worker	12	DOA accident
High	42	♂	Electrician	12	Lung carcinoma
High	43	♂	Policeman	13	Myocard. infarction
High	52	♀	Housewife	6	Breast carcinoma
High	55	♂	Outdoor vendor	6	DOA accident
High	61	♂	Laborer	6	Colon Carcinoma
High	67	♀	Housewife	7	Cervical Carcinoma
High	83	♀	Housewife	7	Arrhythmia

Mexico City, the world's third largest urban area, has some of the worst air quality in the world.(Marley, 2006)

Table 2. Clinical data for subjects experiencing long-term UFP-pollution exposure. The primary causes of death included accidents resulting in immediate death (death on arrival, DOA), arrhythmias, myocardial infarctions, and carcinomas like: gastric, lung, colon, breast, and cervical cancers.(modified after Calderón-Garcidueñas et al., 2004) Insert: Aerial view of Mexico City revealing reduced visibility due to nano-particles from combustion sources.

2. Epigenetics

Since the onset of the initial reports addressing modification and associated inheritable changes of the DNA due to air pollution in animal studies,(Somers et al., 2004; Samet et al., 2004), research activities tried to tackle detrimental health effects related to nano-particle aerosol exposure.(Bacarelli 2009; Nawrot & Adcock, 2009) It soon became evident that the persistent effects of air pollution have a more pronounced effect on the phenotype rather than on the genotype.(Yauk et al., 2007; Vineis & Husgafvel-Pursiainen, 2005; Mahadevan et al., 2005) These findings support the already pursued hypothesis that mutagenic volatile chemicals adsorbed onto airborne particle pollutants induce somatic and germ-line mutations. While it was not always apparent why and how these mutations do occur, recent evidence suggests that a sensitive, easy to modulate layer (via methylation, acetylation, phosphorylation, etc.) is characterized by the so-called epigenome.(Mathers et al., 2010)

The epigenome, in its literal sense, can be regarded as a molecular sleeve sitting on top (*epi-*) of the genome. Without altering the DNA sequence itself, it enables or blocks the readout of the underlying genetic information.(Nadeau et al., 2010; Blum et al., 2010) In fact, many disorders, related either to epi- or genetic mutations can lead to similar or even congruent phenotypes, with the only difference that mutations of the genome are irreversible, whereas epigenomic changes in theory are plastic, thus of non-permanent nature. So far, the relationships between the genome and the epigenome have broadened the spectrum of molecular events, which are related to human diseases. These can be induced *de novo* or inherited, genetic or epigenetic, and most interestingly, some events are influenced by environmental factors. The findings that environmental factors, such as exposure to environmental stimuli (diet, toxins, and even stress) alter the epigenome provide insight to a broad spectrum of disorders. In a bee-hive for example, worker-bees and queens share the same genetic material, yet the differences among them does not consist in altered genetic information, but in the phenotype, fertility, size and life expectancy. The key ingredient that makes a larvae to develop into a worker-bee or a queen is purely based on its nutrition – that is workers receive mainly pollen and honey while queens are fed with royal jelly.(Haydak, 1970) This nutritional difference induces epigenetic modifications that determine the accessibility of genes for their expression (see schematic Figure 7). Similarly, it was possible to show that nutrition during embryonic development affects adult metabolism in humans and other mammals, via persistent alterations in DNA methylation. In particular, dietary supplementations have unintended deleterious influences on the establishment of epigenetic gene regulation.(Waterland & Jirtle, 2003) Even pure physico-environmental stimuli unveiled their epigenetic effects on stem cells that have been exposed to THz radiation. The non-destructive mode revealed athermal effects, like changes in cellular function in which some genes were activated, while others were suppressed.(Alexandrov et al., 2011) These are just some of many investigations that identified environmental factors as modulators of the epigenome and provide perspectives for developing interventions that might decrease the risk of developmental abnormalities, chronic inflammation, cancer, and neuropsychiatric disorders.(Zoghbi & Beaudet, 2007)

The basics of epigenetics are rooted in bio-physico-chemical processes, which can be considered as bookmarks placed into the book of life. Indeed, faulty epigenetic regulations are regularly behind chronic diseases. This can even extend towards a deactivation of specific, fully competent tumor suppression genes.(Rodríguez-Paredes & Esteller, 2011)

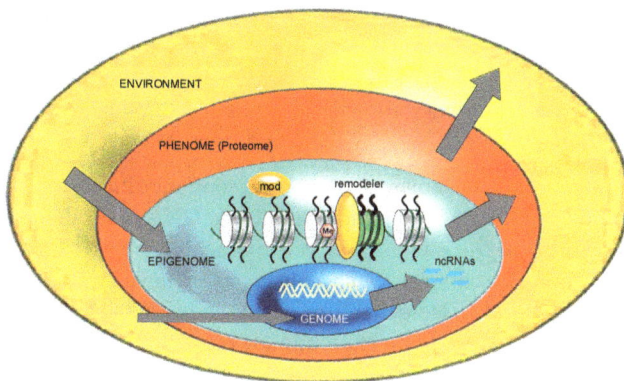

Fig. 7. The epigenome plays an essential role together with the genotype and environmental factors in determining phenotypes. Biochemical reactions affecting gene expression and genome stability include DNA methylation (Me), chromatin-remodelling complexes, covalent histone modifications (mod), the presence of histone variants, or non-coding regulatory RNAs (ncRNAs). The greyish arrows indicate the line and strength of progression.

2.1 Modulating the epigenome

As shown in Figure 8, enabling or inhibiting ribosomal activity to access genetic information is regulated by three major pathways. The first involves direct chemical modifications at DNA-level, the second regards the modification of histone proteins that are closely related with associated gene loci, whereas the third is mediated via various activities of non-coding RNAs.(Zoghbi et al, 2007) A central epigenetic regulatory mechanism comprises methylation of cytosine – whereby DNA-methyl-transferases attaches a methyl-group from the cofactor S-adenosyl-methionin (SAM) onto the cytosine atom C5. DNA methylation mostly takes place at cytosine-guanine-nucleotide. Numerous copies of these dinucleotide sequences are

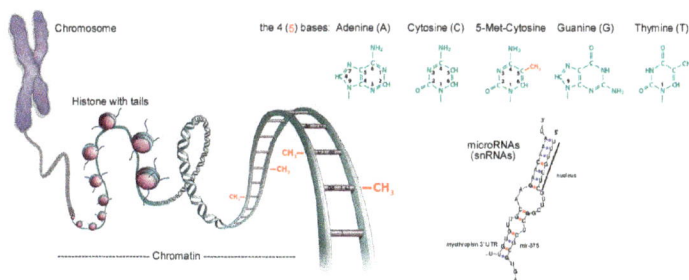

Fig. 8. Supercoiling of DNA via histone mediated proteins triggers a cascade of condensation steps that yield the highly packed mitotic chromosome. Highlighted are the various levels of epigenetic modulation. Methylation at base-level (5-Met-Cytosine) represses gene activity and boosts chromatin condensation, histone-tail modification (e.g. di-methylation, acetylation, etc.) and alters DNA-wrapping thereby inducing de-/activation. Small nuclear RNA (snRNA).(modified after Walker & Gore, 2011 & Qiu, 2006).

located within CpG-islands that constitute the promoter region of genes. If methylation takes place at these sites, it mostly shuts down the corresponding genes.(Esteller 2002) In case of hyper-methylation of promoter regions – as often encountered in tumor suppressor genes – a permanent shut down of these regions is the obvious result, which causes the affected cell to degenerate or mutate into pre-cancerogenous conditions.(Yang et al., 2005)

Another crucial epigenetic mechanism concerns histone-modification and addresses the dense packing of the DNA-filament. If stretched out, the DNA of a cell would cover at least two meters in total length. Supercoiling of the filament enables its packing into a cell nucleus of 10 to 100 μm in diameter. Such packing is achieved via small basic proteins, termed histone cores. Some 147 base-pairs wrap around a histone-octamer, which consists of two H2A/H2B-dimers as well as a H3/H4-tetramer.(Luger et al., 1997) This protein-DNA-unit is known as a nucleosome. Upon DNA-attachment of the linking histone H1, supercoiling occurs in a cascading manner until chromatin is present in its most condensed form: the chromosome (Figure 8). Epigenetics on the histone-level affects the N-terminal tails that protrude from the histone-octamer (see Figure 8). In particular, this regards the basic amino acids lysine and arginine but also serine and threonine. Histone-modifying enzymes append or dislodge specific amino acids. These modifications include methyl-, acetyl- and phosphoryl-groups as well as larger molecules, such as ubiquitin or ADP-ribose.(Biel et al., 2005) In analogy to the genetic code, one refers here to the histone code, as modifications of the basic histone proteins convey specific information.(Strahl & Allis, 2004) Modification of the N-terminal tails loosen the chromatin's density that enables genes to be readily accessible and transcribed. The opposite effect regards densification of the chromatin and concomitant inhibition of gene readability. In addition to the aforementioned modifications, there are also numerous non-histone proteins that can be biochemically modified, likewise yielding de- or activation of corresponding genes. This regards in particular the tumor-suppressor protein p53, which can be deactivated through deacetylation of histone deacetylase-1 (HDAC1).(Luo et al., 2000) Furthermore, a surprisingly large number of RNAs neither functions as messenger, transfer or ribosomal RNAs, and are thus called non-coding RNAs (ncRNAs). Such RNAs regulate gene expression on various levels, including chromatin modification, transcription, RNA modification, RNA splicing, RNA stability and translation. Among them, small interfering RNAs (siRNAs) and microRNAs (miRNAs) are most prominent. Both regulate gene expression through the RNA interference (RNAi) pathway. More than 1% of predicted genes in higher eukaryotic genomes and up to 30% of protein-encoding genes are assumed to be subjected to miRNA regulation. In addition, miRNAs cooperate with transcription factors (TFs) to control gene expression.(Yu, 2007)

Although all cell types within an organism share identical genetic material, they perform different tasks. Task sharing requires a cell-specific readout of genes, which is realized by biochemical markers. Here, epigenetics controls the fate of progeny cells after mitotic division. i.e. lung specific stem cells yield differentiated lung cells although their genome would potentially enable differentiation into any kind of cell. This kind of cellular memory requires to be passed on to the progeny cells. Thus, cellular "learning" predominantly occurs in two ways: either via "bookmarking" or via "paramutation".(Hollick et al., 1997, Sarge & Park-Sarge, 2005) While the former regards cellular regeneration during the fetal stage all the way through the adulthood, the latter is a characteristic feature for the embryonic phase.

2.2 Adult epigenome

Bookmarking transmits cellular memory (i.e. patterns of cellular gene expression) via mitosis to somatic progeny cells of the same type. Throughout one's lifespan, tissue-specific stem cells are responsible for the development and regeneration of entire organs, such as skin, lung, gut, blood system, etc. In order to meet this task, these stem cells, besides revealing an extensive self-renewal potential, encompass also pluripotency. These properties give rise to all cell types of an organ that differentiate to multipotent progenitors with gradually restricted developmental potential. These progenitors subsequently undergo commitment to one of several lineages and then differentiate along the selected pathway into a functionally specialized cell type of that organ.(Fisher, 2002) In other words, stem cells and resulting progenitors as well as specialized tissue cells share the same genome. Yet, the lower the ranking within cellular ontogenesis, the more genes have to be silenced in order to fulfil the requirements that match organ function. Practically, a healthy somatic epithelial lung stem-cell divides to yield a progeny cell that becomes epigenetically tagged in such a way as to provide a specialized cell. This differentiated cell becomes part of the cellular consortium that constitutes an ensemble yielding the lung with all its physiological functions. Under normal physiological conditions, it would be senseless to differentiate into a cell linage other than e.g. specialized epithelial lung cells. This kind of epigenetic tagging (e.g. epithelial lung cell-linage) is stable and heritable such that a mitotically dividing cellular system gives rise to more cells that correspond to the overall phenotype.(Tost, 2008) Studies based on monozygotic twins with similar epigenomes during early years of life revealed remarkable differences in methylated DNA and acetylated histones during later stages of life. This underlines the temporal metastability of the epigenome. (Fraga et al., 2005)

Now, how is epigenetics related to nano-aerosol exposure? As highlighted in Figure 9, environmental exposure of any kind acts as a modulator to the metastable epigenome.(Anway et al., 2005) Metastability affects responsiveness to oxidative stress (Figure 6) and as such

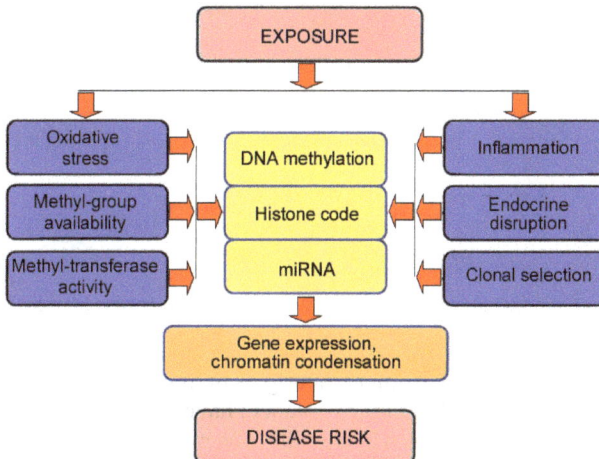

Fig. 9. Potential mechanism linking environmental exposures to epigenetic effects. These effects include DNA methylation, histone codes and miRNA expression. The associated changes modify chromatin organization and condensation, gene expression and ultimately disease risks.(modified after Baccarelli & Bollati, 2009)

renders the organism more susceptible to cardio-vascular as well as respiratory effects of air pollution. The resulting adverse health effects include generation of oxidative stress, inflammation as well as morbidity.(Bollati & Baccarelli, 2010) It was possible to demonstrate that the underlying mechanism regards methylation of the promoter region of the iNOS (inducible nitric-oxide synthase) – found to be suppressed in foundery workers who were exposed to UFPs.(Tarantini et al., 2009) The same authors also reported demethylation effects induced by long-term exposure to particle mass (PM_{10}) exposure in younger individuals individuals. These effects resemble demethylation patterns that are typically observed in old age. Histone modifications have been observed in workers who experienced long-term exposure to nano-sized aerosols at smelters. Both acetylation of the histone H3K9 and demethylation of H3K4 increased by about 15% after a >21 year exposure near the smelter.(Cantone et al., 2011)

Another key concept for the developmental origin of diseases comprises a transfer of the acquired predisposition onto subsequent generations without further environmental impacts (see Figure 10). Such trans-generational epigenetic inheritance is responsible for wider effects, such as a fast-track pathway of adaptation to environmental stress. This improves survival until either more stable genetic changes can provide better adaptations or the environment reverts to the previous status.(Finnegan, 2002)

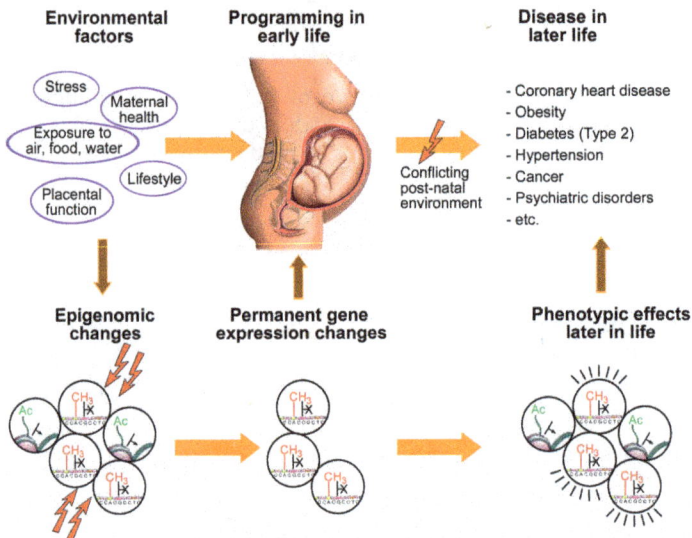

Fig. 10. Epigenotype model of developmental origins of disease. Environmental factors acting in early life (from conception to early infancy) have consequences, which become manifest as an altered risk for diseases in later life. The mother conveys a forecast of the post-natal environment to the genome of the unborn. This includes modifications to its metabolism, whole body physiology and growth trajectory to maximize its chances of post-natal survival. These adaptations become detrimental if the environmental conditions after birth differ from those of the fetal stage. Ac - Histone acetylation/active genes; CH_3 - DNA methylation/silent genes.(modified after Sandovici et al., 2008)

2.3 Embryonic epigenome

Paramutation in comparison to bookmarking, regards the quasi "inheritance" of gene-characteristics (allelic interactions) that are "remembered" and expressed in later generations (e.g. via the germ cell linage). Paramutation occurs when certain control alleles impose an epigenetic imprint on susceptible (paramutable) inferior alleles. The epigenetic imprint is inherited through meiosis and persists even after the interacting alleles have segregated. The observation of heritable but reversible changes in gene expression, is evidence for non-Mendelian genetics, and apparently also in mammalian systems.(Chanlder & Stam, 2004) Paramutation fulfils the criteria for a parental identity mark or "imprint" because it can be established either in the sperm or the oocyte by *de novo* methyl-transferases that act only in one gamete. Once established, it can be stably propagated at each embryonic cell division by a maintenance methyl-transferase, and it can also be erased in the germ line to reset the imprint in the next generation, either by passive demethylation or possibly through the action of a demethylase.(Barlow & Bartolomei, 2008).

The newly fertilized egg or zygote is unique since no other cell has the potential to develop into an entire organism. In order to achieve that, epigenetic marks of both oocyte (female) and sperm (male) are usually efficiently reprogrammed at fertilization, so that upon fertilization the embryonic genome becomes totipotent – an essential property of the zygote.(Surani & Reik, 2007) At the first mitosis after fertilization, most histone marks are quite similar on the maternal and paternal chromosomes.(Santos et al, 2005) To yield the totipotent zygote, dramatic DNA demethylation of the parental genome must be induced by "active demethylation".(Morgan et al., 2005) Hence, gene expression depends on the origin of inheritance. This implies that at an imprinted diploid locus, there is unequal expression of the maternal and paternal alleles. So much so, that in each generation, the parent-specific imprinting marks have to be erased, reset, and maintained. It is obvious that imprinted loci are somewhat prone to errors that may occur during these processes. Indeed, at such an early stage of development, erroneous imprinting in genes, which encode proteins involved in DNA methylation, binding to methylated DNA, and histone modifications may contribute to the fast-growing class of human disorders affecting the epigenome.(Zoghbi & Beaudet, 2007) After further cell divisions, the embryonic stem cells differentiate into roughly 200 different cell lineages, whereby totipotency is gradually downregulated to pluripotency.(Tada et al., 1997; Tada et al., 2001; Cowan et al., 2005) The latter give rise to adult stem cells that generate tissue specific cells.(Surani & Reik, 2007). Tissue-specific stem cells are responsible for the development and regeneration of entire organs, such as skin, gut, and blood system throughout life. To comply with this task, stem cells encompass two unique properties: (i) an extensive self-renewal potential that enables them to propagate in their uncommitted state; (ii) pluripotency that gives rise to all cell types of an organ by differentiation to multipotent progenitors with gradually restricted developmental potential. These progenitors subsequently undergo commitment to one of several cell lineages and then differentiate along the selected pathway into a functionally specialized cell type of the organ.(Fisher, 2002) Transcription factors reprogram the expression of large sets of genes. They act indirectly by (i) affecting gene expression programs (antagonizing other transcription factors through protein-protein interaction) and directly by (ii) the control of gene transcription (recruiting coactivators or corepressors with histone-modifying or chromatin-remodeling activities to regulatory DNA elements). Hence, the activity of a gene is influenced by the local DNA methylation pattern, the state of histone modifications, the nuclear position of the gene relative to repressive heterochromatin domains, and the architecture of the gene locus.[Vickaryous & Hall, 2006] In any case, epigenetic imprinting associated with the in- and/or activation of genes implies that changes acquired during gametogenesis are not only passed on but also extend into embryonic development.[Jaenisch & Gurdon, 2007] In animal studies it was possible to correlate nano-size particle exposure of a smelter to hypermethylation of sperm-DNA that persisted into the next generation even though the filial generation was no longer exposed to this aerosol.[Yauk et al., 2008] Since epigenotypic flexibility regards plasticity during fetal and embryonic development and extends well into the post-natal phase, it is not surprising that the epigenome contributes not only to developmental human disorders, but also to post-natal and even adult diseases.[Jaenisch & Gurdon, 2007] With reference to Table 2, it becomes evident that chronic disease processes can be readily attributed to a long-term exposure to environmental nano-aerosols.

2.4 Chronic diseases and cancer

It is well established that complex diseases, such as heart diseases, diabetes, obesity, Alzheimer's disease, schizophrenia, and bipolar disorder etc. result from the interplay between (epi-)genetic and environmental factors. The interaction with nano-aerosol exposure not only conditions the immune system, thereby inducing to allergic reactions, but also affects the epigenome, which leads to neuro-degenerative disorders or even towards malignancy. It has also been proposed that epigenetic mechanisms explain significant fluctuations of the phenotype between individuals, or even dramatic changes in the incidence of some diseases over short periods of time, such as the rapid increase of asthma incidence in the population.[Petronis, 2001; Bjornsson et al, 2004] Examples include increased methylation levels at the estrogen receptor-alpha and the estrogen receptor-beta gene promotors in proliferating human aortic smooth muscle cells and in atherosclerotic cardiovascular tissues, respectively [Ying et al, 2000; Kim et al, 2007] It has also been demonstrated that

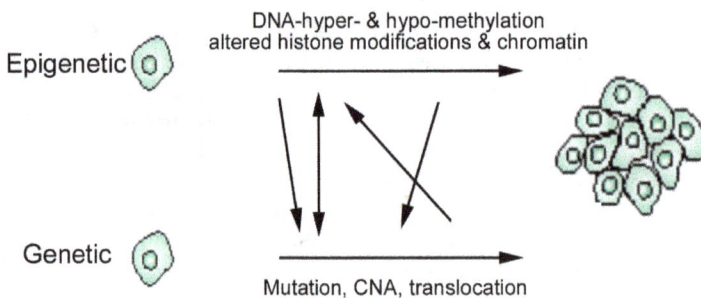

Fig. 11. Interacting pathways to cancer - a mechanism-based model of the pathogenesis of human cancer. Sporadic cancers, which comprise 90-95% of all cancers, almost uniformly exhibit both genetic and epigenetic defects. As suggested by the vertical arrows, these mechanisms show substantial interaction. That is, epigenetic events can cause genetic events, and *vice versa*. Depending on the cancer type, each mechanism can be operative early, late or continuously during the development of the tumor (horizontal arrows). CNA: copy number alteration, including gain, loss and amplification.[modified after Costello & Brena, 2008]

workplace exposure to nano-particles and their associated volatile chemicals induce global demethylation especially of retro-transposons in LINE and SINE sequences. Lowered LINE-1 methylation in peripheral blood leukocytes is a predictor of incidence and mortality from ischemic heart disease (IHD) and stroke.(Baccarelli et al., 2010a, Baccarelli et al., 2010b) Blood samples screened for LINE-demethylation in exposed individuals (e.g. traffic wardens) can be up to 5% lower than in non-exposed subjects.(Bollati et al., 2007) This induces not only premature aging, increased risk of IHD and stroke, but also paves the way to chronic pathology, such as cancer. As stated before, numerous investigations of the epidemiology of cancer reveal that only 5 to 10% of breast, prostate or bowel cancer and 1-2% of melanoma cases are attributable to genetic mutations, while the large bulk does not involve an inherited predisposition at all.(Barlow-Stewart et al., 2007)

Several studies indicate an age-dependent decrease of global DNA methylation pattern, yet evidence suggests that there might be site-specific hyper-methylation involved in cancerogenesis.(VanHelden 1989, Cooney 1993; Rampersaud, 2000) Given the large body of data linking altered DNA methylation to cancer risk or progression, obviously epigenetic changes contribute to the age-related increase in cancer risk.(Mays-Hoopes, 1989; Issa, 1994) The role of diet as a contributing factor in controlling global methylation and its relationship to cancer development has already been illustrated in several cases.(Ingrosso et al., 2003; Waterland & Jirtle, 2003) Thereby epigenetic changes take place despite the pre-systemic metabolic effect of the liver. Since this first-pass effect of the liver is almost inexistent during inhalation, it is apparent that modifications to the epigenome in response to airborne stimuli are even more pronounced and thus directly linked to chronic exposure to airborne pollution. The synergism of nano-aerosols and VOC exposure is known to reduce epigenetic imprinting.(Bollati et al., 2007) As already outlined, suppressed LINE-1 methylation by just 10% (from 82.4% to around 72.6%) increases the associated risk of cancer by a factor of seven.(Zhu et al., 2011) Although aging is the major risk factor associated with cancer development, epigenetic modifications occurring earlier in life underline the important role of the environment in various cancer types. Epigenetic changes induced by environmental factors in the pool of progenitor stem cells of each tissue might be the earliest events during carcinogenesis. It is assumed that the epigenetic progenitor model provides a plausible explanation for both the age dependency and environmental sensitivity of associated cancer risks.(Feinberg et al., 2006). This multistep process implies that an environmental insult induces changes in the epigenome during the early environment at particularly sensitive time-windows during developmental plasticity.(Ozanne & Constincia, 2007) Once environmentally induced epigenetic changes are established, they will be maintained throughout many cell divisions by the epigenetic signalling procedure. The maintenance of such altered epigenetic states leads to stable alterations in gene expression with physiological adapted consequences. Epigenetic programming defects can become irreversible if aberrant organ growth and differentiation ensues as a consequence of an acute response to a transient environmental insult.(Costello & Brena, 2007) Nano-aerosols in combination with benzene exposure for instance are known to induce a significant reduction in LINE-1 and Alu methylation. Both are related to acute myelogenous leukemia.(Snyder, 2002) Likewise, airborne benzene was also associated with hypermethylation in protein p15 and hypomethylation of the cancer-antigen gene MAGE-1.(Bollati et al., 2007)

Since the involvement of proteins in epigenetic pathways is tightly regulated, perturbations at a given level – e.g. through loss-of-function – inevitably will cause human disorders. This implies that epigenetic imprints do also affect transcription, RNA splicing, and protein modifications.(Zoghbi et al., 2007) Evidence is given by studies employing both animals (Liu et al., 1997; Waterland & Jirtle, 2003; Weaver et al., 2004; Wolff et al. 1998) as well as humans (Albert el al., 2005; Bottiglieri et al. 1994; Reynolds et al. 1984;), emphasize the crucial role of epigenetic modulation and the onset of a chronic disease pattern.

From all stated facts, it is obvious that the organism is capable to adapt to a vast range of environmental exposures. The resulting epigenetic modifications are stable and heritable via cell divisions for a given lifespan and affect the phenotypic appearance both on the cellular as well as on the organismic level.(Thaler, 1994) Figure 12 presents a dynamic map that highlights the feedback-cycles of involuntarily as well as deliberate environmental exposure. Thus, epigenetics makes it possible to associate a given exposure related lifestyle to the corresponding phenotype. Hence, long-term environmental exposure must be epigenetically manifest also within the germ cell linage. In this case environmentally related stress-information is passed on to progeny. However, inherited epigenetic imprints are not entirely permanent, as paramutating effects have been documented to persist for up to three generations before they are lost again without altering the sequence of the DNA itself. (Jablonka, 2001)

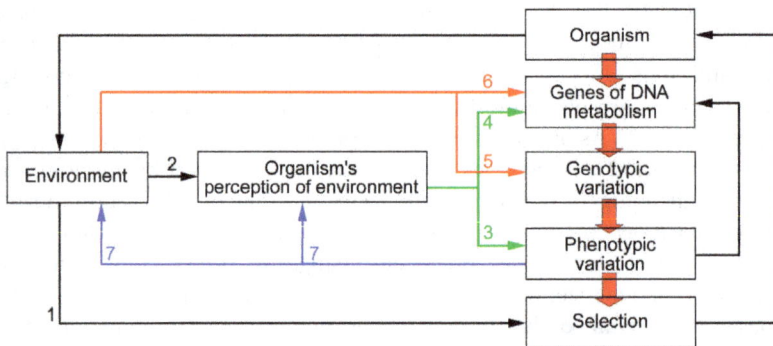

Fig. 12. Environmental influences, exposure to constituents in drinking water, consumed food, inhaled air along with stress and emotions, epigenetically modify genes, without altering the nucleotide sequence. The numbers above outline hierarchical interdependences of eco-systemic relationship in which the organism is embedded in: (1) The environment is the proximate agent of selection. (2) Organism perceives environment. (3) Organismic perception acts on physiology. (4) Organisms modifies genetic metabolism. (5) Environmental impact on DNA. (6) Environmental interaction with genes. (7) Organismic modification of environmental interaction.(modified after Thaler, 1994)

3. Conclusion

Exposure to anthropogenically released nano-aerosols originating mainly from incomplete combustion processes fully unfold their toxicological potential – particularly in congested areas. These particles have multitude ways to enter the body, including adsorption via the skin, the olfactoric nerve bundles (inducing neurodegenerative diseases), deposition within the respiratory tract (directly related to respiratory diseases), ingestion of cleared particle laden mucus (in combination with the former mode of entry, responsible of the wider organismic health problems). Redistribution of the cellularly absorbed particle load throughout the organism is achieved via the blood circulatory and the lymphatic system. Toxicity itself not only depends on the nature of the particle (solubility and hydrophobicity) but also on the surface structure, onto which volatile substances and radionuclides can adsorb. Since nano-sized particles exert more severe distressing effects on the cellular level than their coarser siblings, these aerosols significantly contribute to chronic disease patterns and epigenetic imprinting, which enables acquired lifestyle-related stress-response patterns to be passed on to subsequent generations. Due to the fact that these particles access secondary target organs along with their effects on the organismic level, they definitely will attain further toxico-medical attention in the near future. The availability of more and more products containing designed and engineered nano-particles and fibers, contributes to this dilemma as awareness of associated risks and benefits have not yet led to regulatory guidelines in order to limit side-effects of improper production methods, inadequate usage and irresponsible disposal of these materials. While the exact risk of aerosolized particles is still cumbersome to define, current scientific evidence already stresses the adverse effects of long-term exposure as it tilts the balance towards the emergence of so called "civil-society-related" diseases that so far were either considered harmless or not yet associated to these environmental stressors. While it is not always possible to assign a single detrimental aerosolized agent to a particular disease, the evidence given so far indicates that xenobiotic nano-aerosols along with the adsorbed cocktail of semi-/volatile organic compounds should be considered as promoters and modulators in the emergence of chronic diseases. This interrelation has scarcely been considered in the past. Since epigenotypic flexibility regards plasticity during embryonic development, the post-natal phase, and well into adulthood, it is not surprising that epigenetic imprinting due to airborne nano-aerosol exposure not only contributes to developmental disorders, but also to post-natal and even adult human diseases. This perspective will disproportionally challenges our existing medicare system. This Lamarckian-type of inheritance is achieved via various processes and involves in particular DNA-methylation, histone modifications, mRNA silencing and other regulatory interference mechanisms. Although it is well known that anthropogenic aerosols exert their effects on the cellular, tissue, organ, and organismic level, interference with the phylo-onto-genetic patterns, currently investigated in the field of epigenetics, open a new chapter to these issues. This line of argument may point towards a new understanding of health and disease, whereby the latter should just be regarded as an organismic proxy indicator in the attempt to attain a new organismic steady state. Hence, extended stress exposure (as is the case of environmental nano-aerosols) contain distressing agents that shift the constantly fluctuating homeostatic balance into new oscillating instabilities. However, the difference between these states lies in the fact that in the latter stages affected individuals increasingly feel physically less fit than in the former from which they have been kicked out of.

4. Acknowledgements

The author wants to express his appreciation to Prof. Hanno Stutz from the University of Salzburg (Division of Chemistry and Bioanalytics, Department of Molecular Biology) for substantial input and improvements of the text. This book-chapter builds upon a previously published mini-review available by Madl & Hussain (2011).

This publication was supported by the "*Stiftungs- und Förderungsgesellschaft* on behalf of the Paris Lodron University of Salzburg, Austria.

5. References

Alessandrini F., Schulz H., Takenaka S., Lentner B., Karg E., Behrendt H., Jakob T. (2006). Effects of ultrafine carbon particle inhalation on allergic inflammation of the lung. J. Allergy Clin. Immun., Vol.117: 824-830.

Alessandrini F., Beck-Speier I., Krappmann D., Weichenmeier I., Takenaka S., Karg E., Kloo B., Schulz H., Jakob T., Mempel M., Behrendt H. (2009). Role of Oxidative Stress in Ultrafine Particle-induced Exacerbation of Allergic Lung Inflammation. Am. J. Respir. Crit. Care Med., Vol.179: 984-991.

Alexandrov B.S., Rasmussen K.Ø., Bishop A.R., Usheva A., Alexandrov L.B., Chong S., Dagon Y., Booshehri L.G., Mielke C.H., Phipps M.L., Martinez J.S., Chen H.T., Rodriguez G. (2011). Non-thermal effects of terahertz radiation on gene expression in mouse stem cells. Biomedical Optics Express, Vol.2(9): 2680-2689.

Anway M.D., Cupp A.S., Uzumcu M., Skinner M.K. (2005). Epigenetic transgenerational actions of endocrine disrupters and male fertility. Science, Vol.308(5727): 1466-1469.

Avogbe P.H., Ayi-Fanou L., Autrup H., Loft S., Fayomi B., Sanni A., Vinzents P., Møller P. (2005). Ultrafine particulate matter and high-level benzene urban air pollution to oxidative DNA damage. Carcinogenesis, Vol.26(3): 613-620.

Baccarelli A., Cassano P.A., Litonjua A., Park S.K., Suh H., Sparrow D., Vokonas P., Schwartz J. (2008). Cardiac autonomic dysfunction: effects from particulate air pollution and protection by dietary methyl nutrients and metabolic polymorphisms. Circulation, Vol.117(14): 1802-1809.

Baccarelli A. (2009). Breathe deeply into your genes: genetic variants and air pollution effects. Am. J. Respir. Crit. Care. Med., Vol.179(6):431-432.

Baccarelli A., Bollati V. (2009). Epigenetics and environmental chemicals. Curr. Opin. Pediatr., Vol.21(2): 243-251.

Baccarelli A., Rienstra M., Benjamin E.J. (2010a). Cardiovascular epigenetics: basic concepts and results from animal and human studies. Circ. Cardiovasc. Genet., Vol.3(6): 567-573.

Baccarelli A., Wright R., Bollati V., Litonjua A., Zanobetti A., Tarantini L., Sparrow D., Vokonas P., Schwartz J. (2010b). Ischemic heart disease and stroke in relation to blood DNA methylation. Epidemiology, Vol.21(6):819-828.

Barlow-Stewart K., Dunlop K., Reid V., Saleh M. (2007). The Australian Genetics Resource Book. Centre for Genetics Education. Fact Sheet No. 48, 49, 50, 51; avaialable online (accessed in Feb. 2012): www.genetics.edu.au/factsheet

Barlow D.P., Bartolomei M.S. (2007). Genomic Imprinting in Mammals. In: Epigenetics, Ch.19. Allis C.D., Jenuwein T., Reinberg D., Caparros M.L. eds., EPIGENETICS. Cold Spring Harbor Laboratory Press, New York - USA.

Baron B.A., Willeke K. (2001). Aerosol Measurement Principles, Techniques and Applications, 2nd ed.; Ch.25: p.781-782; Wiley-Interscience, New York, USA.

Baylin S.B. (1997). DNA Methylation: Tying It All Together: Epigenetics, Genetics, Cell Cycle, and Cancer. Science, Vol.277(5334): 1948-1949.

Biel M., Wascholowski V., Giannis A. (2005). Epigenetik – ein Epizentrum der Genregulation: Histone und histonmodifizierende Enzyme. Angew. Chem., Vol.117(21): 3248-3280.

Bird A. (2007). Perceptions of epigenetics. Nature, Vol.447(7143), 396-398.

Bizzozero O.A., DeJesus G., Bixler H.A., Pastuszyn A. (2005). Evidence of nitrosative damage in the brain white matter of patients with multiple sclerosis. Neurochem. Res., Vol.30(1): 139-149.

Bollati V., Baccarelli A., Hou L., Bonzini M., Fustinoni S., Cavallo D., Byun H.M., Jiang J., Marinelli B., Pesatori A.C., Bertazzi P.A., Yang A.S. (2007). Changes in DNA methylation patterns in subjects exposed to low-dose benzene. Cancer Res., Vol.67(3): 876–880.

Bollati V., Baccarelli A. (2010). Environmental epigenetics. Heredity, Vol.105(1):105-112.

Blum J.L., Hoffman C., Xiong J.Q., Zelikoff J.T. (2010). Exposure of Pregnant Mice to Cadmium Oxide (CdO) Nanoparticles (NP) Poses a Risk to the Developing Offspring. Biology of Reproduction, Vol.83: 295.

Bottiglieri X., Hyland K., Reynolds E.H. (1994). The clinical potential of adenomethionine (S-adenosylmethionine) in neurological disorders. Drugs, Vol.48(2): 137-152.

Brook R.D., Franklin B., Cascio W., Hong Y., Howard G., Lipsett M., Luepker R., Mittleman M., Samet J., Smith S.C., Tager I. (2004). Air pollution and cardiovascular disease: a statement for healthcare professionals from the expert panel on population and prevention science of the American Heart Association. Circulation, Vol.109(21): 2655–2671.

Calderón-Garcidueñas L., Reed W., Maronpot R., Henriquez-Roldán C., Delgado-Chavez R., Calderón-Garcidueñas A., Dragustinovis I., Franco-Lira M., Aragón-Flores M., Solt A.C., Altenburg M., Torres-Jardón R., Swenberg J.A. (2004). Brain Inflammation and Alzheimer's-Like Pathology in Individuals Exposed to Severe Air Pollution. Toxicol. Pathol., Vol.32(6): 650-658.

Cantone L., Nordio F., Hou L.F., Apostoli P., Bonzini M., Tarantini L., Angelici L., Bollati V., Zanobetti A., Schwartz J., Bertazzi P.A., Baccarelli A. (2011). Inhalable Metal-Rich Air Particles and Histone H3K4 Dimethylation and H3K9 Acetylation in a Cross-sectional Study of Steel Workers. Environ. Health Perspect., Vol. 119(7): 964-969.

Casals E., Vazquez-Campos S., Bastus N.G., Puntes V. (2008). Distribution and potential toxicity of engineered inorganic nanoparticles and carbon nanostructures in biological systems. Trends Analyt. Chem., Vol.27(8): 672-683.

Chan R.C.-F., Wang M., Li, N., Yanagawa Y., Onoé K., Lee J.J., Nel A.E. (2008). Pro-oxidative diesel exhaust particle chemicals inhibit LPS-induced dendritic cell responses involved in T-helper differentiation. J. Allergy Clin. Immunol., Vol. 118(2): 455-465.

Chandler V.L., Stam M. (2004). Chromatin conversations: Mechanisms and implications of paramutation. Nat. Rev. Genet., Vol. 5(7): 532-544.

Chang C. (2010). The immune effects of naturally occurring and synthetic nanoparticles. J. Autoimmun., Vol.34(3): J234-J246.

Cheng Y.W., Lee W.W., Li C.H., Kang J.J. (2004). Genotoxicity of motorcycle exhaust particles in vivo and in vitro. Toxicol. Sci., Vol.81(1): 103-111.

Cooney C.A. 1993. Are somatic cells inherently deficient in methylation metabolism? A proposed mechanism for DNA methylation loss, senescence and aging. Growth Dev. Aging, Vol.57(4): 261-273.

Costa D.L., Dreher K.L. (1997). Bioavailable Transition Metals in Particulate Matter Mediate Cardiopulmonary Injury in Healthy and Compromised Animal Models. Environ. Health Perspect., Vol.105(5): 1053-1060.

Costello J.F., Brena R.M. (2007). Cancer Epigenetics. In Epigenetics, Ch.12. Tost J. (ed.). Caister Academic Press Norfolk - UK.

Cowan C.A., Atienza J., Melton D.A., Eggan K. (2005). Nuclear reprogramming of somatic cells after fusion with human embryonic stem cells. Science Vol. 309: 1369-1373.

DeNicola G., Karreth F.A., Humpton T.J., Gopinathan A., Wei C., Frese K., Mangal D., Yu K.H., Yeo C.J., Calhoun E.S., Scrimieri F., Winter J.M., Hruban R.H., Iacobuzio-Donahue C., Kern S.E., Blair I.A., Tuveson D.A. (2011). Oncogene-induced Nrf2 transcription promotes ROS detoxification and tumorigenesis. Nature, Vol.475(7354): 106-109.

Devouassoux G., Saxon A., Metcalfe D.D., Prussin C., Colomb M.G., Brambilla C., Diaz-Sanchez D. (2002). Chemical constituents of diesel exhaust particles induce IL-4 production and histamine release by human basophils. J. Allergy Clin. Immunol., Vol.109(5): 847-853.

Dockery D.W., Ware, J.H., Ferris B.G., Speizer, F.E., Cook N.R., Herman S.M. (1982). Change in pulmonary function in children associated with air pollution episodes. J. Air Pollut. Control. Assoc., Vol.32(9): 937-942.

Dockery D.W., Pope C.A., Xu X.P., Spengler J.D., Ware J.H., Fay M.E., Ferris B.G., Speizer F.E. (1993). An Association between Air Pollution and Mortality in Six U.S. Cities. N. Engl. J. Med., Vol.329: 1753-1759.

Donaldson K., Li X.Y., MacNee W. (1998). Ultrafine (Nanometre) Particle Mediated Lung Injury. J. Aerosol Sci., Vol.29(5-6): 553-556.

Donaldson K. (2009). The toxicology of nanoparticles. Queens Medical Research Institute. University of Edinburgh; avaialable online (accessed in Feb. 2012): www.morst.govt.nz/Documents/work/nanotech/Ken-Donaldson--Toxicology-of-Nanoparticles.pdf

Dybdahl M., Risom L., Møller P., Autrup H., Wallin H., Vogel U., Bornholdt J., Daneshvar B., Dragstedt L.O., Weimann A., Poulsen H.E., Loft S. (2003). DNA adduct formation and oxidative stress in colon and liver of Big Blue® rats after dietary exposure to diesel particles. Carcinogenesis, Vol. 24(11): 1759-1766.

Elder A.C., Gelein R., Finkelstein J.N., Cox C., Oberdörster G. (2000). Pulmonary inflammatory response to inhaled ultrafine particles is modified by age, ozone exposure, and bacterial toxin. Inhal. Toxicol., Vol.12(Suppl 4): 227-246.

Esteller M. (2002). CpG Island Hypermethylation and tumor suppressor genes: a booming present, a brighter future. Oncogene, Vol. 21(35): 5427-5440.

Feinberg A.P., Ohlsson R., Henikoff S. (2006). The epigenetic progenitor origin of human cancer. Nat. Rev. Genet. 7(1): 21-33.

Fahy O., Tsicopoulos A., Hammad H., Pestel J., Tonnel A.-B., Wallaert B. (1999). Effects of diesel organic extracts on chemokine production by peripheral blood mononuclear cells. J. Allergy Clin. Immunol., Vol.103(6): 1115-1124.

Fraga M.F., Ballestar E., Paz M.F., Ropero S., Setien P., Ballestar M.L., Heine-Suner D., Cigudosa J.C., Urioste M., Benitez J., Boix-Chornet M., Sanchez-Aguilera A., Ling C., Carlsson E., Poulsen P., Vaag A., Stephan Z., Spector T.D., Wu Y.Z., Plass C.,

Esteller M. (2005). Epigenetic differences arise during the lifetime of monozygotic twins. Proc. Natl. Acad. Sci., Vol.102(30): 10604-10609.

Galli S.J., Tsai M., Piliponsky A.M. (2008). Review: The development of allergic inflammation. Nature, Vol.454(7203): 445-454.

Geiser M., Kreyling W.G. (2010). Deposition and biokinetics of inhaled nanoparticles. Part. Fiber Toxicol., Vol.7(2): 1-17.

Guo H., Morawska L., He C.R., Zhang Y.F., Ayoko G.A., Cao M. (2010). Characterization of particle number concentrations and PM2.5 in a school: influence of outdoor air pollution on indoor air. Environ. Sci. Pollut. Res., Vol.17(6): 1268-1278.

Hautot D., Pankhurst Q.A., Khan N., Dobson J. (2003). Preliminary evaluation of nanoscale biogenic magnetite in Alzheimer's disease brain tissue. Proc. Biol Sci., Vol.270(suppl. 1): S62-S64.

Haydak M.H. (1970). Honey bee nutrition. Ann. Rev. Entomol., Vol. 15: 143-156.

Hoek G., Brunekreef B., Goldbohm S., Fischer P., van den Brandt P.A. (2002). Association between mortality and indicators of traffic-related air pollution in the Netherlands: a cohort study. Lancet, Vol.360(9341): 1203-1209.

Hofmann W. (2011). Modelling inhaled particle deposition in the human lung — A review. J. Aerosol Sci., Vol.42(10): 693-724.

Hollick J.B., Dorweiler J.E. Chandler V.I. (1997). Paramutation and related allelic interactions. Trends Genet., Vol.13(8): 302-308.

Howe H.A., Bodian D. (1940) Portals of entry of poliomyelitis virus in the chimpanzee. Proc. Soc. Exp. Biol. Med., Vol.43: 718–721.

Hunt, A., Abraham, J.L., Judson, B., Berry, C.L. (2003). Toxicologic and Epidemiologic Clues from the Characterization of the 1952 London Smog Fine Particulate Matter in Archival Autopsy Lung Tissues. Environ. Health Perspect., Vol.111(9): 1209-1214.

Hussain M., Madl P., Khan A. (2011). Deposition and biokinetics of inhaled nanoparticles and the emergence of contemporary diseases. Part-I. theHealth, Vol.2(2): 51-59.

Ingrosso D., Cimmino A., Perna A.F., Masella L., De Santo N.G., De Bonis M.L., Vacca M., D'Esposito M., D'Urso M., Galletti P., Zappia V. (2003). Folate treatment and unbalanced methylation and changes of allelic expression induced by hyperhomocysteinaemia in patients with uraemia. Lancet, Vol.361(9370): 1693-1699.

Jablonka E, Lamb M.J. (2002). The Changing Concept of Epigenetics. Ann. N.Y. Acad. Sci., Vol.981: 82–96.

Jaenisch R., Gurdon J. (2007). Nuclear Transplantation and the Reprogramming of the Genome. In: Busslinger M. Tarakhovsky A. (2007). Epigenetic Control of Lymphopoiesis. In Epigenetics, Ch.21. Aliis C.D., Jenuwein T., Reinberg D., Caparros M.L. eds., (2007). Cold Spring Harbor Laboratory Press, New York - USA.

Jones P.A. (2001). Death and methylation. Nature, Vol.409(6817): 141-144.

Kim J., Kim J.Y., Song K.S., Lee Y.H., Seo J.S., Jelinek J., Goldschmidt-Clermont P.J., Issa J.P. (2007). Epigenetic changes in estrogen receptor beta gene in artherosclerotic cardiovascular tissues and in-vitro vascular senescence. Biochim. Biophys. Acta, Vol.1772(1): 72-80.

Kocbach A., Namork E., Schwarze P.E. (2008). Pro-inflammatory potential of wood smoke and traffic-derived particles in a monocytic cell line. Toxicology, Vol. 247: 123-132.

Kreyling W.G. (2003). Translocation of ultrafine solid combustion particles into the vascular and the central nervous system. 7th ETH Conference on Combustion Generated Particles. Zurich – CH.

Kumagai Y., Arimoto T., Shinyashiki M., Shimojo N., Nakai Y., Yoshikawa T., Sagai M. (1997). Generation of reactive oxygen species during interaction of diesel exhaust particle components with NADPH-cytochrome P450 reductase and involvement of the bioactivation in the DNA damage. Free Rad. Biol. & Med., Vol.22(3): 479-487.

Limbach L.K., Li Y., Grass R.N., Brunner T.J., Hintermann M.A., Muller M., Gunther D., Stark W.J. (2005). Oxide Nanoparticle Uptake in Human Lung Fibroblasts: Effects of Particle Size, Agglomeration, and Diffusion at Low Concentrations; Environ. Sci. Technol., Vol.39(23): 9370-9376.

Liu D., Diorio J., Tannenbaum B., Caldji C, Francis D., Freedman A., Sharma S., Pearson D., Plotsky P.M., Meaney M.J. (1997). Maternal care, hippocampal glucocorticoid receptors, and hypothalamic-pituitary-adrenal responses to stress. Science, Vol.277(5332): 1659-1662.

Lubitz S., Schober W., Pusch G., Effner R., Klopp N., Behrendt H., Buters J.T.M. (2009). Polycyclic Aromatic Hydrocarbons from Diesel Emissions Exert Proallergic Effects in Birch Pollen Allergic Individuals Through Enhanced Mediator Release from Basophils. Environ. Toxicol., Vol.25(2): 188-197.

Luger K., Mäder A.W., Richmond R.K., Sargent D.F., Riebmond T.J. (1997). Crystal structure of the nucleosome core particle at 2.8A resolution. Nature, Vol.389(6648): 251-260.

Luo J., Su F., Chen D., Shiloh A., Gu W. (2000). Deacetylation of p53 modulates its effect on cell growth and apoptosis. Nature, Vol.408(6810): 377-381.

Madl P. (2003). Instrumental Development and Application of a Thermodenuder. MSc Thesis. Queensland University of Technology / University of Salzburg. Available online (accessed in Feb. 2012): http://biophysics.sbg.ac.at/exotica/thesis-piero.pdf.

Madl P. (2005). The silent sentinels – the demise of tropical coral reefs, Ch.3. Bufus-Series Vol 32 to 35. Available online (accessed in Feb. 2012): http://biophysics.sbg.ac.at/reefs/reefs.htm.

Madl P. (2009). Anthropogenic Environmental Aerosols: Measurements and Biological Implications. PhD Thesis. University of Salzburg. Available online (accessed in Feb. 2012): http://biophysics.sbg.ac.at/exotica/thesis-piero2.pdf.

Madl P., Hussain M. (2011). Lung deposition predictions of airborne particles and the emergence of contemporary diseases, Part-II. theHealth J., Vol.2(3): 101-107.

Mahadevan B., Keshava C., Musafia-Jeknic T., Pecaj A., Weston A., Baird W.A. (2005). Altered gene expression patterns in MCF-7 cells induced by the urban dust particulate complex mixture standard reference material 1649a. Cancer Res., Vol.65(4): 1251-1258.

Marley N.A. (2006) Researchers to Scrutinize Megacity Pollution during Mexico City Field Campaign. UCAR, Boulder (CO) - USA. Image Mexico City available online (accessed in Feb. 2012): http://www.ucar.edu/news/releases/2006/images/mexico.jpg

Mathers J.C., Strathdee G., Relton C.L. (2010). Induction of epigenetic alterations by dietary and other environmental factors, Ch.1. In: Herceg Z., Ushijima T.; Epigenetics and Cancer, Part B., Academic Press, San Diego (CA) – USA.

Mays-Hoopes L.L. (1989). Age-related changes in DNA methylation: Do they represent continued developmental changes? Int. Rev. Cytol., Vol.114: 181-220.

Morgan H.D., Santos R, Green K., Dean W., and Reik W. (2005). Epigenetic reprogramming in mammals. Hum. Mol. Genet., Vol.14: R47-R58.

Nadeau K., McDonald-Hyman C., Noth E. M., Pratt B., Hammond S. K., Balmes J. (2010). Ambient air pollution impairs regulatory T-cell function in asthma. J. Allergy Clin. Immunol., Vol.126(4): 845-852.

Nagashima M., Kasai H., Yokota J., Nagamachi Y., Ichinose T., Sagai M. (1995). Formation of an oxidative DNA damage, 8-hydroxydeoxyguanosine, in mouse lung DNA after intratracheal instillation of diesel exhaust particles and effects of high dietary fat and beta-carotene on this process. Carcinogenesis, Vol.16(16): 1441-1445.

Nawrot T.S., Adcock I. (2009). The detrimental health effects of traffic-related air pollution. A role for DNA methylation? Am. J. Resp. Crit. Care Med., Vol.179(7): 523-524.

Nel A.E., Diaz-Sanchez D., Ng D., Hiura T., Saxon A. (1998). Enhancement of allergic inflammation by the interaction between diesel exhaust particles and the immune system, J. Allergy Clin. Immunol., Vol.102(4): 539-554.

Oberdörster G., Sharp Z., Atudorei V., Elder A., Gelein R., Kreyling W., Cox C. (2004). Translocation of Inhaled Ultrafine Particles to the Brain. Inhal. Toxicol., Vol.16(6-7): 437-445.

Oberdörster G., Oberdörster E., Oberdörster J. (2005a). Nanotoxicology: An emerging discipline evolving from studies of ultrafine particles. Environ. Health Perspect. Vol.113(7): 832-839.

Oberdörster G., Oberdörster E., Oberdörster J. (2005b). Nanotoxicology: An emerging discipline evolving from studies of ultrafine particles - Supplemental Web Sections; Environ. Health Perspect., Vol.113(7): 830-831.

Oberdörster G., Stone V., Donaldson K. (2007). Toxicology of nanoparticles: A historical perspective. Nanotoxicol., Vol.1(1): 2-25.

Oberdörster G., Utell M.J. (2008). Source Specific Health Effects of Fine/Ultrafine Particles. Annual Progress Report, Rochester Particle Center; available online (accessed in Feb. 2012): http://www2.envmed.rochester.edu/envmed/PMC/anrep08.pdf

Ormstad H., Namork E., Gaarder P.I., Johansen B.V. (1995). Scanning electron microscopy of immunogold labeled cat allergens (Fel-d1) on the surface of airborne house dust particles. J. Immunol. Methods, Vol.187(2): 245-251.

Ozanne S.E., Constancia M. (2007). Mechanisms of disease: the developmental origins of disease and the role of the epigenotype. Nat. Clin. Pract. Endocrinol. Metab., Vol.3(7): 539-546.

Perez L., Tobias A., Querol X., Künzli N., Pey J., Alastuey A., Viana M., Valero N., González-Cabré M., Sunyer J. (2008). Coarse Particles from Saharan Dust and Daily Mortality. Epidemiology, Vol.19(6): 800-807.

Petronis A. (2001). Human morbid genetics revisited: relevance of epigenetics. Trends Genet., Vol.17(3): 142-146.

Poole J.A., Alexis N.E., Parks C., MacInnes A.K., Gentry-Nielsen M.J., Fey P.D., Larsson L., Allen-Gipson D., Von Essen S.G., Romberger D.J. (2008). Repetitive organic dust exposure in vitro impairs macrophage differentiation and functions. J. Allergy Clin. Immunol., Vol.122(2): 375-382.

Pope C.A., Burnett R.T., Thun M.J., Calle E.E., Krewski D., Ito K., Thurston G.D. (2002). Lung Cancer, Cardiopulmonary Mortality, and Long-term Exposure to Fine Particulate Air Pollution. JAMA, Vol.287(9): 1132-1141.

Qiu J. (2006). Epigenetics: Unfinished symphony. Nature, Vol.441(7090): 143-145.

Raab C., Simkó M., Nentwich M., Gazsó A., Fiedeler U. (2010). How Nanoparticles Enter the Human Body and Their Effects There. Institute of Technology Assessment of the Austrian Academy of Sciences. Nanotrust Dossiers. No 003/11.

Rampersaud G.C., Kauwell G.P., Hutson A.D., Cerda J.J., Bailey L.B. (2000). Genomic DNA methylation decreases in response to moderate folate depletion in elderly women. Am. J. Clin. Nutr., Vol.72(4): 998-1003.

Reynolds E.H., Carney M.W., Toone B.K. (1984). Methylation and mood. Lancet, Vol.2(8396): 196-198.

Riedl M., Diaz-Sanchez D. (2005). Biology of diesel exhaust effects on respiratory function. J. Allergy Clin. Immunol., Vol.115(2): 221-228.

Ring J., Krämer U., Schäfer T., Behrendt H. (2001). Why are allergies increasing?. Curr. Opin. Immun., Vol.13: 701-708.

Rodríguez-Paredes M., Esteller M. (2011). Cancer epigenetics reaches mainstream oncology. Nat. Med., Vol.17(3): 330–339.

Rückerl R., Ibald-Mulli A., König W., Schneider A., Wölke G., Cyrys J., Heinrich J., Marder V., Frampton M., Wichmann H.E., Peters A. (2006). Air pollution and markers of inflammation and coagulation in patients with coronary heart disease. Am. J. Respir. Crit. Care Med., Vol.173(4): 432–441.

Samet J.M., DeMarini D.M., Malling H.V. (2004). Do airborne particles induce heritable mutations? Science, Vol.304(5673): 971-972.

Sandovici L., Smith N.H., Ozanne S.E., Constancia M. (2008). The Dynamic Epigenome: The Impact of the Environment on Epigenetic Regulation of Gene Expression and Developmental Programming. In: Tost J. (2008). Epigenetics. Caister Academic Press Norfolk - UK.

Santos R., Peters A.H., Otte A.P., Reik W., Dean W. (2005). Dynamic chromatin modifications characterise the first cell cycle in mouse embryos. Dev. Biol., Vol.280(1): 225-236.

Sayre L.M., Zelasko D.A., Richey P.L., Perry G., Salomon R.G., Smith M.A. (1997); 4-Hydroxynonenal-derived advanced lipid peroxidation end-products are increased in Alzheimer's disease. J. Neurochem., Vol.68(5): 2092-2097.

Sarge K.D., Park-Sarge K. (2005). Gene bookmarking: keeping the pages open. Trends Biochem. Sci., Vol. 30(11): 605-610.

Seppa N. (2000). Silencing the BRCA1 gene spells trouble. Science News, Vol. 157(16): 247.

Shannon C., Weaver W. (1949). The mathematical theory of communication. Univerity of Illinous Press – IL, USA; p.5

Shibata N., Nagai R., Uchida K., Horiuchi S., Yamada S., Hirano A., Kawaguchi M., Yamamoto T., Sasaki S. and Kobayashi M. (2001). Morphological evidence for lipid peroxidation and protein glycoxidation in spinal cords from sporadic amyotrophic lateral sclerosis patients. Brain Res., Vol.917(1): 97-104.

Snyder R. (2002). Benzene and leukemia. Crit. Rev. Toxicol. Vol. 32(3): 155–210.

Somers C.M., McCarry B.E., Malek F., Quinn J.S. (2004). Reduction of particulate air pollution lowers the risk of mutations in mice. Science, Vol.304(5673): 1008-1010.

Strahl B.D., Allis C.D. (2000). The language of covalent histone modifications. Nature, Vol.403(6765): 41-45.

Stobbs N.T. (1952). Foggy Day in Decemeber 1952 - Nelson's Column in December. London image: The Great Smog of 1952. Available online (accessed in Feb. 2012): http://www.geograph.org.uk/photo/765606.

Solomon W.R. (2002). Airborne pollen: A brief life. J. Allergy Clin. Immunol., Vol.109(6): 895-900.

Surani M.A., Reik W. (2007). Germ Line and Pluripotent Stem Cells. In: Epigenetics, Ch.20. Aliis C.D., Jenuwein T., Reinberg D., Caparros M.L. eds., (2007). Cold Spring Harbor Laboratory Press, New York - USA; p.377-395.

Tada M., Tada T, Lefebvre L., Barton S.C., Surani M.A. (1997). Embryonic germ cells induce epigenetic reprogramming of somatic nucleus in hybrid cells. EMBO J., Vol.16(21): 6510-6520.

Tada M., Takahama Y., Abe K., Nakatsuji N., Tada T. (2001). Nuclear reprogramming of somatic cells by in vitro hybridization with ES cells. Curr. Biol., Vol.11(19): 1553-1558.

Takenaka H., Zhang K. Diaz-Sanchez D., Tsien A., Saxon A. (1995). Enhanced human IgE production results from exposure to the aromatic hydrocarbons from diesel exhaust: Direct effects on B-cell IgE production, J. Allergy Clin. Immunol., Vol.95(1 Pt.1): 103-115.

Tarantini L., Bonzini M., Apostoli P., Pegoraro V., Bollati V., Marinelli B., Cantone L., Rizzo G., Hou L.F., Schwartz J., Bertazzi A.P., Baccarelli A. (2009). Effects of particulate matter on genomic DNA methylation content and iNOS promoter methylation. Environ. Health Perspect., Vol.117(2): 217-222.

Taylor P.E., Flagan R.C., Valenta R., Glovsky M. (2002). Release of allergens as respirable aerosols: A link between grass pollen and asthma. J. Allergy Clin. Immunol., Vol.109(1): 51-56.

Thaler D.S. (1994). The evolution of genetic intelligence. Science, Vol.264(5156): 224-225.

Tortora G.J., Grabowski S.R. (1996). Principles of Anatomy and Physiology, 8th ed. Harper Collins, Menlo Park, CA – USA.

Tost J. (2008). Epigenetics. Preface. Caister Academic Press, Norfolk - UK; p.303-386.

Traidl-Hoffmann C., Jakob T., Behrendt H. (2009): Determinants of allergenicity. J. Allergy Clin. Immunol., Vol. 123(3): 558-566.

VanHelden E.G., VanHelden P.D. (1989). Age-related methylation changes in DNA may reflect the proliferative potential of organs. Mutat. Res. Vol.219(5-6): 263-266.

Vladimir V. (2001). Reactive Oxygen Species, Water, Photons and Life. Riv. Biol., Vol.103(2-3): 321-342.

Vickaryous M.K., Hall B.K. (2006). Human cell type diversity, evolution, development, and classification with special reference to cells derived from the neural crest. Biol. Rev., Vol.81(3): 425-455.

Vineis P., Kusgafvel-Pursiainen K. (2005). Air pollution and cancer: biomarker studies in human populations. Carcinogenesis, Vol.26(11): 1846-1855.

Vostal J.J. (1980). Health aspects of diesel exhaust particulate emissions. Bull N.Y. Acad. Med., Vol.56(9): 914-934.

Walker D., Gore A.C. (2011). Transgeneretional neuroendocrine disruption of reproduction. Nat. Rev. Endocrinol., Vol.7(4): 197-207.

Ware J.H., Ferris, B.G., Dockery D.W, Spengler J.D., Stram D.O. (1986). Effects of ambient sulfur oxides and suspended particles on respiratory health of preadolescent children. Am. Rev. Respir. Dis., Vol.133(5): 834-842.

Warner J.A. (2000). Controlling indoor allergens. Pediatric Allergy and Immunology, Vol. 11: 208-219.

Waterland R.A., Jirtle R.L. (2003). Transposable elements: targets for early nutritional effects on epigenetic gene regulation. Mol. Cell. Biol., Vol.23(15): 5293-5300.

Weaver I.C., Cervoni N., Champagne F.A., D'Alessio A.C., Sharma S., Seckl J.R., Dymov S., Szyf M., Meaney M.J. (2004). Epigenetic programming by maternal behavior. Nat. Neurosci., Vol.7(8): 847-854.

Willett W.C. (2002). Balancing Life-Style and Genomics Research for Disease Prevention. Science, Vol.296(5568): 695-698.

Wolff G.L., Kodell R.L., Moore S.R., Cooney C.A. (1998). Maternal epigenetics and methyl supplements affect agouti gene expression in Agouti mice. FASEB J., Vol.12(11): 949-957.

Yazdanbakhsh M., van den Biggelaar A., Maizels R.M. (2001). Th2 responses without atopy: immunoregulation in chronic helminth infections and reduced allergic disease. Trends Immunol., Vol.22(7): 372-377.

Yauk C., Polyzos A., Rowan-Caroll A., Somers C.M., Godshalk R.W., Schooten F.J., Berndt M.L., Pobribny I.P., Koturbash I., Williams A., Douglas G.R., Kovalchuk O. (2008). Germ-line mutations, DNA damage and hypermethylation in mice exposed to particulate air pollution in an urban / industrial location. Proc. Natl. Acad. Sci. USA, Vol.105(2):605-610.

Yang B., House M.G., Guo M., Herman J.G., Clark D.P. (2005) Promoter methylation profiles of tumor suppressor genes in intrahepatic and extrahepatic cholangiocarcinoma. Mod. Pathol., Vol.18(3):412-420.

Ying A.K., Hassanain H.H., Roos C.M., Smiraglia D.J., Issa J.J., Michler R.E., Caligiuri M., Plass C., Goldschmidt-Clermont P.J. (2000). Methylation of the estrogen receptor-alpha gene promoter is selectively increased in proliferating human aortic smooth muscle cells. Cardiovasc. Res., Vol.46(1): 172-179.

Yip M. (2003). Exposure Assessment in a Busway Canyon. MSc Thesis. University of Salzburg. Available online (accessed, in February 2012): http://biophysics.sbg.ac.at/exotica/thesis-mari.pdf

Yu Z.B. (2007). Non-coding RNAs in gene regulation. In: Epigenetics, Ch 7. Tost J. (ed.). Caister Academic Press Norfolk - UK.

Zhang J., Perry G., Smith M.A., Robertson D., Olson S.J., Graham D.G., Montine T.J. (1999). Parkinson's disease is associated with oxidative damage to cytoplasmic DNA and RNA in substantia nigra neurons. Am. J. Pathol., Vol.154(5):1423-1429.

Zhu Z.Z., Sparrow D., Hou L., Tarantini L., Bollati V., Litonjua A.A., Zanobetti A., Vokonas P., Wright R.O., Baccarelli A., Schwartz J. (2011). Repetitive element hypomethylation in blood leukocyte DNA and cancer incidence, prevalence, and mortality in elderly individuals: the Normative Aging Study. Cancer Causes Control., Vol.22(3):437-447.

Zoghbi H.Y., Beaudet A.L. (2007). Epigenetics and Human Disease. Ch.21. Reinberg D., Allis C.D., Jenuwein T., eds. Cold Spring Harbor Laboratory Press, New York - USA.

16

Air Pollution and Cardiovascular Diseases

Najeeb A. Shirwany and Ming-Hui Zou*
*Section of Molecular Medicine, Department of Medicine,
Department of Biochemistry and Molecular Biology,
University of Oklahoma Health Science Center, Oklahoma City, OK
USA*

1. Introduction

Two parallel observations have historically linked poor air quality to human disease. The first of these is the recognition that substances in inspired air can pose health risks, and the second is the view that growing industrialization at the global level has contributed to deteriorating air quality. In the case of the latter, environmental scientists, economists and urban planning experts have spoken and written about the impact of industrial growth, income and urban development in the context of an "EKC" relationship (Environmental Kuznets Curve; "inverted U shaped" curve). This concept is predicated on the fact that as human activities related to industry and urban growth increase, an initial and sharp deterioration in air quality ensues. Subsequently, as income levels in a society inevitably rise, regulation, awareness and increasing attitudes of social and environmental responsibility intervene and the air quality standards improve [1]. However, since many emerging economies and industrial powers (such as China and Brazil) find themselves on the left-hand limb of the curve, the consequent impact of their industrial growth and urban expansion contribute to the aggregate decline in global air quality and pollution.

2. Historical perspective

It is believed that in 1872 Smith published the first scientific report of air pollution [2]. This and subsequent studies have laid the foundation for the scientific examination of pollutants as hazardous components of breathable air and its impact on human populations. In subsequent decades, particularly in the 20th century, several major incidents came to prominence which underscored the importance of air pollution in human health. For example, in 1930 a combination of high atmospheric pressure and mild winds created a heavy fog in Belgium. It is estimated that about 60 deaths were attributable, directly or indirectly to this significant fog event. Later investigations revealed that trapped potent pollutants from chimney exhausts created a toxic cloud that resulted in these fatalities [2]. Seven years later in 1948, an industrial accident caused 20 deaths with thousands of acute illnesses reported because of the smelting plant in Pennsylvania[3]. Another severe event occurred in London in 1952 when pollutants from the use of stoves as well as from

---

*Corresponding Author

industrial plants nearly paralyzed the city. It is estimated that there was a 48% increase in hospital admissions and 163% increase in respiratory illnesses resulting in direct hospital admissions. This event was soon followed by significant increase in numbers of deaths from respiratory illnesses. Such incidents and other related events prompted the establishment of clean air and air quality acts in the United States in 1963 and 1967.

3. Ambient air particulate pollutants and their classification

Particulate pollution in inspired air comprises coarse and fine particles of various sizes. Particles that are considered significant, usually have an aerodynamic diameter (AD) of between 2.5 and 10 μ (PM10). Finer particles are those that are less than 2.5 μ (PM2.5), while ultrafine particles have aerodynamic diameters of less than 0.1 μ (UFPs). The chemical nature and composition of these particles exhibits tremendous diversity and depends on numerous geographical, meteorological and source specific variables. In a general sense, ambient particles include inorganic components such as sulfates, nitrates, ammonium, chloride, and trace metals. In addition, organic materials, crystalline compounds, and biological components are also observed.

The sources of particulate air pollutants can be human, biological but nonhuman, and natural. PM10 particles generally relate to human activities and come from dust, burning of wood, construction and demolition sites, and from sites involved in mining operations. For the size of particle, natural sources include windblown dust and wildfires. Finer particles are generally generated by gaseous materials when they convert to particulate phases during combustion of fuel and industrial activities. PM2.5 sources include power plants, oil refineries, metal processing facilities, automobile exhaust, residential fuel combustion, as well as wildfires. The primary sources of UFP is automobile tailpipe emissions from a variety of vehicles, including aircraft and marine vessels [2].

With regard to human health, of particular importance are particles that are equal to or less than 10 μ in diameter, because these ultimately enter the lung parenchyma[4]. As mentioned above, particles of 10 μ aerodynamic diameters and smaller can be further divided based on size and biological effect. *Coarse* particles range in size from 2.5 to 10 μ, *fine* particles are less than 2.5 μ in diameter, and *ultrafine* particles are less than 0.1 μ. Because of their very small size, particles that are between 0.1 and 2.5 μ are deeply inhaled into the lung parenchyma. If these inhaled particles get deposited in the alveoli they enter the pulmonary circulation and are presumed to also continue to the systemic circulation. This indirect observation has been the presumed mechanism through which particulate pollution impacts the cardiovascular system, which has been reported in a series of major EP immunological and observational studies[5-7]. Recently, considerable research and attention has been given to ultrafine particles as well. These UFPs are less than .1 μ in aerodynamic diameter and are usually attributable to combustion processes from the burning of fossil fuels, for example. These ultrafine particles tend to be short-lived, because they agglomerate and coalesce into larger particles rather quickly[8]. However, these pollutants exhibit a very high rate of deposition in human alveoli and account for a major proportion of the actual numbers of particles in the lung. They also have a high surface area-mass ratio, and this potentially leads to enhanced biological toxicity[9].

4. Air pollution and disease: Evidence from laboratory studies

While a large number of epidemiological studies have clearly implicated particulate air pollutants in the etiology of human cardiovascular disease, controversy remains regarding the underlying biological mechanisms that might explain this phenomenon. Several distinct mechanisms have been hypothesized:

1. *Inflammation, oxidative stress, and endothelial function.* It has been established that inhaled particles can induce inflammation in the lung parenchyma, either directly or by generating free oxygen radicals and increasing oxidative stress, particularly by activating NAD(P)H oxidase[10]. Activation of this oxidase in turn triggers intracellular signaling pathways such as those of MAP kinases and oxidation sensitive transcription factors, such as NFκB and API which control the expression of several genes coding for proinflammatory factors. These include but are not limited to, cytokines, particularly, interleukin 1β, interleukin 6, and interleukin 8. In addition, TNFα and granulocyte-macrophage colony-stimulating factor (GM-CSF), chemokines, and adhesion molecules are also robustly expressed by the signaling systems[11, 12]. These signaling events are critically linked to macrophage and epithelial cell function at the level of pulmonary alveoli and bronchi. Once particulate material inspired air reaches these locations, in systemic as well as local inflammation ensues. These inflammatory changes are evidenced by acute rises in C-reactive protein and fibrinogen as well as increased viscosity of plasma. Stimulation of bone marrow with leukocytosis and circulating immature polymorphonuclear neutrophils, and activated platelets and other procoagulant factors are also involved in this process[10]. It has been demonstrated, for example, that exposure to exhaust from diesel engines contains ingredients that attenuate the acute release of tissue plasminogen activator (t-PA) by the endothelium, resulting in diminished endogenous fibrinolytic capacity[10]. In blood vessels, particulate matter, causes, endothelial dysfunction by inhibiting the formation of nitric oxide, and stimulating the production of endothelin 1, angiotensin II, and thromboxane A2[13, 14]. Angiotensin II, in turn, contributes further to oxidative stress, by increasing the generation of superoxide and through the enhancement of NAD(P)H oxidase activity. These are some of the possible explanations for alterations in endothelium-dependent vasomotor activity induced by particulate pollution. Several studies have demonstrated endothelium-independent alterations as well, which are also related to mechanisms of vasoactivity, such as through the stimulation of sympathetic ennervation as well as by the direct stimulation of angiotensin II AT1 receptors. Cumulatively, these alterations contribute to the development and progression of atherosclerosis, the destabilization of atherosclerotic plaques, and promotion of ischemia antithrombotic states[15]. Experimental models have confirmed that atherosclerosis progresses quite rapidly and plaques have a greater vulnerability to rupture in laboratory animals exposed to PM2.5 and PM10 air pollution over a period of several weeks or several months. In fact, C-reactive protein, which is an acute phase reactant, has been shown to express in greater quantities in the presence of particulate pollution. CRP is a known cardiovascular risk factor because it facilitates its uptake of lipids by macrophages and the expression has been associated with increased fragility atherosclerotic plaques leading to destabilization[16]. Investigators have found a direct correlation between the severity of coronary and aortic atherosclerosis and the number of alveolar macrophages that phagocytose PM10 over several weeks exposure in rabbits who have heritable hyperlipidemia[10].

Several other studies have found that ultrafine particles alongside soluble components can also reach the systemic circulation directly and quickly lead to oxidative stress and inflammation in the heart and arterial vessels, resulting in endothelial dysfunction, without necessarily inducing pulmonary inflammation[17]. Acute exposure to particles of 2.5 μ diameter has also been shown to associate with high levels of circulating markers of lipid and protein oxidation.

Investigators have also determined that the inflammatory reaction to inhaled particulate material is less due to their mass or volume, but rather is a result of the chemical composition and surface area. Fine and ultrafine or even nanoscale particles have greater surface areas in proportion to their mass compared with other particles. The larger surface area of particles results in greater oxidative stress, which itself is consequent upon the larger number of reactive groups present on the surface leading to greater synthesis of reactive oxygen species[18].

Researchers have found that the pro-atherogenic effects of particulate pollution inhalation, in ApoE deficient mice, results in higher susceptibility to atherosclerotic lesions with more significant and more extensive lesions following particulate exposure compared with controls. It has been determined that the greater oxidative effect of ultrafine particles is related to the higher organic carbon content. In addition, it has also been suspected that particles of this small size modulate intracellular calcium concentrations by interfering with the opening of calcium channels on cell membranes, a phenomenon which itself indirectly leads to the synthesis of reactive oxygen species in the cell[13].

The effect of increased oxidative stress, consequent upon exposure to ultrafine particles, leads to mitochondrial dysfunction as well. This increases the production of superoxide radical and activation of p53, a transcription factor which modulates programmed cell death, through the release of pro-apoptotic factors like cytochrome C and AIF.

2. *Autonomic dysfunction*: The autonomic nervous system is also particularly susceptible to the effect of particulate material in the systemic circulation. For example, autonomic dysfunction has been demonstrated after stimulation of pulmonary nerve receptors, following exposure to 10 and 2.5 μ particulate materials. These changes were shown to be either as a direct effect of the particles, or by the instigation of local oxidative stress and inflammation in the lung parenchyma leading to reflex increases in heart rate, reduced variability of the heart rate, as well as heart rate rhythm anomalies[13].

    It has been postulated that the fine and ultrafine particles may also lead to the rhythm anomalies in the heart by virtue of their effects on ion channels in cardiac myocytes [13]. Alternatively, disruption of cardiac autonomic function may also be related to the effect of these particles on oxidative homeostasis in the cardiovascular regulatory nuclei in the central nervous system[19]. Intriguingly, the reverse has also been demonstrated in that autonomic dysregulation can itself trigger cardiac oxidative stress[20].

3. *Dysregulation of intravascular thrombotic system*: inhalation of particulate matter has been shown to enhance arterial thrombosis and coagulation[21]. In this regard, empirical work has resulted in conflicting data. Some studies have shown a positive correlation between particulate matter inhalation and thrombotic dysregulation, while others have not[17]. However, it has been hypothesized that, against the background of vulnerable atherosclerotic plaques in individuals who already have these lesions, disturbances of the thrombotic milieu, are likely to trigger arterial thrombosis, ischemic events, and

catastrophic rupture, leading to embolism. Studies have shown, that increases in fibrinogen and blood viscosity, elevated CRP, increased platelet reactivity, altered levels of coagulation factors, disturbed histamine levels, enhanced interleukin 6 dependent signaling, expression of adhesion molecules, and attenuated release of fibrinolytic factors all appear to be mechanistic underpinnings of the impact of particulate matter on the cardiovascular system[22, 23]. One aspect of these mechanistic theories remains unresolved. It is not known, what relative roles are played by systemic inflammation, disturbed autonomic balance, and the effect of blood-borne mediators of thrombosis, on overall cardiovascular health and disease[8].

4. *PM and heart failure:* exposure to particulate air pollution has also been linked to an increased risk of heart failure as well as hospital admissions resulting from heart failure[24]. It is believed that both pro-ischemic, and dysrhythmic effects of particulate exposure could be responsible for these phenomena. However, a more recent study in mice has also shown that the deposition of particulate material in lung parenchyma can impair the ability of the alveoli to clear fluid because of reduced membrane Na-K-ATPase activity[25]. In these experiments, this effect was abrogated by the use of antioxidants, suggesting that oxidative stress, might play a primary role in such phenomena.

5. *Evidence implicated PM in blood pressure regulation:* in animal studies, evidence has been accumulating that relates particulate air pollution exposure to induced changes in blood pressure. For example, in a Sprague-Dawley rat model, where, angiotensin II was employed to induce hypertension, exposure to concentrated 2.5 μ the particulate pollution for 10 weeks, caused prolonged blood pressure compared with control groups[26]. In this study, aortic vasoconstriction in response to particulate exposure was potentiated with exaggerated relaxation to the Rho-kinase inhibitor Y-27632. Investigators in this paper also demonstrated an increase in ROCK-1 messenger RNA levels and superoxide production in animals exposed to PM, suggesting that even short-term exposure to PM can induce hypertension via superoxide-mediated upregulation of the Rho/ROCK signaling pathway. In other studies, in Murine models of PM exposure, angiotensin II, infusion in conjunction with a rho kinase antagonist, potentiation of the hypertensive phenotype was also reported[27]. Other studies have also shown that exposure to 2.5 μ particulate material increases angiotensin II-induced cardiac hypertrophy, collagen deposition, cardiac RhoA activation, and vascular RhoA activation, suggesting that cardiovascular health effects are consequences of air pollution[4].

5. Air pollution and disease: Evidence from epidemiological studies

A large number of studies have focused on the acute effects of air pollution. In one study, which was carried out in as many as 29 European cities with 43 million inhabitants demonstrated that for each 10 μg per cubic meter increase in 10 μ particulate matter, cardiovascular mortality rose by 69%[28]. In the United States, a survey of 90 cities with 15 million participants revealed a short-term increase in cardiopulmonary mortality of .31% for each 10 μg per cubic meter increase in PM10 when measured over a 24-hour period[29]. Many other studies have also demonstrated significant rises of between .8% and .7% respectively in hospital admissions from heart failure and ischemic heart disease for every 10 μg per cubic meter rise in PM10[30]. The studies have also shown an increase of 1.28% and 4.5%, respectively, and risk of heart failure and acute coronary syndromes for every 10 μg per cubic meter rise in PM2.5[30]. Furthermore, links have also been discovered between short-term rises in PM 2.5 and the incidence of myocardial infarction, occurring within a few hours. Following exposure, ST-segment depression occurs during exercise testing in patients with stable coronary disease, increased heart rates, enhanced incidence of arrhythmias in several different studies[31]. Overall, it has been estimated that between 60,000 and 350,000 sudden cardiac deaths in the United States occur which can be attributed to particulate air pollution.

Investigators have found that exercise-induced myocardial ischemia is exacerbated when humans are exposed to diesel exhaust products which are mainly composed of ultrafine particles, in concentrations similar to those found in heavy traffic in large urban centers (300 μg per cubic meter)[22]. Links have also been demonstrated between pollution from fine particles found in motor vehicle emissions and ST-segment depression recorded via Holter ECG monitoring[32]. In addition, a higher incidence of ischemic and hemorrhagic stroke has also been reported with higher mortality and more hospital admissions, directly connected to short-term increase in airborne PM 10[33]. In a study undertaken in nine US cities of individuals older than 65 years demonstrated that there was a strong association with ischemic stroke, with a rise of 1.03% in hospital admissions for every 23 μg per cubic meter increase in PM 10[34]. In this study an association was also found between specific gaseous co-pollutants and increase in ischemic stroke (carbon monoxide [CO] nitrous oxide [NO2] and sulfur dioxide [SO2]). Indeed, several others have reported links between levels of these gaseous pollutants and mortality or hospitalization rates due to stroke, as well as re-hospitalization in survivors of myocardial infarction[10]. In Germany during the 1985 air pollution episode, epidemiological observations revealed that plasma viscosity, heartrate, and concentrations of C-reactive protein, were increased during this episode[35]. In the United States, in the city of Boston, nitrogen dioxide in the atmosphere and PM2.5 were associated with life-threatening cardiac arrhythmias, leading to the need for drug interventions, including the implantation of cardioverter defibrillators. In addition PM2.5 concentrations were noted to be higher in the hours and days before onset of myocardial infarction in a large group of patients[36, 37]. In a study of individuals older than 65 years, a positive association between stroke mortality and the concentration of fine particles was also demonstrated. In this study, a rise of 6.9% for each interquartile increase in PM2.5 on the day of death, and a 7.4% increase in the 24 hours prior to death, was demonstrated[10]. Such observations have been made on other continents and in other countries as well outside the strict confines of the Western hemisphere. For example, in studies in Shanghai, levels of PM2.5 have been reported to influence daily overall cardiopulmonary mortality, this effect was not observed for particles smaller than 2.5 μ[10]. In other Asian countries as well, rise in urban air pollution and the rise of cardiovascular morbidity has been well documented. In order to understand the effect of air pollution on human health on the vast Asian continent and in particular in countries where a sixth to a fifth of humanity reside (namely China and India), the Health Effects Institute (a Boston based, U.S. non-profit health research corporation) funded the large PAPA (Public Health and Air Pollution in Asia) study. The first phase of this project was undertaken in Thailand from 1999 to 2003, in Hong Kong in China from 1996 to 2004, and in Shanghai and Wuhan in China between 2001 and 2004 [38]. This study documented several common pollutants such as NO_2, SO_2 and PM<10μ and their effect on cardiovascular and respiratory mortality. One interesting conclusion of the investigators (speculative) was that Asian populations might be exposed to outdoor air pollution to a larger extent than Western cohorts because they tend to spend more time outside than indoors while Westerners have more access to air conditioning which tends to mitigate pollution with the use of filters and recirculated ventilation. In the initial data published in the PAPA study, Wuhan in mainland China exhibited highest concentrations of PM10 and O_3, while Shanghai has the highest concentrations of NO_2 and SO_2. In comparison with cities of comparable size in the U.S. (analyzing data from the National Morbidity and Mortality and Air Pollution Study {NMMAPS}), the concentrations of PM10 and SO_2 were found to be much higher for cities included in the PAPA study. For example, the concentrations of PM10 in PAPA had means of 52-142 µg/m3 versus 33 µg/m3 in NMMAPS. In the case of NO_2 and O_3, the results were similar in that U.S. had lower concentrations than those in China [38]. When these results were correlated with mortality figures for cardiovascular and respiratory diseases, predictable patterns emerged. Cause of death ratios were the highest in Wuhan (4:1), followed by Shanghai (3:1) with lower figures documented for Bangkok and Hong Kong [38].

Banerjee et al have recently reported from India that exposure to poor air quality can alter hematological and immunological parameters negatively. In their study, these investigators reported results from 2218 individuals residing in the large urban metropolis of New Delhi ranging in age from 21-65 years who were exposed to vehicular exhaust (the main polluter of air quality in Asia in general and in India in particular). The authors found the prevalence of hypertension 4 fold higher than matched controls (Bannerjee M et al, Int J Hyg Environ Health 2011 Sep 16 Epub). Platelet P-selctins were significantly upregulated in this cohort while CD4+ T-helper cells and CD19+ B cells were found to be depleted and CD56+ NK cells were upregulated. These changes in the immune profile is positively correlated with hypercoagulable states and higher cardiovascular risk[39]. These findings have been effectively reproduced in other Indian studies as well. For example, Barman and colleagues have recently published data from Lucknow (a city in Northern India) in which PM and pollutant heavy metals were also linked to elevated risk of cardiovascular risk [40]. Even earlier studies have shown strong correlation between air pollution levels and cardiovascular risk. Nautiyal et al reported in a study completed in the Indian state of Punjab (a pilot study) demonstrated positive correlations between angina pectoris and PM10 pollution [41].

Aside from cardiovascular mortality, other parameters of cardiovascular function have also been correlated with greater air pollution. For example, in studies from central Europe, a rise in blood pressure at times of greater air pollution has been reported. In the MONICA study from Germany, a significant rise in blood pressure was noted in relation with particulate air pollution even after adjustment for other cardiovascular risk factors[42]. Significant elevations of diastolic blood pressure in 23 normotensive individuals following two hours of exposure to PM2.5 were reported by Urch and colleagues[43]. In Brazil,, monocyte levels also appear to significantly influence systolic, diastolic blood pressure levels per quartile of monocyte concentration. In this setting, sulfur dioxide levels were also noted to affect blood pressure, validating the importance of gaseous co-pollutants[44]. Intriguingly, a large volume of data also suggests that the deleterious effects of particulate air pollution can be aggravated by the presence of cofactors such as diabetes, obesity, hypertension, chronic pulmonary disease, and previous cardiovascular disease, as well as an additive effect of advancing age[45, 46].

A consensus seems to be gathering between clinicians and scientists that the adverse cardiovascular effects of air pollution depend not only on the concentrations of these materials but also on the length of exposure. Prolonged exposure appears to have a cumulative effect as well as a stronger impact and more persistent consequences then shorter exposure. For example, a decrease in PM2.5 over a period of eight years, was shown to significantly attenuate the overall cardiovascular and pulmonary mortality by Laden et al[47].

6. Issues of current and future research focus

An unresolved question is whether the threshold concentrations of particulate air pollution exist below which the risk to the general population dissipates or becomes nonexistent. The importance of this idea is that if such thresholds can be identified, then governmental and private endeavors to reduce air pollution can be pragmatically set to identifiable goals beyond which no further public health benefits would accrue. Some of these issues are now beginning to be analyzed based on epidemiological data. The Health Effects Institute (HEI), conducted a health study beginning in 1996, which is called the National Morbidity, Mortality, and Air Pollution Study (NMMAPS). Subsequent analysis of the study found no evidence of a critical threshold for PM10 in daily all-cause and cardiorespiratory mortality[48]. However, a threshold of about 50 $\mu g/m^3$ was estimated for non-cardiorespiratory causes of death. These and similar analyses suggest that the threshold for acute effects of ozone on lung function changes are likely to be below 100 $\mu g/m^3$/hour maximum.

Several time-series studies have shown a link between day-to-day variations in air pollution concentrations and the rate of deaths per day as well as rates of hospital admissions, however, more detailed correlation remains unclear. For example, it is not certain, by how many days, weeks, or months, such events are increased from baseline[49]. For example, Brunekreef and Holgate have suggested that if deaths occurred just a few days earlier than would have occurred without air pollution, the public health significance of these correlations would be much less severe than if mortality was reduced by months or years[49]. In contrast, effect estimates have been shown to increase with increasing duration of exposure to air pollution, which suggests that there is a stronger effect on mortality in comparison with associations between day-to-day variations in air pollution and deaths. Other data has also shown that many deaths associated with air pollution occur outside hospital settings, which further supports the notion that these individuals were often not terminally ill[50].

Another confounding aspect of the relationship between air pollution and cardiovascular disease in general is to tease out the difference between time spent indoors from that spent outdoors. This is because empirical evidence suggests that indoor pollutant concentrations differ both qualitatively and quantitatively from that found out-of-doors. This has become the basis of criticisms in that it has been questioned if measurement of air pollution on the outside without taking into account exposure indoors, is a valid method of assessing exposure to air pollutants. Thus one study found that for particulate matter and gases, there was no appreciable association between the day-to-day variation in personal exposure to nitrogen dioxide, sulfur dioxide, and ozone[51]. In this study, ambient PM2.5, nitrogen dioxide, sulfur dioxide, and ozone were closely associated with personal PM2.5, strongly suggesting that gaseous and PM 2.5 concentrations outdoors act as a surrogate for personal exposure to PM2.5[51].

Because of these unresolved questions and large costs associated with reducing air pollution, questions regarding the relationship between air pollution and health have become an area of considerable debate in recent decades. Early studies have been criticized for the analytical approach and lack of adequate controls for confounding variables such as weather etc., and US cohort studies have been critiqued for inadequate confounder and co-pollutant controls as well[49]. Several re-analyses have been performed on these older studies and the HEI has itself partnered with the US automobile industry as well as the federal government to attempt to resolve this important debate. In one reanalysis called the Philadelphia time-series study, as well as several others revealed new insights into the role of weather-related variables as well as that of spatial association between air pollution, mortality, and other confounding variables[49].

7. Conclusions

A wide range of experimental and epidemiological studies have established that air pollution is an important determinant of cardiovascular risk and that it can influence more traditional risk factors. It has been shown that alterations by air pollution, specially by fine and ultrafine particles, significantly contribute to the long-term development and progression of atherosclerosis, promotion of atherosclerotic plaques and their instability, and acute cardiovascular events such as stroke, myocardial infarction, arrhythmias, and sudden cardiac death[10]. However, several key questions remain. With rapid developments in molecular biology, proteomics, and genomics, these questions will likely be clarified within the context of complex biological mechanisms involved in cardiovascular injury and their interaction with particulate air pollution and gaseous air pollution. Thus it is likely, that with increasing understanding of the clinical significance of cardiovascular effects of air pollution, a dual approach of abating air pollution as well as using traditional medical tools and pharmaceutical strategies will, in the future, help in abrogating cardiovascular risk and reducing the incidence of cardiovascular pathology in human communities.

8. References

[1] Hettige H MM. Industrial pollution in economic development: the environmental Kuznets curve revisited. *Journal of Development Economics*. 2000;62(2):445-476.

[2] Simkhovich BZ, Kleinman MT, Kloner RA. Air pollution and cardiovascular injury epidemiology, toxicology, and mechanisms. *J Am Coll Cardiol*. 2008;52(9):719-726.

[3] Helfand WH, Lazarus J, Theerman P. Donora, Pennsylvania: an environmental disaster of the 20th century. *Am J Public Health*. 2001;91(4):553.

[4] Sun Q, Hong X, Wold LE. Cardiovascular effects of ambient particulate air pollution exposure. *Circulation*. 2010;121(25):2755-2765.

[5] Dockery DW, Pope CA, 3rd, Xu X, Spengler JD, Ware JH, Fay ME, Ferris BG, Jr., Speizer FE. An association between air pollution and mortality in six U.S. cities. *N Engl J Med*. 1993;329(24):1753-1759.

[6] Hoek G, Brunekreef B, Goldbohm S, Fischer P, van den Brandt PA. Association between mortality and indicators of traffic-related air pollution in the Netherlands: a cohort study. *Lancet*. 2002;360(9341):1203-1209.

[7] Pope CA, 3rd, Burnett RT, Thurston GD, Thun MJ, Calle EE, Krewski D, Godleski JJ. Cardiovascular mortality and long-term exposure to particulate air pollution:

epidemiological evidence of general pathophysiological pathways of disease. *Circulation.* 2004;109(1):71-77.

[8] Brook RD, Franklin B, Cascio W, Hong Y, Howard G, Lipsett M, Luepker R, Mittleman M, Samet J, Smith SC, Jr., Tager I. Air pollution and cardiovascular disease: a statement for healthcare professionals from the Expert Panel on Population and Prevention Science of the American Heart Association. *Circulation.* 2004;109(21):2655-2671.

[9] Daigle CC, Chalupa DC, Gibb FR, Morrow PE, Oberdorster G, Utell MJ, Frampton MW. Ultrafine particle deposition in humans during rest and exercise. *Inhal Toxicol.* 2003;15(6):539-552.

[10] Nogueira JB. Air pollution and cardiovascular disease. *Rev Port Cardiol.* 2009;28(6):715-733.

[11] Bhatnagar A. Environmental cardiology: studying mechanistic links between pollution and heart disease. *Circ Res.* 2006;99(7):692-705.

[12] Franchini M, Mannucci PM. Short-term effects of air pollution on cardiovascular diseases: outcomes and mechanisms. *J Thromb Haemost.* 2007;5(11):2169-2174.

[13] Bai N, Khazaei M, van Eeden SF, Laher I. The pharmacology of particulate matter air pollution-induced cardiovascular dysfunction. *Pharmacol Ther.* 2007;113(1):16-29.

[14] Nurkiewicz TR, Porter DW, Barger M, Castranova V, Boegehold MA. Particulate matter exposure impairs systemic microvascular endothelium-dependent dilation. *Environ Health Perspect.* 2004;112(13):1299-1306.

[15] Suwa T, Hogg JC, Quinlan KB, Ohgami A, Vincent R, van Eeden SF. Particulate air pollution induces progression of atherosclerosis. *J Am Coll Cardiol.* 2002;39(6):935-942.

[16] Donaldson K, Stone V, Seaton A, MacNee W. Ambient particle inhalation and the cardiovascular system: potential mechanisms. *Environ Health Perspect.* 2001;109 Suppl 4:523-527.

[17] Brook RD. Cardiovascular effects of air pollution. *Clin Sci (Lond).* 2008;115(6):175-187.

[18] Behndig AF, Mudway IS, Brown JL, Stenfors N, Helleday R, Duggan ST, Wilson SJ, Boman C, Cassee FR, Frew AJ, Kelly FJ, Sandstrom T, Blomberg A. Airway antioxidant and inflammatory responses to diesel exhaust exposure in healthy humans. *Eur Respir J.* 2006;27(2):359-365.

[19] Chahine T, Baccarelli A, Litonjua A, Wright RO, Suh H, Gold DR, Sparrow D, Vokonas P, Schwartz J. Particulate air pollution, oxidative stress genes, and heart rate variability in an elderly cohort. *Environ Health Perspect.* 2007;115(11):1617-1622.

[20] Rhoden CR, Wellenius GA, Ghelfi E, Lawrence J, Gonzalez-Flecha B. PM-induced cardiac oxidative stress and dysfunction are mediated by autonomic stimulation. *Biochim Biophys Acta.* 2005;1725(3):305-313.

[21] Ghio AJ, Hall A, Bassett MA, Cascio WE, Devlin RB. Exposure to concentrated ambient air particles alters hematologic indices in humans. *Inhal Toxicol.* 2003;15(14):1465-1478.

[22] Mills NL, Tornqvist H, Gonzalez MC, Vink E, Robinson SD, Soderberg S, Boon NA, Donaldson K, Sandstrom T, Blomberg A, Newby DE. Ischemic and thrombotic effects of dilute diesel-exhaust inhalation in men with coronary heart disease. *N Engl J Med.* 2007;357(11):1075-1082.

[23] Mills NL, Tornqvist H, Robinson SD, Gonzalez M, Darnley K, MacNee W, Boon NA, Donaldson K, Blomberg A, Sandstrom T, Newby DE. Diesel exhaust inhalation causes vascular dysfunction and impaired endogenous fibrinolysis. *Circulation.* 2005;112(25):3930-3936.

[24] Wellenius GA, Schwartz J, Mittleman MA. Particulate air pollution and hospital admissions for congestive heart failure in seven United States cities. *Am J Cardiol.* 2006;97(3):404-408.

[25] Mutlu GM, Snyder C, Bellmeyer A, Wang H, Hawkins K, Soberanes S, Welch LC, Ghio AJ, Chandel NS, Kamp D, Sznajder JI, Budinger GR. Airborne particulate matter inhibits alveolar fluid reabsorption in mice via oxidant generation. *Am J Respir Cell Mol Biol.* 2006;34(6):670-676.

[26] Sun Q, Yue P, Ying Z, Cardounel AJ, Brook RD, Devlin R, Hwang JS, Zweier JL, Chen LC, Rajagopalan S. Air pollution exposure potentiates hypertension through reactive oxygen species-mediated activation of Rho/ROCK. *Arterioscler Thromb Vasc Biol.* 2008;28(10):1760-1766.

[27] Ying Z, Yue P, Xu X, Zhong M, Sun Q, Mikolaj M, Wang A, Brook RD, Chen LC, Rajagopalan S. Air pollution and cardiac remodeling: a role for RhoA/Rho-kinase. *Am J Physiol Heart Circ Physiol.* 2009;296(5):H1540-1550.

[28] Zanobetti A, Schwartz J, Samoli E, Gryparis A, Touloumi G, Peacock J, Anderson RH, Le Tertre A, Bobros J, Celko M, Goren A, Forsberg B, Michelozzi P, Rabczenko D, Hoyos SP, Wichmann HE, Katsouyanni K. The temporal pattern of respiratory and heart disease mortality in response to air pollution. *Environ Health Perspect.* 2003;111(9):1188-1193.

[29] Samet JM, Dominici F, Curriero FC, Coursac I, Zeger SL. Fine particulate air pollution and mortality in 20 U.S. cities, 1987-1994. *N Engl J Med.* 2000;343(24):1742-1749.

[30] Pope CA, 3rd, Muhlestein JB, May HT, Renlund DG, Anderson JL, Horne BD. Ischemic heart disease events triggered by short-term exposure to fine particulate air pollution. *Circulation.* 2006;114(23):2443-2448.

[31] Toren K, Bergdahl IA, Nilsson T, Jarvholm B. Occupational exposure to particulate air pollution and mortality due to ischaemic heart disease and cerebrovascular disease. *Occup Environ Med.* 2007;64(8):515-519.

[32] Chuang KJ, Coull BA, Zanobetti A, Suh H, Schwartz J, Stone PH, Litonjua A, Speizer FE, Gold DR. Particulate air pollution as a risk factor for ST-segment depression in patients with coronary artery disease. *Circulation.* 2008;118(13):1314-1320.

[33] Hong YC, Lee JT, Kim H, Kwon HJ. Air pollution: a new risk factor in ischemic stroke mortality. *Stroke.* 2002;33(9):2165-2169.

[34] Wellenius GA, Schwartz J, Mittleman MA. Air pollution and hospital admissions for ischemic and hemorrhagic stroke among medicare beneficiaries. *Stroke.* 2005;36(12):2549-2553.

[35] Peters A, Doring A, Wichmann HE, Koenig W. Increased plasma viscosity during an air pollution episode: a link to mortality? *Lancet.* 1997;349(9065):1582-1587.

[36] Peters A, Dockery DW, Muller JE, Mittleman MA. Increased particulate air pollution and the triggering of myocardial infarction. *Circulation.* 2001;103(23):2810-2815.

[37] Ackermann-Liebrich U, Leuenberger P, Schwartz J, Schindler C, Monn C, Bolognini G, Bongard JP, Brandli O, Domenighetti G, Elsasser S, Grize L, Karrer W, Keller R, Keller-Wossidlo H, Kunzli N, Martin BW, Medici TC, Perruchoud AP, Schoni MH,

Tschopp JM, Villiger B, Wuthrich B, Zellweger JP, Zemp E. Lung function and long term exposure to air pollutants in Switzerland. Study on Air Pollution and Lung Diseases in Adults (SAPALDIA) Team. *Am J Respir Crit Care Med.* 1997;155(1):122-129.

[38] Wong CM, Ou CQ, Chan KP, Chau YK, Thach TQ, Yang L, Chung RY, Thomas GN, Peiris JS, Wong TW, Hedley AJ, Lam TH. The effects of air pollution on mortality in socially deprived urban areas in Hong Kong, China. *Environ Health Perspect.* 2008;116(9):1189-1194.

[39] Banerjee M, Siddique S, Mukherjee S, Roychoudhury S, Das P, Ray MR, Lahiri T. Hematological, immunological, and cardiovascular changes in individuals residing in a polluted city of India: A study in Delhi. *Int J Hyg Environ Health.*

[40] Barman SC, Kumar N, Singh R, Kisku GC, Khan AH, Kidwai MM, Murthy RC, Negi MP, Pandey P, Verma AK, Jain G, Bhargava SK. Assessment of urban air pollution and it's probable health impact. *J Environ Biol.*31(6):913-920.

[41] Nautiyal J, 3rd, Garg ML, Kumar MS, Khan AA, Thakur JS, Kumar R. Air pollution and cardiovascular health in Mandi-Gobindgarh, Punjab, India - a pilot study. *Int J Environ Res Public Health.* 2007;4(4):268-282.

[42] Ibald-Mulli A, Stieber J, Wichmann HE, Koenig W, Peters A. Effects of air pollution on blood pressure: a population-based approach. *Am J Public Health.* 2001;91(4):571-577.

[43] Urch B, Silverman F, Corey P, Brook JR, Lukic KZ, Rajagopalan S, Brook RD. Acute blood pressure responses in healthy adults during controlled air pollution exposures. *Environ Health Perspect.* 2005;113(8):1052-1055.

[44] de Paula Santos U, Braga AL, Giorgi DM, Pereira LA, Grupi CJ, Lin CA, Bussacos MA, Zanetta DM, do Nascimento Saldiva PH, Filho MT. Effects of air pollution on blood pressure and heart rate variability: a panel study of vehicular traffic controllers in the city of Sao Paulo, Brazil. *Eur Heart J.* 2005;26(2):193-200.

[45] Dubowsky SD, Suh H, Schwartz J, Coull BA, Gold DR. Diabetes, obesity, and hypertension may enhance associations between air pollution and markers of systemic inflammation. *Environ Health Perspect.* 2006;114(7):992-998.

[46] Peel JL, Metzger KB, Klein M, Flanders WD, Mulholland JA, Tolbert PE. Ambient air pollution and cardiovascular emergency department visits in potentially sensitive groups. *Am J Epidemiol.* 2007;165(6):625-633.

[47] Laden F, Schwartz J, Speizer FE, Dockery DW. Reduction in fine particulate air pollution and mortality: Extended follow-up of the Harvard Six Cities study. *Am J Respir Crit Care Med.* 2006;173(6):667-672.

[48] Daniels MJ, Dominici F, Samet JM, Zeger SL. Estimating particulate matter-mortality dose-response curves and threshold levels: an analysis of daily time-series for the 20 largest US cities. *Am J Epidemiol.* 2000;152(5):397-406.

[49] Brunekreef B, Holgate ST. Air pollution and health. *Lancet.* 2002;360(9341):1233-1242.

[50] Schwartz J. What are people dying of on high air pollution days? *Environmental research.* 1994;64:26-35.

[51] Sarnat JA, Schwartz J, Catalano PJ, Suh HH. Gaseous pollutants in particulate matter epidemiology: confounders or surrogates? *Environ Health Perspect.* 2001;109(10):1053-1061.

17

Particulate Matter and Cardiovascular Health Effects

Akeem O. Lawal and Jesus A. Araujo
Division of Cardiology, Department of Medicine,
David Geffen School of Medicine,
University of California, Los Angeles, CA
USA

1. Introduction

Several studies have shown that exposure to air pollution leads to important adverse health effects resulting in increased morbidity and mortality (Brook, Franklin et al. 2004; Bhatnagar 2006; Brook, Rajagopalan et al. 2010). According to the World Health Organisation, air pollution constitutes the 13th leading cause of mortality in the world (WHO 2009). Increased mortality is mostly due to increased cardiovascular diseases in the exposed population, particularly those of ischemic nature (Pope, Burnett et al. 2004). While air pollutants are composed of a mixture of particulate matter (PM) and gases such as carbon monoxide, ozone, sulphur oxide and nitrogen oxide, recent studies have shown that the particulate matter component of air pollution is mainly responsible for the cardiovascular health effects (Araujo and Nel 2009). This chapter will mostly focus on the links between PM, atherosclerosis and ischemic heart disease.

2. Particulate matter and sources of exposure

2.1 Classification of particulate matter

Particulate matter is constituted by compounds of varying sizes, numbers, chemical composition, and derived from various sources. They are mainly classified according to their size and divided into the following categories: i) Thoracic particles, with an aerodynamic diameter less than 10 micrometers ($< 10~\mu m$), ii) Coarse particles, with an aerodynamic diameter greater than 2.5 micrometers and less than 10 micrometers ($PM_{2.5-10}$), iii) Fine particles, less than 2.5 micrometers and iv) Ultrafine particles (UFP), less than 0.1 micrometers ($<0.1~\mu m$) (Table1) (U.S.EPA 2004).

2.2 Sources of PM emissions

The characterization of PM emissions is rather complicated by limitations in the measuring instrumentation. In addition, emissions from the same sources change with time and operating conditions. What is actually released by the source of the emissions may not be what is found in the atmosphere at varying distances from the sources, since particulates undergo a variety of chemical and physical transformations after they have been released into the atmosphere (Robinson, Donahue et al. 2007). Particulates are emitted directly from various sources such as fossil fuel combustion (primary particles) like diesel and gasoline exhaust particles or are formed from gases through chemical reactions involving atmospheric oxygen (O_2), water vapor (H_2O), free radicals such as nitrate (NO_3^-) and hydroxyl ($\cdot OH$) radicals, reactive species such as ozone (O_3), organic gases from anthropogenic and natural sources and pollutants such as nitrogen oxides (NOx) and sulfur dioxide (SO_2) (U.S.EPA 2004).

As shown in Table 1, various sources are involved in the generation of particles via different mechanisms that result in particles with distinct lognormal modes in the particle-size distributions by number and volume (nucleation, Aitken, accumulation and coarse modes). Nucleation and Aitken modes contain particles derived from the combustion of fossil fuels and are due to the nucleation of gas-phase compounds to form condensed-phase species in particles that are newly formed with little chance to grow (nucleation mode) or in newly formed particles in the process of coagulation (Aitken mode). These modes of generation are responsible for the ultrafine particles, emitted indeed from the combustion of fuels such as diesel and gasoline used in the operation of motor vehicles, aircrafts and ships. In the accumulation mode, particles grow in size and accumulate either by coagulation or by condensation, responsible for the generation of the bigger fine particles that are also emitted

Particles	Sources	Mode of generation	Aerodynamic diameter (μm)	Atmospheric half-life
Ultrafine particles (UFP)	Combustion of fossil fuels (gasoline & diesel) and emissions from mobile sources (aircrafts, ships, motor vehicles)	Fresh emissions, secondary photochemical reactions (nucleation mode)	< 0.1	Minutes to hours
Fine particles (FP)	Residential fuel combustion, power plants, tailpipe and brake emissions and Oil refineries	Condensation, coagulation conversion of gas-to-particle (accumulation mode)	<2.5	Days to weeks
Coarse particles (CP)	Suspension from construction, plant and animal fragments, disturbed soil (mining, farming)	Evaporation of sprays, suspension of dusts, mechanical disruption (crushing, grinding, abrasion of surfaces)	2.5-10	Minutes to hours
Thoracic particles (TP)	-	-	<10	-

Source: Modified from Araujo (Araujo and Nel 2009).

Table 1. Size fractions of PM emissions into the atmosphere.

from fuel combustion as well as power plants, oil refineries, tail pipe and brake emissions. The coarse mode contains coarse particles (PM $_{2.5-10}$) derived from the suspension of material from construction, disturbed soil as well as plant and animal fragments. Most of the coarse particles in the PM$_{10}$ are emitted into the atmosphere from the process and open dust sources of fugitive emissions. The process sources are associated with industrial operations such as rock crushing while the open dust sources are those that produce non-ducted emissions of solid particles formed from forces of wind or machinery acting on exposed material. The latter include particulate emissions from industrial sources associated with the open transport, storage, transfer of raw, intermediate and waste aggregate materials and nonindustrial sources such as unpaved roads and parking lots, paved streets and highways, agricultural tilling and heavy construction activities (WRAP 2004). Different sources involved in the generation of ambient particulate can be classified depending on their mobility, into mobile vs. non-mobile sources, as shown in table 2.

Sources	Percentage contribution (% total/ % within category)
Mobile Sources *	28%
1. On-road Mobile Sources	10%
i. Heavy Gasoline Vehicles	3%
ii. Light Gasoline Trucks	10%
iii. Cars& Motorcycles	15%
iv. Diesel Vehicles	72 %
2. Non-road Mobile Sources	18%
i. Railroads	7%
ii. Aircraft	7%
iii. Marine	10%
iv. Gasoline Equipment	20%
v. Diesel Equipment	56%
Not Mobile Sources *	72%
PM $_{2.5}$ Emissions by Source Sector **	
Road Dust	21.5%
Industrial Processes	12.1%
Electricity Generation	11.5%
Fires	9.2%
Residential Wood combustion	8.5%
Waste Disposal	6.2%
Non-road Equipment	6.0%
Fossil fuel combustion	4.8%
On road vehicles	3.0%
Solvent Use	0.2%
Fertilizer and Livestock	0.03%
Miscellaneous	17.0%

* Based on data from the 1999 National Air Quality and Emissions Trends Report (U.S.EPA 1999).
** Based on data from the 2005 National Summary of Particulate matter emissions by source sector (U.S.EPA 2005).

Table 2. Different sources of PM$_{2.5}$ emissions into the atmosphere.

2.2.1 Mobile sources

Emissions from mobile sources contribute significantly to measured levels of particulate matter. These emissions mainly consist of fine and ultrafine particles. According to the report of the 1999 National Emissions by Source in several US states (USEPA 1999), the mobile sources include: 1) On-road sources, which account for ~ 10% of the total emissions in US and include emissions from Cars and Motorcycles (15%), Diesel vehicles (72%), Light Gasoline Trucks (10%) and Heavy Gasoline Vehicles (3%), 2) Non-road sources, which accounts for 18% of total emissions in US including emissions from Marine (10%), Diesel (56%) and Gasoline Equipment (20%) (Table 2). Indeed, emissions from diesel vehicles accounted for about three-quarter of the on-road emissions (USEPA 1999). Importantly, particles derived from motor vehicles seem to be especially potent in increasing cardiovascular mortality (Laden, Neas et al. 2000; Pekkanen, Peters et al. 2002) and cardiac hospital admissions (Janssen NA 2002).

2.2.2 Non-mobile sources

Emissions from Non-mobile sources, also known as the Point Sources, contribute in a larger degree to the total PM emitted into the atmosphere throughout the US (Table 2) (USEPA 1999). Point sources are stationary, large and identifiable sources of emissions that release pollutants into the atmosphere. According to the National Emissions statistics of $PM_{2.5}$ throughout the US in 1999, non-mobile sources include large industries such as manufacturing plants, power plants, fires, residential wood combustion, waste disposal, non road equipment, paper mills and refineries (table 2) (USEPA 1999). Additional sources include biogenic non-anthropogenic sources such as trees and vegetation, oil and gas seeps and microbial activity (USEPA 1999).

3. Particulate matter, duration of exposure and cardiovascular health effects

Exposure to air pollutants has been associated with increased risk for adverse cardiovascular health effects, mostly attributed to those in the particulate phase. Indeed, ambient particulate matter (PM) has been linked to a whole variety of atherothrombotic endpoints (Ghio, Hall et al. 2003; Pope, Muhlestein et al. 2006; Baccarelli, Martinelli et al. 2009). Although epidemiological studies with PM_{10} and $PM_{2.5}$ show that exposure to these particles are associated with adverse cardiovascular effects such as myocardial infarction and stroke death, greater effects are associated with the smaller particle sizes ($PM_{2.5}$) (Analitis, Katsouyanni et al. 2006; Pope 2006; Zanobetti and Schwartz 2009). Thus, the risk of cardiovascular mortality posed by PM_{10} exposure could be in a good degree due to its fraction of particles < 2.5 μm in diameter ($PM_{2.5}$). According to the World Health Organization, exposure to PM may be responsible for an estimate of about 800,000 excess deaths worldwide in each year as a result of myocardial infarction, arrhythmias or heart failure (WHO 2002). Several epidemiological studies have shown that the increased risk in cardiovascular morbidity and mortality occurs as a result of both short- and long-term exposures to PM as summarized in Table 3.

Duration of exposure	Type of Study	Study	No. of subjects	Exposure variable	Outcome variable	Major Findings	Ref.
Short-term exposure studies	Time Series	NMMAPS	~50 million	PM_{10}	Daily CP mortality	20 $\mu g/m^3$ increases in PM 10 caused 0.6% increased in daily CP mortality in adults from 20 to 100 US cities and hundreds of countries.	(Dominici, Zeger et al. 2000; Dominici, McDermott et al. 2003; Peng, Dominici et al. 2005; Dominici, Peng et al. 2007)
		NCHS	NCHS data on 112 US cities	$PM_{2.5}$	Daily CV mortality	10 $\mu g/m^3$ increase in 2-day averaged PM 2.5 caused 0.98% and 0.85% increase in daily all cause and CV mortality respectively	(Baccarelli, Martinelli et al. 2009)
		APHEA2	~43 million	PM_{10}	Daily CV mortality	20 $\mu g/m^3$ increases in PM 10 caused a 1.5% increased in daily CV mortality in adults from 29 European cities.	(Katsouyanni 2003; Analitis, Katsouyanni et al. 2006)
		APHENA	NMMAPS, APHEA2 & Canadian studies	PM_{10}	Daily CV mortality	10 $\mu g/m^3$ increase in PM 10 caused 0.2 to 0.6% increased in daily all-cause mortality in individuals from the US (90 cities), Europe (22 cities) and Canada (12 cities), mostly in individuals>75-year old.	(Samoli, Peng et al. 2008)
	Case cross-over	IHCS	~12865	$PM_{2.5}$	Ischemic coronary events	10 $\mu g/m^3$ increases in $PM_{2.5}$ caused a 4.5% increase in daily acute ischemic coronary events in patients in UTAH. Patients with pre-existing CAD have significant increased frequency.	(Pope, Muhlestein et al. 2006)
		MI Registry in Augsburg (data from KORA)	691	Traffic	MI	Exposure to traffic for 1 h increases the relative risk for an MI by 2.92 before the event.	(Peters, von Klot et al. 2004)

Duration of exposure	Type of Study	Study	No. of subjects	Exposure variable	Outcome variable	Major Findings	Ref.
Long-term exposure studies	Prospective Cohort	American Cancer Society II (extended analysis)	~500000	$PM_{2.5}$	CV mortality	10 µg/m³ increase long-term exposure to PM $_{2.5}$ in a 16-year study caused 12% increased in risk for CV death. The largest cause of mortality was due to Ischemic heart disease and smaller numbers of people died from arrhythmias and heart failure	(Pope, Burnett et al. 2004)
		Harvard Six Cities (extended analysis)	8096	$PM_{2.5}$	CV mortality	10 µg/m³ increase long-term exposure to PM $_{2.5}$ in a 28-year study of six US cities residents show a 1.28% increased in relative risk for CV. Significant reduction in CV mortality was observed in some cities and this correlate with decrease in PM $_{2.5}$	(Laden, Schwartz et al. 2006)
		NHS	66250	PM_{10}	Acute events of CV and CV mortality	10 µg/m³ increase in long-term exposure to PM $_{10}$ caused a 43% increased in fatal CAD disease among nurses from northeastern US	(Puett, Schwartz et al. 2008)
		Woman's Health Initiative	65893	$PM_{2.5}$	CV acute events and CV mortality	10 µg/m³ increases in long-term exposure to PM $_{2.5}$ caused a 24% and 76% increase in CV acute events and CV mortality respectively in healthy post-menopausal women in 36 US cities.	(Miller, Siscovick et al. 2007)

NMMAPS, National Morbidity, Mortality and Air Pollution Study; APHEA2, Air Pollution and Health: A European Approach; IHCS, Intermountain Heart Collaborative Study; APHENA, Air Pollution and Health: A Combined European and North American Approach study; CV, Cardiovascular; CP, Cardiopulmonary; MI, Myocardial Infarction; KORA, Cooperative Health Research in the Region of Augsburg; NCHS, National Center for Health Statistics. Modified from (Araujo and Brook 2011)

Table 3. Selected studies showing a relationship between exposure to PM of different lengths of duration and increased cardiovascular morbidity and/or mortality.

3.1 Short-term exposures

Although the mechanisms involved in the generation of PM-mediated adverse cardiovascular health effects are still not clear, several studies have shown that short-term exposures to PM results in cardiovascular systemic effects (Table 3). Thus, in a study carried out in eight European cities, zero to one-day exposure to PM_{10} and black smoke associated with a significant increase in cardiac hospital admission in subjects of all ages and ischemic heart disease in people over 65 years (Le Tertre, Medina et al. 2002). Results from the National Morbidity, Mortality, and Air Pollution study (NMMAPS) and the Air Pollution and Health: A European Approach (APHEA2) involving ~50 and ~43 million subjects respectively found that there was 0.6 and 1.5 % increase in daily cardiopulmonary mortality rate for every 20 µg/m³ increase in PM_{10}, respectively (Dominici, McDermott et al. 2003; Analitis, Katsouyanni et al. 2006). Several studies have also shown that exposure to $PM_{2.5}$ result in similar or even larger magnitudes of association. For instance, in the Intermountain Heart Collaborative Study (IHCS) involving 12,865 subjects, 10 µg/m³ increases in ambient PM $_{2.5}$ associated with a daily increase of 4.5% in acute ischemic coronary events. Studies have also shown that short-term exposures to PM associated with increased risk for hospitalization for cardiovascular diseases. Dominici et al showed among 11.5 million U.S medicare enrollees > 65 year old that admission rates for all cardiovascular causes increased in association with 10 µg/m³ increases in $PM_{2.5}$, leading to an increase of 0.44% in ischemic heart, 1.28% in heart failure and 0.81% in cerebrovascular diseases (Dominici, McDermott et al. 2003). There is also evidence that exposure to UFP associates with increased risk for cardiac hospital admissions, as discussed in section 6.

3.2 Long-term exposures

A number of studies have also shown an association between long-term exposures to PM and cardiovascular morbidity and mortality (Table 3). Thus, in an extended analysis (16-year follow-up) of a cohort survival study from the American Cancer Society, 10 µg/m³ increases in long-term exposure to PM $_{2.5}$ associated with a 12% increase in CV deaths, most of which was due to ischemic heart disease, with a smaller percentage due to arrhythmias and heart failure (Table 3) (Pope, Hansen et al. 2004). This study suggests that long-term PM exposures carry greater cardiovascular risk as compared with the magnitude of risk observed in studies evaluating short-term exposures. An even stronger correlation was observed in the Women Health Initiative Study involving 65,893 post menopausal women, which showed 24% and 76% increase in cardiovacular acute events and mortality, respectively (Miller, Siscovick et al. 2007).

4. Particulate matter and cardiovascular ischemic effects

It is noteworthy that PM-mediated enhancement of cardiovascular morbidity and mortality is mostly due to the promotion of ischemic events, such as myocardial infarction. The fact that PM exposures associate with cardiovascular ischemic endpoints both in the short-term as well as long-term exposure studies suggests that PM activates various pathways resulting in both short- as well as long-term effects.

Indeed, several studies have shown that short-term (hours to days) exposures to PM significantly enhance cardiovascular risk (Peters, Frohlich et al. 2001; Miller, Siscovick et al. 2007). Although the mechanisms by which exposure to PM induces cardiac ischemic events such as myocardial infarctions are not yet well defined, several studies have suggested that PM can be associated with hemodynamic, hemostatic as well as cardiac rhythm alterations that may account for some of the PM-induced acute cardiac events. In support of this observation, there are significant associations between exposure to ambient PM and acceleration of heart rate, diminished heart rate variability (Pope and Kalkstein 1996; Peters, Doring et al. 1997), ventricular fibrillation, increased number of therapeutic intervention in patients with implanted cardiovascular-defibrillators (Peters, Liu et al. 2000) and increased plasma viscosity (Peters, Doring et al. 1997).

Many studies have also reported a strong correlation between long-term exposure to PM $_{2.5}$ and cardiac ischemic endpoints as shown above, which strongly suggests that PM-enhancement of atherosclerosis could be one of the pathogenic mechanism(s). Indeed, several studies involving human subjects have shown significant associations between $PM_{2.5}$ exposure and atherosclerosis. Thus, in a cross-sectional study carried out in 798 individuals in Los Angeles by Kunzli et al, carotid intima-medial thickness (CIMT) was found to increase by 5.9% for every 10 μg/m³ rise in $PM_{2.5}$ levels (Kunzli, Jerrett et al. 2005). Same group found an acceleration in the annual rate progression of CIMT among individual living within 100 m of a highway which was more than twice the population mean progression (Kunzli, Jerrett et al. 2010). In another study, Hoffman et al found an association between coronary artery calcification (CAC) scores, an index of coronary atherosclerosis, and long-term residential exposure to high traffic (Hoffmann, Moebus et al. 2007). They found that in 4,494 participants, subjects living within 101-200 m, 51-100 or less than 50 m from a major road showed 8%, 34% and 63% increase respectively in the probability of having CAC compared with subjects living >200 m. Furthermore, Diez Roux et al reported a 1-3% increase in CIMT per 21 μg/m³ increase in PM_{10} or 12.5 μg/m³ increase in $PM_{2.5}$ in subjects exposed to PM_{10} over long-term (20-year means and 2001 mean) and 20-year $PM_{2.5}$ respectively (Diez Roux, Auchincloss et al. 2008). Likewise, in a related study, Allen et al reported an increase risk for aortic calcification with $PM_{2.5}$ exposure (Allen, Criqui et al. 2009). All these studies confirm that exposure to PM triggers-off the development of atherosclerosis in humans, which may be a major mechanism how long-term exposure to PM leads to enhanced cardiovascular ischemic endpoints.

Three putative "general mediating" pathways (Fig.1) have been proposed for the cardiovascular effects of air pollutants such as: 1) Autonomic nervous system imbalance, 2) Induction of pumonary and thereby systemic inflammation/oxidative stress via "spill-over" of mediators (e.g. cytokines, activated white cells/platelets) into the systemic circulation, 3) Access of particles or specific chemical constituents into the systemic circulation which thereby cause direct affects upon the heart and vasculature (Araujo and Brook 2011) (Fig.1). Some of the PM-mediated short-term effects may be related to pathway #1. On the other hand, the occurrence of long-term effects strongly suggests that PM would promote atherosclerosis, likely via a combination of pathway #2 and #3. However, while it has been proposed that UFP could access the systemic circulation (pathway #3), clear confirmation of this possibility is still lacking.

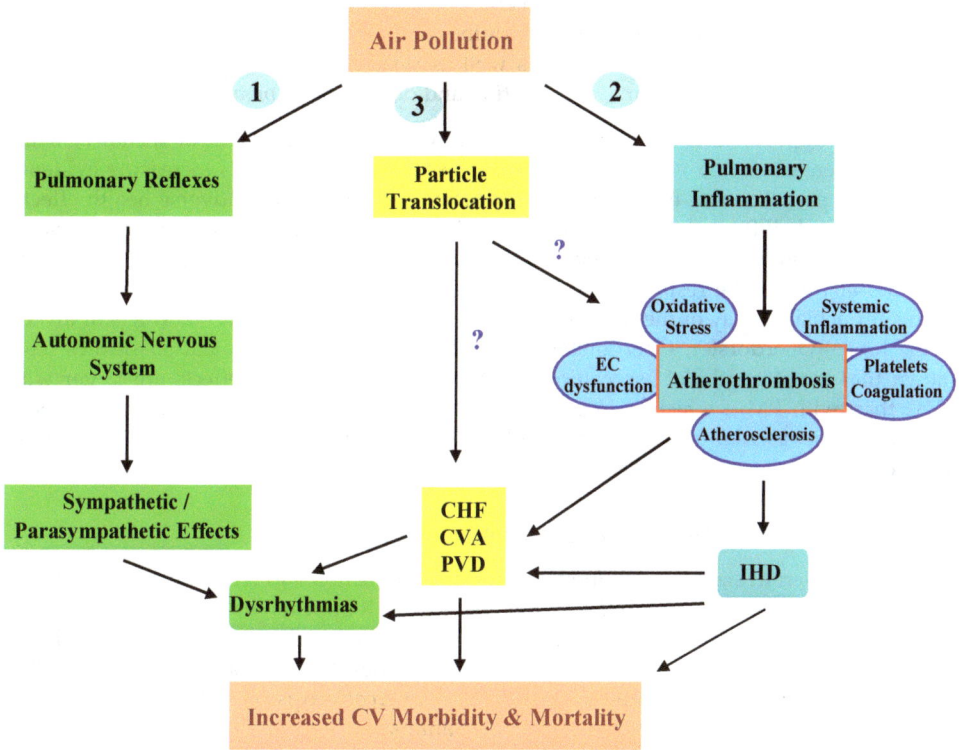

Fig. 1. **Possible mechanisms that link exposure to PM with cardiovascular diseases**. Three mediating pathways are proposed. Air pollution can; 1) Activate the autonomic nervous system resulting in the alteration of heart rate viability and induction of dysrhythmias, 2) Enhance atherothrombotic processes that can lead to development of pulmonary oxidative stress and inflammation with systemic "spill-over" of inflammatory mediators, 3) Particles and/or their chemical constituents can translocate directly to the systemic circulation. CHF= Congestive heart failure, CVA= Cerebrovascular accident, PVD= Peripheral vascular disease, EC= Endothelial cells, IHD= Ischemic heart disease. Taken from (Araujo and Brook 2011).

It is possible then that the predominant mechanism(s) mediating PM-enhancement of cardiovascular ischemic events could be different among studies that evaluate short- vs. long-term effects. In addition, particulate of different size fractions as well as originating from different sources could exhibit significant differences in their ability to induce systemic vascular effects as will be discussed in the next two sections below.

5. What determines PM-related cardiovascular toxicity?

5.1 Particle size

The ability of PM to induce adverse health effects in humans may be a function of a number of factors, some of which are likely to determine its deposition in the respiratory tract and ability to induce both local pulmonary as well as systemic vascular effects. These factors include, among others, size, shape, composition and density of ambient particles

Particle size appears to be an important determinant of PM toxicity. For instance, while PM_{10} has been associated with ischemic cardiovascular events, there is increasing evidence that smaller particles may be responsible for most of cardiovascular adverse health effects. Thus, $PM_{2.5}$ and UFP are thought to be more toxic than larger particles, which may be partly based on their ability to access deeper portions into the lungs. These particles can pass the proximal airway of the respiratory system (throat and larynx) and get deposited into the tracheobronchial airway of the lungs or in the gas exchange region (alveolar ducts or alveoli of the lungs)(Oberdorster, Oberdorster et al. 2005). However, as population studies have shown that the strengths of association of PM with cardiovascular effects are larger with $PM_{2.5}$ than with PM_{10}, which is in support of the notion that a small particle size would facilitate cardiovascular toxicity, there is still a paucity of epidemiological studies in relation to the cardiovascular effects of UFP.

5.2 Importance of chemical composition and redox potential

Particles of different sizes have different chemical composition and redox potential that may affect their ability to act through the various pathways. It appears that PM ability to trigger and/or enhance reactive oxygen species (ROS) production, resulting in tissue oxidative stress, may be of key importance in cardiovascular tissues. For instance, PM-mediated ROS formation may be central in the enhancement of atherosclerosis by PM, likely due to the promotion of systemic prooxidant and proinflammatory effects. Since atherosclerosis is characterized by lipid deposition and oxidation in the arterial wall, factors that promote lipid oxidation and retention in the artery wall may exacerbate the pathogenic process. These lipids are derived from plasma low-density lipoprotein (LDL) particles that travel into the arterial wall and get trapped in the subendothelial space where they can be oxidatively modified (Steinberg 1997; Araujo, Barajas et al. 2008). Several studies using different PM size fractions (PM_{10}, $PM_{2.5}$, UFP) to expose different animal models (apoE[-/-] mice, LDL-R[-/-] mice and hyperlipidemic rabbits) via different modes of exposure (inhalation of CAPs, oropharyngeal/intratracheal instillation) converge to demonstrate that PM exposure leads to enhanced atherosclerotic lesions or plaques with altered composition, suggesting that the associations encountered in epidemiological studies are very likely to be causal (Araujo 2011a).

Experimental animal data also supports the notion that particulate of the smallest size may be able to induce larger proatherogenic effects. Araujo et al reported that exposure of apoE[-/-] mice to concentrated ambient particles (CAPs) in the fine and ultrafine-size ranges resulted in a significant increase of lipid peroxidation in the liver and triggering of a Nrf2-regulated antioxidant response in mice exposed to ultrafine particulate, likely to be indicative of the greater levels of oxidation induced by ultrafines. Stronger prooxidant effects led to bigger atherosclerotic plaques in mice exposed to ultrafines (Araujo, Barajas et al. 2008). The greater toxicity of UFP could be due to their greater content in prooxidant organic chemicals, greater bioavailability for those reactive compounds due to their larger surface-to-mass ratio and/or greater lung retention (Araujo and Nel 2009). Indeed, UFPs contained twice as much organic carbon (OC) as FP did, accompanied by an increase in the relative content of polyaromatic hydrocarbon (PAH), which may have been one of the factors why UFP induced greater enhancement of aortic atherosclerosis than $PM_{2.5}$ did (Araujo, Barajas et al. 2008). In addition, transition metals have also been largely implicated in the PM hazardous health effects (Ntziachristos, Froines et al. 2007; Ayres, Borm et al. 2008). While there is a large experimental evidence that points out towards these various candidates, there is no substantial evidence that links PAH contents with adverse cardiovascular effects as of yet.

5.3 Factors determining susceptibility

PM cardiovascular toxicity could also be influenced by different susceptibility to PM-mediated systemic effects. Indeed, epidemiological, toxicological and controlled human exposure studies have examined whether the health effects of PM are modulated by preexisting conditions such as coronary artery disease, congestive heart failure, obesity, age and diabetes among others. Several reports identify obesity as a likely susceptibility factor. For instance, short-term exposure to PM has been shown to associate with a reduction in heart rate variability (Schwartz, Litonjua et al. 2005) and higher levels of plasma inflammatory markers such as IL-6 and C-reactive protein in a greater degree among obese individuals (Dubowsky, Suh et al. 2006). Studies of the Veteran's Normative Aging and Women's Health Initiative Cohorts also show that long-term exposure to PM associated with an increase in inflammatory markers and cardiovascular events in individuals with body mass index (BMI) ≥ 25 kg/m^2 compared with < 25 kg/m^2 (Zeka, Sullivan et al. 2006; Miller, Siscovick et al. 2007). In addition, it appears that preexistent congestive heart failure increases the susceptibility to PM-induced cardiovacular events and mortality (Bateson and Schwartz 2004; Pope 2006). Additional studies are required to confirm these findings and better dissect potential susceptibility factors.

6. Relationship between sources of emission and cardiovascular health effects

While different sources of emission may generate pollutants with different physicochemical characteristics, the mechanisms involved in PM-mediated enhancement of cardiovascular morbidity and mortality have been assumed to be common to all particulate derived from various sources (Figure 1). However, this may be an oversimplification of a rather complex relationship that has not been well studied yet, partly due to the lack of understanding about the critical toxic components of air pollution that are responsible for the cardiovascular effects.

PM from different sources may trigger and/or enhance ROS formation which may lead to tissue oxidative stress, inflammation, endothelial dysfunction, resulting in the formation of atherosclerotic plaques (Figure 2). The particular source of emission is to play an important role in determining particle size and chemical composition. It should also be noted that some elements in PM are more source-specific than others and therefore, the extent of PM induction of cardiovascular effects may largely depend on the specific source of PM. For instance, elements such as iron and zinc are mostly derived from steel production while nickel is linked to oil combustion (Li Z 2004; Zheng L 2004; Qin Y 2006). Some elements might be site-specific. For instance, coal and oil combustion at power plants are the main source of sulphur in Canada. Therefore, better understanding of the sources and type of emissions may help to elucidate the particle components responsible for cardiovascular toxic effects.

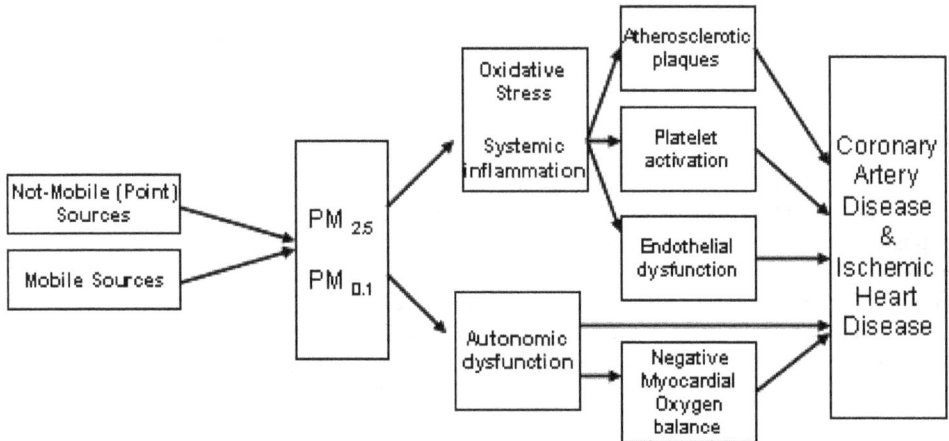

Fig. 2. **Mechanisms involved in the development of ischemic heart disease generated from PM emitted from different sources.** PM, emitted from different sources, may induce various systemic effects and modulate various pathways resulting in enhanced ischemic heart disease events.

One clear example about the difficulties to study the influence of the source of exposures on health effects is in relation to the particle size. Numerous studies have shown that combustion-derived particles from mobile sources are associated with increased cardiovascular morbidity and mortality. Studies have shown that $PM_{2.5}$ exhibits a larger association with cardiovascular endpoints than PM_{10}. These associations have been established mostly using metrics based on PM mass. However, there has been a notion that even smaller particles such as ultrafines would be more active to induce cardiovascular effects based on experimental in-vitro and animal evidence (Schulz, Harder et al. 2005; Araujo, Barajas et al. 2008) but supportive epidemiological evidence is somewhat weaker. Exposure to UFP comes from both mobile as well as non-mobile sources but capturing their toxicity is rather difficult since they are very dependant on the proximity to the source and the conventional particle mass or particle number-based metrics are inadequate to fully capture their toxic potential. Therefore, dissecting the actual contribution of the source of PM to cardiovascular events is difficult as no adequate parameters have so far been used to assess the contribution of UFP at the point of exposure in relation to the point of emission. Despite the short-comings related to the use of total particle number concentration to estimate UFP concentration, studies have shown that exposure to UFP significantly associates with increased hospital admissions due to acute myocardial infarction (Lanki, Pekkanen et al. 2006; von Klot, Peters et al. 2005) as well as heart failure (Belleudi, Faustini et al.; Dominguez-Rodriguez, Abreu-Afonso et al. 2011 et al.), summarized in table 4. Unfortunately, UFP assessment was only made based on particle number concentration whereas a more comprehensive characterization of the UFP and the particular source of emissions would have been very informative.

Study	Number of cardiac hospital admissions	Environmental exposure parameters	Types of cardiac admissions	Mayor findings
(von Klot, Peters et al. 2005)	6655 first hospital readmissions	PNC PM_{10} Gases (CO, NO_2, O_3)	Acute MI Angina pectoris, HF Dysrhythmia	Cardiac readmissions increased by 2.1% and 2.6% per each increase of 10 µg/m³ of PM_{10} and 10 000 particles/cm³, respectively
(Lanki, Pekkanen et al. 2006)	26 854 first MI admissions	PNC PM_{10} Gases (CO, NO_2, O_3)	Acute MI	Hospitalization for first MI increased by 0.5% per each increase of 10 000 particles/cm³ (lag 0). Associations were greater among fatal events and subjects < 75 years
(Belleudi, Faustini et al. 2010)	90 056 cardiac hospital admissions	PNC PM_{10} $PM_{2.5}$	HF ACS Other cardiac causes	HF and ACS increased by 2.4% and 2.3% respectively per each increase of 10 µg/m³ of PM2.5 (lag 0). HF increased by 1.7% per each increase of 9392 particles/cm³ (lag 0)
(Dominguez-Rodriguez, Abreu-Afonso et al. 2011)	3229 hospital admissions	PNC PM_{10} $PM_{2.5}$ PM_1 Gases (CO, SO_2, NO_2, O_2)	HF ACS	UFP found to be a risk factor for HF admissions compared to admissions for ACS (odds ratio = 1.4)

ACS, acute coronary syndromes; HF, heart failure; MI, myocardial infarction; PM, particulate matter; PNC, particle number concentration; UFP, ultrafine particles. Modified from (Araujo 2011b).

Table 4. Studies linking exposure to ultrafine particles and cardiac hospital admissions.

The assessment of Nickel content exemplifies another situation where a thorough characterization of the source of emissions might prove to be helpful. Nickel (Ni) is an important metal produced from the combustion of fossil fuels. Ni content in $PM_{2.5}$ has been identified as an indicator of oil combustion not only from mobile sources such as vehicles (Brook2004) and ships (Ying, Yue et al. 2009) but also from industries such as cement production, asbestos mining and milling, iron and steel foundries, municipal waste sludge incineration and cooling tower. Exposure to Nickel, Lead and Sulfur have been reported to significantly correlate with all-cause mortality in a study involving 6 US cities (Laden, Neas et al. 2000). Ni has also been reported to induce coronary vasoconstriction, decreased heart rate variability, increased incidence of arrhythmias, increased expression of cardiac cytokines IL-6 and TGF-β, and monocytic cell infiltration in animal models (Rubanyi and Kovach 1980). Lippman et al found that exposure of apoE-/- mice to $PM_{2.5}$, containing Ni at an average concentration of 43 ng/m^3 for 6h/day, 5d/wk for 6 months in Tuxedo, NY resulted in acute changes in heart rate, heart rate variability and enhanced atherosclerosis (Lippmann, Ito et al. 2006). It turns out that elevated concentrations of Ni in that area were in relation to Ni emitted from the International Nickel Company at Sudbury Ontario at a distance of more than 800 km away from NYC (Lippmann, Ito et al. 2006). Exposures to a similar $PM_{2.5}$ mass but with different Ni contents that would determine different degrees of atherosclerosis would certainly substantiate the role of Ni metal in PM-mediated atherogenesis.

7. Conclusions

Several conclusions can be drawn about the cardiovascular effects of PM as follows: a) Short and long-term exposure to PM associates with various cardiovascular endpoints, especially of ischemic nature, likely in a causal manner, b) PM-induction of systemic prooxidative and proinflammatory effects appears to be key in the development of atherosclerosis, c) Particle size and chemical composition are important determinants of PM cardiovascular toxicity, d) Thorough characterization of the sources of exposure may help to identify toxic components of air pollution.

Dissecting the actual contribution of each source of emissions to the CV effects would involve a more comprehensive characterization of the type of emissions that would need to go beyond PM mass-based metrics and include qualitative analysis. No single source is likely to account for all PM-mediated cardiovascular events in a dense urban setting and instead, various sources may contribute in different degrees in different locations. Therefore, to better assess the contribution of each source of emission to cardiovascular events, consideration should be given to the type of emission source, composition of the PM derived from the source of emissions and the proximity of the emission source to contact, among other factors.

8. Funding

Writing of this chapter was supported by the National Institute of Environmental Health Sciences, National Institutes of Health (RO1 Award ES016959 to Jesus A. Araujo).

9. References

Allen, R. W., M. H. Criqui, et al. (2009). "Fine Particulate Matter Air Pollution, Proximity to Traffic, and Aortic Atherosclerosis." *Epidemiology*(20): 254-264.

Analitis, A., K. Katsouyanni, et al. (2006). "Short-term effects of ambient particles on cardiovascular and respiratory mortality." *Epidemiology* 17(2): 230-3.

Araujo, J. (2011a). "Particulate air pollution, systemic oxidative stress, inflammation, and atherosclerosis." *Air Quality, Atmosphere & Health* 4(1): 79-93.

Araujo, J. A. (2011b). "Are ultrafine particles a risk factor for cardiovascular diseases?" *Rev Esp Cardiol* 64(8): 642-5.

Araujo, J. A., B. Barajas, et al. (2008). "Ambient particulate pollutants in the ultrafine range promote early atherosclerosis and systemic oxidative stress." *Circ Res* 102(5): 589-96.

Araujo, J. A. and R. D. Brook (2011). Cardiovascular Effects of Particulate-Matter Air Pollution: An Overview and Perspectives *Environmental Cardiology: Pollution and Heart Disease*. A. Bhatnagar. Cambridge, U.K., Royal Society of Chemistry. Issues in Toxicology No.8: 76-104.

Araujo, J. A. and A. E. Nel (2009). "Particulate matter and atherosclerosis: role of particle size, composition and oxidative stress." *Part Fibre Toxicol* 6: 24.

Ayres, J. G., P. Borm, et al. (2008). "Evaluating the Toxicity of Airborne Particulate Matter and Nanoparticles by Measuring Oxidative Stress Potentialâ€"A Workshop Report and Consensus Statement." *Inhalation Toxicology* 20(1): 75-99.

Baccarelli, A., I. Martinelli, et al. (2009). "Living near major traffic roads and risk of deep vein thrombosis." *Circulation* 119(24): 3118-24.

Bateson, T. F. and J. Schwartz (2004). "Who is sensitive to the effects of particulate air pollution on mortality? A case-crossover analysis of effect modifiers." *Epidemiology* 15(2): 143-9.

Belleudi, V., A. Faustini, et al. "Impact of fine and ultrafine particles on emergency hospital admissions for cardiac and respiratory diseases." *Epidemiology* 21(3): 414-23.

Bhatnagar, A. (2006). "Environmental cardiology: studying mechanistic links between pollution and heart disease." *Circ Res* 99(7): 692-705.

Brook, R. D., B. Franklin, et al. (2004). "Air pollution and cardiovascular disease: a statement for healthcare professionals from the Expert Panel on Population and Prevention Science of the American Heart Association." *Circulation* 109(21): 2655-71.

Brook, R. D., S. Rajagopalan, et al. (2010). "Particulate Matter Air Pollution and Cardiovascular Disease. An Update to the Scientific Statement From the American Heart Association." *Circulation*: CIR.0b013e3181dbece1.

Chen, Q., M. G. Espey, et al. (2005). "Pharmacologic ascorbic acid concentrations selectively kill cancer cells: action as a pro-drug to deliver hydrogen peroxide to tissues." *Proc Natl Acad Sci U S A* 102(38): 13604-9.

Diez Roux, A. V., A. H. Auchincloss, et al. (2008). "Long-term exposure to ambient particulate matter and prevalence of subclinical atherosclerosis in the Multi-Ethnic Study of Atherosclerosis." *Am J Epidemiol* 167(6): 667-75.

Dominguez-Rodriguez, A., J. Abreu-Afonso, et al. (2011). "Comparative study of ambient air particles in patients hospitalized for heart failure and acute coronary syndrome." *Rev Esp Cardiol* 64(8): 661-6.

Dominici, F., A. McDermott, et al. (2003). Mortality among residents of 90 cities. In Revised Analyses of Time-Series Studies of Air Pollution and Health. Boston, MA, Health Effects Institute: 9-24.

Dominici, F., A. McDermott, et al. (2003). "National maps of the effects of particulate matter on mortality: exploring geographical variation." *Environ Health Perspect* 111(1): 39-44.

Dominici, F., R. D. Peng, et al. (2007). "Does the effect of PM10 on mortality depend on PM nickel and vanadium content? A reanalysis of the NMMAPS data." *Environ Health Perspect* 115(12): 1701-3.

Dominici, F., S. L. Zeger, et al. (2000). "A measurement error model for time-series studies of air pollution and mortality." *Biostatistics* 1(2): 157-75.

Dubowsky, S. D., H. Suh, et al. (2006). "Diabetes, obesity, and hypertension may enhance associations between air pollution and markers of systemic inflammation." *Environ Health Perspect* 114(7): 992-8.

Ghio, A. J., A. Hall, et al. (2003). "Exposure to concentrated ambient air particles alters hematologic indices in humans." *Inhal Toxicol* 15(14): 1465-78.

Hoffmann, B., S. Moebus, et al. (2007). "Residential exposure to traffic is associated with coronary atherosclerosis." *Circulation* 116(5): 489-96.

Janssen NA, S. J., Zanobetti AH et al (2002). "Air conditioning and source-specific particles as modifiers of the effect of PM(10) on hospital admissions for heart and lung disease." *Environ Health Perspect* 110: 43-9.

Katsouyanni, K. (2003). "Ambient air pollution and health." *Br Med Bull* 68: 143-56.

Kunzli, N., M. Jerrett, et al. (2010). "Ambient air pollution and the progression of atherosclerosis in adults." *PLoS One* 5(2): e9096.

Kunzli, N., M. Jerrett, et al. (2005). "Ambient air pollution and atherosclerosis in Los Angeles." *Environ Health Perspect* 113(2): 201-6.

Laden, F., L. M. Neas, et al. (2000). "Association of fine particulate matter from different sources with daily mortality in six U.S. cities." *Environ Health Perspect* 108(10): 941-7.

Laden, F., J. Schwartz, et al. (2006). "Reduction in fine particulate air pollution and mortality: Extended follow-up of the Harvard Six Cities study." *Am J Respir Crit Care Med* 173(6): 667-72.

Lanki, T., J. Pekkanen, et al. (2006). "Associations of traffic related air pollutants with hospitalisation for first acute myocardial infarction: the HEAPSS study." *Occup Environ Med* 63(12): 844-51.

Le Tertre, A., S. Medina, et al. (2002). "Short-term effects of particulate air pollution on cardiovascular diseases in eight European cities." *J Epidemiol Community Health* 56(10): 773-9.

Li Z, H. P., Husain L, Qureshi S, Dutkiewicz VA, Schwab JJ (2004). "Sources of fine particle composition in New York City." *Atmos Environ* 38: 6521–6529.

Lippmann, M., K. Ito, et al. (2006). "Cardiovascular effects of nickel in ambient air." *Environ Health Perspect* 114(11): 1662-9.

Miller, K. A., D. S. Siscovick, et al. (2007). "Long-term exposure to air pollution and incidence of cardiovascular events in women." *N Engl J Med* 356(5): 447-58.

Ntziachristos, L., J. Froines, et al. (2007). "Relationship between redox activity and chemical speciation of size-fractionated particulate matter." *Particle and Fibre Toxicology* 4(1): 5.

Oberdorster, G., E. Oberdorster, et al. (2005). "Nanotoxicology: an emerging discipline evolving from studies of ultrafine particles." *Environ Health Perspect* 113(7): 823-39.

Pekkanen, J., A. Peters, et al. (2002). "Particulate air pollution and risk of ST-segment depression during repeated submaximal exercise tests among subjects with coronary heart disease: the Exposure and Risk Assessment for Fine and Ultrafine Particles in Ambient Air (ULTRA) study." *Circulation* 106(8): 933-8.

Peng, R. D., F. Dominici, et al. (2005). "Seasonal analyses of air pollution and mortality in 100 US cities." *Am J Epidemiol* 161(6): 585-94.

Peters, A., A. Doring, et al. (1997). "Increased plasma viscosity during an air pollution episode: a link to mortality?" *Lancet* 349(9065): 1582-7.

Peters, A., M. Frohlich, et al. (2001). "Particulate air pollution is associated with an acute phase response in men; results from the MONICA-Augsburg Study." *Eur Heart J* 22(14): 1198-204.

Peters, A., E. Liu, et al. (2000). "Air pollution and incidence of cardiac arrhythmia." *Epidemiology* 11(1): 11-7.

Peters, A., S. von Klot, et al. (2004). "Exposure to traffic and the onset of myocardial infarction." *N Engl J Med* 351(17): 1721-30.

Pope, C. A., 3rd, R. T. Burnett, et al. (2004). "Cardiovascular mortality and long-term exposure to particulate air pollution: epidemiological evidence of general pathophysiological pathways of disease." *Circulation* 109(1): 71-7.

Pope, C. A., 3rd, M. L. Hansen, et al. (2004). "Ambient particulate air pollution, heart rate variability, and blood markers of inflammation in a panel of elderly subjects." *Environ Health Perspect* 112(3): 339-45.

Pope, C. A., 3rd and L. S. Kalkstein (1996). "Synoptic weather modeling and estimates of the exposure-response relationship between daily mortality and particulate air pollution." *Environ Health Perspect* 104(4): 414-20.

Pope, C. A., 3rd, J. B. Muhlestein, et al. (2006). "Ischemic heart disease events triggered by short-term exposure to fine particulate air pollution." *Circulation* 114(23): 2443-8.

Pope, C. A., 3rd, J. B. Muhlestein, et al (2006). ""Ischemic heart disease events triggered by short-term exposure to fine particulate air pollution."" *Circulation* 114(23): 2443-8.

Puett, R. C., J. Schwartz, et al. (2008). "Chronic particulate exposure, mortality, and coronary heart disease in the nurses' health study." *Am J Epidemiol* 168(10): 1161-8.

Qin Y, K. E., Hopke PK (2006). "The concentrations and sources of PM2.5 in the metropolitan New York City." *Atmos Environ* 40 (Suppl): 312–332.

Robinson, A. L., N. M. Donahue, et al. (2007). "Rethinking organic aerosols: semivolatile emissions and photochemical aging." *Science* 315(5816): 1259-62.

Rubanyi, G. and A. G. Kovach (1980). "Cardiovascular actions of nickel ions." *Acta Physiol Acad Sci Hung* 55(4): 345-53.

Samoli, E., R. Peng, et al. (2008). "Acute effects of ambient particulate matter on mortality in Europe and North America: results from the APHENA study." *Environ Health Perspect* 116(11): 1480-6.

Schulz, H., V. Harder, et al. (2005). "Cardiovascular effects of fine and ultrafine particles." *J Aerosol Med* 18(1): 1-22.

Schwartz, J., A. Litonjua, et al. (2005). "Traffic related pollution and heart rate variability in a panel of elderly subjects." *Thorax* 60(6): 455-61.

Steinberg, D. (1997). "Low density lipoprotein oxidation and its pathobiological significance." *J Biol Chem* 272(34): 20963-6.

U.S. EPA (1999). National Air Quality andEmissions Trends Report, 1999. USA, U.S. Environmental Protection Agency Office of Air Quality Planning and Standards Emissions Monitoring and Analysis Division Air Quality Trends Analysis Group Research Triangle Park, North Carolina 27711: 79-237.

U.S. EPA (2004). Air quality criteria for particulate matter (Final Report, Oct 2004), U.S. Environmental Protection Agency, Washington, DC, EPA 600/P-99/002aF-bF.

U.S. EPA (2005). National Summary of Particulate matter emissions by source sector. USA, U.S. Environmental Protection Agency, Research Triangle Park, North Carolina 27711. Accessed at www.epa.gov/air/urbanair/6poll.html on December 2010.

von Klot, S., A. Peters, et al. (2005). "Ambient air pollution is associated with increased risk of hospital cardiac readmissions of myocardial infarction survivors in five European cities." *Circulation* 112(20): 3073-9.

WHO (2002). Reducing Risks, Promoting Healthy Life. Geneva, Switzerland, World Health Organization. (http://www.who.int/healthinfo/global_burden_disease/GlobalHealthRisks_report_full.pdf).

WHO (2009). Global health risks: Mortality and burden of diseases attributable to selected major risks. Geneva, World Health Organization. (http://www.who.int/healthinfo/global_burden_disease/GlobalHealthRisks_report_full.pdf).

WRAP (2004). Definitions of Dust, Western Regional Air Partnership.October 21. www.nmey.state.nm.us/aqb/documents/FDHandbook_06.pdf.

Ying, Z., P. Yue, et al. (2009). "Air Pollution and Cardiac Remodeling: A Role For RhoA/Rho-kinase." *Am J Physiol Heart Circ Physiol*: 01270.2008.

Zanobetti, A. and J. Schwartz (2009). "The effect of fine and coarse particulate air pollution on mortality: a national analysis." *Environ Health Perspect* 117(6): 898-903.

Zeka, A., J. R. Sullivan, et al. (2006). "Inflammatory markers and particulate air pollution: characterizing the pathway to disease." *Int J Epidemiol* 35(5): 1347-54.

Zheng L, H. P., Husain L, Qureshi S, Dutkiewicz VA, Schwab JJ, Drewnick F, Demerjian KL (2004). "Sources of fine composition in New York City." *Atmos Environ.* 38: 6521–6529.

Permissions

The contributors of this book come from diverse backgrounds, making this book a truly international effort. This book will bring forth new frontiers with its revolutionizing research information and detailed analysis of the nascent developments around the world.

We would like to thank Dr. Mukesh Khare, for lending his expertise to make the book truly unique. He has played a crucial role in the development of this book. Without his invaluable contribution this book wouldn't have been possible. He has made vital efforts to compile up to date information on the varied aspects of this subject to make this book a valuable addition to the collection of many professionals and students.

This book was conceptualized with the vision of imparting up-to-date information and advanced data in this field. To ensure the same, a matchless editorial board was set up. Every individual on the board went through rigorous rounds of assessment to prove their worth. After which they invested a large part of their time researching and compiling the most relevant data for our readers. Conferences and sessions were held from time to time between the editorial board and the contributing authors to present the data in the most comprehensible form. The editorial team has worked tirelessly to provide valuable and valid information to help people across the globe.

Every chapter published in this book has been scrutinized by our experts. Their significance has been extensively debated. The topics covered herein carry significant findings which will fuel the growth of the discipline. They may even be implemented as practical applications or may be referred to as a beginning point for another development. Chapters in this book were first published by InTech; hereby published with permission under the Creative Commons Attribution License or equivalent.

The editorial board has been involved in producing this book since its inception. They have spent rigorous hours researching and exploring the diverse topics which have resulted in the successful publishing of this book. They have passed on their knowledge of decades through this book. To expedite this challenging task, the publisher supported the team at every step. A small team of assistant editors was also appointed to further simplify the editing procedure and attain best results for the readers.

Our editorial team has been hand-picked from every corner of the world. Their multi-ethnicity adds dynamic inputs to the discussions which result in innovative outcomes. These outcomes are then further discussed with the researchers and contributors who give their valuable feedback and opinion regarding the same. The feedback is then collaborated with the researches and they are edited in a comprehensive manner to aid the understanding of the subject.

Apart from the editorial board, the designing team has also invested a significant amount of their time in understanding the subject and creating the most relevant covers. They scrutinized every image to scout for the most suitable representation of the subject and create an appropriate cover for the book.

The publishing team has been involved in this book since its early stages. They were actively engaged in every process, be it collecting the data, connecting with the contributors or procuring relevant information. The team has been an ardent support to the editorial, designing and production team. Their endless efforts to recruit the best for this project, has resulted in the accomplishment of this book. They are a veteran in the field of academics and their pool of knowledge is as vast as their experience in printing. Their expertise and guidance has proved useful at every step. Their uncompromising quality standards have made this book an exceptional effort. Their encouragement from time to time has been an inspiration for everyone.

The publisher and the editorial board hope that this book will prove to be a valuable piece of knowledge for researchers, students, practitioners and scholars across the globe.

List of Contributors

Selahattin Incecik and Ulaş Im
Istanbul Technical University, Department of Meteorology, Maslak, Istanbul, Turkey
University of Crete Department of Chemistry, Environmental Chemical Processes Laboratory (ECPL) Voutes, Heraklion, Crete, Greece

Diofantos G. Hadjimitsis, Kyriacos Themistocleous and Argyro Nisantzi
Cyprus University of Technology, Cyprus

Bang Quoc Ho
Institute of Environment and Resources, Vietnam National University in HCM City, Vietnam

Florentina Villanueva, José Albaladejo, Beatriz Cabañas, Pilar Martín and Alberto Notario
Castilla La Mancha University, Spain

Daniela Buske
Federal University of Pelotas, Pelotas, RS, Brazil

Marco Tullio Vilhena and Bardo Bodmann
Federal University of Rio Grande do Sul, Porto Alegre, RS, Brazil

Tiziano Tirabassi
Institute ISAC, National Research Council, Bologna, Italy

Dante L. Cáceres and Sergio A. Alvarado O.
Division of Epidemiology, School of Public Health, Faculty of Medicine, University of Chile, Chile
Grups de Recerca d'Amèrica i Àfrica Llatines (GRAAL), Unitat de Bioestadística, Facultat de Medicina, Universitat Autònoma de Barcelona, Barcelona, Spain

Claudio Z. Silva
Division of Epidemiology, School of Public Health, Faculty of Medicine, University of Chile, Chile

N.F. Elansky, I.B. Belikov, O.V. Lavrova, A.I. Skorokhod and R.A. Shumsky
A.M. Obukhov Institute of Atmospheric Physics RAS, Russia

C.A.M. Brenninkmeijer
Max Planck Institute for Chemistry, Germany

O.A. Tarasova
World Meteorological Organization, Switzerland

Mohd Zamri Ibrahim, Marzuki Ismail and Yong Kim Hwang
Department of Engineering Science, Faculty of Science and Technology, Universiti Malaysia Terengganu, Kuala Terengganu, Malaysia

M. Maatoug, K. Taïbi, A. Akermi, M. Achir and M. Mestrari
Faculty of Natural Sciences, Ibn Khaldoun University, Tiaret, Algeria

Nebojša Topić
Luka Koper, d.d., Koper, Slovenia

Matjaž Žitnik
Jožef Stefan Institute, Ljubljana and Faculty of Mathematics and Physics, University of Ljubljana, Slovenia

Haroldo F. de Campos Velho, Eduardo F.P. da Luz and Fabiana F. Paes
Laboratory for Computing and Applied Mathematics (LAC), National Institute for Space Research (INPE) São José dos Campos (SP), Brazil

Débora R. Roberti
Department of Physics, Federal University of Santa Maria (UFSM) Santa Maria (RS), Brazil

Yara de Souza Tadano
State University of Campinas – Sao Paulo, Department of Mechanical Engineering, Brazil

Cássia Maria Lie Ugaya and Admilson Teixeira Franco
Federal University of Technology – Paraná, Department of Mechanical Engineering, Brazil

Kyriaki Kitikidou and Lazaros Iliadis
Democritus University of Thrace, Department of Forestry and Management of the Environment and Natural Resources, Orestiada, Greece

Yungling Leo Lee
Institute of Epidemiology and Preventive Medicine, College of Public Health, National Taiwan University, Taipei, Taiwan

Guang-Hui Dong
Department of Biostatistics and Epidemiology and Department of Environmental and Occupational, School of Public Health, China Medical University, Shenyang, PR China

Pierre Madl
Division of Physics and Biophysics, Department of Materials Research and Physics University of Salzburg, Austria

Najeeb A. Shirwany and Ming-Hui Zou
Section of Molecular Medicine, Department of Medicine, Department of Biochemistry and Molecular Biology, University of Oklahoma Health Science Center, Oklahoma City, OK, USA

Akeem O. Lawal and Jesus A. Araujo
Division of Cardiology, Department of Medicine, David Geffen School of Medicine, University of California, Los Angeles, CA, USA